Contents at-a-glance

FRONT COVER

At a modern day's end, the sun sinks from the eye's view into a plume of air-borne pollution. This view is over St. Louis, Missouri, USA; the plume will drift northeastward to the edge of the North American continent, then onwards over the Atlantic Ocean

Introduction to Environmental Chemistry

Nigel J Bunce
University of Guelph

Wuerz Publishing Ltd
Winnipeg, Canada

phone (204) 453-7429
fax (204) 453-6598

Wuerz Publishing Ltd
Winnipeg, Canada

(204) 453-7429
fax (204) 453-6598

Introduction to Environmental Chemistry
Nigel J. Bunce
ISBN 0-920063-50-0 paperback

Printed in Canada

Contents

ERRATUM

Due to an unfortunate error in manufacturing ...

pages "ix" and "xi" were switched.

We are sorry for any inconvenience.

Preface

The design of this first year program in chemistry is a little different from the conventional first year chemistry curriculum. The major chemical principles are covered: stoichiometry, chemical energetics, equilibria, rates of chemical reactions, and electrochemistry. In order to allow time for the discussion of more applied topics, certain aspects of chemistry had to be left out. The most important of these is atomic and molecular structure (electronic structures of atoms, Lewis structures, hybridization etc). Structural aspects of chemistry are thus down-played, although it is assumed that you have some knowledge of these topics from high school chemistry. For example, it is assumed that you are familiar with the difference between ionic and covalent bonding.

The text contains no coverage of organic chemistry except in two contexts: combustion of fuels and the chemistry of chlorofluorocarbons.

I am very grateful to Drs. George Ferguson, Bob McCrindle and especially Allan Colter for commenting on an earlier draft of these notes. The remaining errors and omissions are, of course, mine; I would welcome any suggestions for improving the clarity and usefulness of what you see here. Most of all, I hope that you derive something of the pleasure from your study of environmental chemistry that I have gained from mine.

Nigel Bunce
Department of Chemistry and Biochemistry,
University of Guelph,
July 1993

To Caroline

Photo Credits

Front Cover	Ted Spiegel
page 4	all 4 photos, Ted Spiegel
facing page 144	a, Ted Spiegel; b, Bill Eppridge; c,d, Ted Spiegel
facing page 145	both photos, Ted Spiegel
page 232	*Environ. Sci. Technol.* **27** (4) 628 (1993)
facing page 240	a,b, Ted Spiegel; c, Mark Edwards; d, Mark Richards
facing page 241	Kathleen Manscill
facing page 272	all 4 photos, Ted Spiegel
facing page 273	both photos, Tom Pantages
facing page 385	top photos, Walt Lane
	bottom photos, Ted Spiegel
page 394	Rene Dionne
facing page 400	a, J Proctor; b, Mark Edwards; c,d, Herbert Girardet
facing page 401	photo at top, Mark Edwards; photo at bottom, Magda Havas
Back Cover	Ted Spiegel

Photographers' addresses

Ted Spiegel:
R.D. #2, Box 353A, South Salem NY, 10590 (USA)

Mark Edwards &
Herbert Girardet:
Still Pictures, 199A Shooters Hill Road, Blackheath, London SE3 8UL (UK)

Bill Eppridge &
Mark Richards:
Dot Pictures, 253 West 73rd Street #3B, New York NY, 10023 (USA)

Tom Pantages:
34 Centre, Concord NH, 03301 (USA)

Kathleen Manscill:
Librarian, Great Smoky Mountains National Park, Gatlinburg TN (USA)

Rene Dionne:
c/o Inco Ltd, Copper Cliff ON (Canada)

J Proctor:
Salisbury Cathedral, Salisbury (UK)

Walt Lane:
c/o Librarian, Gettysburg National Military Park, Gettysburg PA (USA)

Magda Havas:
Trent University, Peterborough ON (Canada)

1 INTRODUCTION TO ENVIRONMENTAL CHEMISTRY

1.1 Introduction

The goal of this book is to introduce you both to fundamental chemical principles and to show how these principles apply to the chemistry of the environment. Many of the issues which confront society today and which we encounter through the news media are chemical in nature. Indeed, the word chemical carries very negative connotations to many people, and there is a widespread misconception that chemicals are limited in definition to substances which are both manufactured by the chemical industry and at the same time harmful to people and/or to the environment. Such, of course, is not the case. Everything in the material world, natural or synthetic, is a chemical substance. Many synthetic substances are of great value to modern society - drugs, paints, adhesives; even the clothes you are now wearing are in all probability wholly or partly synthetic. Nevertheless, it is true that there are some chemical substances - again, natural as well as synthetic - which can cause problems, and we shall meet some of these in this book. Problem substances are commonly referred to as **pollutants**, although it is useful to keep in mind the concept that a pollutant is a substance in the wrong place, at the wrong time, or in the wrong amount.

It would be a mistake to equate "environmental chemistry" with "pollution chemistry". The chemistry of the clean, or unpolluted, environment is worthy of study in its own right; moreover, our study of the problems caused by specific pollutants requires us to understand the operation of the unpolluted environment in order to know why a particular substance represents a pollution problem.

In this book we shall study the chemical principles at work in the environment - which, of course, are the same principles that govern chemical processes in the laboratory. In order to discuss specific environmental issues, we shall need to introduce certain industrial processes, which currently or previously have been associated with problems. However, I hope to dispel the view that "environmental pollution just keeps getting worse and worse". Today, our society is much more aware of pollution and much less tolerant of it than in the past. Awareness has come about largely because of the advances in analytical chemistry over the past 40 years; the parts per million and parts per billion that we read about in popular magazines today were undetectable in the past. People had no choice but to live with the pollution they did not know about. Rachel Carson's book *Silent Spring*, published in 1962, was one of the first exposés of the effects of the indiscriminate use of synthetic chemicals on the natural environment.

Chemical innovation can solve environmental problems or reduce their impact. Consider mercury pollution from chlor-alkali plants. This was the top environmental issue in 1970. We never hear about it today. Why? Because of the development of mercury-free, membrane cells for the electrolysis of brine. In another area, enormous research efforts are going on internationally to find replacement substances for the chlorofluorocarbons (CFCs) which are to be phased out before the end of this century under the terms of an international agreement known as the Montréal Protocol.

Closer to my home, Canadian technology is involved in the complete redesign of Inco's smelting operation at Sudbury, Ontario so as to meet the 1994 targets for reduced emission of acidic gases, at a cost of almost $500 million. Union Carbide Canada has developed the Can-Solv process for reducing acidic emissions from coal burning power plants. I hope that learning about the chemistry behind these and other issues will convince you that the study of chemistry is an important and worthwhile part of the study of environmental science.

1.2 Compartments of the environment

In order to consider the chemistry of the environment it is useful to divide, in our minds, the environment into different compartments, or **reservoirs**:

Definition

*the solid land, or **lithosphere**; oceans, rivers and lakes: the **hydrosphere**, and finally the **atmosphere**.*

The chemical processes that take place in the different compartments can be modelled in the laboratory, and are subject to the same constraints in terms of energetic feasibility and rate of progress as reactions studied in the laboratory.

Atmospheric chemistry is represented by chemical processes in the gas phase, while the chemistry of the hydrosphere is the chemistry of aqueous solutions. The chemistry of the lithosphere is the most complicated; it includes both reactions occurring in the solid state and reactions occurring on the surfaces of particles, at a gas-solid or a solid-liquid interface. For this reason, this book will focus mainly on gas phase and aqueous phase reactions.

It is important to remember the role of biological processes in shaping the environment of our planet. To take one simple example, the presence of 21% oxygen in the Earth's atmosphere is entirely due to the biochemical process of photosynthesis by algae and green plants. The Earth's atmosphere is highly oxidizing as a result, and the ultimate fate of both natural and synthetic trace components of the atmosphere is (chemical) oxidation to simple substances such as carbon dioxide and water. These atmospheric reactions could not proceed in the absence of oxygen. The presence of life on Earth, and the chemical reactions which are undertaken on a huge scale by living organisms, make the chemistry of the Earth's environment quite different from the chemistry of the other planets in the solar system.

One factor which we shall consider repeatedly in this book is the importance of **sunlight** as an energy source in the environment. Light is a form of energy, and under the proper conditions light energy can be harnessed to drive chemical reactions. Such reactions are called **photochemical reactions**. The input of light energy permits chemical reactions to be driven energetically uphill - in other words, light can make possible chemical reactions which would fail in its absence. A familiar example is photosynthesis, the overall reaction for which is given below.

$$6\ CO_2 + 6\ H_2O \rightarrow C_6H_{12}O_6 + 6\ O_2$$

Photosynthesis cannot proceed in the absence of light. In fact, the spontaneous reaction in the absence of light is respiration, the reverse of photosynthesis.

$$C_6H_{12}O_6 + 6\ O_2 \rightarrow 6\ CO_2 + 6\ H_2O$$

The physical and chemical processes which deposit a substance into a particular compartment are known as **sources** or **source reactions**. For example, the flow of rivers is a source of both water and dissolved salts into the oceans. Emissions from coal-fired electricity generating plants are a source of acidic gases such as sulfur dioxide into the atmosphere.

Environmental scientists often make the distinction between **point sources** and **non-point sources** in discussing pollution problems. A point source means that the emission occurs from a defined location, whereas non-point sources are by definition dispersed. For example, coal fired electricity generating stations are point sources of nitrogen oxides into the atmosphere. Automobiles and trucks, residential woodstoves, and forest fires are non-point sources of these same nitrogen oxides into the air. Point sources of pollution are easier in principle to regulate than non-point sources, because there are fewer of them.

Conversely, processes which remove a substance from a given reservoir are known as **sinks**. Deposition in the form of sea shells (calcium carbonate) is an important sink for dissolved calcium in the oceans; oxidation to carbon dioxide and water is the chief sink for methane in the atmosphere.

The lithosphere, the hydrosphere, and the atmosphere

The examples just given include cases where the source or sink is another chemical species within the same reservoir, and others where there is transport from or to a different reservoir.

1.3 Residence times

Definition

***Residence times** are defined as follows:*

$$\textit{Residence time} = \frac{\textit{amount of substance in the reservoir}}{\textit{rate of inflow to, or outflow from, reservoir}}$$

The numerator of this equation will have units such as grams, moles, moles per liter, or tonnes, while the denominator will have units such as grams per second, moles per minute, moles per liter per second, or tonnes per year etc. In all cases, the units should be chosen to be compatible, so that the units representing amount of substance cancel, and the final quantity has the units of time.

Example: *The concentration of lead in the blood of an adult male was 140 micrograms per liter ($\mu g\ L^{-1}$), and the blood volume was 4.8 L. The net transfer of lead into this person's bones was 7.5 $\mu g\ day^{-1}$, and his net excretion rate was 24 micrograms per day ($\mu g\ day^{-1}$). Calculate the residence time of lead in this person's blood.*

Answer: *Total amount of lead in the reservoir (the person's blood):*

$= 140\ \mu g\ L^{-1} \times 4.8\ L = 672\ \mu g$

Total rate of loss of lead out of the blood (to bone plus excretion):

$= 7.5\ \mu g\ day^{-1} + 24\ \mu g\ day^{-1} = 31.5\ \mu g\ day^{-1}$

$$\text{Residence time} = \frac{\text{total lead in the blood}}{\text{rate of loss from bloodstream}}$$

$$= 672\ \mu g / 31.5\ \mu g\ day^{-1} = 21\ days$$

The answer is given to two significant figures: see Section 1.7.3.

Residence times may cover the whole possible range, from very short to very long. For example, oxygen atoms are formed by the action of sunlight in the upper atmosphere (Chapter 15). They are extremely reactive, and their residence time in the atmosphere is measured in fractions of a second. At the other end of the scale, the residence time of dissolved ions such as Na^+ in the oceans is estimated to be as long as 70 million years.

The residence time is also sometimes known as the **lifetime**. There is possible confusion between the term lifetime (or residence time) and the **half-life**. By definition, the half-life is the time taken for the amount of substance entering the reservoir to reach half its maximum value, or the amount of substance leaving the reservoir to fall to half its initial value. The lifetime turns out to be about 40% greater than the half-life (for explanation, see Section 7.6). However, both are of the same order of magnitude.

1.4 Ionic and covalent bonding

This section is intended as review only. If you are already familiar with this topic, please skip to the next section.

Ionic compounds are composed of charged particles called **ions** (atoms or groups of atoms carrying one or more positive or negative charges). Positively charged ions are called cations, and negatively charged ions are called anions. In the solid state, ionic compounds are held together by electrostatic attraction between the cations and anions. In aqueous solution, the ions separate from each other, a process called **dissociation**, and each cation and anion moves freely through the solution, surrounded by a sheath of water molecules. Metallic elements have a tendency to give up electrons (which are themselves negatively charged) to form cations, while non-metals have the tendency to acquire electrons to form anions.

Types of compounds

Ionic compounds (salts)

KBr (potassium bromide, K^+ and Br^-)
NH_4Cl (ammonium chloride, NH_4^+ and Cl^-)
$CaCl_2$ (calcium chloride, Ca^{2+} and 2 Cl^-)
$NaCl$ (sodium chloride, Na^+ and Cl^-)
$CaCO_3$ (calcium carbonate, Ca^{2+} and CO_3^{2-})

Covalent compounds

CH_4 (methane)
$CH_3(CH_2)_6CH_3$ (octane)
CCl_4 (carbon tetrachloride)
$C_6H_{12}O_6$ (glucose)

The position of an element in the periodic table often gives the clue as to the numerical charge of the corresponding cation or anion. This is because the number of **valence electrons** (those available for bonding) is equal to the number of the periodic group to which the element belongs.

Periodic group I	Li, Na, K, Rb, Cs	+1
Periodic group II	(Be)*, Mg, Ca, Sr, Ba	+2
Periodic group VI	O, S, (Se)*	-2
Periodic group VII	F, Cl, Br, I	-1

* Be, Se, and the elements below Se form very few ionic compounds

Although we shall not study the details of bonding in this book, many readers will recall that the possession of an electron count equal to that of a noble gas element (periodic group VIII) confers unusual stability on an ion. This concept explains the data above; for example, calcium (periodic group IIA) can lose two electrons to form Ca^{2+} with electron count the same as that of the noble gas argon; likewise fluorine (periodic group VII) can gain one electron to form F^- with the same electron count as the noble gas neon. These simple rules do not apply directly to the transition or "d block" elements.

Covalent bonding is common when non-metallic elements bond together. Covalent bonding involves the sharing of one or more pairs of electrons between atoms. In the majority of stable covalent compounds, the various atoms share sufficient electrons that each is surrounded by the same number of electrons as one of the noble gases. Thus in the water molecule (H_2O), the oxygen atom (periodic group VI) gains a share in two extra electrons by sharing them with the two hydrogen atoms, and thereby is surrounded by eight electrons like the noble gas neon. Likewise in the covalent compound CCl_4, the carbon atom (periodic group IV) gains a share in four extra electrons (one from each chlorine), and each chlorine atom (periodic group VII) gains a share in one extra electron provided by carbon.

We can emphasize the ionic or covalent nature of the bonding in a given compound by the way we write the formula. For example, we could emphasize the ionic nature of sodium chloride (NaCl) by writing it as Na^+Cl^-, or we could show water to be a covalent substance by writing H-O-H instead of H_2O. Lines between atoms are conventionally recognized as representing covalent links between the atoms concerned.

Ionic and covalent compounds typically have very different properties. Ionic compounds are generally able to conduct electricity in the molten form and in aqueous solution. Typically they have high melting points and boiling points, because of the strong electrostatic attraction between cations and anions. Water is a good solvent for many ionic compounds (but not all, see Chapter 12) because the sheath of water molecules surrounding the ion stabilizes the ions to a comparable extent as the electrostatic attraction in the crystal. Substances which dissociate completely in water to give ions are called **strong electrolytes**, the term electrolyte signifying that the solution conducts electricity. Few other solvents have as much stabilizing capacity for ions, and as a result, ionic compounds have very little solubility in liquids such as hydrocarbons (*e.g.* gasoline), alcohols, and other organic solvents.

Covalent compounds tend to have the opposite properties from ionic substances. Covalent compounds do not conduct electricity (they are insulators); they tend to have low melting points and boiling points, because they are composed of individual molecules, whose mutual attraction is less than that of oppositely charged ions. Although it is dangerous to over-generalize, they tend to have appreciable solubility in organic solvents such as hydrocarbons and alcohols, but very little solubility in water. However, water is a rather reactive solvent, and many covalent substances dissolve in water as a result of chemical reaction. One example is hydrogen chloride (HCl), a gas which consists of covalent molecules at room temperature. When it dissolves in water it is converted to hydrochloric acid, which is a strong electrolyte.

$$HCl(gas) + H_2O(liquid) \quad \rightarrow \quad H_3O^+ + Cl^- \quad \text{(both solvated ions)}$$

In other cases, the covalent substance only partially dissociates into ions; the solution conducts electricity only feebly, and is called a **weak electrolyte**. Hydrogen fluoride (HF) is an example of a weak electrolyte in water.

Selected properties of ionic and covalent compounds

		MP	BP	Conductivity**	Solubility water	hydrocarbons	alcohols
ionic compounds							
KBr		735	1435	150	++	-	-
NH_4Cl		340	520	150	++	-	+
$CaCl_2$		780	>1600	135	+++	-	+
NaCl		800	1410	125	++	-	-
covalent compounds							
CH_4	methane	-182	-164	~0	-	+	-
CH_3CH_2OH	alcohol	-115	80	~0	∞	∞	∞
$CH_3(CH_2)_6CH_3$	octane	-55	125	~0	-	∞	∞
$C_6H_{12}O_6$	glucose	150	-	~0	+++	-	+

* d, decomposes at high temperatures
** "Conductivity" is Equivalent Conductance in gm equiv cm^{-3}

solubility:
-- less than 0.1 g/100 mL
\+ 0.1 to 10 g/100 mL
++ 10 to 100 g/100 mL
+++ over 100 g/100 mL

In still other cases, the covalent compound reacts with water and is thereby changed into completely different substance(s). An example is given by phosphorus pentachloride, PCl_5.

$$PCl_5 + 4\ H_2O \quad \rightarrow \quad H_3PO_4 + 5\ HCl$$

The aqueous solution does not contain PCl_5 at all, but instead a mixture of phosphoric and hydrochloric acids.

1.5 Chemical nomenclature

This section is intended as review only. If you are already familiar with this topic, please skip to the next section.

Chemical substances are named so as to provide an unambiguous way of communicating the identity of a given substance, and equally, unambiguous rules for translating a name into a chemical formula, and *vice versa.*

The simplest forms of matter from the chemist's perspective are the chemical **elements**, whose names and symbols appear in the Periodic Table (inside front cover). Each element contains only one kind of atom. Elements may occur in nature as single atoms (*e.g*, Ar), as molecules (*e.g.*, N_2), or as virtually infinite assemblies of atoms (*e.g.*, carbon in the form of diamond, and metals). About 80% of all the chemical elements are metals. Some elements occur in more than one form, called **allotropes** (*e.g.*, molecular oxygen, O_2, and ozone, O_3).

Compounds consist of more than one type of atom. Compounds may be **ionic** or **molecular**. The simplest compounds contain only two kinds of atom and are called binary compounds, the prefix "bi" signifying two. Binary compounds can always be identified because the name ends in "ide": example, calcium fluoride.

Binary ionic compounds are held together by the electrostatic attraction of oppositely charged **ions**. Binary ionic compounds are recognized by being composed of a metal (the cation) and a non-metal (the anion). For example, sodium chloride is composed of sodium cations (Na^+) and chloride anions (Cl^-). In any stable ionic substance the sum of positive charges must be balanced by the sum of negative charges. By convention, the cation name always precedes the anion name: thus, copper (cation) sulfate (anion). Ionic compounds, binary or otherwise, are also known as **salts**; most salts are strong electrolytes - they dissociate completely in aqueous solution.

Examples:	*calcium chloride, $CaCl_2$*	*1 Ca^{2+} and 2 Cl^-*
	zinc sulfide, ZnS	*1 Zn^{2+} and 1 S^{2-}*
	aluminum oxide, Al_2O_3	*2 Al^{3+} and 3 O^{2-}*

If the element can form more than one type of ion, the charge on the ion (generally a cation) is indicated by a roman numeral.

Examples:	*copper(I) oxide, Cu_2O*	*2 Cu^+ and 1 O^{2-}*
	iron(III) chloride	*1 Fe^{3+} and 3 Cl^-*

Binary molecular compounds are held together by covalent bonds, in which the atoms share one or more pairs of electrons. Molecular compounds are generally formed when both partners are non-metals. In order to indicate the relative numbers of atoms, the following multiplying prefixes are used.

- mono 1 (usually omitted)
- di 2
- tri 3
- tetra 4
- penta 5
- hexa 6

Examples:

bromine trifluoride	BrF_3
silicon tetrachloride	$SiCl_4$
tetrasulfur tetranitride	S_4N_4
carbon monoxide	CO

In this last example, note both the use of the prefix "mono" (to avoid confusion with carbon dioxide, CO_2), and the contraction of monooxide to monoxide (to make the word easier to pronounce).

A few very common compounds are known by **trivial** (non-systematic) names. We should learn the following trivial names.

trivial name		formula
• water	dihydrogen oxide	H_2O
• ammonia	trihydrogen nitride	NH_3
• methane	tetrahydrogen carbide	CH_4

More complex ionic compounds are composed of polyatomic ions in which the ion itself is held together by covalent bonds. For example, sodium sulfate Na_2SO_4 is composed of 2 Na^+ ions for every 1 sulfate (SO_4^{2-}) ion, in which the sulfur atom is bonded covalently to the four oxygen atoms. On the next page is a list of commonly encountered polyatomic ions.

***Examples** of names of compounds containing polyatomic ions:*

- $(NH_4)_2SO_4$ *ammonium sulfate*
- $Na_2S_2O_3$ *sodium thiosulfate*
- K_2HPO_4 *dipotassium hydrogen phosphate*

	ammonium NH_4^+	
thiosulfate $S_2O_3^{2-}$	nitrite NO_2^-	nitrate NO_3^-
hypochlorite ClO^-	chlorite ClO_2^-	chlorate ClO_3^-
perchlorate ClO_4^-		
sulfite SO_3^{2-}	sulfate SO_4^{2-}	hydrogen sulfate HSO_4^-
carbonate CO_3^{2-}	hydrogen carbonate (common name, bicarbonate) HCO_3^-	
phosphate PO_4^{3-}	hydrogen phosphate HPO_4^{2-}	dihydrogen phosphate H_2PO_4-
chromate CrO_4^{2-}	dichromate $Cr_2O_7^{2-}$	permanganate MnO_4^-

Some salts crystallize from water with water molecules bound into the crystal. Such crystals are called **hydrates**. The water of hydration (also called water of crystallization) is included in the name, as shown in the following examples.

$CaSO_4.2H_2O$	calcium sulfate dihydrate
$Ni(NO_3)_2.6H_2O$	nickel(II) nitrate hexahydrate

In cases where the commonly encountered substance has water of hydration, the term **anhydrous** (without water) may be used to emphasize that the substance in question lacks water of hydration.

Example: *$CuSO_4.5H_2O$* — *copper(II) sulfate pentahydrate (commonly termed "bluestone").*

$CuSO_4$ — *copper(II) sulfate; this could (but does not have to) be called anhydrous copper(II) sulfate, to emphasize that it lacks water of crystallization.*

1.6 SI measurements

This section is intended as review only. If you are already familiar with this topic, please skip to the next section.

It is assumed that students using this book will be familiar with SI metric measurements. In this book, we shall encounter the following SI base, or fundamental, units.

• mass	kilogram, kg
• length	meter, m
• time	second, s
• temperature	kelvin, K
• amount of chemical substance	mole, mol

Derived units are made up by combining SI fundamental units. Some examples are:

• force	newton, N = kg m s^{-2}
• pressure	pascal, Pa = N m^{-2} = kg m^{-1} s^{-2}
• energy	joule, J = N m = kg m^2 s^{-2}
• density	kg m^{-3}

We shall also make considerable use of units which are outside the strict SI system.

• mass	tonne (metric ton), t	1 t = 10^6 g = 1000 kg
• volume	liter, L	1 L = 1 dm^3 = 10^{-3} m^3
• Celsius temperature	degree Celsius, °C	t (°C) = T (K) - 273.15
• pressure	atmosphere, atm	1 atm = 101.325 Pa

In Section 1.2, we noted that residence times could cover almost every conceivable value, large or small. Such a wide range of values, whether they be times, rates, masses of material, equilibrium constants, is a characteristic of environmental chemistry. Quantities tend either to be very large (what is the total mass of ozone in the stratosphere?) or very small (what is the concentration of polychlorinated biphenyls in the water of Lake Ontario?). In order to avoid the need to carry exponents in every step of every calculation, SI provides for **multiplying prefixes** to be used to indicate quantities substantially larger or smaller than the fundamental unit. These multiplying prefixes can be used with any SI base or derived unit, and are also used with some of the non-SI units, notably the liter. The multiplying prefixes are given below.

Commonly used prefixes for large quantities, with examples:

- kilo, k 10^3 one thousand kmol, kPa
- mega, M 10^6 one million Mg[1], MPa
- giga, G 10^9 one billion GJ
- tera, T 10^{12} one trillion Tg

It is very unusual to use the large multiplying prefixes with the time unit the second, because of the well established common units of time: minutes (min), hours (h), days (d), and years (yr), even though these units are not recognized as SI. Large multiplying prefixes are almost never used with the liter in scientific work (although dairy farmers will be familiar with the hectoliter = 100 L). No multiplying prefixes, large or small, are commonly used with the atmosphere or Celsius temperature. The only common large multiplying prefixes used with the tonne are kilotonne (= 1000 t, 10^9 g), and megatonne (= 1 million tonnes, 10^{12} g).

Commonly used prefixes for small quantities, with examples:

- deci, d 10^{-1} dm (the only common example is the dm^3 = 1 L)
- centi, c 10^{-2} cm (the only common example is the centimeter)
- milli, m 10^{-3} mmol, mJ
- micro, μ 10^{-6} μs, μL
- nano, n 10^{-9} nm, ng
- pico, p 10^{-12} pmol, ps

Note the examples of the use of small multiplying prefixes with the second and liter.

It is never correct to use more than one multiplying prefix.

1 kkg = 1 Mg; 1 mμmol = 1 nmol; 1 nkm = 1 μm

1.7 Chemical calculations

This section is intended as review only. If you are already familiar with this topic, please skip to Chapter 2.

---

[1] Be careful not to confuse the megagram with the chemical symbol for magnesium, since both look the same. The context will show you which is intended.

1.7.1 Qualitative and quantitative aspects of chemistry

Chemistry is a science that has both **qualitative** and **quantitative** aspects. This is true whatever branch of chemistry we discuss - a study in a research laboratory, an industrial process, or environmental chemistry. Qualitative aspects of chemistry involve *identifying* substances: *what* was the product of a chemical reaction; *what* is the impurity in a commercial product; *what* is the contaminant present in a sample of drinking water? Advances in qualitative analysis allow chemists to determine unambiguously the identities and the molecular structures of increasingly complicated molecules.

Quantitative aspects of chemistry centre around *how much* material is present. *How much* product can be formed from a given amount of reactant; *how much* of a toxic impurity is present in a sample of animal feed? Advances in quantitative analysis allow chemists to analyse for ever smaller amounts of substance.

Both qualitative and quantitative analysis are needed to study chemical processes. You have to know *what* is present before you can determine *how much*, or even to decide what method to use to determine how much. The appropriate methods of qualitative analysis tend to vary greatly with the substance being identified, and we shall usually assume in this book that some suitable method has already been used to identify the substances of interest.

The quantitative aspects of chemistry involve calculations. Chemical calculations can be intimidating to some students. However, this need not be so. The calculations in this book require little more than a familiarity with basic addition, subtraction, multiplication, and division, together with the ability to use logarithms and exponents. Besides, all these functions are available on scientific calculators. Thus the actual calculations present no real difficulty. Where difficulty is likely to arise is in deciding how to tackle a problem: in other words, in formulating a strategy for working the calculation. The worked examples in this book will focus on the issue of *how* to do the problem, as much as actually working the problem. Try to take note of the comments provided as clues in the worked examples. Look at the information you have had provided, and try to decide how to use it to extract the information you need.

1.7.2 Dimensional analysis

Units provide great assistance in working chemical problems. Let's take a simple example. The density of a solution of sodium bromide is measured to be 1.24 grams per cubic centimeter. What volume of solution is required to provide 55 grams of sodium bromide solution?

Two pieces of information were given: the density of the solution, and the mass of sodium bromide solution required. The information to be calculated is the volume of solution.

Hopefully we remember that density, mass and volume are interrelated (the relationship is actually density = mass/volume). The trouble is: do we multiply or divide, and what do we multiply or divide with what? This is where the units are helpful.

Since density = mass/volume, it follows that volume = mass/density.

The problem is worked as follows:

$$\begin{aligned}\text{volume required} &= \text{mass of NaBr solution/density of solution}\\ &= \frac{55\ \text{g}}{1.24\ \text{g cm}^{-3}}\\ &= 44\ \text{cm}^3\end{aligned}$$

Notice that the units grams cancelled, leaving cm^{-3} in the denominator (same as cm^3 in the numerator). These units, cm^3, are units of volume, so our calculation is dimensionally correct. The inclusion of the units, as well as the numerical values, in a calculation is called **dimensional analysis**.

Let's see what would have happened if we had combined the quantities mass and density in other possible ways (which are incorrect).

First wrong approach:

$$\begin{aligned}\text{volume required} &= \text{mass of NaBr solution x density of solution}\\ &= 55\ \text{g x}\ 1.24\ \text{g cm}^{-3}\\ &= 68\ \text{g}^2\ \text{cm}^{-3}\end{aligned}$$

Second wrong approach:

$$\begin{aligned}\text{volume required} &= \text{density of solution/mass of NaBr solution}\\ &= \frac{1.24\ \text{g cm}^{-3}}{55\ \text{g}}\\ &= 0.023\ \text{cm}^{-3}\end{aligned}$$

Neither $g^2\ cm^{-3}$ nor cm^{-3} are units of volume (cm^{-3} is the unit of $volume^{-1}$). Therefore neither of our wrong approaches can be correct, because both gave an answer which was dimensionally incompatible with the quantity needed, namely a volume.

Rule: *Always include the units at every step in every calculation*

1.7.3 Significant figures

In the previous calculation (the correct one!), I obtained the following result with my calculator.

$$\text{volume required} = \text{mass of NaBr solution/density of solution}$$
$$= \frac{55\ \text{g}}{1.24\ \text{g cm}^{-3}}$$
$$= 44.35483871\ \text{cm}^3$$

However, I only reported the answer as 44 cm^3. Why?

The answer to this question is bound up with the precision of measurements. The mass of sodium bromide solution required was given as 55 g, from which we must conclude that it must have been larger than 54 g and smaller than 56 g. Whether it was actually 55.0000 g, or 55.27 g, or 54.9826 g we cannot say. The value 55 g is said to be precise to two significant figures.

Likewise, the density of the solution was reported as 1.24 g cm^{-3}: somewhere between 1.23 g cm^{-3} and 1.25 g cm^{-3}. By the same reasoning, this value had a precision of three significant figures. If we were simply to record the answer as given on the calculator as 44.35483871 cm^3, the implication would be that we are sure of the exactness of the answer to between 44.35483870 cm^3 and 44.35483872 cm^3, which is clearly not what we intend. Therefore, the precision of any calculated quantity is **not** something that we can obtain directly from our calculator; it must be determined by us. This section considers how we make such a determination.

How many significant figures are implied by a given number?

Examples:

	Sig. Figs
21.26	*4*
0.00648	*3 (the zeros simply mark the position of the decimal point)*
600.0	*4 (the zeros were not needed just to locate the decimal point)*
2000	*? (can't tell whether or not the zeros are significant)*

Frequently, the use of exponential (scientific) notation is helpful in specifying the number of significant figures. The first three values above are 2.126×10^1; 6.48×10^{-3}; and 6.000×10^2 in scientific notation; however, the last one is ambiguous. Is it 2×10^3, 2.0×10^3; 2.00×10^3, or 2.000×10^3? We simply cannot tell. Reporting quantities in scientific notation avoids all possible ambiguity.

Rounding up and down

Suppose I want to report an answer to three significant figures when the number on my calculator has more than three digits. How should I proceed? The rule is to truncate when the following digit is less than 5, and to round up when the following number is 5 or greater.

Examples:

21.216	*becomes 21.2*
0.0034871	*becomes 0.00349*

We now review how calculations are carried out with the proper number of significant figures.

Multiplication and division

These are the easiest. Simply count the number of significant figures in each quantity being multiplied or divided. The proper number of significant figures in the answer is the same as the datum with the smallest number of significant figures.

Example:

$$\frac{2.9486 \times 0.0428 \times 83.867}{0.53 \times 100.778} = 0.1981568699 = 2.0 \times 10^{-1}$$

The data point with the smallest number of significant figures is 0.53 (2 sig. figs.).

Addition and subtraction

Here the rule is slightly different; you have to pay attention to the location of the decimal point. If the numbers are in exponential notation, you may have to "normalize" them to the same exponent.

Example 1: *12.556 + 0.002 + 102.5 + 0.1472 = ?*

Let's write this out in full

$$\begin{array}{r} 12.556 \\ 0.002 \\ 102.5 \\ 0.1472 \\ \hline 115.2052 \end{array} \Rightarrow 115.2$$

Everything beyond the first decimal place is insignificant, because this is the precision of the number 102.5.

Example 2: *104.8231 - 104.7397 = ?*

$$\begin{array}{r} 104.8231 \\ 104.7397 \\ \hline 0.0834 \end{array}$$

Notice the loss of precision from 7 significant figures to 3. This commonly happens; precision is frequently lost in subtraction, especially in the subtraction of numbers of similar value. A typical example is in measuring the mass of a small amount of substance in a relatively massive container. The three lines of this example might well be:

mass of sample + container	*104.8231 grams*
mass of empty container	*104.7397 grams*
mass of sample	*0.0834 grams*

Example 3: $2.56 \times 10^{-3} + 5.387 \times 10^{-4} + 6.2 \times 10^{-2} = ?$

$$\begin{array}{l} 0.00256 \\ 0.0005387 \\ 0.062 \\ \hline 0.065 \end{array} \quad \text{or} \quad \begin{array}{rl} 2.56 & \times 10^{-3} \\ 0.5387 & \times 10^{-3} \\ 62 & \times 10^{-3} \\ \hline 65 & \times 10^{-3} = 6.5 \times 10^{-2} \end{array}$$

Logarithms and exponents

Remember that a logarithm is not like an ordinary number. In the statement $\log_{10}(x) = 12.35$, the numbers preceding the decimal point serve only to fix the power of 10: they are not significant[2]. Only the numbers following the decimal point are significant.

Rule: *The number of decimal places in the logarithm is the same as the number of significant figures in the actual number.*

Therefore if $\log_{10}(x) = 12.35$, $x = 2.238721139 \times 10^{12} \Rightarrow 2.2 \times 10^{12}$. This is especially important in dealing with pH, Chapter 11.

Similar considerations apply to natural logarithms and exponents. For example, if $y = 6.772 \times 10^3$ (4 significant figures), $\ln(y) = 8.820551743 \Rightarrow 8.8206$ (4 decimal places).

Carrying significant figures through a calculation

Should you round off at each stage? Opinions vary; personally, I think this is a waste of time. I usually carry all the figures displayed in the calculator from stage to stage in the calculation, and round off at the end. When writing down an intermediate answer, I often write it with one more than the proper number of significant figures, but this is a matter of choice. What matters is proper rounding at the end. There is one rule however; *never carry fewer than the correct number of significant figures*. Once you have lost precision, you cannot get it back.

Quantities that are exact

Compare the following statements.

1. I am 2 m tall.
2. I have 2 sisters.

The first quantity (2 m) is precise to 1 significant figure; the second is exact (an integral number), since I could not possibly have 2.1 sisters. The same distinction arises in chemical calculations. Consider the well-known reaction between hydrogen and oxygen to form water.

$$2\ H_2 + O_2 \rightarrow 2\ H_2O$$

---

[2] The technical term for the figures preceding the decimal point is the **characteristic** of the logarithm; the figures following the decimal point are called the **mantissa**.

Exactly two hydrogen molecules react with exactly one oxygen molecule.

Question: *How many molecules of hydrogen are needed to react with 3.592×10^4 molecules of oxygen?*

Answer:

Number of H_2 molecules:

$$= 3.592 \times 10^4 \text{ molecules of } O_2 \times \frac{(2 \text{ molecules of } H_2)}{(1 \text{ molecule of } O_2)}$$

$$= 7.184 \times 10^4 \text{ molecules of } H_2$$

The answer is precise to 4 significant figures (the precision of the datum), not one significant figure (the molar ratio) because the 2 molecules of H_2 per molecule of O_2 is an integral number.

Problems

Section 1.1

1. Which of the following represents a source of pollution by lead?
   a) lead sheeting on a cathedral roof
   b) lead electrodes in your car battery
   c) lead ions in your drinking water
   d) discarded lead paint in the city dump
   e) a solution of lead nitrate in the laboratory

2. The maximum allowable concentration of copper in Ontario drinking water is 1.0 micrograms per liter (1.0 $\mu g\ L^{-1}$). Which of the following would be polluted?
   a) drinking water containing 0.002 $\mu g\ L^{-1}$ of copper
   b) drinking water containing 2.0 $\mu g\ L^{-1}$ of copper
   c) cooling water for a steel making plant containing 5.0 $\mu g\ L^{-1}$ of copper
   d) a mineral rock containing 3.2% copper by weight

3. What are your thoughts on whether the following water samples should be considered polluted?
   a) water from an underground aquifer (your well perhaps) into which seepage from an industrial chemical dumpsite has flowed
   b) a stream in a remote area of northern Ontario which has dissolved mercury salts which are naturally present in the underlying rock
   c) a rural well into which run off from fertilizer or manure has seeped from a neighbouring farm
   d) a mountain stream just downstream of the carcass of a dead mountain sheep.

Section 1.2

1. Which of the following are point sources of air pollution? Pollutants shown in parentheses.
   a) the cars in Toronto during the morning rush hour (nitrogen oxides and hydrocarbons)
   b) a lead smelter where used car batteries are recycled (lead)
   c) a coal-fired electricity generating station (acidic gases)
   d) the booster rockets of a space shuttle (nitrogen oxides)
   e) automobile air conditioners (chlorofluorocarbons)
   f) a municipal garbage incinerator (nitrogen oxides and dioxins).

2. Which of the following are non-point sources of water pollution? (Pollutants are given in parentheses.)
   a) agricultural run-off from farms (fertilizers)
   b) leachate from a city dump (toxic solvents)
   c) the sewers from a manufacturer of chlorinated hydrocarbons
   d) dust from city streets (lead)
   e) tailings from an abandoned zinc mine (metals, such as Zn, Pb, Cu)

Section 1.3

1. The residence time of the substance tetrachloroethylene in the rat is about 8 hours. Calculate the amount of tetrachloroethylene in a rat if the measured rate of excretion averages 24 mg per day.

2. Data on mercury poisoning show that a constant intake of mercury in the diet of 5 mg per day leads to the accumulation of 550 mg of mercury in all compartments of the body (blood, bone, brain, etc.). Calculate the residence time of mercury in the body.

3. The chlorofluorocarbon CFC-11 has an estimated lifetime of 70 years in the atmosphere. Calculate the mass of CFC-11 that would be present in the atmosphere if the release of this substance continued at the 1980's level of about 300,000 tonnes per year.

4. The mass of nitrogen in the atmosphere is 4×10^{18} kg, and its sinks from the atmosphere include (i) biological nitrogen fixation by bacteria, 2×10^{11} kg yr^{-1}; (ii) production of NO in thunderstorms, 7×10^{10} kg yr^{-1}; (iii) chemical synthesis of ammonia, 5×10^{10} kg yr^{-1} (all data refer to loss of nitrogen). Calculate the residence time of nitrogen in the atmosphere.

5. The residence time of the water in Lake Erie is 2.7 years. The concentration of phosphorus in Lake Erie in 1983 was 15 $\mu g\ L^{-1}$ and the volume of the lake is 484 km^3. Calculate the average rate of input of phosphorus into Lake Erie at that time.

Section 1.4

1. Give the charges on the common ions formed by the following elements: Cl, Ca, O, K, Mg, Na, Br, S.

2. Predict whether the following compounds are likely to be ionic or covalent: CO_2, $MgCl_2$, C_3H_8, BF_3, KBr, CaO, NH_3.

Section 1.5

1. Name the following ionic compounds:

a) $CaCO_3$	b) $MgCl_2$
c) $Na_2Cr_2O_7$	d) $CuSO_4.5H_2O$
e) CdS	f) $Al_2(SO_4)_3$
g) $FeCl_3.6H_2O$	h) $NaClO_2$
i) $KHCO_3$	j) $(NH_4)_3PO_4$

2. Name the following covalent compounds:

a) Br_2O	b) AsF_5
c) S_2Cl_2	d) P_2S_3
e) SiH_4	f) CCl_4
g) ClO_2	h) NCl_3

3. Write chemical formulae corresponding to the following names:
iron(II) carbonate
potassium hydrogen sulfate
sodium carbonate monohydrate
copper(II) chloride
ammonium dichromate
sodium chlorate
magnesium perchlorate
manganese(II) sulfate
sodium thiosulfate pentahydrate

Section 1.6

1. Give the following in terms of the SI base or derived unit:
Example: 2 cm = 2×10^{-2} m <u>or</u> 0.02 m

a) 5 mm	b) 28 kmol
c) 1.6×10^3 mJ	d) 650 μs
e) 8.4×10^4 μg	f) 0.004 kPa
g) 27 GJ	h) 26,000 μL
i) 18 pg	j) 2.6×10^4 Mmol

Section 1.7

1. How many significant figures are represented by each of the following numbers?

a) 12.6	b) 0.0024
c) 14.600	d) 0.08800
e) 12.64 (as a logarithm)	f) exp (5.662)

2. Give each of the values below to the number of significant figures indicated in brackets. Use both conventional and scientific notation.

a) 16.7221	(3)	b) 4.44926	(3)
c) 0.001448	(2)	d) 67.2274	(4)
e) 0.0005512	(1)	f) 100.00	(4)

3. Give the following answers to the correct number of significant figures.
   a) $12.65 + 0.00336 + 107.221 + 0.61 - 74.1122$
   b) $\left[\dfrac{12.65 \times 0.00336 \times 107.221 \times 0.61}{74.1122}\right]$
   c) $21.4469 - 21.4407$
   d) $6.5 \times 10^{-3} + 2.68 \times 10^{-4} + 6.6622 \times 10^{-2}$
   e) $8.732 \times 10^{-4} - 88.561 \times 10^{-5}$
   f) $18.2273/3.6$
   g) $\ln(64.72)$
   h) $\ln(102.6)$
   i) $\exp(-4.662)$
   j) $\text{antilog}_{10}(5.6)$
   k) $3.2\,\pi$

2 STOICHIOMETRY, Part 1: Mole Relationships

2.1 The mole concept

Stoichiometry is the study of the mass/mass relationships between one reactant and another, or between reactant and product in a chemical reaction - for example, how much of compound X is needed to react completely with a given amount of compound Y. Scientific interest in stoichiometry dates back over two centuries to the Law of Conservation of Mass, the Law of Constant Composition and the Law of Multiple Proportions[1]. These Laws could only be formulated once chemical substances were studied quantitatively, and it was this quantitative perspective that led to the emergence of chemistry as a separate scientific discipline. From the foregoing ideas eventually developed the concept of the **mole** as the fundamental unit of the amount of chemical substance - that substances react with each other in simple *molar* ratios, which are not immediately obvious from the corresponding mass ratios. Today the mole concept is so important that the mole has been selected as one of the seven fundamental quantities in the SI scheme of units. The mole is the SI quantity for "amount of chemical substance", symbol n. The equation below illustrates the use of this symbol to specify the amount of water in a particular sample.

$n(H_2O) = 1.28$ mol

---

[1] Law of Conservation of Mass: Matter is neither created nor destroyed in chemical reactions; the total mass of all products of a chemical reaction is equal to the total mass of all reactants.
Law of Constant Composition: the ratios by mass of the elements in a pure chemical substance never vary.
Law of Multiple Proportions: when elements combine to give two or more binary compounds, their combining ratios by mass bear a simple proportional relationship to each other.
Example: CO and CO_2; the mass ratios O/C are 1.333 and 2.667 respectively.

This means "the amount of the chemical substance water is 1.28 mol". The identity (name or formula) of the substance in question is placed in parentheses, or occasionally as a subscript. Note that the units are included.

Example: *Write an equation to indicate the presence of 2.3 micromoles of hemoglobin.*

Answer: *n(hemoglobin) = 2.3 μmol.*

Notice that it does not matter that we do not know the chemical formula of hemoglobin. Either a name or a chemical formula (or even a convenient abbreviation) can be put in the bracket.

2.2 The mole defined

Definition

The ***mole*** *is defined as the amount of substance which contains the same number of "entities" as the number of atoms in exactly 12 grams of the common isotope of carbon.*

Let's look at this definition in a little detail. First, the common isotope of carbon is the one of mass 12 *i.e.*, ^{12}C. The definition has the effect of making the mass of 1 mole of this isotope exactly 12 grams, *i.e.*, 12.0000.... grams. Second, the mole is defined in terms of the number of particles. However many atoms of ^{12}C there might be in 12.00000 grams of this isotope, there will be the same number of atoms or molecules in 1 mole of any other substance. The actual number of atoms in 12 grams of ^{12}C is 6.022×10^{23} (Avogadro's number): this is an experimental number, and not a value obtained from theory. It follows that 1 mole of any substance contains 6.022×10^{23} of the appropriate formula units. Therefore there is always a constant number of entities in a mole[2].

The "entities" referred to may be atoms, ions, or molecules, but they must be specified. It is correct to refer to a mole of sodium (atoms), a mole of water (molecules), a mole of sulfate ions (SO_4^{2-}), or even a mole of a so-called "non-stoichiometric" compound such as copper hydride (which has the formula $CuH_{0.9}$). However, you must be careful to avoid ambiguity. Even the term "a mole of oxygen" is ambiguous, unless you specify whether it is a mole of oxygen atoms (16 grams of oxygen) or a mole of oxygen molecules (32 grams of oxygen). It follows that you cannot speak meaningfully of a mole of substance unless you know the chemical formula.

---

[2] An analogy for this definition of the mole would be to define a dozen in the following way: a dozen is the same number of objects as the number of eggs in a carton. You would have to count the eggs in the box, in order to discover that there are 12 objects in a dozen.

A problem with the mole concept is that Avogadro's number is inconceivably large. There are 5 billion people on Earth, but this would only amount to 8×10^{-15} mol of us - always supposing we could write a chemical formula! By the same token, the masses of individual atoms and molecules are inconceivably small. Each of the 6.022×10^{23} atoms of ^{12}C in a mole weighs only 2×10^{-23} gram. It is because atoms and molecules are so incredibly small that chemistry seems to many people to be a very abstract subject. Atoms and molecules are outside the range of our senses - we cannot see one or weigh one. Put another way, the small sizes of these particles mean that the numbers of atoms and molecules in macroscopic (that is, detectable to the senses) amounts of matter are unimaginably large, and can only be discussed in terms of scientific notation. This places a very real barrier between chemistry and the non-scientist. (The scientist still cannot really imagine the number of entities in a mole, but learns to accept it!)

Example: *Calculate the total number of atoms in 0.05 mol of water (about 1 mL).*

Answer: $n(H_2O) = 0.05$ *mol*

number of water molecules $= 0.05 \text{ mol} \times (6.022 \times 10^{23} \text{ molecules mol}^{-1})$

$= 3 \times 10^{22}$ *molecules*

number of atoms $= 3 \times 10^{22} \text{ molecules} \times (3 \text{ atoms}/1 \text{ molecule})$

$= 9 \times 10^{22}$ *atoms*

2.3 Molar mass

The **molar mass** is defined as the mass in grams of 1 mol of substance. The molar mass of any compound is readily calculated from tables (the molar masses of all the elements are given in the Periodic Table at the front of this book), provided that you know the chemical formula of the substance. We use the symbol **M** for molar mass, with the name or formula of the substance in parenthesis or as a subscript, *e.g.*, $\mathbf{M}(CO_2) = 44.0103$ g mol^{-1} means that the molar mass of carbon dioxide is 44.0103 grams per mole; likewise $\mathbf{M}_{HCl} = 36.461$ g mol^{-1} means that the molar mass of hydrogen chloride is 36.461 grams per mole.

The molar masses of the elements reported in tables are the weighted averages of all the naturally occurring isotopes of the element in question. For example, the molar mass of natural carbon is 12.0115 g mol^{-1}, and not 12.0000 g mol^{-1}, because natural carbon contains small amounts of other isotopes besides the commonest one, ^{12}C.

Example: *The element boron has two natural isotopes, ^{10}B, 19.78%, molar mass 10.0129 g mol^{-1}, and ^{11}B, 80.22%, molar mass 11.0093 g mol^{-1}. Calculate the molar mass of natural boron.*

Answer: *Since the molar mass of natural boron is a weighted average, we can write:*

$$
\begin{aligned}
\mathbf{M}(B) &= 0.1978 \times \mathbf{M}(^{10}B) + 0.8022 \times \mathbf{M}(^{11}B) \\
&= (0.1978 \times 10.0129\ g\ mol^{-1}) + (0.8022 \times 11.0093\ g\ mol^{-1}) \\
&= 10.81\ g\ mol^{-1}
\end{aligned}
$$

The molar mass of a molecular or ionic substance is obtained as the sum of the molar masses of all its elements. For example, the molar mass of carbon dioxide CO_2 is 12.0115 + (2 x 15.9994) = 44.0103 g mol^{-1}, and the molar mass of sulfate ion (SO_4^{2-}) is 32.06 + (4 x 15.9994) = 96.06 g mol^{-1}. Note in the latter example that the mass of the extra two electrons that give the sulfate ion its 2- charge is ignored; the mass of an electron is insignificant compared with the masses of the atoms.

2.4 Calculation of number of moles given the mass of substance

The conversion factor which links mass with the number of moles of substance involves the molar mass - reasonably enough, since the units of molar mass are grams per mole.

Definition

molar mass = *mass (g)/amount of substance (mol)*

Example: *Calculate the number of moles in 27 grams of ethyl alcohol, C_2H_6O.*

Answer: *First calculate the molar mass of ethyl alcohol. Note that since the mass in grams was given only to two significant figures, it is not necessary to calculate the molar mass to great precision.*

$$
\begin{aligned}
\mathbf{M}(C_2H_6O) &= (2 \times 12.0\ g\ mol^{-1}) + (6 \times 1.0\ g\ mol^{-1}) + (1 \times 16.0\ g\ mol^{-1}) \\
&= 46.0\ g\ mol^{-1} \\
n(C_26H_6O) &= 27\ g/(46.0\ g\ mol^{-1}) = 0.59\ mol \qquad \text{(2 significant figures)}
\end{aligned}
$$

Comment: Many students have difficulty deciding whether the mass of substance should be divided or multiplied by the molar mass, or which quantity is divided by which. To make this easier, **always** include the units in your calculations, and make sure that they cancel such that the answer has the right units. In the example above:

$$n(C_2H_6O) = 27 \text{ g x } \left[\frac{1 \text{ mol}}{46.0 \text{ g}}\right] = 0.59 \text{ mol}$$

The answer is in moles, as it should be. Notice that dividing by 46.0 g mol $^{-1}$ is the same as multiplying by 1 mol per 46.0 g.

Earlier, we defined the mole in terms of convenient amounts of substance for laboratory experiments. However, the amounts of substance we encounter in practical situations are often very large (how much water is there in the oceans?) or very small (how much dioxin is present in a liter of human milk?). Fortunately, we can use the SI prefixes to avoid the need to carry large positive and negative exponents.

Example: *Soil at an industrial site is contaminated with 1.7 μg trichlorophenol per gram of soil. The molar mass of trichlorophenol is 198 g mol $^{-1}$. Calculate the number of moles of trichlorophenol (TCP) in a 25 g sample of this soil.*

Answer:

Mass of TCP in 25 g soil = 25 g soil x 1.7 μg TCP per g soil
= 42.5 μg TCP (carrying one extra sig. figure)
n(TCP) = 42.5 μg/198 g mol $^{-1}$
= 0.21 μmol

Note that if the mass is in μg, the amount of substance automatically works out in μmol. To see this more clearly, the last step of the calculation is rewritten below.

$$n(\text{TCP}) = 42.5 \times 10^{-6} \text{ g}/198 \text{ g mol}^{-1} = 2.1 \times 10^{-7} \text{ mol}$$

This is the same as 0.21 x 10 $^{-6}$ mol, or 0.21 μmol. The same principle will automatically apply to any SI prefix.

Example: *Calculate the number of moles of copper in 5,500 tonnes of ore which has a copper content of 1.48%. (Remember that 1 tonne = 1000 kg = 1 Mg)*

Answer: *Assume that 5,500 t implies 2 significant figures, and carry one extra figure through the calculation till the end.*

mass of Cu = 5,500 tonnes ore x (1.48 tonnes Cu/100 tonnes ore)
= 81.4 tonnes
*n(Cu) = 81.4 Mg/**M**(Cu) = 81.4 Mg/63.6 g mol $^{-1}$*
= 1.3 Mmol (or 1.3 x 10 ^{6} mol)

Again, the last step in the calculation is written out in full below.

n(Cu) = 81.4 Mg/63.6 g mol $^{-1}$ = 81.4 x 10 ^{6} g/63.6 g mol $^{-1}$
= 1.3 x 10 ^{6} mol = 1.3 Mmol

2.5 Molecular formula and percentage composition

One of the first things we need to know about an unknown substance is its molecular formula. The molecular formula tells us the numbers of each kind of atom in one molecule of the compound. For example, the molecular formula of benzene is C_6H_6, telling us that each benzene molecule contains six carbon atoms and six hydrogen atoms. The molecular formula is not necessarily the same as the simplest formula, which merely gives the **ratio** of the atoms in the molecule. For example, the ratio of carbon atoms to hydrogen atoms in benzene is 6:6, which is the same as 1:1, and so the simplest formula of benzene is CH.

Two steps are needed to determine the molecular formula of a substance: (i) obtain the simplest formula; (ii) obtain the molar mass, in order to find out how many units of the simplest formula are present in each molecule.

2.5.1 Simplest formula

The simplest formula of any substance is obtained by chemical analysis, in which the analyst converts one of the elements in the substance into a well defined form, such as a solid of well defined composition. For example, all the barium in a sample might be converted into barium sulfate, which can be filtered, dried, and weighed. The amount of barium sulfate (mol) can then be calculated from the mass of the solid. In another example, all the iron in a sample might be converted to a highly coloured derivative (called ferrous phenanthroline), and the concentration of iron determined from the intensity of the colour, using an instrument called a spectrometer.

Whatever analytical method, direct weighing or an instrumental method, is chosen, it is essential that the conversion of the chosen element into its derivative can be carried out quantitatively. The practice of analytical chemistry has been to develop procedures that are reliable for this purpose.

The amount of the element analysed is usually reported as a percent by mass, since at this stage, the molecular formula of the compound is not yet known. The sample is analysed element by element until all its mass has been accounted for (*i.e.*, the sum of all the percentages by mass reaches 100%). From this information, we can determine the **simplest formula**, sometimes called the **empirical formula**[3], as shown in the example at the top of p. 31.

Analytical methodology is very well developed for the analysis of organic (carbon-containing) compounds. The chemistry involves the complete oxidation of a known mass of the compound to CO_2 and H_2O, since all organic compounds contain carbon, and virtually all also contain hydrogen.

---

[3] "Empirical" means found by experiment, as opposed to deduced from a theory.

Example: *A compound is known to contain only phosphorus and chlorine. A 0.1176 g sample of the compound is dissolved in water, and $AgNO_3$ solution is added. This causes all the chlorine in the sample to be precipitated as AgCl. The AgCl is filtered, dried, and weighed: mass 0.3680 g. Calculate the simplest formula of the compound.*

Answer: *We work backwards from the information we have (the mass of AgCl) to the information we want (the formula of the compound). The link between these pieces of information is that* all *the chlorine in the original compound was converted into AgCl. Furthermore, since the original compound contained only P and Cl, any mass not accounted for as Cl must have been present as P.*

$$M(AgCl) = M(Ag) + M(Cl) = 107.87 + 35.45 = 143.32\ g\ mol^{-1}$$
$$n(AgCl) = 0.3680\ g/143.32\ g\ mol^{-1} = 2.568 \times 10^{-3}\ mol$$

From the formula of silver chloride,

$$n(Cl) = n(AgCl) = 2.568 \times 10^{-3}\ mol$$
$$mass(Cl) = 2.568 \times 10^{-3}\ mol \times 34.45\ g\ mol^{-1} = 0.09102\ g$$

Therefore, by difference:

$$mass(P) = 0.1176 - 0.09102 = 0.02658\ g$$

Now find the molar ratio:

$$n(P) = 0.02658\ g\ /\ 30.97\ g\ mol^{-1} = 8.582 \times 10^{-4}\ mol$$
$$n(Cl) = n(AgCl) = 2.568 \times 10^{-3}\ mol\ \text{(calculated earlier)}$$
$$n(Cl)/n(P) = 2.568 \times 10^{-3}\ mol/8.582 \times 10^{-4}\ mol = 2.992$$

This value is very close to 3, suggesting the mole ratio n(Cl) : n(P) = 3 : 1. Hence the simplest formula of the compound is PCl_3.

These products (CO_2 and H_2O) are measured quantitatively. All the carbon atoms of the sample are converted to CO_2, and all the hydrogen atoms to H_2O. Although many organic compounds also contain oxygen, it is not possible to analyse quantitatively for oxygen in this way, and so the percent by mass of oxygen is usually taken as the difference between 100% and the sum of the percentages by mass of all other elements. Today, the complete analysis is routinely carried out using less than 1 mg of the sample.

Example: *Analysis of a compound known to contain only C, H, and O gave the following result: C, 71.42%; H, 9.52%. What was the simplest formula of the compound?*

Answer: *The two percentages reported add up to 80.94%. Therefore, the compound also contained oxygen, 19.06%.*

The mole ratio C: H: O can be obtained by realizing that the percentages are equal to the masses of C, H, and O in a 100 g sample of the compound.

$$n(C) = \text{mass C/molar mass C} = 71.42\ g/12.01\ g\ mol^{-1} = 5.947\ mol$$
$$n(H) = \text{mass H/molar mass H} = 9.52\ g/1.008\ g\ mol^{-1} = 9.44\ mol$$
$$n(O) = \text{mass O/molar mass O} = 19.06\ g/16.00\ g\ mol^{-1} = 1.191\ mol$$

Therefore the ratio C: H: O is 5.947 : 9.44 : 1.191. Dividing through by the smallest of these numbers gives the ratio C : H : O as 4.99 : 7.93 : 1, and so the most likely simplest formula is C_5H_8O.

2.5.2 *Molecular formula*

It is not possible to determine the molecular formula on the basis of elemental analysis alone. Consider this series of compounds:

CH_2O $C_2H_4O_2$ $C_3H_6O_3$

All these substances have the same elemental composition: C, 40.0; H, 6.7; O, 53.3%. On the basis of elemental analysis they are indistinguishable, and so the analytical data C, 40.0; H, 6.7; O, 53.3% could apply equally to any of them. In order to deduce the molecular formula from the simplest formula, another piece of information is required: this is the **molar mass**. Most commonly today, the molar mass of the compound is obtained with the aid of a complex analytical instrument called a mass spectrometer[4].

Example: *Analysis of the highly toxic substance dioxin for carbon, hydrogen, and chlorine gave the following result: C, 44.72; H, 1.25; Cl, 44.10%. The molar mass of dioxin was found to be 322 g mol^{-1}. Determine the molecular formula of dioxin.*

Answer: *The percentages reported add up to 90.07%. Therefore, the compound also contained oxygen, 9.93%.*

The mole ratio C: H: Cl: O can be obtained as in the previous example by realizing that the percentages are equal to the corresponding masses in a 100 g sample of dioxin.

$n(C) = \text{mass C/molar mass C} = 44.72\ g/12.01\ g\ mol^{-1} = 3.723\ mol$

$n(H) = \text{mass H/molar mass H} = 1.25\ g/1.008\ g\ mol^{-1} = 1.240\ mol$

$n(Cl) = \text{mass Cl/molar mass Cl} = 44.10\ g/35.45\ g\ mol^{-1} = 1.244\ mol$

$n(O) = \text{mass O/molar mass O} = 9.93\ g/16.00\ g\ mol^{-1} = 0.621\ mol$

Dividing through by the smallest of these numbers, as before, gives the ratio C : H : Cl : O as 5.999 : 1.998 : 2.005 : 1, and so the most likely simplest formula is $C_6H_2Cl_2O$.

The molar mass corresponding to the simplest formula is:

$$(6 \times 12.0) + (2 \times 1.0) + (2 \times 35.5) + (1 \times 16.0) = 161\ g\ mol^{-1}$$

Therefore the number of units of the simplest formula in the molecular formula is 322 g mol^{-1}/161 g mol^{-1} = 2. Hence the molecular formula of dioxin is $C_{12}H_4Cl_4O_2$.

---

[4] Although the working of a mass spectrometer is beyond the scope of this course, we may note here that the sample is introduced into the mass spectrometer in the gaseous form and, under the influence of a beam of high energy electrons, its molecules (M) are ionized into cations (M^+). In one common form of mass spectrometer, these cations are subjected to a magnetic field, which causes their trajectory to follow a curved path inside the instrument. The larger the molar mass, the less curved the trajectory for a given strength of the magnetic field. The molar mass of M^+, and hence that of M, can be deduced from the strength of the magnetic field needed to cause the ions to follow a trajectory of fixed curvature, and thence pass into a detector.

2.6 Solutions

Many reactions take place in solution, with aqueous solutions being most commonly encountered. In this section, we show that the mole concept is equally as applicable to substances in solution as it is to pure elements and compounds.

Concentration is always a ratio between the amount of substance and the amount of the matrix in which it is present. A very useful measure of concentration, the **molar concentration**, is the amount of substance per unit volume; its units are moles per liter.

$$\text{conc(X)} = \frac{\text{moles of X}}{\text{volume of solution, L}}$$

Many books use the symbol $\underline{M}$ for molar concentration, but we shall avoid doing so in this textbook, because of the possible confusion with **M** for molar mass. Instead we shall use mol L^{-1} or similar units. Other units of concentration include grams per liter, moles per kilogram, and grams per 100 grams (commonly called weight percent).

Notice that if we have two solutions, one containing one mole of solute in 1 L of water, the other one containing 1 μmol of solute in 1 μL of water, they both have the same concentration (1 mol L^{-1}), even though the amounts of substance (moles) differ by a factor of 1 million! In the same way, if you pour orange juice from the same jug into both a small wine glass and a large beer stein, the concentrations in the wineglass and the beer stein are the same, even though the total amounts of solution differ greatly.

From the equation above, we see that the number of moles of solute are obtained thus:

$$n(\text{X}) = \text{c(X)} \times \text{V(soln)}$$

Example: *How many moles of sodium ions are contained in 25 mL of a 4.6 mmol* L^{-1} *solution of sodium sulfate,* Na_2SO_4*?*

Answer:

$$\begin{aligned} n(Na_2SO_4) &= c(Na_2SO_4) \times V(soln) \\ &= (4.6\ mmol\ L^{-1}) \times (25 \times 10^{-3}\ L) \\ &= 0.115\ mmol \end{aligned}$$

However, there are 2 mol Na^+ *for each 1 mol* Na_2SO_4*:*

$$n(Na^+) = n(Na_2SO_4) \times (2\ mol\ Na^+/1\ mol\ Na_2SO_4) = 0.23\ mmol$$

2.7 Concentrations in parts per million, parts per billion, etc

Environmental concentrations are often reported in units such as parts per million (ppm), parts per billion (ppb) etc, especially in the news media. These units are used to describe the concentrations of substances, usually contaminants, at very low levels. *They are defined by mass, not by moles.* A definition by mass is an advantage if the chemical composition of the contaminant is unknown or if the contaminant is present in a solid. These units are applicable both to solids and liquids.

One part per million, ppm, means 1 part by mass in a million, 1 in 10^6. It could therefore be 1 gram in 1 million grams, 1 microgram in a gram, 1 milligram per kilogram, etc, etc. (To get a feel for 1 ppm, think of sharing one Aspirin, or ASA, tablet (about 300 mg), among five 60 kg adults.) Thus a soil contaminated with 5 ppm lead contains 5 μg Pb per gram soil, or 5 grams per tonne. It would not be practical to describe the concentration in terms of moles of lead per liter of soil.

In the specific case of water as solvent, 1 ppm is the same as 1 mg L^{-1}.

density of water = 1 g cm^{-3} = 1 kg L^{-1}

1 ppm = 1 mg kg^{-1} = 1 mg kg^{-1} x 1 kg L^{-1} = 1 mg L^{-1}

One part per billion, ppb, means 1 part by mass in a billion, 1 in 10^9. It could therefore be 1 gram in 1 billion grams, 1 microgram per kg, 1 milligram per tonne, etc. (To get a feel for 1 ppb, think of dissolving one small crystal of salt from the salt shaker (about 1 μg for a crystal 0.1 mm x 0.1 mm x 0.1 mm) in 3 servings of soup.) Drinking water contaminated with 1 ppb mercury contains 1 μg mercury per liter.

One part per trillion, ppt, means 1 part by mass in a trillion, 1 in 10^{12}. It could therefore be 1 gram in 1 trillion grams (same as 1 gram per million tonnes), 1 microgram per tonne, etc. To get a feel for 1 ppt, try to imagine dissolving that same small crystal of salt in 100 cases of beer. Great Lakes water contaminated with 1 ppt polychlorinated biphenyls (PCBs) contains 1 ng PCB per liter.

Example*: Calculate (i) the mass (ii) the number of moles of dioxin (molar mass 322 g mol^{-1}) in 150 mL of human milk which contains 12.2 ppt dioxin. Assume milk has a density of 1 g/mL.*

Part (i): *Moles are not involved in this calculation, so we shall not need to use the molar mass.*

	conc of dioxin	=	*12.2 x 10^{-12} g dioxin/1 g milk (same as 12.2 pg dioxin/1 g milk)*
	mass of dioxin	=	*(12.2 pg dioxin/1 mL milk) x 150 mL milk*
		=	*1830 pg (same as 1.83 ng)*
Part (ii):	*n(dioxin)*	=	*1830 pg/322 g mol^{-1}*
		=	*5.7 pmol*

One must be very careful not to become confused when the units ppm, ppb, and ppt are used. Consider the value 3 ppb as the concentration of a contaminant in drinking water. The value 3 ppb corresponds to 3 $\mu g\ L^{-1}$ in water. Suppose this quantity were reported as 3000 ppt; it is still the same quantity, since 3000 ppt is the same as 3 ppb; however, the number used is much larger, and anyone's first reaction would be to think that the sample must be very contaminated. Conversely, the value 0.003 ppm looks like a very small quantity, suggesting a very low level of contamination, but still refers to the same 3 $\mu g\ L^{-1}$. The point of this paragraph is to emphasize that the units chosen for concentration can be used to manipulate your impression of the level of contamination of an environmental sample.

2.8 Dilution

Frequently we need to know how much of a "stock" solution needs to be taken to produce a less concentrated solution having a known concentration. The key point to realize is that **the amount of solute in the diluted solution is exactly the same as the amount of solute taken from the stock solution**. This has to be true, because you only add water (just as you only add water when you dilute frozen concentrated orange juice - the concentration is what changes, not the amount of substance). The idea is seen in the following examples, and is equally applicable to concentrations in ppm etc, as to concentrations in moles per liter.

Example 1: *Calculate the volume of a 1.2 mol L^{-1} hydrochloric acid solution needed to prepare 150 mL of 0.18 mol L^{-1} hydrochloric acid solution.*

Answer: *First calculate the amount of substance needed in the final, diluted solution.*

$$n(HCl) = (150 \times 10^{-3}\ L) \times (0.18\ mol\ L^{-1}) = 2.7 \times 10^{-2}\ mol$$

$$\text{or: } n(HCl) = (150\ mL) \times (0.18\ mol\ L^{-1}) = 27\ mmol$$

$$n(HCl\ in\ stock\ solution) = 2.7 \times 10^{-2}\ mol$$

$$V(stock\ solution) = n(HCl)/c(HCl) = 2.7 \times 10^{-2}\ mol/1.2\ mol\ L^{-1} = 2.25 \times 10^{-2}\ L = 23\ mL$$

Thus, to get the desired solution, you must take 23 mL of the stock solution, and add water till the final volume is 150 mL.

Example 2: *What volume of a stock solution of NaF, containing 1.0% fluoride ion by weight, would be needed to produce 1.0 million liters of fluoridated water having fluoride concentration 1.2 ppm?*

Answer: *Again calculate the amount of substance needed in the final, diluted solution. In this example, we work in terms of mass, not moles. Recall that in aqueous solution, 1 ppm = 1 mg L^{-1}.*

$$\text{mass of } F^- = (1.0 \times 10^6 \text{ L}) \times (1.2 \text{ mg } F^-/1 \text{ L solution})$$
$$= 1.2 \times 10^6 \text{ mg} \quad \text{(same as 1.2 kg)}$$

The initial, stock solution was 1.0% F^- by weight; that is, 1.0 g F^- per 100 g (or 100 mL) water. So we need to know how much stock solution to take in order to obtain the requisite 1.2 kg.

$$V(\text{stock solution}) = \text{mass}(F^-)/c(F^-)$$
$$V(\text{stock}) = 1.2 \text{ kg}/(1.0 \text{ g}/100 \text{ mL})$$

Let's make the concentration units more convenient for cancellation:

$$1.0 \text{ g}/100 \text{ mL} = 10.0 \text{ g L}^{-1} = 1.0 \times 10^{-2} \text{ kg L}^{-1}$$
$$V(\text{stock}) = 1.2 \text{ kg}/(1.0 \times 10^{-2} \text{ kg L}^{-1}) = 1.2 \times 10^2 \text{ L}$$

Therefore, water must be added to 1.2×10^2 L of the stock solution until the final volume is 1.0×10^6 L. Actually, in this example, the volume of concentrated solution is so small compared with the final volume that you could simply pour 1.2×10^2 L of stock solution into 1 million liters of water, without introducing any significant error.

Dilution is a very useful technique in the laboratory, where we frequently make a whole series of solutions from a given stock solution. The advantage is that the solute only needs to be weighed out once to make the stock solution; all other operations can be done quickly using pipettes. Look at the following two examples, and then make sure that you understand how the dilutions were calculated.

Example 1: *We have a 100 ppm stock solution of the pesticide malathion, and need to prepare a series of less concentrated solutions in order to calibrate an analytical instrument. The solutions required are to be 100 mL each of 10, 20, 40, 60, 80, 100 ppm malathion. The calculated amounts of stock solution and water are given in the table below. Make sure that you understand how the values in the table are calculated.*

conc required	volume of stock soln	volume of water
10 ppm	10 mL	90 mL
20	20	80
40	40	60
60	60	40
80	80	20
100	100	none

Example 2 illustrates "serial dilution", which is done when we need to prepare a series of solutions of widely differing concentration. Suppose that this time we want to prepare, from the same 100 ppm stock solution of malathion, 100 mL each of a series of solutions that are 10, 1, 0.1, 0.01, and 0.001 ppm. If we proceeded as in Example 1, the volumes of stock solution being diluted to 100 mL would get as small as 0.001 mL. This is difficult to do accurately. Instead, we use serial dilution as follows.

- *Solution A (10 ppm) is 10 mL of stock solution diluted to 100 mL*
- *Solution B (1 ppm) is 10 mL of Solution A diluted to 100 mL*
- *Solution C (0.1 ppm) is 10 mL of Solution B diluted to 100 mL*
- *Solution D (0.01 ppm) is 10 mL of Solution C diluted to 100 mL*
- *Solution E (0.001 ppm) is 10 mL of Solution D diluted to 100 mL*

The chief disadvantage of serial dilution is the "propagation of errors". Any error in making Solution B from Solution A is "propagated", or carried through, into every subsequent solution. Thus Solution E combines the accumulated errors in the preparation of Solutions A through D.

Problems

Section 2.1

1. Write equations to represent the following statements.
   a) the amount of carbon dioxide is 0.227 moles
   b) the amount of iron(II) chloride is 67 micromoles
   c) the amount of methane is 84.6 kilomoles
   d) the amount of bromine trifluoride is 2.62 millimoles
   e) the mass of water is 28 grams
   f) the mass of copper is 67.3 kilograms
   g) the volume of mercury is 12.44 milliliters
   h) the pressure of molecular nitrogen is 77 kilopascals
   i) the volume of gaseous argon is 47 cubic meters
   j) the pressure of sulfur dioxide is 14.27 atmospheres

Section 2.2

1. Calculate the number of moles corresponding to:
   a) 2.6×10^{20} atoms of helium
   b) 7.22×10^{27} molecules of methane
   c) 48 molecules of water
   d) 2,000,000,000 sulfate ions

2. Calculate the number of:

   a) molecules in 0.27 moles of carbon dioxide
   b) molecules in 43 pmol of dioxin
   c) atoms in 7.32 x 10^{-1} μmol of ammonia (count all the atoms)
   d) ions in 5.82 mmol of calcium chloride (count all the ions)
   e) water molecules in 14.72 nmol of copper(II) sulfate pentahydrate

3. Tell whether the following statements are defined or undefined.

   a) 0.2 mol of molecular nitrogen
   b) 5.6 mol of atomic hydrogen
   c) 2.4 mmol of ozone
   d) 0.13 μmol of chlorine
   e) 6.4 kmol of iron

Sections 2.3 - 2.4

1. Calculate the mass in grams of 0.38 mol of the element scandium.

2. Calculate the number of chlorine atoms in 28.6 g of chloroform $CHCl_3$.

3. Calculate i) the number of moles, and ii) the mass in grams represented by 3.72 x 10^{21} molecules of ethyl alcohol C_2H_6O.

4. Calculate the mass in grams of 0.112 moles of chlorate ions.

5. Calculate the mass of one atom of iodine.

6. Calculate the number of hydrogen atoms in 17.2 g of aniline C_6H_7N.

7. Calculate the number of moles of oxygen atoms in 3.402 kg of ozone (O_3).

8. Calculate the mass in micrograms of 1.17 x 10^{16} molecules of acetylene, C_2H_2.

9. Calculate the mass of sulfur in 28.62 mg of ammonium persulfate $(NH_4)_2S_2O_8$.

10. Natural carbon has molar mass 12.01115 g mol^{-1}. It consists of two principal isotopes: ^{12}C, and ^{13}C which has M = 13.00336 g mol^{-1}. What percentage of natural carbon is ^{12}C?

11. Elemental chlorine contains two isotopes: ^{35}Cl (M = 34.96885 g mol^{-1}, 75.77%) and ^{37}Cl (24.23%). The molar mass of natural chlorine is 35.453 g mol^{-1}. Calculate the molar mass of ^{37}Cl.

Section 2.5

1. Calculate the percentages by mass of all elements in sodium dihydrogen phosphate, NaH_2PO_4.

2. Calculate the empirical formula of a compound analysing for Cr, 33.57%; Cl, 45.78%; remainder oxygen.

3. A 7.3 mg sample of an organic compound was oxidized completely in oxygen, and gave 22.9 mg CO_2 and 9.4 mg H_2O. The molar mass was 83 ± 2 g mol^{-1}. What was its molecular formula?

4. A 1.000 g sample of sulfur ore was converted quantitatively to sulfate ion (SO_4^{2-}) and all the sulfate was precipitated with excess $BaCl_2$. The resulting precipitate of $BaSO_4$ weighed 7.002 g. Calculate the percent by mass of sulfur in the ore.

5. Calculate the molecular formula of aspirin, given that the compound analyses as C, 60.0; H, 4.44% (remainder oxygen) and that the molar mass is 180 g mol^{-1}.

6. Calculate the percentages by mass of all elements in saccharin, $C_7H_5NO_3S$.

7. A mothproofing substance analysed for C, 49.0; H, 2.7; Cl, 48.3%.
   a) What was its simplest formula?
   b) The molar mass of the compound was found to be 147 g mol^{-1}; what was the molecular formula?

8. A 2.2606 g sample of a hydrated calcium sulfate $CaSO_4.xH_2O$ was heated to constant weight, where x is an unknown number. The residue, which was pure $CaSO_4$, weighed 1.7874 g. What was the formula of the hydrate?

9. A 3.06 mg sample of organic compound X gave 7.86 mg CO_2, 1.83 mg H_2O, and 0.22 mg N_2 upon combustion. Its mass spectrum showed a peak indicating M^+ = 394 g mol^{-1}. What was its molecular formula?

10. Combustion analysis of 2.818 mg of peroxyacetyl nitrate, which is present in polluted air and contains C, H, N and O, gave 2.049 mg CO_2, 0.629 mg H_2O; in a separate analysis 3.70 mg of the compound was converted to 0.521 mg NH_3. What is the simplest formula of this compound?

Sections 2.6 - 2.7

1. Calculate the concentration of NaCl in a solution made up by dissolving 0.1270 g of pure NaCl in enough water to make 250.0 mL of solution. Express your answer in mol L^{-1} and in ppm.

2. Calculate the mass of ascorbic acid (vitamin C, $C_6H_8O_6$) that you would need to weigh out to obtain 100.0 mL of a solution of concentration 2.5 mmol L^{-1}.

3. What is the concentration (moles per liter) of cadmium in a waste water stream that is contaminated by 3.6 ppm of $CdCl_2$?

4. The maximum allowable concentration of copper in drinking water is 1.0 ppm. Assuming an intake of 2.0 L of drinking water per day, what is the maximum intake of copper from drinking water to which a person is exposed?

5. What is the maximum mass of caffeine $C_8H_{10}N_4O_2$ that could be extracted from 250.0 g of coffee beans which contained 485 ppm of caffeine?

6. What volume of 4.76 mmol L^{-1} $CaCl_2$ solution would be needed to provide 1.00×10^{-2} moles of chloride ions?

7. The concentration of phosphorus in the water of Lake Ontario was 10 ppb in 1989.

   (a) Calculate this in μmol L^{-1};
   (b) If all the phosphorus is assumed to be present as phosphate ion, calculate the mass of phosphate per cubic meter of water. Assume a precision of two significant figures.

Section 2.8

1. Starting with a solution of NaCl of concentration 0.12 mol L^{-1}, explain how you would prepare 150 mL of solution whose concentration of NaCl was 7.6 mmol L^{-1}.

2. Starting with a solution of Na_2SO_4 of concentration 0.272 mmol L^{-1}, explain how you would prepare 4.00 L of a solution whose concentration of Na^+ was 5.88 μmol L^{-1}.

3. A student makes the following serial dilutions of a 165 ppm solution of $BaCl_2$.

   (a) 5.0 mL diluted to 20.0 mL;
   (b) 5.0 mL of solution (a) diluted to 20 mL;
   (c) 5 mL of solution (b) diluted to 25 mL.

   What was the concentration of $BaCl_2$ in solution (c)?

4. Calculate the concentration of Hg^{2+} in a solution formed by diluting 2.5 mL of 17 ppm Hg^{2+} to 80 mL.

5. Starting with a solution containing 50 ppm F^-, explain how you would prepare 10.0 mL each of the following solutions: 25 ppm, 20 ppm, 10 ppm, 5 ppm.

6. Starting with a solution containing 10.0 mmol L^{-1} $CdCl_2$, explain how you would prepare 10.0 mL each of the following solutions 1.0 mmol L^{-1}, 0.10 mmol L^{-1}, 10 μmol L^{-1}, 1.0 μmol L^{-1} Cd^{2+}.

STOICHIOMETRY, Part 2: Chemical Equations and Chemical Reactions

3.1 Chemical equations

A balanced chemical equation is the chemist's shorthand for conveying a great deal of information in a standard format. Consider the information in this well known equation.

$$2\ H_2 + O_2 \rightarrow 2\ H_2O$$

1. The equation identifies the **reactants** (H_2 and O_2) and **product(s)** (H_2O) of the reaction. The arrow always points from the reactants to the products.

2. The equation also tells us the combining ratios of the reactants and products, and these ratios can be either **ratios of molecules** or **ratios of moles**. In this case for example, we see that 2 moles of molecular hydrogen react with one mole of molecular oxygen. Equally, 2 molecules of molecular hydrogen react with one molecule of molecular oxygen. We also see that two molecules of water are formed for each molecule of oxygen that reacts, and equally, two moles of water are formed for each mole of oxygen that reacts. Balanced chemical equations thus convey much more information than unbalanced ones: they convey the **stoichiometry** of the reaction. "Balanced" means that the total numbers of atoms and (if relevant) charges are the same on both sides of the equation. Here is a more complicated example.

$$MnO_4^- + 5\ Fe^{2+} + 8\ H^+ \rightarrow Mn^{2+} + 5\ Fe^{3+} + 4\ H_2O$$

The numbers of atoms on each side of the equation are:

Mn, 1 O, 4 Fe, 5 H, 8

The numbers of charges also balance:

reactants: $MnO_4^-(-1) + Fe^{2+}(5 \times +2) + H^+(8 \times +1) = +17$

products: $Mn^{2+}(+2) + Fe^{3+}(5 \times +3) = +17$

We could use this equation to say that 1 mol of permanganate ion reacts with 5 mol of Fe^{2+} (or that 1 permanganate ion reacts with 5 Fe^{2+} ions); that 8 H^+ ions are used up for every 5 Fe^{3+} ions that are produced, etc.

Extra information can be obtained from a chemical equation if the physical states of the reactants and products are included. Compare the two examples below: the first indicates that gaseous water (steam) is being formed, the second that the product is liquid water. We shall see that this distinction is very important when we consider the energetics of chemical reactions in Chapter 5.

$$2\ H_2(g) + O_2(g) \rightarrow 2\ H_2O(g)$$

$$2\ H_2(g) + O_2(g) \rightarrow 2\ H_2O(\ell)$$

3.2 Mass and mole relations in chemical reactions

The advantage of a balanced chemical equation is that it allows us to determine how much of one reactant is needed to react with another, or how much product is formed from a given amount of reactant, and so on. The heart of such problems lies in relating the number of moles of the different substances to each other through the use of the coefficients of the balanced chemical equation. For example, in relating mass of reactant to mass of product, our strategy involves the following steps.

Scheme 1

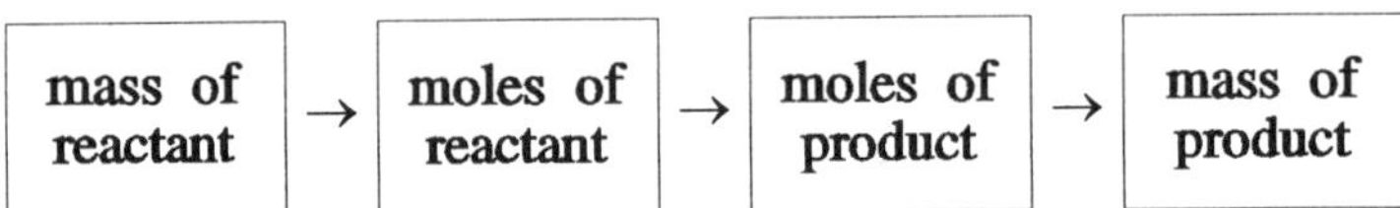

The simplest example would be a reaction where all the mole ratios are 1:1, but exactly the same approach is taken when the mole ratios are more complex.

Example 1: *Calculate the mass of carbon dioxide released to the atmosphere when 155 tonnes of limestone (calcium carbonate) are roasted in the production of cement. The equation is given below.*

$$CaCO_3 \rightarrow CaO + CO_2$$

Answer: *We note that 1 mol $CaCO_3$ affords 1 mol CO_2. Scheme 1 indicates that this calculation requires three steps.*

Step 1: *Calculate molar amount of $CaCO_3$, using the molar mass of $CaCO_3$ (100 g mol^{-1}).*

$n(CaCO_3) = 155$ Mg $CaCO_3/(100$ g $mol^{-1}) = 1.55$ Mmol

Step 2: *Determine the mole ratio from the balanced chemical equation - in this case 1:1*

$n(CO_2) = n(CaCO_3) = 1.55$ Mmol

Step 3: *Calculate mass of CO_2, using molar mass of CO_2 (44 g mol^{-1})*

mass of $CO_2 = 1.55$ Mmol x 44 g $mol^{-1} = 68$ Mg (same as 68 t)

Example 2: *Calculate the mass of lime ($Ca(OH)_2$) that would be needed to neutralize completely 26 kg of phosphoric acid (H_3PO_4) in an aqueous waste stream, assuming the stoichiometry below.*

$$2\ H_3PO_4 + 3\ Ca(OH)_2 \rightarrow Ca_3(PO_4)_2 + 6\ H_2O$$

Answer: *First calculate the molar masses of the relevant substances, in this case H_3PO_4, 98g mol^{-1} and $Ca(OH)_2$, 74g mol^{-1}. The calculation then proceeds as in the previous example.*

Step 1: *Calculate molar amount of H_3PO_4:*

$n(H_3PO_4) = 26$ kg/98 g $mol^{-1} = 0.265$ kmol

Step 2: *Determine the mole ratio from the balanced chemical equation.*

$n(Ca(OH)_2) = n(H_3PO_4)$ x (3 mol $Ca(OH)_2$/2 mol H_3PO_4)
$= 0.265$ kmol x (3 mol $Ca(OH)_2$/2 mol H_3PO_4)
$= 0.398$ kmol

Step 3: *Calculate mass of lime needed.*

mass($Ca(OH)_2$) $= 0.398$ kmol x 74 g $mol^{-1} = 29$ kg

The balanced chemical equation can also be used to "work backwards" to determine how much reactant would be needed to form a given amount of product. The next example is of interest because the substance involved, chlorine dioxide, is used in some communities to disinfect drinking water (Chapter 10). It is used as an alternative to chlorine; in Ontario, several communities switch temporarily from chlorine to chlorine dioxide when they encounter taste or odour problems with the water. To place this problem in perspective, note that the average person drinks about 2 L of water per day, but that a typical household consumes about 1 m^3 (1000 L) of water daily, mostly for bathing, laundry, flushing toilets etc.

Example: *What mass of sodium chlorite ($NaClO_2$) would be needed to prepare enough chlorine dioxide (ClO_2) in order to disinfect 1 million L of drinking water with 2.2 ppm of ClO_2? The equation is given below:*

$$10\ NaClO_2 + 5\ H_2SO_4 \rightarrow 8\ ClO_2 + 5\ Na_2SO_4 + 2\ HCl + 4\ H_2O$$

Answer: *From the definition of ppm in aqueous solution (1 ppm (aq) = 1 mg L^{-1}), the mass of ClO_2 to be prepared is:*

$$\text{mass}(ClO_2) = (1.0 \times 10^6 L) \times (2.2\ mg\ L^{-1}) = 2.2 \times 10^6 mg \qquad \text{(same as 2.2 kg)}$$

We now proceed as in Scheme 1. Molar masses are $\mathbf{M}(ClO_2) = 67.5\ g\ mol^{-1}$;

$$\mathbf{M}(NaClO_2) = 90.5\ g\ mol^{-1}.$$

Step 1: *Calculate $n(ClO_2)$:*

$$n(ClO_2) = 2.2\ kg/67.5\ g\ mol^{-1} = 3.26 \times 10^{-2}\ kmol \Rightarrow 32.6\ mol$$

Step 2: *Determine the mole ratio from the balanced chemical equation.*

$$n(NaClO_2) = n(ClO_2) \times (10\ mol\ ClO_2/8\ mol\ ClO_2)$$
$$= 32.6\ mol \times (10\ mol\ ClO_2/8\ mol\ ClO_2) = 40.7\ mol$$

Step 3: *Calculate the mass of sodium chlorite needed.*

$$\text{mass}(NaClO_2) = 40.7\ mol \times 90.5\ g\ mol^{-1} = 3.69 \times 10^3\ g \Rightarrow 3.7\ kg$$

3.3 Sequential reactions

It is not always possible to proceed from the reactants to a desired product by way of a single chemical reaction. Often, a sequence of chemical transformations is needed in order to effect a particular chemical change. In that case, it is necessary to follow the amounts of substances through the sequence of reactions step by step. A similar procedure is followed to determine how much reactant is needed to make a certain amount of product, only this time you work backwards from the product to the reactant step by step. These ideas are illustrated in the next two examples: the commercial production of nitric acid and the smelting of zinc ore to make the metal.

Example 1: *How many moles of HNO_3 could theoretically be produced from 1000 mol H_2, without considering recycling of the NO formed in the last step of the process? The reactions are:*

$$N_2 + 3\ H_2 \rightarrow 2\ NH_3$$
$$4\ NH_3 + 5\ O_2 \rightarrow 4\ NO + 6\ H_2O$$
$$2\ NO + O_2 \rightarrow 2\ NO_2$$
$$3\ NO_2 + H_2O \rightarrow 2\ HNO_3 + NO$$

Answer: *We must follow the number of moles, reaction by reaction.*

Reaction 1: $n(NH_3) = 1000\ mol\ H_2 \times (2\ mol\ NH_3/3\ mol\ H_2) = 667\ mol\ NH_3$

Reaction 2: $n(NO) = n(NH_3) = 667\ mol$

Reaction 3: $n(NO_2) = n(NO) = 667\ mol$

Reaction 4: $n(HNO_3) = 667\ mol\ NO_2 \times (2\ mol\ HNO_3/3\ mol\ NO_2) = 445\ mol$

Example 2: *What mass of zinc sulfide would be needed to produce 1500 t of zinc according to the following reactions?*

$$2\ ZnS + 3\ O_2 \rightarrow 2\ ZnO + 2\ SO_2$$
$$ZnO + C \rightarrow Zn + CO$$

Answer: *Work backwards from the information given (mass of Zn) to that required (mass of ZnS).*

$n(Zn) = 1500\ Mg/65.4\ g\ mol^{-1} = 22.9\ Mmol$

Reaction 2: $n(Zn) = n(ZnO) = 22.9\ Mmol$

Reaction 1: $n(ZnS) = n(ZnO) = 22.9\ Mmol$

Calculate mass of ZnS:

$\text{mass } ZnS = 22.9\ Mmol \times \mathbf{M}(ZnS) = 22.9\ Mmol \times 97.5\ g\ mol^{-1}$

$= 2.2 \times 10^3\ Mg \Rightarrow 2300\ t$

3.4 Limiting reactants

So far, we have not considered that a chemical reaction might stop short of completion because of a shortage of one of the reactants. Let's go back to our example of the combination of elemental hydrogen and oxygen to form water.

$$2\ H_2 + O_2 \rightarrow 2\ H_2O$$

Let us imagine introducing 1 mol of O_2 into a flask along with various quantities of H_2 and initiating the reaction between them. Here are some possibilities.

H_2, mol	O_2, mol	Result
1.0	1.0	Reaction stops after 1.0 mol H_2O is formed; all the H_2 is used up, but 0.5 mol O_2 is left over
2.0	1.0	Reaction goes to completion with 2.0 mol H_2O formed; no H_2 or O_2 is left over
3.0	1.0	Reaction stops after 2.0 mol H_2O are formed; all the O_2 is used up, but 1 mol H_2 is left over

In the first case, all the hydrogen is used up and the reaction stops because it runs out of hydrogen; we would say that hydrogen is the **limiting reactant** and that oxygen was present **in excess** (some was left over). Correspondingly in the third reaction oxygen is the limiting reactant and hydrogen is present in excess. Only in the second reaction are the amounts of reactants exactly balanced to follow the stoichiometry of the reaction.

The example just given was chosen so that the limiting reactant would be easy to identify. The next example shows a general method of identifying the limiting reactant.

Example: *The reaction of iron(II) with dichromate ions in acid solution follows the stoichiometry below.*

$$6\ Fe^{2+} + 14\ H^+ + Cr_2O_7^{2-} \rightarrow 6\ Fe^{3+} + 2\ Cr^{3+} + 7\ H_2O$$

What is the maximum amount of Cr^{3+} that can be produced when 0.352 mmol $FeCl_2$, 0.858 mmol $HClO_4$, and 0.0599 mmol $K_2Cr_2O_7$ react together?

Answer:

Step 1: We first have to recognize that $FeCl_2$ is the source of the Fe^{2+} in the equation, and that likewise $HClO_4$ is the source of the H^+, and that $K_2Cr_2O_7$ is the source of the $Cr_2O_7^{2-}$.

Step 2: Determine limiting reactant:

In order to determine the limiting reactant, we must put the number of moles of each substance on an equal footing by dividing the number of moles of each reactant by its coefficient in the equation. *The following argument shows that this is reasonable. Consider the relationship between Fe^{2+} and $Cr_2O_7^{2-}$; 6 Fe^{2+} are consumed for one $Cr_2O_7^{2-}$. The Fe^{2+} ions are used up six times as fast, so we divide $n(Fe^{2+})$ by 6 in order to put the amounts of Fe^{2+} and $Cr_2O_7^{2-}$ on an equal basis.*

$$n(Fe^{2+})/6 = 0.352/6 = 0.0587$$

$$n(H^+)/14 = 0.858/14 = 0.0613$$

$$n(Cr_2O_7^{2-})/1 = 0.0599$$

The smallest number represents the limiting reactant: in this case Fe^{2+}.

Step 3: Determine yield of product:

This is obtained from the amount of the limiting reactant and the balanced chemical equation. Note that you use the actual amount of the limiting reactant, not the number you obtained in the determination of which reactant was limiting.

$$n(Cr^{3+}) = n(Fe^{2+}) \times (2\ mol\ Cr^{3+}/6\ mol\ Fe^{2+})$$

$$= 0.352\ mol\ Fe^{2+} \times (2\ mol\ Cr^{3+}/6\ mol\ Fe^{2+}) = 0.117\ mol$$

3.5 Percentage yield and material balance

3.5.1 Percentage yield

It would be convenient for us if every reaction proceeded to give a single product with 100% efficiency, and that the product could be isolated from the reaction mixture without loss. Unfortunately, that is rarely the case. There may be side reactions, in which other products form; the initial products of the reaction may react further (to give "secondary products"); or there may be losses of product during its isolation and purification. We define a **theoretical yield** as the yield expected according to the balanced chemical equation, and an **actual** or **experimental yield**, which is the amount of product actually obtained. The **percentage yield** is defined as follows.

% yield = 100 x (experimental yield/theoretical yield)

Definition

Example: *The fermentation of glucose to produce ethyl alcohol proceeds as follows.*

$$C_6H_{12}O_6 \rightarrow 2\ C_2H_6O + 2\ CO_2$$

*When 1.56 kg of glucose (**M** = 180 g mol^{-1}) was fermented and the product distilled, the yield of ethyl alcohol (**M** = 46.0 g mol^{-1}) was 685 g. What was the percentage yield in this preparation?*

Answer: *We are given the experimental yield of ethyl alcohol (685 g), and can calculate the theoretical yield from the balanced chemical equation.*

$$n(C_6H_{12}O_6) = 1.56 \times 10^3\ g/180\ g\ mol^{-1} = 8.67\ mol$$

From the equation,

$$n(C_2H_6O) = n(C_6H_{12}O_6) \times (2\ mol\ C_2H_6O/1\ mol\ C_6H_{12}O_6)$$
$$= 17.3\ mol$$
$$\text{theoretical mass } C_2H_6O = 17.3\ mol \times 46.0\ g\ mol^{-1} = 797\ g$$
$$\%\ yield = 100 \times (\text{experimental yield/theoretical yield})$$
$$= 100 \times (685/797) = 85.9\%$$

A concept that many people find difficult is to calculate how much reactant is needed when the reaction does not proceed in 100% yield. One method of solving such problems is to write down an "experimental" equation alongside the "theoretical" one. The theoretical equation is the one that follows the nominal stoichiometry. The experimental one gives the coefficients of the substances as might be obtained experimentally, taking into account losses of material in the various manipulations of reagents and the fact that sometimes, not all the reactant proceeds to the desired product - some is diverted to byproducts. Two examples are given to illustrate this concept.

Example 1: *How many moles of glucose would be needed to prepare 5.7 mol ethyl alcohol by fermentation if the percent yield in the reaction is 78%? The theoretical equation was given in the previous example.*

Answer:

Theoretical equation: $C_6H_{12}O_6 \rightarrow 2\ C_2H_6O + 2\ CO_2$

Experimental equation: $C_6H_{12}O_6 \rightarrow 1.56\ C_2H_6O$ + *other products*

Note: theoretically, 1 mol glucose → 2 mol alcohol. In practice, 1 mol glucose gives (78% of 2 mol C_2H_6O), which is 1.56 C_2H_6O mol.

We can now determine $n(C_6H_{12}O_6)$ directly:

$$n(C_6H_{12}O_6) = 5.7\ mol\ C_2H_6O \times (1\ mol\ C_6H_{12}O_6/1.56\ mol\ C_2H_6O) = 3.2\ mol$$

Example 2: *$POBr_3$ may be prepared by the following reaction sequence.*

$$P_4(s) + 10\ Br_2(g) \rightarrow 4\ PBr_5(s) \qquad \text{yield, 93 \%}$$

$$P_4O_{10}(s) + 6\ PBr_5(s) \rightarrow 10\ POBr_3(s) \qquad \text{yield, 81 \%}$$

Calculate the yield in moles of $POBr_3$, starting with 0.24 mol phosphorus (P_4) and unlimited Br_2 and P_4O_{10}.

Answer: *Write "experimental" equations for the two reactions.*

$$P_4(s) + 10\ Br_2(g) \rightarrow 3.72\ PBr_5(s) \qquad \text{yield, } 0.93 \times 4 = 3.72$$

$$P_4O_{10}(s) + 6\ PBr_5(s) \rightarrow 8.1\ POBr_3(s) \qquad \text{yield, } 0.81 \times 10 = 8.1$$

From reaction 1:

$$n(PBr_5) = 0.24 \text{ mol } P_4 \times (3.72 \text{ mol } PBr_5/1 \text{ mol } P_4) = 0.89 \text{ mol}$$

From reaction 2:

$$n(POBr_3) = 0.89 \text{ mol } PBr_5 \times (8.1 \text{ mol } POBr_3/6 \text{ mol } PBr_5) = 1.2 \text{ mol}$$

Percentage yield is an extremely important concept in industry, where thousands of tonnes of a given product may be manufactured each year. An increase in the percentage yield of even 1% is valuable in that it generates more product for sale, and correspondingly fewer byproducts to dispose of. The first point is important because even a fraction of 1% yield is a great deal of material when thousands of tonnes are being manufactured annually. The second is important as tighter regulations determine what disposal practices are acceptable, and increasingly, acceptable disposal requires additional processing and hence additional cost.

Example: *A manufacturer produces 50,000 tonnes of terephthalic acid ($C_8H_6O_4$) annually by oxidation of p-xylene (C_8H_{10}). The product sells for \$5.87/kg. What is the additional income obtained if the yield in the reaction can be increased from 84.8 to 86.3%?*

Answer: *The increased yield is 1.5%, which represents 880 t = 50,000 t x (86.3/84.8).*

Increased income = 880 t x (10^3 kg/1 t) x \$5.87 kg^{-1} = \$5,200,000.

3.5.2 Material balance

To obtain a material balance means to determine quantitatively the fate of all the reactants in a chemical reaction or industrial process. When a material balance has been obtained, the sum of the percentage yields of all products, including unreacted starting materials, will approach 100%. This concept is important in environmental chemistry, where frequently it is toxic trace products of a reaction that are of the greatest concern. The environmental chemist will therefore seek to obtain the best possible material balance in order to determine how much material is unaccounted for. Likewise, an industrial chemist will need to have the best possible material balance in order to identify byproducts of the main reaction, and hence to determine whether they can be recycled into useful product, and if they cannot, what separation and disposal procedures will be required.

Example: *A chemist attempts to prepare chloromethane CH_3Cl by chlorinating methane.*

$$CH_4 + Cl_2 \rightarrow CH_3Cl + HCl$$

From a mixture of 1.03 mol methane and 0.17 mol chlorine the following products were identified: unreacted methane, 0.82 mol; chloromethane (CH_3Cl), 0.11 mol; dichloromethane (CH_2Cl_2), 0.02 mol; HCl, 0.15 mol. What was the material balance, based on (a) methane (b) chlorine?

Answer:

(a) The products derived from methane (containing carbon) are unreacted methane (0.82 mol), chloromethane (0.11 mol), and dichloromethane (0.02 mol) for a total of 0.95 mol.

$$\text{Material balance} = 0.95 \text{ mol}/1.03 \text{ mol} = 0.92 \Rightarrow 92\%$$

(b) The products containing chlorine are chloromethane (0.11 mol), dichloromethane (2 x 0.02 mol) and HCl (0.15 mol). Note that dichloromethane contains 2 Cl atoms per molecule. Total Cl recovered = 0.11 + 0.04 + 0.15 = 0.30 mol. In calculating the material balance, remember that the Cl_2 used as reactant contained 2 Cl atoms per molecule, hence there were 2 x 0.17 or 0.34 mol Cl in the reactants.

$$\text{Material balance} = 0.30 \text{ mol}/0.34 \text{ mol} = 0.88 \Rightarrow 88\%$$

Notice that the material balance differs according to the reactant upon which the calculation is based.

3.6 Chemical reactions in solution

So far, we have considered only reactions between pure compounds. However, molar relationships can be applied to reactions in solution, by making use of the correlation between amount of substance (mol) and the molar concentration of a solution.

$$n(\text{X}) = \text{conc (mol L}^{-1}) \times \text{Volume (L)}$$

3.6.1 Mole relationships in solution

Calculations are carried out in just the same way as for pure substances. All the concepts of the previous sections — limiting reactants, sequential reactions, yields — apply equally to substances in solution as to pure substances. The next example is typical.

Example: *Calculate the mass of calcium carbonate formed by the reaction of 200 mL of 0.0124 mol* L^{-1} $CaCl_2$ *solution with 50 mL of 0.0722 mol* L^{-1} Na_2CO_3 *solution, according to the equation below:*

$$CaCl_2(aq) + Na_2CO_3(aq) \rightarrow CaCO_3(s) + 2\ NaCl(aq)$$

Answer: *We assume that the calcium carbonate precipitates completely. Notice that we are provided with concentration and volume for both reactants, and can therefore determine the number of moles of each of them. This is therefore a limiting reactant problem.*

$$n(CaCl_2) = c \times V = 0.0124\ mol\ L^{-1} \times 0.200\ L = 2.48 \times 10^{-3}\ mol$$

$$n(Na_2CO_3) = 0.0722\ mol\ L^{-1} \times 0.050\ L = 3.61 \times 10^{-3}\ mol$$

This is a 1:1 reaction, and $CaCl_2$ *is present in the smaller amount;* $CaCl_2$ *is the limiting reactant.*

$$n(CaCO_3) = n(\text{limiting reactant}) = 2.48 \times 10^{-3}\ mol$$

$$\text{mass } CaCO_3 = n(CaCO_3) \times \mathbf{M}(CaCO_3)$$

$$= 2.48 \times 10^{-3}\ mol \times 100.2\ g\ mol^{-1}$$

$$= 0.248\ g$$

3.6.2 Net ionic equations

Look again at the equation given in the previous problem:

$$CaCl_2(aq) + Na_2CO_3(aq) \rightarrow CaCO_3(s) + 2\ NaCl(aq)$$

All the substances are ionic. When we examine what ions are present in solution we see the following (all ions are (aq)):

$$Ca^{2+} + 2\ Cl^- + 2\ Na^+ + CO_3^{2-} \rightarrow 2\ Na^+ + 2\ Cl^- + CaCO_3(s)$$

The reactant ions are Ca^{2+}, 2 Cl^-, 2 Na^+, and CO_3^{2-}. The product ions are 2 Na^+ and 2 Cl^-. The $CaCO_3$ has precipitated; it is no longer in solution.

We therefore notice that the 2 Na^+ and the 2 Cl^- ions did not change their chemical identity during the reaction. They could therefore be omitted from the equation, since they do not change chemically: they are called **spectator ions.** Leaving out the spectator ions, we can write a **net ionic equation**.

$$Ca^{2+}(aq) + CO_3^{2-}(aq) \rightarrow CaCO_3(s)$$

The net ionic equation includes only the chemical species that change their chemical identity during the reaction, and omits the spectator ions.

The question immediately arises, "which substances are ionic?" or alternatively, "which substances are strong electrolytes?". This question is answered with the aid of a few simple rules.

1. All **salts** in aqueous solution dissociate into ions when dissolved in aqueous solution. Salts are ionic compounds: cation-anion combinations, such as the metal cation (and NH_4^+) derivatives of families such as chlorides, bromides, iodides, nitrates, sulfates etc. (Review the Table of Common Anions from Chapter 1.) Insoluble salts — those which are present principally as solids — are written as the electrically neutral formula, rather than as dissociated ions in aqueous solution. $CaCO_3$ in the last example was a case in point. For the moment, we shall have to be told which salts are insoluble. Later (Chapter 12) we shall learn which salts are soluble and which are not.

2. The **strong acids** (of which HCl, HNO_3, $HClO_4$, and H_2SO_4 are the commonest) dissociate completely into ions in water; all other acids remain principally in the undissociated form. Thus HCl(aq) → H^+(aq) + Cl^-(aq), but acetic acid (a weak acid, Chapter 11) stays principally in the undissociated form in aqueous solution.

3. The **alkali metal hydroxides** (Na, K, Rb, Cs) and the hydroxides of **Ca, Sr,** and **Ba** dissociate into ions; all other metal hydroxides are insoluble.

4. **Covalent substances** such as CO_2, H_2O, N_2, NO, and NH_3 do not dissociate.

In writing a net ionic equation, it is important to remember that the atoms and charges must balance on each side of the equation, just as in any other chemical equation. Another way of saying this is that the same number of spectator ions must be omitted from each side of the equation.

Example: *Write the net ionic equation for the following reaction:*

$$KIO_3 + 5\ KI + 6\ HCl \rightarrow 6\ KCl + 3\ I_2 + 3\ H_2O$$

Answer: *The following substances dissociate (ions are shown):*

$K^+\ IO_3^-$; $K^+\ I^-$; $H^+\ Cl^-$; $K^+\ Cl^-$

I_2 and H_2O are molecular, and do not dissociate.

The full equation is:

$$K^+(aq) + IO_3^-(aq) + 5\ K^+(aq) + 5\ I^-(aq) + 6\ H^+(aq) + 6\ Cl^-(aq)$$
$$\rightarrow 6\ K^+(aq) + 6\ Cl^-(aq) + 3\ I_2 + 3\ H_2O(\ell)$$

We omit the spectator ions, which are 6 K^+ and 6 Cl^-.

The net ionic equation is:

$$IO_3^-(aq) + 5\ I^-(aq) + 6\ H^+(aq) \rightarrow 3\ I_2 + 3\ H_2O(\ell)$$

3.7. Chemical analysis

Successful chemical analysis requires that the amount of the substance being analysed (the **analyte**) can be determined quantitatively without interference from other components of the material under analysis. The methods of carrying out quantitative analysis are limited only by the imaginations of chemists in finding chemical reactions having these characteristics. The two classical methods of analysis are **gravimetric analysis** and **volumetric analysis**. They are generally simple to carry out and require no sophisticated - and hence expensive - equipment. Introductory courses in chemistry therefore rely on classical methods of analysis.

Unfortunately, classical methods lack sensitivity, and are not easily applicable to the analysis of solutions having analyte concentrations < 1 ppm. This is a problem in environmental analysis, where the analyst is often concerned about the presence of contaminants at the ppb level, or even lower. Such analyses require instrumental methods.

3.7.1 Gravimetric analysis

Gravimetric analysis means analysis by weight, and so the analyte must be converted to some solid substance which can be isolated and weighed. Gravimetric analysis is the most direct form of analysis, since the amount of substance (moles) in the material which is weighed is obtained in one calculation by use of the molar mass. A key requirement for successful gravimetric analysis is that the substance being precipitated has extremely low solubility in water. If the water solubility is significant, some of the analyte will be left in solution and the mass of the precipitate will be underestimated. In the example below, the precipitate is silver chloride. AgCl is very well suited for gravimetric analysis, because its solubility in water is less than 2 mg L^{-1} (See example, next page).

A point to remember in carrying out gravimetric analysis (though the same consideration applies to other methods of quantitative analysis) is that the analyte must be the limiting reactant with the other reactant in excess. For example, in the precipitation of AgCl just considered, if insufficient chloride ion were present to precipitate all the silver as AgCl, the measured weight would be an underestimate.

Example: *A silver ore is analysed by dissolving 7.1649 g of the ore in concentrated acid, and then re-precipitating the silver as AgCl(s) upon addition of a solution containing chloride ion. The dried precipitate weighs 0.1047 g. Calculate the percent by weight silver in the ore.*

Answer: *This is another application of Scheme 1.*

mass AgCl → n(AgCl) → n(Ag) → mass Ag

$n(AgCl) = 0.1047\,g/\mathbf{M}(AgCl) = 0.1047\,g/143.32\,g\,mol^{-1}$

$= 7.305 \times 10^{-4}\,mol$

Since all the silver was precipitated as AgCl, n(Ag) = n(AgCl)

$mass\ Ag = n(Ag) \times \mathbf{M}(Ag) = 7.305 \times 10^{-4}\,mol \times 107.88\,g\,mol^{-1}$

$= 7.881 \times 10^{-2}\,g$

$\%\ by\ mass\ Ag = 100 \times (7.881 \times 10^{-2}\,g/7.1649\,g) = 1.100\,\%$

The disadvantage of gravimetric analysis is that it is time-consuming. Precipitates must be carefully collected, filtered and dried (and re-dried!) to constant weight. If the precipitate is not completely dry, any water remaining adds to the total mass, and the mass of the precipitate is overestimated. Also, gravimetric analysis is labour intensive, and therefore not well suited to applications such as routine testing or quality control.

3.7.2 Volumetric analysis (Titrations)

Volumetric analysis involves the reaction together of substances in solution. As the name implies, it is the volume of the solutions that is measured. Generally, a fixed volume of the analyte solution is placed in a flask using a pipette, and the reagent solution (usually called the **titrant**) is added from a burette. The concentration of the titrant solution will have been determined beforehand. The burette is marked with a volume scale so that at any moment the volume of titrant added may be measured. Provided that the analyte and the titrant react with each other to completion just enough titrant will have been added to consume the analyte exactly at the end of the titration. This is called the **equivalence point**. The volume of titrant needed to reach the equivalence point is measured, and since its concentration is already known, the number of moles of titrant can be calculated.

Successful titration requires that the titrant and the analyte react with each other completely. Titration is a useful method of analysis in that the procedure is rapid and inexpensive. With only a little practice, the analyst should be able to obtain results reproducible to within at least ± 1%.

An experimental consideration is that very frequently the analyte solution and the titrant solution are both colourless, and some means must be found to signal the moment when the reaction is complete (the **end point** of the titration). A common means of determining the end point is by the use of a coloured **indicator** solution. The indicator has the property of changing colour at the end point. If the indicator is well chosen, the end point of the titration (when the indicator changes colour) will closely match the equivalence point (the point when the stoichiometry of the titration reaction is exactly satisfied).

3.7.2.1 Acid-base titrations

Let's illustrate the concept of volumetric analysis by reference to a titration between hydrochloric acid and sodium hydroxide. Assume the hydrochloric acid is the titrant (known concentration, placed in the burette), and the sodium hydroxide solution is the analyte (unknown concentration, placed in the flask). A few drops of phenolphthalein are added as the indicator; this substance is a dye which has a red colour in alkaline solution and is colourless in acidic solution.

At the beginning of the titration, sodium hydroxide is in the flask and the solution is red. HCl is added in small portions from the burette. As the end point approaches the colour of the solution fades, and the end point is taken as the permanent decolourization of the solution. The burette is then read to determine the volume of acid needed to reach the equivalence point.

3.7.2.2 Iodometric titration

The example just given (HCl + NaOH) is an **acid-base** titration. Acid-base titration is applicable to a wide variety of acids and bases, although the indicator must be selected carefully according to the properties of the acid and the base. This issue will be addressed in Chapter 11.

Another commonly used method is **iodometric titration**, which is useful when the analyte has oxidizing properties. Iodometric titration makes use of sequential reactions, as shown in the following example for the analysis of residual chlorine in drinking water. The chlorine is present in the water as a disinfectant, see Chapter 10. The procedure involves *(i)* adding a source of iodide ion in acidic solution, whereupon the iodide ion is oxidized quantitatively to I_2 by the analyte; *(ii)* titration of the I_2 against a solution of sodium thiosulfate of known concentration. In step *(i)*, the iodide ion is added in excess, so that Cl_2 is the limiting reactant.

$$Cl_2 + 2\,I^- \rightarrow 2\,Cl^- + I_2$$

$$I_2 + 2\,Na_2S_2O_3 \rightarrow Na_2S_4O_6 + 2\,NaI$$

Example: *In the titration described in Section 3.7.2.1, 25.00 mL of sodium hydroxide solution were placed in the flask and the concentration of HCl was 0.1352 mol L^{-1}. The volume of HCl needed to decolourize the phenolphthalein solution was 16.72 mL. Calculate the concentration of the sodium hydroxide solution. The chemical reaction is:*

$$HCl + NaOH \rightarrow NaCl + H_2O$$

Answer: *Calculate the number of moles of HCl (the titrant), and relate this to the number of moles of NaOH (the analyte) by means of the chemical equation.*

$$n(HCl) = c(HCl) \times V(HCl) = 0.1352\ mol\ L^{-1} \times 16.72\ mL = 2.261\ mmol$$

From the equation, $n(HCl) = n(NaOH) = 2.261\ mmol$

$$c(NaOH) = n(NaOH)/V(NaOH) = 2.261\ mmol/25.00\ mL$$

$$= 9.042 \times 10^{-2}\ mol\ L^{-1}$$

Follow-up question: *What is the net ionic equation for the reaction between HCl and NaOH?*

Answer: *Identify the ionic substances on both sides of the equation.*

Reactants: $H^+\ Cl^-$ *and* $Na^+\ OH^-$

Products: $Na^+\ Cl^-$ *and* H_2O *(undissociated)*

Therefore the Na^+ *and* Cl^- *ions are spectator ions. The net ionic equation is:*

$$H^+(aq) + OH^-(aq) \rightarrow H_2O(\ell)$$

In fact, the same net ionic equation can be written for the reaction of any strong acid with any strong base (see Chapter 11).

3.7.2.3 Back titration

Occasionally, analysts make use of a special technique called **back titration** in which a known, but excess amount of chemical reagent is added to the analyte. Since the analyte is the limiting reactant, some of the other reagent is left over, and the amount left over is determined by titration (in a sense by titrating *back* to the equivalence point). The amount of analyte is calculated from the amount of the reagent that was used up, and hence not available to take part in the back titration. Back titration is used in cases where the analyte is for some reason unstable in solution (for example, it volatilizes easily from the solution) or where the analyte is a solid, and it would be difficult to tell exactly when the last of it dissolved.

Example of iodometric titration: *A 100.0 mL sample of drinking water is treated with excess KI in acidic solution, and the liberated iodine is titrated against 1.236 mmol* L^{-1} $Na_2S_2O_3$ *solution, of which 7.58 mL are needed to react with all the* I_2*. Calculate the concentration of chlorine in the water, in ppm.*

Answer: *As always, work back from the information given (the concentration and volume of* $Na_2S_2O_3$*) to that required (the concentration of* Cl_2 *in the original solution.)*

n($Na_2S_2O_3$) = c x V = 1.236 mmol L^{-1} x 7.58 mL = 9.37 μmol (Footnote 1)

From the equations:

(i) n(I_2) = 9.37 μmol $Na_2S_2O_3$ x (1 mol I_2/2 mol $Na_2S_2O_3$)
= 4.68 μmol

(ii) n(Cl_2) = n(I_2) = 4.68 μmol

Initial concentration of Cl_2 = n(Cl_2)/V(water) = 4.68 μmol/0.100 L
= 46.8 μmol L^{-1} = 4.68×10^{-5} mol L^{-1}

Recall that in aqueous solution 1 ppm = 1 mg L^{-1}:

c(Cl_2) = 4.68×10^{-5} mol L^{-1} x **M**(Cl_2)
= (4.68×10^{-5} mol L^{-1}) x (70.9×10^{3} mg mol^{-1})
= 3.32 mg L^{-1} = 3.32 ppm.

Example of back titration: *A 0.1639 g sample of limestone (impure calcium carbonate) was dissolved in 50.00 mL of 0.07553 mol* L^{-1} *hydrochloric acid. The HCl left over after the reaction was titrated against 0.02237 mol* L^{-1} *NaOH solution, of which 26.98 mL were needed to reach the end point. Calculate the percent of* $CaCO_3$ *in the limestone sample. The equations are:*

$$CaCO_3(s) + 2\,HCl(aq) \rightarrow CaCl_2 + CO_2(g) + H_2O(\ell)$$
$$HCl(aq) + NaOH(aq) \rightarrow NaCl(aq) + H_2O(\ell)$$

Answer: The total amount of HCl added was 0.07553 mol L^{-1} x 50.00 mL = 3.777 mmol.

The amount of HCl left over (= n(NaOH)) = 0.02237 mol L^{-1} x 26.9 mL = 0.6035 mmol.

Therefore the amount of HCl used up by reaction with $CaCO_3$ is given by the difference:

n(HCl used up) = 3.777 - 0.604 = 3.173 mmol.

From equation 1, n($CaCO_3$) = 3.373 mmol HCl x (1 mol $CaCO_3$/2 mol HCl) = 1.586 mmol.

Mass of $CaCO_3$ = 1.586 mmol x **M**($CaCO_3$) = 1.586 mmol x 100.2 g mol^{-1} = 0.1590 g.

Percent of $CaCO_3$ in the limestone = 100 x (0.1590 g/0.1639 g) = 97.00%

---

1 milli (m, 10^{-3}) x milli (m, 10^{-3}) = micro (μ, 10^{-6})

3.7.3 Instrumental analysis

Instrumental analysis is less direct than gravimetric or volumetric methods of analysis in that the instrument is configured so that an electrical signal (resistance, voltage, current) generated in an appropriate detector is a known function of the concentration of the analyte. Consequently, the concentration of the analyte is obtained by measuring the proper electrical signal. Usually this is done by the construction of a **calibration curve**, in which the electrical signal is measured for each of a series of known concentrations of the analyte and a graph is made of Electrical Signal *vs.* Concentration of Analyte. The signal due to the "unknown" concentration of analyte is then measured, and its concentration read off the graph.

Briefly, here are some of the advantages of instrumental methods of analysis.

1. Generally much more sensitive than classical methods, allowing the analysis of much more dilute samples.

2. Since the output is an electrical signal, instrumental analysis is amenable to on-line analysis, and to computer control (automated analysis).

3. Frequently much more rapid than "wet chemical" methods.

However, nothing comes without a price; modern analytical instruments are often very expensive to install, ranging from a few hundred dollars (pH meter) to hundreds of thousands of dollars (mass spectrometer).

One common and relatively inexpensive piece of analytical equipment is the **UV-visible** spectrometer, in which the concentration of analyte is related to its capacity to absorb a beam of light that is shone upon the solution. With the proper configuration of the instrument, the Absorbance of light is directly proportional to the concentration of the analyte. Absorbance is related logarithmically to the intensity of the radiation transmitted through the solution and hence incident upon the detector:

$$\text{Absorbance} = -\log_{10}\{\text{Fraction of light transmitted through solution}\}$$

Hence the calibration curve for a UV-visible spectrometer is a straight line, Absorbance *vs*. Concentration.

It is scarcely an exaggeration to say that instrumental methods of analysis are responsible for the emergence of and interest in environmental chemistry. The tiny traces of contaminants which of are concern to society today were unmeasurable only a generation ago. Instrumental methods now exist for **separating** complex mixtures into their components so that each can be identified and quantitated free of interference by other components of the matrix. Other instruments then quantitate the separated materials at levels of sensitivity unheard of as recently as 1960. In the 1950s and early 1960s, 1 ppm was considered "trace analysis". Today, we regularly see reports of quantitative analysis at the ppb, ppt levels and even lower.

What is the ultimate sensitivity possible in trace analysis? A somewhat facetious answer would be that the ultimate goal is to detect a single molecule. That is not attainable. Let's consider the example of dioxin (**M** = 322 g mol $^{-1}$) as our analyte.

Example: *How many molecules of dioxin are detected when 1 pg of dioxin is analysed by mass spectrometry? (This is about the current limit of detection by this method.)*

Answer:

n(dioxin) = $(1\ pg/322\ g\ mol^{-1}) = 3 \times 10^{-3}\ pmol = 3 \times 10^{-15}\ mol$

number of molecules = $(3 \times 10^{-15}\ mol)(6.022 \times 10^{23}\ molec\ mol^{-1})$

= 2×10^{9} *molecules*

We therefore see that current levels of detection involve billions of molecules. Consequently, there is still great scope for the development of analytical methods of even greater sensitivity. At the present time, we have no idea as to what this ultimate sensitivity may be, as far as routine analyses are concerned. In the purely research setting, it has been possible to detect light emission from as few as 400 molecules - but this was a situation tailored to be as favourable as possible, and therefore not yet a practical possibility.

From the point of view of the environmentalist, the attainment of ever lower limits of detection raises a practical problem. We are reaching the zone of analytical capability where, as has been stated, we can detect virtually any analyte in virtually any sample of interest. People are naturally very concerned to discover that toxic substances (toxic metals such as lead and mercury, pesticides, polychlorinated biphenyls, dioxins) are present in their food and water. The truth is that yes, these substances can now be detected and quantified in all of these matrices. Should we be concerned to discover that our food and water are "full of poisons"? In trying to deal rationally with these problems we must bear in mind another truth first propounded over 400 years ago by the physician Paracelsus: "It is the dose that makes the poison". (Anything is toxic if given in large enough doses.) As environmentalists, it is our concern to determine not only the concentrations of toxic substances present in a particular sample (since we know they *will* be found, given a sufficiently sensitive analytical method), but whether the concentration is large enough to be a cause for concern. This concept is particularly difficult for the general public, who do not realize the connection between dose and toxicity; instead, they are unlikely to see beyond the fact that their food and water may be "full of poisons".

Problems

Section 3.1

1. Balance the following unbalanced chemical equations.

a) $N_2 + H_2 \rightarrow NH_3$

b) $Cu + H_2SO_4 \rightarrow CuSO_4 + SO_2 + H_2O$

c) $C_3H_8 + O_2 \rightarrow CO_2 + H_2O$

d) $FeSO_4 + Na_2Cr_2O_7 + H_2SO_4 \rightarrow Fe_2(SO_4)_3 + Na_2SO_4 + Cr_2(SO_4)_3 + H_2O$

e) $NaClO_3 + H_2SO_4 + SO_2 \rightarrow ClO_2 + NaHSO_4$

f) $N_2O_4 + N_2H_4 \rightarrow N_2 + H_2O$

Section 3.2

1. Calculate the mass of $CuSO_4$ that can be obtained from the reaction of 7.83 g of copper(II) oxide and excess sulfuric acid.

$$CuO + H_2SO_4 \rightarrow CuSO_4 + H_2O$$

2. Calculate the mass of oxygen that can be obtained by reacting 0.100 mol $KMnO_4$ with excess hydrogen peroxide and sulfuric acid by the reaction below.

$$2KMnO_4 + 5H_2O_2 + 3H_2SO_4 \rightarrow K_2SO_4 + 2MnSO_4 + 8H_2O + 5O_2$$

3. Calculate the mass of the explosive TNT ($C_7H_5N_3O_6$) that can be prepared by treatment of 1.2 t of toluene C_7H_8 with excess nitric acid.

$$C_7H_8 + 3HNO_3 \rightarrow C_7H_5N_3O_6 + 3H_2O$$

4. Calculate the mass of oxygen that a person would need to use in order to respire 0.50 kg of sugar $C_{12}H_{22}O_{11}$.

$$C_{12}H_{22}O_{11} + 12\ O_2 \rightarrow 12\ CO_2 + 11\ H_2O$$

5. Calculate the mass of hydrogen chloride that would be needed to convert 18 tonne of methanol CH_3OH to chloromethane.

$$CH_3OH + HCl \xrightarrow{450°C} CH_3Cl + H_2O$$

6. Calculate the mass of fluorine that would be needed to prepare 1500 kg of CFC-12 (CF_2Cl_2) by the reaction below.

$$CCl_4 + 2F_2 \xrightarrow{SbF_5 \text{ (catalyst)}} CF_2Cl_2 + 2ClF$$

Section 3.3

1. White lead $Pb_3(OH)_2(CO_3)_2$ which was formerly used as a pigment in paints can be converted to red lead Pb_3O_4 by the reactions below.

$$Pb_3(OH)_2(CO_3)_2 + 6HNO_3 \rightarrow 3Pb(NO_3)_2 + 2CO_2 + 3H_2O$$

$$Pb(NO_3)_2 \xrightarrow{\text{heat}} PbO + N_2O_5$$

$$6PbO + O_2 \xrightarrow{\text{heat}} 2Pb_3O_4$$

Calculate the mass of white lead needed to prepare 145 g of Pb_3O_4.

2. Nickel is manufactured from its sulfide ore by the reaction sequence below.

(i) $2NiS + 3O_2 \rightarrow 2NiO + 2SO_2$

(ii) $NiO + CO \rightarrow Ni + CO_2$

a) What mass of nickel can be made from 1.0 x 10^3 tonnes of nickel sulfide (1 tonne = 1000 kg)?

b) If a particular ore is 2.73% NiS (remainder inert material) how much ore is needed to make each tonne of nickel?

3. As a byproduct of nickel manufacture, iron pyrite FeS_2 is roasted to give Fe_2O_3 (which can be reduced to produce iron). The SO_2 byproduct can be converted to sulfuric acid. Calculate the maximum yield of H_2SO_4 obtainable from 120 t of FeS_2.

$$4\ FeS_2 + 11\ O_2 \rightarrow 2Fe_2O_3 + 8SO_2$$

$$2\ SO_2 + O_2 \rightarrow 2SO_3$$

$$SO_3 + H_2O \rightarrow H_2SO_4$$

4. Silver(II) oxide is a useful oxidizing agent, which has been used in organic chemistry to produce azobenzene $C_{12}H_{10}N_2$ from aniline C_6H_7N.

$$2AgNO_3 + K_2S_2O_8 + 2KOH \rightarrow 2AgO + 2K_2SO_4 + 2KNO_3 + H_2O$$
$$2AgO + C_6H_7N \rightarrow C_{12}H_{10}N_2 + Ag_2O + 2H_2O$$

Calculate the mass of $AgNO_3$ needed to prepare 2.6 g of $C_{12}H_{10}N_2$

Section 3.4

1. A mixture of H_2 (4.2 g) and O_2 (26.7 g) is subjected to an electric spark. Water forms:

$$2H_2 + O_2 \rightarrow 2H_2O$$

What quantities of all substances (H_2, O_2, and H_2O) are present after the reaction?

2. Phenylsulfur trifluoride is prepared by the reaction below:

$$(C_6H_5)_2S_2 + 6AgF_2 \rightarrow 2C_6H_5SF_3 + 6AgF$$

a) What is the maximum yield (in grams) of phenylsulfur trifluoride that can be prepared from 100 g of $(C_6H_5)_2S_2$ and 178 g AgF_2?

b) Under the above conditions, which reagent is left over, and by how much?

3. What is the maximum yield of SO_3 that could be obtained from 260 kg of SO_2 and 75 kg of O_2?

$$SO_2 + ½O_2 \rightarrow SO_3$$

4. Calculate the amount of hydrochloric acid left over after 1.566 g of calcium carbonate (marble) is dissolved in 100.0 mL of 0.524 mol L^{-1} HCl:

$$CaCO_3 + 2HCl \rightarrow CaCl_2 + CO_2 + H_2O$$

5. Determine the limiting reactant in the following reaction involving $KMnO_4$ (0.155 g), CH_2O, (0.0427 g) and H_2SO_4 (150.0 mL of 0.00272 mol L^{-1}).

$$4KMnO_4 + 5CH_2O + 6H_2SO_4 \rightarrow 2K_2SO_4 + 4MnSO_4 + 5CO_2 + 11H_2O$$

Section 3.5

1. Quicklime CaO is prepared from limestone by the reaction:

   $$CaCO_3 \rightarrow CaO + CO_2$$

   What weight of quicklime can be prepared from 1.00×10^3 kg of limestone that contains 94.6% $CaCO_3$?

2. Nitric oxide is made commercially by oxidation of ammonia:

   $$4NH_3 + 5O_2 \rightarrow 4NO + 6H_2O$$

   a) What mass of oxygen is needed to oxidize 28.2 t of NH_3 to NO?

   b) If 28.2 t of NH_3 afford 44.7 t of NO, what is the percent yield of nitric oxide?

3. Chlorine dioxide ClO_2 is prepared commercially by the following reaction, which is carried out in aqueous solution:

   $$2NaClO_3 + SO_2 + H_2SO_4 \rightarrow 2ClO_2 + 2NaHSO_4$$

   a) Calculate the theoretical yield of ClO_2 obtained when 1.00 kg each of $NaClO_3$, CO_2 and H_2SO_4 are allowed to react.

   b) If the experimental yield of ClO_2 is 83%, what mass of $NaClO_3$ is needed to prepare 500 g of ClO_2?

4. Nitrobenzene $C_6H_5NO_2$ is prepared by the reaction below:

   $$C_6H_6 + HNO_3 \rightarrow C_6H_5NO_2 + H_2O$$

   a) When 5.0 g of C_6H_6 and excess HNO_3 are allowed to react, the yield of $C_6H_5NO_2$ is 6.7 g. Calculate the percent yield.

   b) What mass of C_6H_6 would you need to use to prepare 175 g of $C_6H_5NO_2$ under the same experimental conditions?

5. Pure oxygen can be prepared in the laboratory by decomposing potassium chlorate:

   $$2KClO_3 \rightarrow 2KCl + 3O_2$$

   If the reaction proceeds in 87% yield, calculate the mass of $KClO_3$ required to prepare 0.15 mol of O_2.

Section 3.6

1. a) What volume of 0.118 mol L^{-1} hydrochloric acid would be needed to neutralize exactly 400 cm^3 of 0.0632 mol L^{-1} barium hydroxide solution?

   b) Write the net ionic equation for the reaction which occurs.

2. In aqueous solution, isopropyl alcohol C_3H_8O is oxidized to acetone C_3H_6O by the reaction below.

   $3C_3H_8O + Na_2Cr_2O_7 + 4H_2SO_4 \rightarrow 3C_3H_6O + Na_2SO_4 + Cr_2(SO_4)_3 + 7H_2O$

   a) Write the net ionic equation.

   b) What volume of 0.337 mol L^{-1} $Na_2Cr_2O_7$ would be needed to oxidize 1484 cm^3 of 0.163 mol L^{-1} isopropyl alcohol in the presence of excess H_2SO_4?

3. Zinc dissolves in hydrochloric acid by the reaction:

   $Zn(s) + 2HCl(aq) \rightarrow ZnCl_2(aq) + H_2(g)$

   a) Write the net ionic equation for this reaction.

   b) What is the minimum volume of 0.13 mol L^{-1} hydrochloric acid needed to dissolve 6.4 g of zinc?

4. What volume of 0.300 mol L^{-1} HCl is required to react completely with 26.4 g of washing soda $Na_2CO_3.10H_2O$?

   $Na_2CO_3.10H_2O + 2HCl \rightarrow 2NaCl + CO_2 + 11H_2O$

5. Calculate the maximum mass of lead iodide PbI_2 that can be prepared by mixing 0.100 L^{-1} of 0.124 mol L^{-1} NaI solution and 0.150 L of 0.040 mol L^{-1} $Pb(NO_3)_2$ solution.

Section 3.7.

1. A 0.1172 g sample of a metal oxide MO_2 was reduced to the metal in a stream of hydrogen gas. The mass of metal was 0.1007 g. What was the metal?

2. A binary compound of lead and oxygen has a lead content of 90.7%. When 1.0000 g of this compound is heated to 600°C, oxygen is evolved and a residue of 0.9767 g remains. Write a balanced equation for the reaction which occurs.

3. A 7.78 g sample of a mixture of NaCl and $NaHCO_3$ is heated to 300°C. $NaHCO_3$ decomposes but NaCl is stable.

   $$2NaHCO_3 \rightarrow Na_2CO_3 + H_2O + CO_2$$

   After decomposition, the sample weighs 5.93 g. Calculate the mass percent of NaCl in the original mixture.

4. A mixture of $CaCl_2$ and $CuCl_2$ (1.005 g total) is dissolved in water, and hydrogen sulfide is passed through the solution. Copper sulfide precipitates, but the calcium salt is unaffected.

   $$CuCl_2 + H_2S \rightarrow CuS + 2HCl$$

   The dried precipitate of CuS weighs 0.442 g. What was the percent $CuCl_2$ in the mixture?

5. Iodine pentoxide is used to analyse for carbon monoxide by the reaction below:

   $$5CO + I_2O_5 \rightarrow 5CO_2 + I_2$$
   $$I_2 + 2S_2O_3^{2-} \rightarrow S_4O_6^{2-} + 2I^-$$

   When a 10.8 dm^3 sample of air contaminated with CO is passed over excess I_2O_5, enough I_2 is liberated to react with 8.62 cm^3 of 0.0822 mol L^{-1} $Na_2S_2O_3$ solution. Calculate the concentration of CO in the air sample in the units mg CO per liter.

6. An unknown acid is analysed by titration against 0.1043 mol L^{-1} $Ba(OH)_2$. A 0.7364 g sample of the acid (which has only one acidic hydrogen) required 25.66 mL of the $Ba(OH)_2$ solution for complete neutralization. Calculate the molar mass of the acid.

7. An older method for analysing the Ca^{2+} concentration in blood serum uses the reactions below.

   [1] $Ca^{2+} + (NH_4)_2C_2O_4 \rightarrow CaC_2O_4(s) + 2NH_4^+$
   ammonium oxalate

   [2] $CaC_2O_4(s) + H_2SO_4 \rightarrow H_2C_2O_4 + CaSO_4$

   [3] $5H_2C_2O_4 + 2KMnO_4 + 3H_2SO_4 \rightarrow K_2SO_4 + 2MnSO_4 + 10CO_2 + 8H_2O$

   a) Write net ionic equations for reactions [1] -[3].

   b) Calculate the calcium ion concentration if a 2.0 cm^3 blood serum sample required 0.68 cm^3 of 2.44 x 10^{-3} mol L^{-1} $KMnO_4$ in the final titration.

8. A 25.0 cm^3 sample of triethylaluminum $Al(C_2H_5)_3$ solution, which is used as a catalyst for polymerization of ethylene, was allowed to react with 100 cm^3 of 0.103 mol dm^{-3} HCl. The excess HCl was back titrated against 0.142 mol dm^{-3} ammonia solution, of which 16.75 cm^3 was needed for neutralization. What was the concentration of $Al(C_2H_5)_3$ in the catalyst solution?

$$Al(C_2H_5)_3 + 3HCl \rightarrow AlCl_3 + 3C_2H_6$$
$$HCl + NH_3 \rightarrow NH_4Cl$$

9. A sample of limestone (impure $CaCO_3$) was analysed by the following procedure:

i) dissolution in excess dilute HCl solution

ii) titration against standard NaOH solution

In such an experiment 2.262 g of limestone was dissolved in 100 cm^3 of dilute hydrochloric acid of concentration 0.582 mol dm^{-3}. Exactly ten cm^3 of this solution was withdrawn and titrated against 0.211 mol dm^{-3} sodium hydroxide, of which 6.85 cm^3 was needed to reach the equivalence point. What was the percent $CaCO_3$ in the limestone?

$$CaCO_3 + 2HCl \rightarrow CaCl_2 + CO_2\,(g) + H_2O$$
$$HCl + NaOH \rightarrow NaCl + H_2O$$

10. A water sample at 0 °C was analysed for dissolved oxygen by the reaction sequence below.

[1] $Mn^{2+} + 2OH^- + ½O_2 \rightarrow MnO_2(s) + H_2O$

[2] $MnO_2 + 4H^+ + 2I^- \rightarrow Mn^{2+} + I_2 + 2H_2O$

[3] $I_2 + 2Na_2S_2O_3 \rightarrow Na_2S_4O_6 + 2NaI$

A 50.00 cm^3 water sample was treated with Mn^{2+}; the precipitate of MnO_2 was used to oxidize (excess) I^- to I_2, and the I_2 was titrated against 0.01037 mol L^{-1} $Na_2S_2O_3$, of which 10.64 mL were required. Calculate the oxygen content of the original water sample in ppm.

INDUSTRIAL PROCESSES

4.1 The chemical industry

The term "the chemical industry" covers that sector of industry which provides chemical substances used by society. It would be wrong to think of the chemical industry as a single, monolithic entity; the companies involved in the chemical industry range from huge multinational organizations with world-wide sales in the multi-billion dollar range, down to businesses which offer small quantities of highly specialized products for sale. Chemical commodities which are produced in kilotonne amounts per year are often known as "heavy chemicals" (nothing to do with molar mass!), whereas those produced in small quantities are known as "fine chemicals", especially when produced in a high state of purity.

Since everything in the world is chemical in nature, it is difficult to draw a boundary around "the chemical industry". Traditionally, the following are not considered part of the chemical industry, even though their products are of a chemical nature.

- the mining industry (see Chapter 17)
- the pulp and paper industry
- the petroleum refining industry (that is, the refining of crude oil for fuel). However, the **petrochemicals industry** (see below) is considered part of the chemical industry
- the pharmaceuticals industry (human and veterinary drugs). The purity demands on pharmaceuticals make these products "ultra-fine" chemicals.

Each year, the magazine *Chemical and Engineering News* publishes lists of both the largest U.S. producers of chemicals, and the substances which are produced in the greatest volume[1]. In 1992, 50 companies had sales of over $1 billion in the U.S. The five largest companies for U.S. sales were:

1) Du Pont (U.S.)	$15.5 billion	
2) Dow Chemical (U.S.)	$12.9	
3) Exxon (U.S.)	$10.6	excluding petroleum refining
4) Hoechst Celanese (Germany)	$ 6.5	
5) Monsanto (U.S.)	$ 5.4	

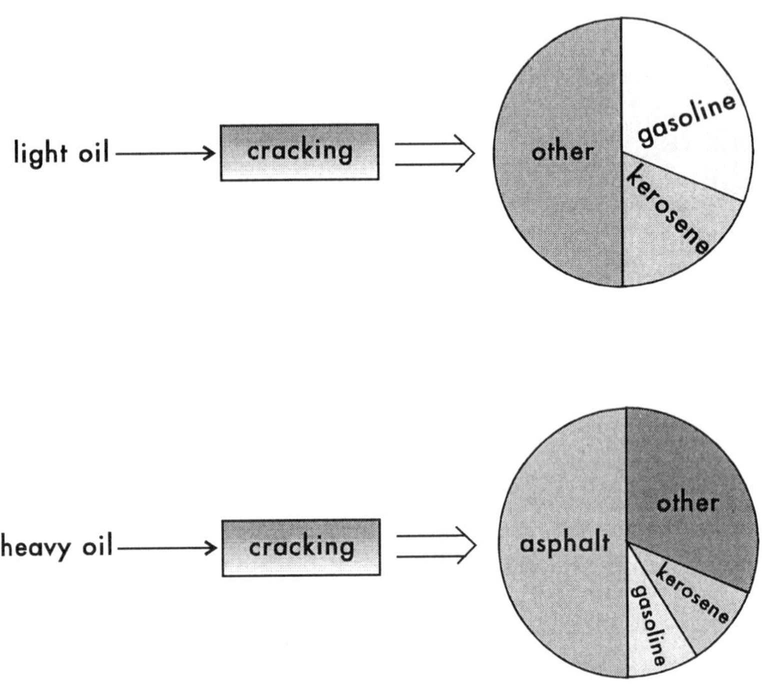

Through catalytic cracking, the petrochemicals industry provides the basic materials for much of the chemical industry.

[1] 1992 data: Top 50 Chemicals, *Chem Eng News*, April 12, 1993; Top 100 Chemical Producers, *Chem Eng News*, May 3, 1993.

4.2 Some high-volume chemical products

In 1992, the ten substances manufactured in the largest amount each had production > 10 million tonnes in the U.S.A. They were:

sulfuric acid	3.8×10^7 tonnes (38 million tonnes)
nitrogen	2.8
oxygen	2.0
ethylene[a]	1.8
ammonia	1.6
lime	1.5
phosphoric acid	1.2
sodium hydroxide	1.1
chlorine	1.0
propylene[a]	1.0

a a petrochemical product

It is difficult to grasp the scale of operations of the heavy chemicals industry. For example, compare the U.S. production of sulfuric acid with the 0.1 mol L^{-1} solution of sulfuric acid commonly found in the university chemistry laboratory.

Example: *Calculate (i) the number of moles of sulfuric acid produced each year by U.S. industry (ii) the volume of 0.1 mol L^{-1} sulfuric acid that could be made from this amount of acid. Express your answers both in scientific notation and in the conventional (non-scientific) format.*

Answer: *(i)*

$$M(H_2SO_4) = 98 \text{ g mol}^{-1}$$

$$n(H_2SO_4) = 3.7 \times 10^7 \text{ Mg}/98 \text{ g mol}^{-1}$$

$$= 3.8 \times 10^5 \text{ Mmol}$$

(380 Gmol or 380,000,000,000mol)

(ii)

$$V(H_2SO_4) = 3.8 \times 10^5 \text{ Mmol}/0.1 \text{ mol L}^{-1}$$

$$= 4 \times 10^6 \text{ ML } (4 \times 10^{12} \text{ L or } 4 \times 10^9 \text{ m}^3)$$

In non-scientific notation, this is 4,000,000,000,000 liters, or 4 trillion liters. This is equivalent to 4 cubic kilometers, or 1 cubic mile, of sulfuric acid solution.

In the following sections, we examine how some of these high volume substances are produced, and their uses.

Sulfuric acid

Sulfuric acid is so well established as the No.1 chemical that it is possible to rank nations as developed, developing, or undeveloped economically as easily by their national annual production of sulfuric acid as by monitoring their GNP (gross national product of goods and services). Canada, for example, with a GNP about 1/10 that of the U.S.A., produces about 4 million tonnes of sulfuric acid annually.

Most sulfuric acid is produced starting with elemental sulfur, which is available very cheaply. One source of sulfur is a large subterranean deposit beneath the states of Texas and Louisiana; it is recovered inexpensively by passing superheated water (about 180 °C, and 6-8 atm pressure) and hot compressed air underground; the sulfur (melting point 113 °C) liquifies and is driven to the surface through metal pipes under the pressure of the gases forced down into the ore. Since sulfur is the only material liquified under these conditions, it is obtained in a high state of purity.

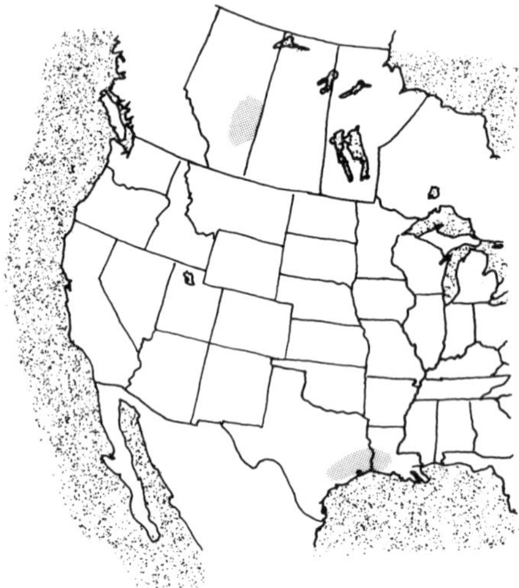

Within North America, large amounts of sulfur come from Alberta, Louisiana, and Texas.

Sulfur is cheaply available in western Canada as a byproduct of the natural gas industry and the synthetic crude oil industry (the Athabasca oilsands project, near Fort McMurray, Alberta). As obtained from the ground, natural gas contains significant quantities of hydrogen sulfide, which has the unpleasant odour of rotting eggs. For this reason, the crude natural gas is known as "sour gas". Furthermore, upon combustion of sour gas, the hydrogen sulfide is oxidized to sulfur dioxide, which is responsible for acid precipitation (Chapter 13). Sour gas is therefore "sweetened" by converting the hydrogen sulfide to sulfur, which is removed as the solid. More sulfur is obtained by sweetening than can be used in Canada for producing sulfuric acid, and so much of the excess is exported.

$H_2S(g) + \frac{1}{2} O_2(g) \rightarrow S(s) + H_2O(g)$

There are three steps in the production of sulfuric acid.

1. production of SO_2
2. oxidation of SO_2 to SO_3
3. conversion of SO_3 to H_2SO_4

1. Starting with elemental sulfur, the dioxide is formed by combustion in air.

$$S(s) + O_2(g) \rightarrow SO_2(g)$$

In Canada, the metal smelting industry is an important alternative source of SO_2. Commercially important metals such as copper, zinc, lead, and nickel occur as sulfides, which are roasted in air as the first step in the production of the metal. Sulfur dioxide is a byproduct of this reaction.

e.g. $2\ ZnS + 3\ O_2 \rightarrow 2\ ZnO + 3\ SO_2$

The release of sulfur dioxide to the atmosphere is a serious contributor to acidic precipitation (Chapter 13), and so it is worthwhile to purify the SO_2 for conversion to sulfuric acid, rather than letting it escape into the atmosphere.

2. Sulfur dioxide is oxidized catalytically to sulfur trioxide, a reaction which occurs in the gas phase at a reasonable rate at about 450 °C.

$$SO_2 + \frac{1}{2} O_2 \rightarrow SO_3$$

As discussed further in Chapters 8 and 13, this reaction is an example of an equilibrium process. The conditions of temperature and pressure are chosen so as to maximize the yield of sulfur trioxide, while maintaining a rapid rate of reaction.

3. In principle, sulfur trioxide can be converted to sulfuric acid by reaction with water.

$$SO_3 + H_2O \rightarrow H_2SO_4$$

However, this reaction is rather slow, and it is difficult to ensure that all the SO_3 gets trapped when SO_3 gas is passed into water. Failure to trap all the SO_3 would result in the formation of a sulfuric acid fog around the production facility as the escaped SO_3 reacted with moisture in the air. Therefore, for environmental reasons, SO_3 is trapped in concentrated sulfuric acid to give a superconcentrated form of sulfuric acid called **oleum**, which is then diluted with water to give the conventional "concentrated" sulfuric acid (commonly 98% H_2SO_4, 2% water).

$$SO_3 + H_2SO_4 \rightarrow H_2SO_4.SO_3 \qquad \text{(oleum)}$$
$$H_2SO_4.SO_3 + H_2O \rightarrow 2\ H_2SO_4$$

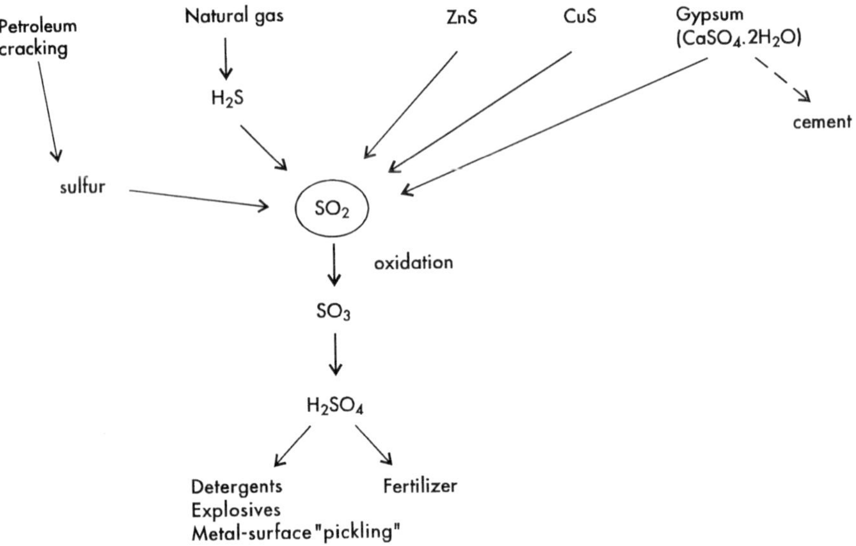

Some sources of SO_2 and uses of H_2SO_4

Because the reactants (sulfur and air) are so cheap, sulfuric acid is an extremely inexpensive commodity chemical, the pure liquid selling for under 20¢ per kg in bulk.

Large amounts of sulfuric acid are used in the production of fertilizer: the formation of "superphosphate" from rock phosphate. Rock phosphate (approximate composition $Ca_3(PO_4)_2$) is very insoluble, and hence unavailable to plants as a nutrient; conversion to calcium hydrogen phosphate increases its solubility and availability.

$$2\ Ca_3(PO_4)_2 + H_2SO_4 \rightarrow 2\ CaHPO_4 + CaSO_4$$

Other important uses of sulfuric acid include the manufacture of synthetic detergents and explosives, as well as in the "pickling" of metal surfaces such as automotive parts before painting. Pickling removes any surface layer of rust (Fe_2O_3), allowing better adhesion between the paint primer and the metal.

$$Fe_2O_3 + 3\ H_2SO_4 \rightarrow Fe_2(SO_4)_3 + 3\ H_2O$$

Nitrogen and Oxygen

These commodities are manufactured inexpensively from air, which is an approximately 4:1 mixture of N_2: O_2 (Section 6.2). The separation of the two elements depends on their different boiling points as liquids: N_2: 77 K (-196 °C); O_2: 90 K (-183 °C).

Gases in general may be liquified by cooling them below the boiling point of the liquid (example: cool steam below 100 °C) or by subjecting the gas to high pressure (example: compression of propane, whose boiling point at atmospheric pressure is -42 °C, but which is sold in metal tanks under pressure at ordinary temperature). It is not possible to liquify air at ordinary temperatures no matter how much pressure is applied (see Chapter 6), and so liquefaction is carried out by cooling air to below -100 °C and then applying pressure. The liquid air thus formed is then subjected to **fractional distillation**: the liquid is allowed to boil and the vapours to rise up a large, cooled fractionating tower, where they repeatedly condense and vaporize. This achieves equilibrium between vapour and liquid. Nitrogen, which is more volatile, is drawn as a gas from the top of the tower, and oxygen is withdrawn lower down. The separated oxygen and nitrogen are delivered to customers either as liquids stored in vessels similar to large "Thermos" bottles, or as compressed gases in steel cylinders.

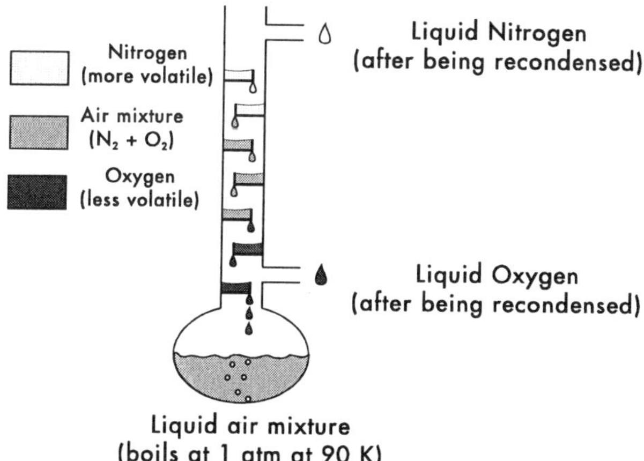

The process of "fractional distillation" permits the separation of liquid nitrogen from liquid oxygen

The major use of nitrogen is conversion to ammonia: see below. Other uses take advantage of the chemical inertness of nitrogen, for example as an unreactive blanket of gas in the machining of reactive metals which might take fire in air, and as a **cryogen**. Cryogens are super-cold materials which are used for cooling purposes, such as the maintenance of superconducting magnets and the storage of biological materials such as frozen bull semen.

The major use of pure oxygen is in the refining of certain metals, notably steel, where oxygen is used in large amounts to remove carbon impurities from impure or "pig" iron (Chapter 17). Smaller amounts are used to produce extremely hot flames in welding applications, for example the oxygen-hydrogen flame, and the oxy-acetylene flame. Burning a fuel in pure oxygen gives a hotter flame than combustion in air, when the oxygen is diluted with nitrogen. Liquid oxygen is only rarely used as a cryogen because liquid nitrogen is much safer. Liquid oxygen is a very powerful oxidizing agent which can react explosively upon contact with many organic materials, including oils and greases, and under some conditions even with wood and plastic materials.

Ethylene and Propylene

Ethylene (C_2H_4, or $CH_2{=}CH_2$) and propylene (C_3H_6, or $CH_3CH{=}CH_2$) are examples of **petrochemicals**, which are manufactured from petroleum. They are also examples of **organic** compounds - that is, compounds whose structures include chains or rings of carbon atoms.

The present major route to ethylene is the catalytic dehydrogenation (dehydrogenation means removal of hydrogen) from ethane (C_2H_6), which is present in natural gas. The principal component of natural gas is methane, CH_4, which is separated from ethane by fractional distillation (the same process by which nitrogen and oxygen are separated).

$$C_2H_6 \rightarrow C_2H_4 + H_2$$

Demand for ethylene is expanding so much that increasingly it is being produced by **catalytic cracking** of the larger hydrocarbons of crude oil. At high temperatures and in the presence of appropriate catalysts, hydrocarbons containing chains of 20 and more carbon atoms can be split apart into smaller hydrocarbons, of which ethylene and propylene are the most important. Ethylene and propylene can then be separated by fractional distillation. The following example illustrates the cracking of a C_{20} hydrocarbon into ethylene, although the chemistry is much more complex in practice.

$$C_{20}H_{42} \rightarrow 10\ C_2H_4 + H_2$$

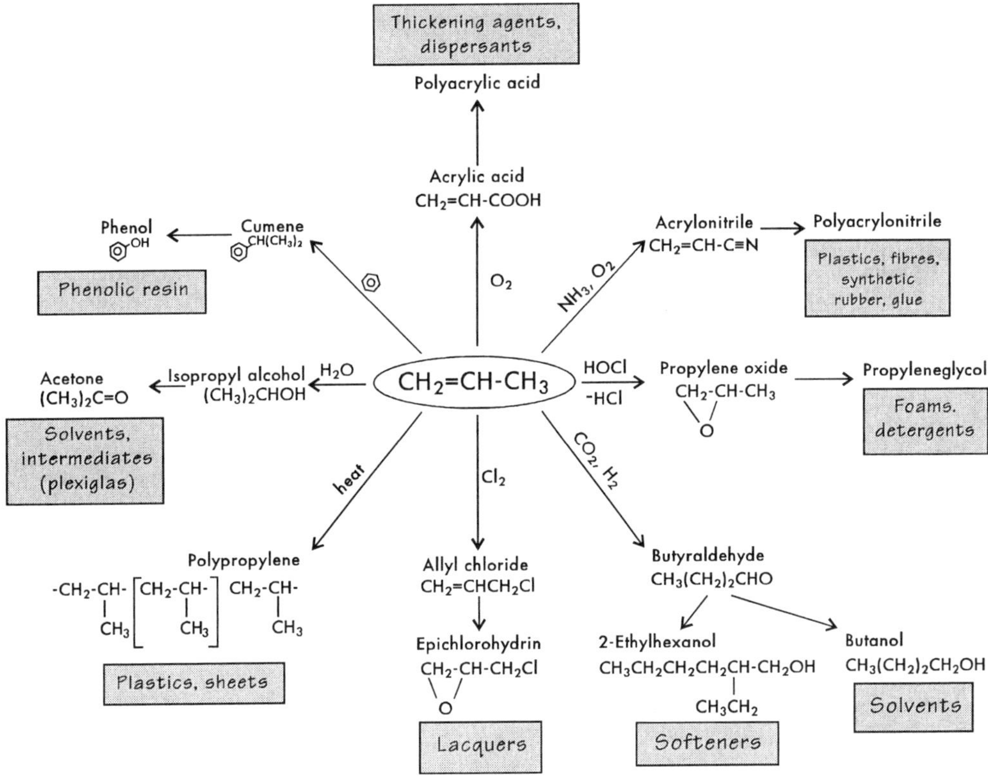

Some of the many uses for propylene

Ethylene and propylene are both used as feedstocks for the manufacture of **polymers**. Polyethylene and polypropylene are produced from ethylene and propylene respectively. In each of these polymers the molecules are formed by the end-to-end joining of thousands of individual molecules of ethylene or propylene.

$$e.g. \quad n\,C_2H_4 \rightarrow \sim CH_2CH_2\text{-}CH_2CH_2\text{-}CH_2CH_2\text{-}CH_2CH_2\sim$$

Polyethylene is used as shrink wrapping for foods, as a low cost plastic for applications such as plastic buckets, and as an electrical insulator. Polypropylene, in addition to these uses, can be formed into fibers from which items such as hard-wearing carpets can be made.

Both ethylene and propylene are the source materials for a wide variety of other organic chemicals. A few examples include ethyl alcohol, ethylene glycol (antifreeze), acetic acid and styrene, all of which are made from ethylene.

An important branch of petrochemicals research is "C_1 chemistry". Natural gas (major component methane, CH_4) is currently a very inexpensive feedstock. (The term "feedstock" implies its use for the purpose of chemical conversion to higher value products, as opposed to simply burning it as a fuel.) Methane is already used as the feedstock for the manufacture of hydrogen (see section immediately following) and also for the production of methanol (CH_3OH), which is used, among other things, as windshield de-icer, and which is also under consideration as an "oxygenated " automotive fuel in the fight against urban air pollution (Section 9.8). A major research thrust in C_1 chemistry is to find ways of joining the C_1 fragments together to produce petrochemicals of more complex structure for the further manufacture of plastics, paints, fibers, and other high value organic compounds.

Ammonia and Nitric Acid

Ammonia is manufactured by the **Haber Process**, developed by the German chemist Fritz Haber just before the outbreak of World War I. The Haber process involves passing nitrogen and hydrogen gases over a metal oxide catalyst at elevated temperature and high pressure. Ammonia is formed directly.

$$N_2 + 3\ H_2 \rightarrow 2\ NH_3$$

The hydrogen needed for this reaction is produced from natural gas by reaction with steam at high temperature.

$$CH_4 + 2\ H_2O \rightarrow CO_2 + 4\ H_2$$

The Haber process has been of inestimable value to agriculture because it permits the inexpensive production of nitrogen-containing fertilizers, using the nitrogen of the air as the source of nitrogen. This reaction is an example of **nitrogen fixation**, by which the unreactive nitrogen of the atmosphere is converted into a chemically more reactive form. Nitrogen is also fixed naturally, both by microorganisms and when air is heated to high temperature, as in a lightning strike (see Chapter 6).

Prior to the introduction of synthetic nitrogen fixation, the only nitrogenous fertilizers were manure and natural deposits of "Chile saltpeter", $NaNO_3$, mined in Chile. This resource was being rapidly depleted in the early 1900s. Ammonia is used in fertilizers by conversion to ammonium sulfate (using sulfuric acid), to ammonium nitrate (using nitric acid), or to urea $CO(NH_2)_2$.

$$CO_2 + 2\ NH_3 \rightarrow CO(NH_2)_2$$

Ammonia (boiling point -33 °C) is easily liquified, and is used in this form (anhydrous ammonia) — see footnote 2 — as a fertilizer for corn. In this application, ammonia is injected directly below the surface of the soil. Liquified ammonia is also a convenient means of shipping ammonia from producer to an off-site user.

Example: *Compare the volume occupied by 1.0 t of ammonia in the liquid form (density at -33 °C = 0.68 kg L^{-1}) and in the gaseous form at 20 °C at 8.5 atm (density = 6.7×10^{-3} kg L^{-1})*

Answer: *1.0 t = 1000 kg*

V(liquid) = 1000 kg/0.68 kg L^{-1} = 1.5×10^3 L (1.5 m^3)

V(gas) = 1000 kg/6.7×10^{-3} kg L^{-1} = 1.5×10^5 L ⇒ 150 m^3

Even at the elevated pressure of 8.5 atm, the gaseous form of ammonia occupies one hundred times the volume of the liquid.

Another major use of ammonia is in the manufacture of nitric acid. Commonly, a nitric acid plant is set up on the same site as an ammonia facility in order to avoid the cost of shipping the ammonia from one location to another.

Nitric acid manufacture involves a sequence of reactions.

$$4\,NH_3 + 5\,O_2 \rightarrow 4\,NO + 6\,H_2O$$
$$2\,NO + O_2 \rightarrow 2\,NO_2$$
$$3\,NO_2 + H_2O \rightarrow 2\,HNO_3 + NO$$

The NO formed in the last step is recycled.

Notice that the manufacture of ammonia and nitric acid involves only very inexpensive substances. Ammonia is made from nitrogen (made by liquifying air) and hydrogen (made from natural gas), and the only other reactants needed for conversion to nitric acid are oxygen (available from air) and water. It is a characteristic of industrial processes that the cheapest possible reactants are always used. For instance most Canadian ammonia and nitric acid plants are located in Alberta, where natural gas is cheaply available.

[2] The term anhydrous (without water) is used to distinguish pure liquid ammonia from household ammonia, which is an aqueous solution of ammonia used for cleaning.

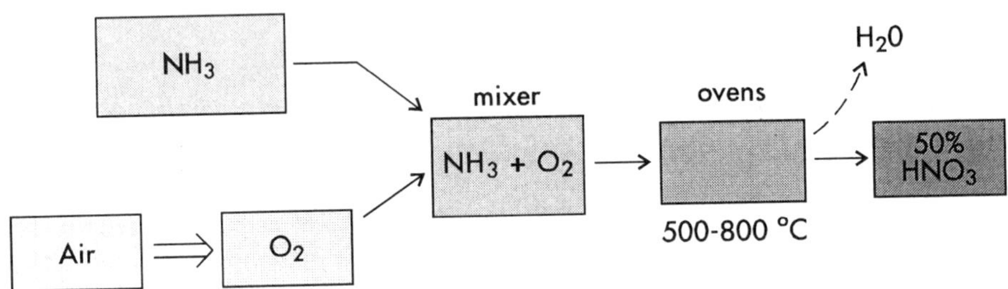

The manufacture of nitric acid from ammonia.

Much industrial research is directed towards optimizing reaction conditions and finding and improving catalysts for specific reaction steps in order to operate at lower temperature or in higher percent yield in order to reduce costs. Profit margins on high volume chemicals are very small, and the difference between profit and loss may hinge upon reducing production costs by a small fraction of a cent per kilogram (which translates to a few dollars per tonne).

The concepts just described are illustrated very well by considering the fate of an older process for fixing nitrogen, the reaction between N_2 and O_2 at high temperature.

$$N_2 + O_2 \rightarrow 2\ NO$$

This reaction is very energy-intensive: it requires a temperature of 2000 °C (compared with about 500 °C for the Haber process), and is an energy-consuming ("endothermic" — see Chapter 5) reaction whereas the Haber process is energy releasing ("exothermic"). Besides, the yield of NO is rather small. For these reasons, this technology was completely superceded by the development of the Haber process.

Lime

The expression "lime" covers two related substances: quicklime, CaO; and slaked lime, $Ca(OH)_2$. They are made by roasting limestone (calcium carbonate) above about 800 °C.

$$CaCO_3 \rightarrow CaO + CO_2$$
$$CaO + H_2O \rightarrow Ca(OH)_2$$

Calcium oxide is produced on a very large scale in the manufacture of cement and glass. Calcium hydroxide finds use as an inexpensive base for neutralizing acids. For example, spent pickling liquors containing unreacted sulfuric acid may be neutralized with lime; likewise, an aqueous spray of lime is used to neutralize acidic emissions from coal burning power stations (Chapter 13).

Sodium Hydroxide and Chlorine

These two commodities are generally produced together, by electrolysis[3] of a solution of brine (aqueous sodium chloride solution). The reaction is usually known as the **chlor-alkali process**: chlor-alkali is short for chlorine + alkali (sodium hydroxide).

$$2\ NaCl + 2\ H_2O \rightarrow 2\ NaOH + Cl_2 + H_2$$

Notice the use of another extremely cheap raw material (rock salt) for this reaction. The major operating cost of the chlor-alkali process is the electricity used for electrolysis.

The stoichiometry of the chlor-alkali process is such that almost equal masses of both commodities are obtained.

Example: *Calculate the masses of chlorine and sodium hydroxide formed by electrolysis of 1.0 t of NaCl.*

Answer:

M(NaCl)	$= 58.5\ g\ mol^{-1}$
n(NaCl)	$= 1.0\ Mg/58.5\ g\ mol^{-1} = 1.7 \times 10^{-2}\ Mmol$
n(NaOH)	$= n(NaCl) = 1.7 \times 10^{-2}\ Mmol$
M(NaOH)	$= 40\ g\ mol^{-1}$
mass(NaOH)	$= 1.7 \times 10^{-2}\ Mmol \times 40\ g\ mol^{-1} = 0.68\ Mg \Rightarrow 0.68\ t$
$n(Cl_2)$	$= 1.7 \times 10^{-2}\ Mmol\ NaCl \times (1\ mol\ Cl_2/2\ mol\ NaCl)$
	$= 8.5 \times 10^{-3}\ Mmol$
$\mathbf{M}(Cl_2)$	$= 71\ g\ mol^{-1}$
$mass(Cl_2)$	$= 8.5 \times 10^{-3}\ Mmol \times 71\ g\ mol^{-1} = 0.60\ Mg \Rightarrow 0.60\ t$

---

3 Electrolysis involves driving a chemical reaction by passing a current of electricity through the reactants. For more detail, see Chapter 16.

Summary: Production of high-volume chemicals

chemical product		origins
sulfuric acid	H_2SO_4	from S(elemental) in underground deposits; as a byproduct of natural gas production.
nitrogen and oxygen	N_2 and O_2	by liquefying air, and separating N_2 from O_2 by distillation.
ethylene and propylene	C_2H_4 and C_3H_6	manufactured from petroleum, mainly through catalytic cracking.
ammonia	NH_3	by heating $N_2 + H_2$ under pressure (the Haber Process).
nitric acid	HNO_3	from $NH_3 + O_2$ (usually manufactured at an ammonia plant).
lime	CaO and $Ca(OH)_2$	from roasted limestone, $CaCO_3$.
sodium hydroxide and chlorine	$NaOH$ and Cl_2	from electrolysis of brine (the chlor-alkali process).

Thus the ratio by mass of NaOH: Cl_2 produced in the chloralkali process is $0.68 : 0.60 = 1.1 : 1$. This is approximately the weight ratio shown in the listing of Top Chemicals at the beginning of this chapter.

The simultaneous production of two high volume chemicals has an interesting economic consequence: whichever of them is in greater demand at any time drives the level of production, and relegates the other to the status of a byproduct. As the price of the scarcer commodity rises, the other is in over-supply and its price falls. For several years the demand for sodium hydroxide has been greater than the demand for chlorine, and while the bulk price of sodium hydroxide is in the range of \$300 per tonne, chlorine can be purchased for as little as \$100 per tonne although they are produced in the same process and in similar quantities.

Demand for chlorine has been in decline for environmental reasons. One major outlet for this commodity has been in bleaching paper. However, concerns about the release of potentially toxic, chlorinated organic byproducts into rivers from pulp mills have led to research into methods of bleaching that require less chlorine: these include the use of chlorine dioxide (ClO_2) or peroxide bleaches, neither of which afford organo-chlorine byproducts. Another major use of chlorine has been in chlorofluorocarbon manufacture, but these substances are being phased out because of concern that they are responsible for depletion of stratospheric ozone (Chapter 15). Other chlorinated solvents such as chloroform, trichloroethylene and tetrachoroethylene are experiencing reduced demand because of concerns about their toxicity; these also employ chlorine in their manufacture.

The reduced demand for chlorine has therefore made the chlor-alkali process less attractive financially than in former years. This has led to the revival of an old process for the manufacture of NaOH without the concurrent formation of Cl_2. This technology uses **soda ash** (sodium carbonate) as starting material. Soda ash is cheaply available in North America in the form of "trona" ($Na_2CO_3.H_2O$), which occurs naturally in large deposits in the U.S. Midwest. The soda ash process involves the treatment of an aqueous solution of sodium carbonate with lime.

$$Na_2CO_3(aq) + Ca(OH)_2(aq) \rightarrow 2\ NaOH(aq) + CaCO_3(s)$$

As we have seen already, lime is made inexpensively from limestone, $CaCO_3$. In principle, it would appear that the $CaCO_3$ could be recycled back into lime, but this is inconvenient in practice, because the $CaCO_3$ is produced as a fine aqueous suspension or "slurry", which is very difficult to free from water. As a result, the disposal of the large volume of slurry is itself a considerable environmental problem.

4.3 Chemical pollution

The chemical industry has had a poor public image in terms of pollution. In the past, the environment appeared to be limitless, and waste materials were dumped into landfills, into rivers, and into the oceans with little thought for what would happen to them. The first regulations to limit emissions from factories and to protect the health of workers date back over a century, but the real impact of environmental pollution became apparent only in the 1960s. This happened because of advances in analytical methods, especially the development of instrumental methods of analysis (Chapter 3). Instrumental analysis made possible the detection and quantitation of part per million, part per billion (and subsequently, even lower) quantities of xenobiotic[4] chemicals in streams and lakes, in wildlife, and even in food and drinking water. This alerted the general public, chemical industry and governments to the finite extent of the environment, and to the need for tighter controls on emissions and improved waste disposal practices.

Today, the major chemical manufacturers in the developed world recognize the importance of proper handling and disposal in terms of their corporate image; in Canada for example, the members of the Canadian Chemical Producers Association adhere to a strict code of ethics called "Responsible Care", which involves responsibility for their products right through to the ultimate consumer; responsibility does not end when the product passes their gates. Legislation on

---

4 "Xenobiotic" = foreign to life, *i.e.*, synthetic, or manmade

emissions requires advanced treatment of aqueous waste streams and, increasingly, air emissions. Industrial waste treatment plants are frequently multi-million dollar facilities employing the latest technology for chemical and biological degradation of xenobiotics.

Serious problems still exist however; some smaller companies remain indifferent to chemical pollution and/or lack the resources to invest in the latest technology for waste treatment. A particular problem concerns the abandoned dump sites which were used in the days before there were strict regulations on waste disposal. Where the site is still owned by the same company, it is usually possible to get the organization to remediate (clean up) the site. Frequently however, the offending company has been long out of business, and the multi-million dollar costs of remediation may fall on the taxpayers. In the U.S., the "Superfund" legislation provides a mechanism for sharing the cost of remediation of some of the worst polluted sites jointly between government and industry.

One of the problems facing chemical companies is that low levels of emission add up to major problems when manufacturing is carried out on a large scale. This may lead to very different perspectives on what is an acceptable level of emission.

Example: *The compound methyl methacrylate ($C_5H_8O_2$, abbreviated MMA) is used to make the acrylic polymer Plexiglas. During manufacture of Plexiglas, up to 0.5% of the MMA (which is rather volatile) may be lost into the atmosphere. Calculate the average daily loss of MMA from a plant which uses 5,000 t yr^{-1} of MMA.*

Answer:

Loss of MMA $= (0.5/100) \times 5{,}000\ t\ yr^{-1} = 2.5 \times 10^1\ t\ yr^{-1}$

daily loss $= 2.5 \times 10^1 t\ yr^{-1}/(365\ d\ yr^{-1}) = 7 \times 10^{-2}\ t\ d^{-1} = 70\ kg\ d^{-1}$

Depending on your point of view, limiting losses of MMA to 70 kg per day may represent an acceptable level of containment (only 0.5% emitted), or a completely unacceptable level of emission (25 tonnes per year).

However, one should always keep in mind that since chemicals are product, chemicals are also profit; therefore companies seek to sell the maximum amount possible to their customers. Emissions to the environment and material sent for waste disposal represent lost profits. It is therefore in any company's interests to minimize these losses.

Example: *The annual U.S. production of styrene (C_8H_8) is 4.1×10^6 tonnes. Assume that when the crude product is distilled, 0.6% of the material is high boiling residue ("still bottoms"). What mass of still bottoms require treatment and/or disposal?*

Answer: *mass* $= 4.1 \times 10^6\ t \times (0.6/100) = 2.5 \times 10^4\ t$

Again we see that while the waste to be disposed of is small in percentage terms, the mass of waste from a large scale industry is formidable.

The following are some of the factors which work to reduce emissions of manufactured substances into the environment.

- reduce waste, maximize profits
- improve corporate image
- public pressure, leading to legislation
- monitoring programs

The last of the factors, monitoring programs, is extremely important, since they force compliance with legislation. Legislation without monitoring is completely ineffective at controlling any industry which is indifferent to environmental concerns. However, it must be realized that even the most stringent legislation and the most comprehensive monitoring cannot abate emissions completely. The issue for society is to determine what are *acceptable* levels of emissions, as well as what can *practically* be achieved and monitored. Setting standards that cannot be achieved or monitored makes no sense. The practicality of emission standards also has to include consideration of what a particular industry can afford to achieve and still remain in business. Hard political decisions often revolve around the competition between high environmental standards and the threatened closure of an industrial operation which may be a major employer.

A final consideration in this context is the ever-changing nature of environmental legislation. A company which met all the legislative requirements of 1970 may find itself today with a site requiring millions of dollars to remediate to today's standards. Today, when an industrial plant is sold, a key consideration is whether the buyer or the seller will be responsible for any environmental problems which may later arise. Sometimes, the buyer purchases only the plant, with the vendor remaining fully liable for environmental consequences of its former operation. The day appears to be fast approaching when a company will never be able to remediate a site and then abandon it completely; a residual responsibility will always remain. This is a particular headache for mining companies (Chapter 17), whose operations lead to the permanent alteration of sites of hundreds or thousands of hectares.

Emission monitoring requires chemical analysis, and we stress again the need for reliable instrumental methods for the detection and quantitation of trace amounts of xenobiotics in water, air, soil, and wildlife. Monitoring may be done by government scientists, or alternatively, industry itself may be required to analyse its own effluents, and to keep records which may be inspected by government regulators. The latter approach is being developed in Ontario, where the MISA program (Municipal and Industrial Strategies for Abatement) places the responsibility on industry for testing for defined pollutants in all water leaving the plant. In the 1990s, as water pollution becomes increasingly controlled, attention is expected to shift towards control and monitoring of air emissions.

Municipal effluents are also covered under the MISA program: sewage treatment plants continue to be significant sources of water pollution. Political expediency and practicality make it easier politically to shut down an offending industrial polluter than a municipal sewage treatment facility, not least because shutting down the plant does not diminish the flow of sewage.

Many people have the impression that pollution by industry is an ever-growing problem. Reasons for this include greater awareness of the environment by the public, in large measure due to the efforts of environmental advocacy groups, and the increasing sophistication of chemical analysis, which allows the detection and quantitation of ever-smaller amounts of xenobiotics. This leads to a *perception* of increasing levels of pollution, when the reverse is actually the case. We are extremely unlikely to see repetitions of the Love Canal (Niagara Falls, New York) fiasco, which has become almost a synonym for industrial chemical pollution. The Love Canal is just one of many abandoned dump sites along the U.S. side of the Niagara River. These date from the first half of this century, when many chemical manufacturers were attracted to the area by cheap hydroelectricity. In those days, chemical wastes were disposed of very casually.

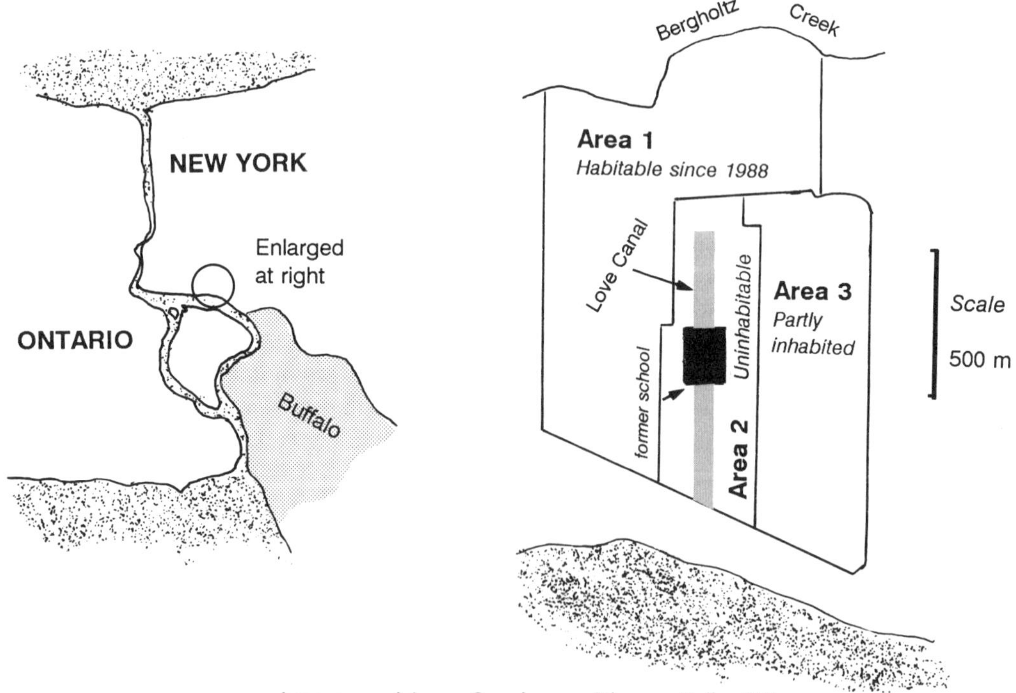

Area around Love Canal, near Niagara Falls, NY.

In the Love Canal incident, thousands of tonnes of toxic chlorinated still bottoms were simply dumped into an open trench, about 1.6 km long, 3 m wide, and varying in depth between 3 and 13 m. The trench had been dug as part of an unsuccessful canal-building project to bypass Niagara Falls in the late 1800s. Chemical dumping began in the 1940s, and when the trench was full it was covered with clay. The environmental problems which followed occurred because the site was subsequently developed residentially, and the homes and a school became contaminated with chemical wastes. This is not an isolated incident; other examples of residential development on former garbage dumps have led to abandonment of homes due to continued seepage of combustible methane into basements - the result of ongoing anaerobic decay. Greater environmental awareness in the developed world today means that society is unlikely to tolerate repetitions of these incidents.

While there is reason for cautious optimism about industrial pollution in the developed world, the outlook is very different in the rapidly industrializing areas of the "Third World". As in Europe and North America a century ago, the early stages of industrialization tend to involve low technology and accompanying high levels of pollution. In North America this is seen very clearly along the border between the U.S. and Mexico.

The higher population of the developing world today, compared with the developed world as it was a century ago, means that industrialization in the Third World could unleash a torrent of pollution far higher than the planet has experienced to date. Something of this kind of scenario has been observed with the opening up of the former Soviet bloc, whose heavy industry is now recognized to be outdated and highly polluting. A severe ecological problem — possibly a disaster — is now evident in the Aral Sea in former Soviet Asia. Diversion of its waters for irrigation has caused this equivalent of the North American Great Lakes to shrink in size. Meanwhile, continued industrial pollution has aborted a once thriving fishery and led to the eradication of several unique aquatic species.

The solution for both developed and developing nations is somehow to transfer advanced, less polluting technology so that the industrial revolution of the 21st century is environmentally less damaging than that which took place in the 19th and early 20th centuries. Immediately the question arises, "Who pays?"; private industry normally develops and patents new technologies, and the development costs are passed on to their customers. Older technology is not covered by patents, which run for 17 years from the date of filing. Older, more polluting technology is therefore more readily accessible to companies in Third World countries than up-to-date technology. Moreover, in a struggling economy, emission control technology may be viewed as an expensive and unnecessary luxury.

Problems

1. a) Calculate in moles the quantity of ammonia produced each year in the U.S. (15 million tonnes).

   b) Canadian industry produces about 25% as much ammonia as the U.S., yet the Canadian population is only 10% that of the U.S. Can you suggest a reason?

2. The Inco Ltd. mine in Sudbury processes about 35,000 t of ore daily. The ore has typical sulfur content of 9.8%. Calculate:

   (a) the mass of SO_2 that would be emitted to the atmosphere if all this ore were roasted without pollution controls;

   (b) the mass of sulfuric acid that could be produced if 43% of the sulfur dioxide were captured and converted to H_2SO_4.

3. Calculate the mass of ethylene (C_2H_4) required to manufacture 2,500 t of polyethylene if the yield of product is 94%.

   $$n\ C_2H_4 \rightarrow \sim\sim\sim\sim (CH_2CH_2)_n \sim\sim\sim\sim$$

4. Ethylene glycol $C_2H_6O_2$ is manufactured from ethylene for use as an antifreeze. Calculate the mass of ethylene required to make the U.S. annual output of ethylene glycol (2.3 million tonnes).

5. The U.S. production of urea (CON_2H_4) is 7.3 million tonnes. Calculate the percentage of U.S. ammonia production (15 million tonnes) which is used to make urea.

   $$CO_2 + 2\ NH_3 \rightarrow CON_2H_4 + H_2O$$

6. In the production of nitric acid, ammonia is oxidized in a multi-step reaction.

   $$4\ NH_3 + 5\ O_2 \rightarrow 4\ NO + 6\ H_2O$$
   $$2\ NO + O_2 \rightarrow 2\ NO_2$$
   $$3\ NO_2 + H_2O \rightarrow 2\ HNO_3 + NO$$

   The NO formed in the last step is recycled.

   a) Calculate the mass of ammonia needed to make the 1991 Canadian output of 0.91×10^6t HNO_3.

   b) What fraction of Canadian ammonia output (3.7 million tonnes) is used to make nitric acid?

7. Canadian production of Cl_2 and NaOH in 1991 were 1.4 and 1.5 million t, respectively. Calculate the dollar value of the Canadian chlor-alkali industry at an approximate bulk price of NaOH, \$300 t^{-1} and Cl_2, \$100 t^{-1}.

8. In 1985, Canadian transportation was responsible for the emission of 138×10^6 t of CO_2 and 7.2×10^6 t of CO into the atmosphere. Calculate the mass of petroleum (assumed composition about $C_{10}H_{20}$) responsible for these emissions.

5 ENERGETICS IN CHEMICAL REACTIONS I: ENTHALPY CHANGES

5.1 Conservation of energy

The study of the energy changes associated with physical and chemical processes is a branch of science called **thermodynamics**. Chemical reactions will only proceed if the products are more stable than the reactants, unless energy is provided to the reaction from an outside source, for example in the form of heat, light or electricity. In the environmental context, sunlight is an important outside energy source.

You may wonder why the energetics of chemical processes might be of concern in the environmental context. Energetics are important in determining what chemical processes might be possible. Consider the question of the fate of, for example, a pollutant in the environment. What happens to it? What is it likely to be converted to? In order to look for probable "secondary pollutants", we would need to think of likely chemical and physical processes which might remove this substance from its current environmental compartment. A calculation of the energetics of the reactions under consideration allows us to determine which are energy-releasing, and hence might be possible in principle. Of course, even though a reaction is favourable from the energetic point of view does not mean that it will actually occur in the environment at a measurable rate; kinetic information (Chapter 7) is needed to decide how fast each of the energetically possible reactions might proceed, and hence which process is actually dominant.

Our study of chemical energetics will be in two parts. In this chapter, we examine **enthalpy** (heat) changes accompanying chemical reactions. However, heat changes are not the only contributors to the overall energy change. Chapter 14 will deal with another aspect of chemical energetics: the **free energy change**, which determines whether a chemical process is **spontaneous** or **nonspontaneous**.

The Law of Conservation of Energy[1] states that energy can neither be created or destroyed; it can only be converted from one form into another. Heat, light, electricity and mechanical work are all forms of energy, and all can be interconverted, although not always with 100% efficiency. In particular, heat cannot be converted completely into any of the other forms of energy; the efficiency of machinery such as turbines, electricity generators, and automobiles is limited by fundamental restrictions on the conversion of heat into other energy forms.

The **First Law of Thermodynamics** is one statement of the law of energy conservation. It states that the energy changes comprise a combination of heat (q) and work (w) terms, eq. [1].

[1] $\Delta E = q + w$

ΔE (in some texts, ΔU) is the change in the **internal energy** of the system. In eq. [1], as in all equations involving energy change, the signs are arranged by convention such that energy released is given a negative sign. Energy is released if ΔE is negative, when q is negative (heat is evolved), and when w is negative (by convention, work done by a system is negative; w would be negative if the chemical reaction produced work: for example, if an explosion forced an expansion of gases).

Rule: *A negative sign always indicates an energy-releasing process.*

Figure 1 illustrates why energy release is given a negative sign; the products of the reaction shown have a lower energy content than the reactants; energy was released when the reactants (higher energy content) were changed into the products (lower energy content). Reactions involving the evolution of heat are termed **exothermic**; reactions which require the input of external heat are called **endothermic**. For the majority of chemical reactions the "w" term in eq. [1] is much smaller than the "q" or heat term, and can often be neglected.

The mechanical work term is only significant in chemical processes where gases are produced or consumed. Under these circumstances, $w = -P\Delta V$, where P and V have their usual meanings of pressure and volume. From the ideal gas

---

[1] The well-known equation $E = mc^2$ extends the concept of energy conservation further, by indicating that matter and energy can be considered equivalent. This equivalence finds practical application in nuclear processes, such as power reactors, nuclear bombs, and the Sun and stars, whose power generation comes form the conversion of matter into heat and light. In chemical reactions, the interchange of matter and energy is so infinitesimal that it can be neglected. If that were not so, there could be no law of conservation of mass during chemical reactions. Recall that the concept of conservation of mass is critical to the concept of stoichiometry.

law (see also Section 6.1):

$$PV = nRT, \text{ so } P\Delta V = \Delta nRT$$

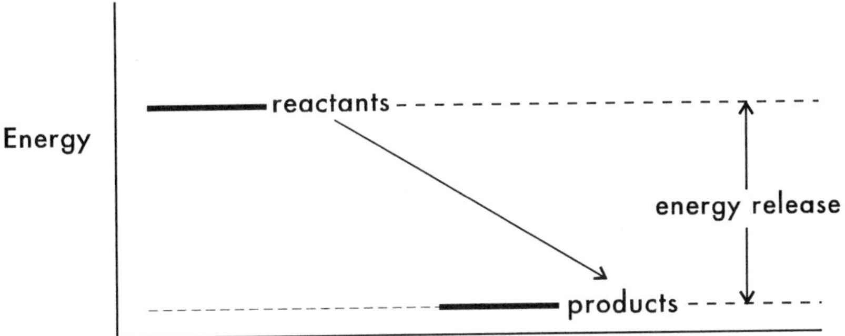

Figure 5.1: Energy release when reactants are converted to products of lower energy content

If necessary the work term can be evaluated by considering the number of moles of *gas* evolved (positive) or absorbed (negative) in the reaction. Only gaseous reactants and products are considered in determining Δn, because the volume changes are negligible in reactions involving only solids and liquids.

Example: *2 NiS(s) + 3 O_2(g) → 2 NiO(s) + 2 SO_2(g)*
In this reaction Δn = -1 (i.e. 3 mol reactant gases, 2 mol product gases).

5.2 Enthalpy changes and thermochemical equations

The enthalpy can be considered as the heat content of a chemical substance or reaction system. Unfortunately, no-one has ever succeeded in determining the enthalpy of any substance absolutely; the best that can be done is to determine the **enthalpy change** ΔH associated with a given chemical or physical process. Restating this important point, chemical reactions involve energy changes; the changes can be measured, even though the absolute magnitudes of the energy contents of reactants and products are unknown.

The enthalpy change is defined more specifically as the heat change that occurs when a physical process or a chemical reaction occurs at constant temperature and pressure; if the pressure is 1 atm we use the superscript zero to indicate a "standard" enthalpy change $\Delta H°$. Under conditions of constant pressure, eq. [1] can be rewritten.

$$\Delta E = q + w = \Delta H° - P\Delta V = \Delta H° - \Delta nRT$$

Thus: $\Delta H° = \Delta E + \Delta nRT$

For all but the most precise work, the same value of $\Delta H°$ can be used for pressures other than 1 atm, *provided that the pressure remains constant during the course of the reaction.* This is important if we want to calculate enthalpy changes of reactions in the upper atmosphere, where $p_{total} << 1$ atm, but still use enthalpy data that have been obtained in the laboratory at 1 atm pressure. Furthermore, the enthalpy change is approximately independent of temperature, provided that all the reactants and products stay in the same physical states, and the heat change is measured when the reactants and the products are at the same temperature. In the most precise work, the small variation of $\Delta H°$ with temperature must be recognized and the temperature is included as a subscript. Thus $\Delta H°_{298}$ is the enthalpy change for a reaction occurring at constant pressure at the constant temperature 298 K (25°C). We shall neglect the variation of $\Delta H°$ with temperature in this book.

Definition

*The **enthalpy change** is the heat change at constant pressure and temperature.*

Let us examine the characteristics of enthalpy change by reference to a particular chemical reaction, the reduction of carbon monoxide to methanol. This reaction is carried out on a large scale industrially, and has $\Delta H° = -90.2$ kJ mol $^{-1}$.

[2] $CO(g) + 2\ H_2(g) \rightarrow CH_3OH(g)$ $\quad\quad \Delta H° = -90.2$ kJ mol $^{-1}$

Equation [2] is an example of a **thermochemical equation**. It is an ordinary balanced chemical equation, which contains two additional pieces of information: the value of the enthalpy change associated with the reaction, and the physical states of matter of each reactant and product. The latter are very important; as we shall see, the enthalpy change depends upon the states of matter of the chemical substances. Therefore no thermochemical equation is complete without inclusion of the states: s (solid), ℓ (liquid), or g (gas) for pure substances, or (aq) for aqueous solutions.

The enthalpy change is part of the stoichiometry of the process, as seen from the units: kJ mol $^{-1}$. This concept carries several important implications.

1. In the reaction above, the evolution of 90.2 kJ mol $^{-1}$ accompanies the production of 1 mol of CH_3OH, but the disappearance of 2 mol H_2. Enthalpy changes in kJ mol $^{-1}$ therefore refer to 1 mol of the equation as written. In thermochemical equations, the coefficients are interpreted as numbers of moles rather than numbers of molecules.

2. The production of 1 mol of methanol releases 90.2 kJ of heat, whereas production of 100 mol of methanol releases 9,020 kJ of heat. Strictly, these heat changes apply to reactions where the products have been allowed to warm or cool to the original temperature of the reactants; this is what is meant by saying that enthalpy changes refer to reactions carried out at constant temperature. The scale-up of reactions from the laboratory to the industrial scale requires careful consideration of how to dispose of the heat of exothermic reactions, because the amount of heat goes up proportionately with the scale of the reaction. Many industrial accidents have occurred over the years due to the failure to cool exothermic reactions sufficiently. An example which made world-wide headlines was an accident in Seveso, Italy in 1976, where a runaway exothermic reaction caused an explosion which contaminated the town with dioxin; nearly 20 years later, the health status of these residents is still being monitored regularly.

3. The enthalpy change ΔH° is a characteristic property of a reaction, and its value depends only on the initial and final states of the system. It therefore follows that the enthalpy change for the reverse reaction (eq. [3]) must be numerically equal to that of reaction [2], but with the sign reversed.

[3] $CH_3OH(g) \rightarrow CO(g) + 2\ H_2(g) \quad \Delta H° = +90.2 \text{ kJ mol}^{-1}$

By the same reasoning, the enthalpy change for the overall reaction is independent of the reaction pathway. For example, reaction [2] can be carried out in two stages, reactions [4] and [5].

[4] $CO(g) + H_2(g) \rightarrow CH_2O(g) \quad \Delta H° = +2.0 \text{ kJ mol}^{-1}$

[5] $CH_2O(g) + H_2(g) \rightarrow CH_3OH(g) \quad \Delta H° = -92.2 \text{ kJ mol}^{-1}$

Notice that if you add up equations [4] and [5] the substance $CH_2O(g)$ drops out on both sides, restoring eq. [2]. The sum of the enthalpy changes of reactions [4] and [5] is the enthalpy change of reaction [2]. This is an example of **Hess' Law of heat summation**, discovered in the middle of the last century: *when you add together a number of thermochemical reactions, you add together the enthalpy changes*. Put another way, the enthalpy change for the conversion of CO(g) to $CH_3OH(g)$ is the same whether you carry out the reaction directly, or whether you take it in two steps.

Figure 2 summarizes the preceding points graphically. The reactants CO(g) and 2 $H_2(g)$ lie 90.2 kJ mol $^{-1}$ above the product, $CH_3OH(g)$. Conversion of $CO(g)$ and 2 $H_2(g)$ into $CH_3OH(g)$ releases 90.2 kJ mol $^{-1}$, and the reverse reaction requires the absorption of 90.2 kJ mol $^{-1}$. The energetics of the overall reaction are independent of whether the reaction is carried out in a single step, or whether CH_2O is formed as an intermediate stage.

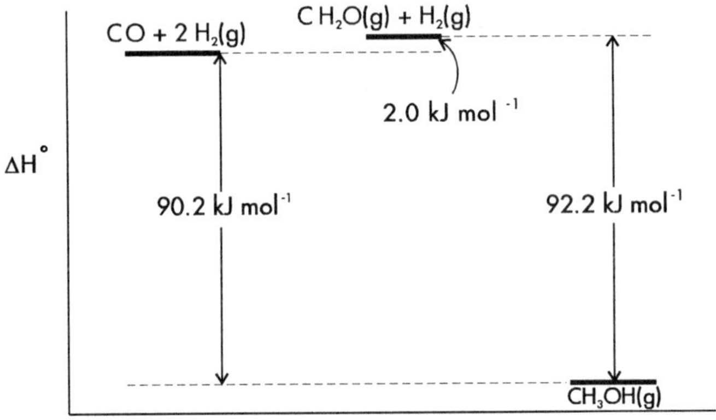

Figure 5.2: Energetics of the conversion of 1 mol carbon monoxide and 2 mol hydrogen into 1 mol methanol.

We noted earlier that enthalpy changes are approximately independent of the temperature at which the reaction is carried out. This is only true provided that the substances all remain in the same physical states. For example, the formation of liquid water from its elements has $\Delta H° = -285.8$ kJ mol $^{-1}$ at 25 °C.

$$H_2(g) + \frac{1}{2} O_2(g) \rightarrow H_2O(\ell) \qquad \Delta H° = -285.8 \text{ kJ mol}^{-1}$$

It would be approximately correct to use the same value of $\Delta H°$ for the formation of liquid water at 100 °C. However, it would not be valid to use this value for the formation of steam at 100 °C.

$$H_2(g) + \frac{1}{2} O_2(g) \rightarrow H_2O(g) \qquad \Delta H° = -241.8 \text{ kJ mol}^{-1}$$

In thermochemical equations, $H_2O(g)$ and $H_2O(\ell)$ are to be considered as different substances, as are the different physical states of all other chemical substances. Therefore, when looking up values of enthalpies in tables, always be sure to check that you have the correct physical form of the substance in question.

The law of heat summation is very useful for calculating the enthalpy changes for reactions for which the enthalpy change is not known experimentally. Suppose you want to know the enthalpy change for the oxidation of nitric oxide

NO by ozone O_3, a reaction which is very important in the polluted urban atmosphere (Section 9.1).

[6] $NO(g) + O_3(g) \rightarrow NO_2(g) + O_2(g)$ $\Delta H° = ?$

You have been able to find data for two other reactions, [7] and [8].

[7] $NO(g) + \frac{1}{2} O_2(g) \rightarrow NO_2(g)$ $\Delta H° = - 53.4$ kJ mol $^{-1}$

[8] $3 O_2(g) \rightarrow 2 O_3(g)$ $\Delta H° = + 284.6$ kJ mol $^{-1}$

Equations [7] and [8] can be manipulated to recreate the desired equation [6]. Equation [7] is the right way round (1 mol NO(g) on the left, 1 mol $NO_2(g)$ on the right); equation [8] links the two substances $O_2(g)$ and $O_3(g)$, but the equation needs to be reversed and rewritten in terms of 1 mol $O_3(g)$ not 2 mol. This means that the value of $\Delta H°$ must be divided by 2 and the sign reversed, eq. [9].

[7] $NO(g) + \frac{1}{2} O_2(g) \rightarrow NO_2(g)$ $\Delta H° = - 53.4$ kJ mol $^{-1}$

[9] $O_3(g) \rightarrow \frac{3}{2} O_2(g)$ $\Delta H° = - 142.3$ kJ mol $^{-1}$

Adding equations [7] and [9] gives equation [6], and yields the desired information, $\Delta H° = - 195.7$ kJ mol $^{-1}$.

5.3 Definition of the standard molar enthalpy of formation

The calculation of reaction enthalpies using Hess' Law is limited in practice because you cannot always find the enthalpy changes for all the reactions you need in order to "recreate" the reaction in which you are interested. Standard enthalpies of formation (symbol $\Delta H°_f$) represent a systematized way of providing the information needed to calculate the enthalpy change for virtually any reaction. Textbooks and reference books such as the *Handbook of Chemistry and Physics* contain extensive compilations of standard enthalpies of formation. A smaller collection is found in Appendix 1 of this book.

Definition

*The **standard molar enthalpy of formation** of any substance is the heat change that occurs when 1 mol of the substance is formed from its elements, and the elements are in their usual physical states at 25°C and 1 atm pressure.*

Notice the four parts to the definition:

1. 1 mol of substance
2. is formed
3. from its elements
4. elements in their usual physical states at 25 °C, 1 atm (*i.e.*, close to ordinary laboratory conditions).

Example: *Write a thermochemical equation having $\Delta H°$ equal to the standard molar enthalpy of formation of solid sodium nitrate.*

Answer:

Step 1: Sodium nitrate is $NaNO_3$. The only substance on the right hand side of the required equation is 1 mol of $NaNO_3$, the substance whose standard molar enthalpy of formation is required.

$$? \rightarrow NaNO_3(s)$$

Step 2: The elements which are constituents of $NaNO_3$ (that is Na, N, and O) are present on the left hand side of the equation.

$$? (Na) + ? (N) + ? (O) \rightarrow NaNO_3(s)$$

Step 3: Now balance the equation. You will need the proper states of each element. Na is a solid; N and O are both gaseous molecules N_2 and O_2 at 25 °C. You need 1 mol of Na, ½ mol of N_2, and 3/2 mol of O_2 to balance the equation. Don't worry that some of the numbers of moles of elements are fractional; the important thing is to get 1 mol of substance on the right hand side.

$$Na + \tfrac{1}{2} N_2(g) + 3/2\ O_2(g) \rightarrow NaNO_3(s)$$

Aside: How do we learn the physical states of all the elements at 25 °C and 1 atm? Actually, it is quite easy.

- Relatively few elements are gases, and they are probably familiar to you. The noble gases are He, Ne, Ar, Kr, Xe, and Rn. Five elements exist as diatomic gases: H_2, N_2, O_2, F_2 and Cl_2.
- Only two elements are liquids at room temperature: Hg and Br_2.
- All other elements are solids at 25 °C, 1 atm. Basically, if you do not recognize an element, it is most probably a solid.

Standard enthalpies of formation of elements

It follows from the definition of standard enthalpy of formation that the standard enthalpy of formation of any element must be zero when it is already in its "standard state", the usual physical state at 25 °C, 1 atm.

Example 1: *What is the standard enthalpy of formation of gaseous Cl_2?*
Answer: *Zero; chlorine occurs as $Cl_2(g)$ at 25 °C and 1 atm.*

Example 2: *Is the standard enthalpy of formation of $Br_2(g)$ equal to zero?*
Answer: *No. Bromine is a liquid at 25 °C and 1 atm. (The standard enthalpy of formation of gaseous Br_2 is actually + 30.9 kJ mol^{-1}).*

Standard enthalpies of formation of aqueous ions

It has already been noted that absolute values of enthalpies cannot be measured. The concept of standard molar enthalpy of formation avoids this problem by defining the standard molar enthalpy of elements of formation as zero when they are in their usual physical states at 25 °C and 1 atm.

For ions in aqueous solution, the analogous convention is that the standard molar heat of formation of H^+(aq) is taken to be zero. For example, suppose we determine the standard molar enthalpy of formation of the strong electrolyte HCl(aq) to be -167.2 kJ mol^{-1}. Unless we make an assumption about $\Delta H°_f$ for one of the ions, we would have no way of dividing up the -167.1 kJ mol^{-1} between H^+(aq) and Cl^-(aq). The convention assigns $\Delta H°_f = 0$ to H^+(aq) and $\Delta H°_f = -167.1$ kJ mol^{-1} to Cl^-(aq). Now when we measure the standard molar enthalpy of formation of, for example, NaCl(aq) as -407.2 kJ mol^{-1} we can apportion this between Cl^-(aq): -167.1 kJ mol^{-1} and Na^+(aq): -240.1 kJ mol^{-1}. Hence the standard molar heat of formation of any aqueous ion can be established, once that of H^+(aq) has been assigned.

5.4 Using standard enthalpies of formation

5.4.1 Calculation of enthalpy changes for chemical reactions

Suppose we want to calculate $\Delta H°$ for the reaction below - thinking (incorrectly as it turns out) that this might be a plausible route for the removal of NO_2 from the atmosphere.

[10] $$2\ NO_2(g) \rightarrow N_2O_4(g) \qquad \Delta H° = ?$$

We look up in standard tables ΔH°_f for $NO_2(g)$ (+33.2 kJ mol $^{-1}$) and $N_2O_4(g)$ (+ 9.2 kJ mol $^{-1}$). How do we proceed?

First we might write out the thermochemical equations which define the standard heats of formation of $NO_2(g)$ and $N_2O_4(g)$.

[11]	$\frac{1}{2} N_2(g) + O_2(g) \rightarrow NO_2(g)$	$\Delta H^\circ = + 33.2$ kJ mol $^{-1}$
[12]	$N_2(g) + 2\ O_2(g) \rightarrow N_2O_4(g)$	$\Delta H^\circ = + 9.2$ kJ mol $^{-1}$

Now we could use Hess' Law. Reverse eq. [11] and multiply by 2; call this equation [13]. Keep eq. [12] as it is.

[13]	$2\ NO_2(g) \rightarrow N_2(g) + 2\ O_2(g)$	$\Delta H^\circ = -\ 66.4$ kJ mol $^{-1}$
[12]	$N_2(g) + 2\ O_2(g) \rightarrow N_2O_4(g)$	$\Delta H^\circ = + 9.2$ kJ mol $^{-1}$

Adding these equations gives the desired eq. [10], $\Delta H^\circ = -\ 57.2$ kJ mol $^{-1}$

By the use of Hess' Law, we have just exploited the concept of standard heats of formation to convert the reactant, 2 $NO_2(g)$ into the product, $N_2O_4(g)$. We used an imaginary reaction sequence in which we first broke the reactant down into its elements and then recombined the elements into the product. This approach worked because the heat change for the overall process is independent of the pathway taken, as explained previously.

Look carefully at the heat changes corresponding to reactions [13] and [12]. They correspond to $\{-\ 2\ \Delta H^\circ_f(NO_2,g)\}$ and $\{\Delta H^\circ_f(N_2O_4,g)\}$ respectively. Thus:

$$\Delta H^\circ(\text{reaction 10}) = \Delta H^\circ_f(N_2O_4,g) - 2\ \Delta H^\circ_f(NO_2,g)$$

This is a specific example of a general rule:

[14] $\Delta H^\circ(\text{reaction}) = \sum \Delta H^\circ_f(\text{products}) - \sum \Delta H^\circ_f(\text{reactants})$

Rule: *The enthalpy change for any reaction can be calculated as the sum of the standard heats of formation of all the products minus the sum of the standard heats of formation of all the reactants.*

This rule is completely general, and follows directly from Hess' Law. It applies to aqueous solutions as well as to pure substances. In dealing with aqueous solutions, you must decide which substances are ionized in order to look up the correct heats of formation.

Example: *Calculate the standard enthalpy change for the reaction below.*

$$CaCO_3(s) + 2\,HCl(aq) \rightarrow CaCl_2(aq) + CO_2(g) + H_2O(\ell)$$

First write the net ionic equation (omit the two Cl^- *ions which appear on both sides of the equation).*

$$CaCO_3(s) + 2\,H^+(aq) \rightarrow Ca^{2+}(aq) + CO_2(g) + H_2O(\ell)$$

Now apply equation [14].

$\Delta H°(reaction) = \sum \Delta H°_f(products) - \sum \Delta H°_f(reactants)$

$= \{\Delta H°_f(Ca^{2+}, aq) + \Delta H°_f(CO_2, g) + \Delta H°_f(H_2O, \ell) - \Delta H\ell°_f(CaCO_3, s) - 2\,\Delta H°_f(H^+, aq)\}$

$= (-542.8) + (-393.5) + (-285.8) - (-1206.9) - (2 \times zero)$

$= -15.2\ kJ\ mol^{-1}$

Always be very careful when looking up the standard enthalpies of formation of substances in tables. If you are careless, you may end up with the wrong substance!

Examples:

$H_2O(\ell)$: $\Delta H°_f = -285.8\ kJ\ mol^{-1}$

$H_2O(g)$: $\Delta H°_f = -241.8\ kJ\ mol^{-1}$

$OH(g)$: $\Delta H°_f = +38.9\ kJ\ mol^{-1}$

$OH^-(aq)$: $\Delta H°_f = -230.0\ kJ\ mol^{-1}$

The hydroxyl radical OH is a completely different substance from the hydroxide ion OH^-. We shall meet OH(g) in our study of atmospheric chemistry (Chapter 9), while OH^- is encountered in solution when we discuss acids and bases (Chapter 11).

5.4.2 Calculating enthalpy changes for physical processes

Like chemical reactions, physical changes are accompanied by changes in heat content. For example, heat is absorbed when water boils at 100 °C, even though its temperature does not change.

$$H_2O(\ell) \rightarrow H_2O(g) \qquad \Delta H° = ?$$

We calculate this enthalpy change just like that for a chemical reaction, using the standard enthalpies of formation of liquid and gaseous H_2O given above[2].

$$\begin{aligned}\Delta H°(\text{reaction}) &= \sum \Delta H°_f(\text{products}) - \sum \Delta H°_f(\text{reactants}) \\ &= \Delta H°_f(H_2O, \ell) - \Delta H°_f(H_2O, g) \\ &= (-241.8) - (-285.8)\ \text{kJ mol}^{-1} \\ &= +\ 44.0\ \text{kJ mol}^{-1}\end{aligned}$$

This enthalpy change is called the heat of vaporization of water.

Learn these terms associated with various physical changes.

Process	Symbol for enthalpy change
Fusion (solid melts to liquid)	$\Delta H°_{fus}$
Vaporization (liquid boils or evaporates to gas)	$\Delta H°_{vap}$
Sublimation (solid evaporates to gas, bypasses liquid)	$\Delta H°_{sub}$
Atomization (substance is converted to gaseous atoms)	$\Delta H°_{at}$
Solution (substance dissolves in solvent)	$\Delta H°_{soln}$

The first four of these processes are always endothermic (positive ΔH°), no matter what the substance. However, heats of solution may be exothermic or endothermic, depending upon the solvent/solute combination. Heat is required to cause a solid to melt, or a liquid to vaporize. From Hess' Law, the heat of sublimation for a given substance will be the sum of its heats of fusion and vaporization.

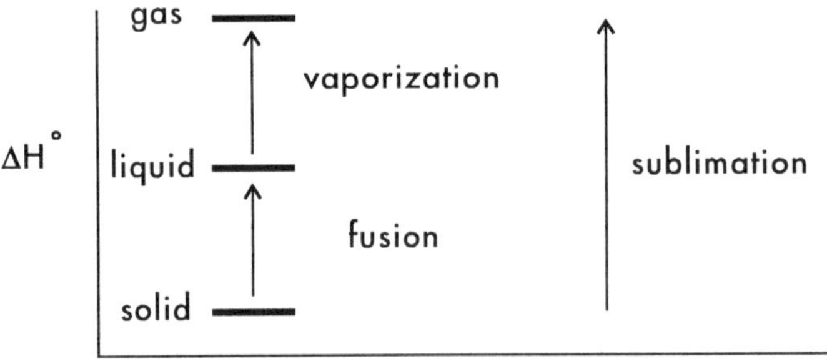

[2] It was mentioned earlier that ΔH° changes slightly with temperature. In this example, the calculated value of the heat of vaporization using standard enthalpies of formation appropriate to 298 K is +44.0 kJ mol^{-1}, in agreement with experimental observations at this temperature. The experimental value at the normal boiling point of water, 100 °C (373 K), is +40.9 kJ mol^{-1}. It is possible to carry out more refined calculations, but we shall not do so in this course.

Example: *Write a balanced thermochemical equation to show that the standard molar heat of fusion of ethanol (C_2H_6O) is 5.0 kJ mol^{-1}.*

Answer:

The process involved is $C_2H_6O(s) \rightarrow C_2H_6O(\ell)$.

Hence the required thermochemical equation is:

$$C_2H_6O(s) \rightarrow C_2H_6O(\ell) \qquad \Delta H° = +\ 5.0\ kJ\ mol^{-1}$$

Conversely, the standard molar heat change when liquid ethanol freezes would be -5.0 kJ mol^{-1}.

Atomization is a rather special process, in which the substance is converted to gaseous atoms; it turns out to be very important in establishing accurate values of bond energies, which are discussed below in Section 5.6. For example, the following reaction has ΔH° equal to the heat of atomization of carbon.

$$C(s, graphite) \rightarrow C(g)$$

Atomization is frequently different from sublimation, in that bonds may be broken. Compare these two reactions.

$I_2(s) \rightarrow I_2(g)$	heat of sublimation of I_2
$\frac{1}{2}\ I_2(s) \rightarrow I(g)$	heat of atomization of solid iodine

5.5 Fuels and combustion reactions

5.5.1 Definition of heat of combustion

As noted already, chemical reactions are accompanied by energy changes. Some chemical reactions are carried out principally to obtain energy: in other words, to convert chemical energy locked up in the structure of the reactants into a more usable form. Such reactants are called **fuels**. All common fuels (wood, oil, coal) are carbon compounds (strictly, coal is mostly elemental carbon), and give up their energy through **combustion**.

Definition

Combustion *is the complete oxidation of a substance through the use of air or O_2; in the case of carbon-containing fuels, the products are CO_2 and H_2O.*

The heat released upon the combustion of 1 mol of a fuel is the standard molar enthalpy of combustion, $\Delta H°_{comb}$. Although fuels burn at high temperatures, so that any H_2O is formed in the gaseous state, tabulated values of $\Delta H°_{comb}$ are corrected to the (constant) temperature of 298 K, when water would be in the form $H_2O(\ell)$.

Example: *Write a thermochemical equation having $\Delta H°$ equal to the standard molar enthalpy of combustion of liquid isooctane, C_8H_{18}, a component of gasoline, for which the standard molar enthalpy of combustion is 5456 kJ mol^{-1}.*

Answer: *We need (i) to write the equation in terms of 1 mol of isooctane, (ii) to remember that the products are $CO_2(g)$ and $H_2O(\ell)$, and (iii) to balance the equation by the use of the stoichiometric number of moles of $O_2(g)$.*

$$C_8H_{18}(\ell) + 12\tfrac{1}{2}\ O_2(g) \rightarrow 8\ CO_2(g) + 9\ H_2O(\ell) \quad \Delta H° = -5456\ \text{kJ mol}^{-1}$$

Note the negative sign; all combustion reactions are exothermic *(otherwise the fuel wouldn't burn).*

5.5.2 *Heat contents of fuels*

The value of $\Delta H°_{comb}$ indicates how much heat can be obtained from a mole of fuel. For most fuels, the major products are CO_2 and H_2O, so the energy gap between the reactants (fuel + O_2,g) and these products determine how much heat is available. The less negative the value of $\Delta H°_f$ for a particular fuel, the more heat is available. The extreme case is when $\Delta H°_f$ of the fuel is positive (that is the breakdown of the fuel into its elements is exothermic, and the fuel is an **endothermic compound**). Acetylene (C_2H_2, $\Delta H°_f$ = + 227 kJ mol^{-1}) is an example of an endothermic compound, and this explains in part the high temperature of the oxyacetylene flame used in welding and metal cutting. The heat liberated on combustion of 1 mol of acetylene, at the temperature of the flame, is equal to the enthalpy of formation of 2 mol CO_2 (g) and 1 mol H_2O (g) plus the 227 kJ mol^{-1} by which the formation of acetylene is endothermic. A comparison between the heat liberated upon combustion of 1 mol each of acetylene and ethane (C_2H_6, $\Delta H°_f$ = -85 kJ mol^{-1}) is shown below.

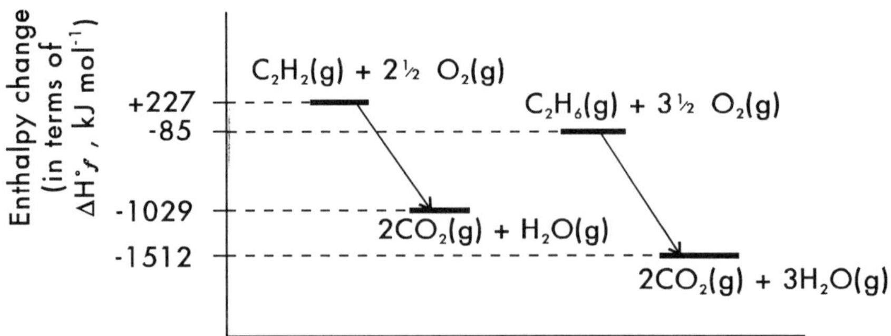

Figure 5.3: A comparison of ΔH_f for acetylene and ethane

As a practical consideration, the **heat content** of a fuel is often more useful than the standard molar heat of combustion. The heat content is the heat of combustion on a weight basis. Some typical values are shown below.

Fuel	Heat content, MJ kg $^{-1}$
$H_2(g)$	130
$CH_4(g)$	50
gasoline, fuel oil	44-48
coal	28-30

Consideration of the heat content of a fuel is useful in transportation applications, where the mass the fuel itself (plus its container) reduces the payload that can be carried, since the fuel must also be transported. For example, although hydrogen gas is a very attractive fuel from the point of view of its own weight, it unfortunately requires a massive container for the pressurized gas, thus offsetting its favourable heat content per kilogram. Various strategies have been considered to take advantage of the high heat content of hydrogen. *Liquid hydrogen* does not have to be pressurized, but has to be extremely well insulated, since its boiling point is -253 °C (20 K). Alternatively, there are various *metallic hydrides* which give up their hydrogen upon gentle warming — for example, being vanadium hydride, which has the approximate composition $VH_{0.6}$. However, the expense of making the metallic hydride adds greatly to the cost of this potential fuel, while the conversion of H_2 to a metal derivative negates the favourable energy-to-mass property of elemental hydrogen.

Fuels in common use today are classified as either **fossil fuels** or **biomass fuels**. While biomass-derived fuels are in principle renewable resources, fossil fuels are a finite resource which cannot be replaced. Fossil fuels are coal (principally elemental carbon), oil (a complex mixture of hydrocarbons, compounds composed of only carbon and hydrogen), and natural gas (principally methane, CH_4). The major biomass fuel is wood, although in some countries certain crops are grown for their fuel value. Ethanol (C_2H_6O) is obtained as a biomass derivative by fermentation of sugar. The sugar is either produced specifically for the purpose, or obtained by hydrolysis of starch (from corn). "Gasohol", sold in some countries, is a blend of ethanol, a renewable resource, with gasoline, a fossil fuel.

Example: *What mass of ethanol can be obtained by fermenting 17,000 mol of glucose ($C_6H_{12}O_6$) if the isolated yield is 73%? The stoichiometric equation is:*

$$C_6H_{12}O_6 \rightarrow 2\ C_2H_6O + 2\ CO_2$$

Answer: *Because the percentage yield is only 73%, 1 mol of $C_6H_{12}O_6$ produces (0.73 x 2 mol), that is 1.46 mol of C_2H_6O, rather than 2 mol as the "theoretical" equation indicates. The "experimental" equation is:*

$$C_6H_{12}O_6 \rightarrow 1.46\ C_2H_6O + \text{other products}$$

Hence: $n(C_2H_6O) = 17\ \text{kmol}\ C_6H_{12}O_6 \times (1.46\ \text{mol}\ C_2H_6O/1\ \text{mol}\ C_6H_{12}O_6)$

$= 25\ \text{kmol}$

Since $\mathbf{M}(C_2H_6O) = 46\ \text{g mol}^{-1}$, *mass of ethanol* $= 25\ \text{kmol} \times 46\ \text{g mol}$

$= 1.2 \times 10^3\ \text{kg} \Rightarrow 1.2\ \text{tonnes}$

Both biomass and fossil fuels produce carbon dioxide upon combustion. Increased industrialization worldwide is accompanied by an increasing use of fuel, and a corresponding increase in emissions of carbon dioxide. As we shall see in Chapter 6, the partial pressure of CO_2 in the atmosphere is increasing and is predicted to be the cause of global warming in the next century. The issue of CO_2 emissions to the atmosphere is relevant to the discussion of chemical energetics because fuels are burned principally for their heat content, and CO_2 is a byproduct of this activity. We can think of a number of major industries where this is so.

- Thermal power stations for electricity generation. The fuel is burned to raise steam, which drives the turbines.
- Smelting of metals. Iron, nickel, and many other metals are produced by reducing the metal oxide by means of coal or coke. For example:

$$2C(s) + O_2\ (g) \rightarrow 2\ CO(g)$$
$$Fe_2O_3(s) + 3\ CO(g) \rightarrow 2\ Fe(\ell) + 3\ CO_2\ (g)$$

  In this equation, carbon monoxide rather than carbon is the actual substance which reduces Fe_2O_3 .

- Production of lime and cement. Here a fuel is burned to produce heat in order to decompose calcium carbonate, and in addition, further CO_2 is produced through the decomposition of limestone, $CaCO_3$.

$$CaCO_3(s) \rightarrow CaO(s) + CO_2(g)$$

5.5.3 Clean and dirty fuels

Burning fuels not only affords CO_2, which is a source of environmental concern, but other byproducts in addition. Coal contains small amounts of sulfur, oxidation of which yields SO_2, which is a major cause of acid rain (Chapter 13). Coal is therefore considered a "dirty" fuel. Petroleum oil generally contains little sulfur, while the major sulfur compound in natural gas (H_2S) is removed by the process of "sweetening", which means catalytic oxidation of the H_2S in the gas to give elemental sulfur. Petroleum products and natural gas are therefore considered to be "clean" fuels. However, all combustion reactions give heat, and lead to the formation of small quantities of nitrogen oxides, which are responsible for photochemical smog (Chapter 9).

$$N_2(g) + O_2(g) \rightarrow 2\ NO(g) \qquad \Delta H^\circ = +\ 180\ kJ\ mol^{-1}$$

This endothermic reaction becomes more efficient as the temperature rises. This is significant because all "heat engines" (machines which use heat as their source of mechanical energy) become more efficient as the temperature rises (Chapter 14). Automobiles and power stations are examples of heat engines which are designed to operate at the maximum temperature practicable. Efficiency of operation therefore conflicts with limiting emissions of NO.

Proponents of a cleaner environment may call for increased use of electricity or for a switch to hydrogen as a fuel which "burns to give only a whiff of steam". Although electricity is clean at the point of use, it is a substantial point source of CO_2, nitrogen oxides, and maybe SO_2 as well, if it is generated thermally from fossil fuels (although these pollutants would not be produced at a hydroelectric or a nuclear generating station). The person who plugs an appliance into the wall is unaware of any pollution which may have been created at the power plant.

Hydrogen represents a similar situation. Hydrogen is not found as the element on Earth, and so the hydrogen must be produced as an "energy currency" from some other source, which in general will be less "clean" than the hydrogen. The proposed sources are *(i)* the high temperature reaction of steam with natural gas or *(ii)* the high temperature reaction of steam with powdered coal or (iii) electrolysis of water. Reaction *(i)* produces 0.25 mol of CO_2 for every mole of H_2 formed. Reaction *(ii)* is endothermic, and the necessary high temperature is maintained by burning some of the coal. Reaction (iii) requires the expenditure of electrical energy. Because the production of hydrogen is not necessarily a "clean" process environmentally, caution should be exercised in regarding hydrogen as the super-clean fuel of the future.

(i) $CH_4(g) + 2\ H_2O(g) \rightarrow CO_2(g) + 4\ H_2(g)$

(ii) $C(g) + H_2O(g) \rightarrow CO(g) + H_2(g)$

(iii) $2\ H_2O(\ell) \xrightarrow{\text{electrical energy}} 2\ H_2(g) + O_2(g)$

Even hydrogen will cause nitrogen oxides to be formed when it burns, unless pure oxygen is used for combustion rather than air. The reason is that the $N_2/O_2/NO$ reaction occurs whenever air is heated - and air always contains nitrogen.

5.6 Bond dissociation energies

The "bond dissociation energy" can be thought to represent the "strength" of a covalent bond.

Definition

*The **Bond Dissociation Energy (BDE)** is the enthalpy change for the dissociation of the bond into its constituent atoms or free radicals when the dissociation occurs in the gas phase.*

Note particularly that the bond dissociation energy is restricted to the dissociation of covalent bonds, and to their dissociation into neutral particles. Since the bond dissociation energy is the enthalpy change associated with *breaking* the bond, bond dissociation enthalpies are always positive quantities. The magnitude of the bond dissociation energy is a measure of bond strength. Appendix 2 contains a listing of average dissociation energies of commonly encountered bonds.

Example: *$O_2(g) \rightarrow 2\ O(g)$ $\Delta H° = 498\ kJ\ mol^{-1}$*
Hence the bond dissociation energy of the O-O bond in molecular oxygen is 495 kJ mol^{-1}. Notice that this equation also defines the standard enthalpy of formation of two moles of atomic oxygen; the bond dissociation energy of $O_2(g)$ is exactly twice the heat of formation of O(g).

Example: *Which of the following reactions has ΔH° equal to the bond dissociation energy of the bond that is broken?*

A: $H\text{-}Cl(g) \rightarrow H^+(aq) + Cl^-(aq)$

B: $H\text{-}Cl(g) \rightarrow H^+(g) + Cl^-(g)$

C: $H\text{-}Cl(g) \rightarrow H(g) + Cl(g)$

Answer: *C. Equations A and B both involve ionic dissociation, and in addition, equation A has the products in aqueous solution. Only equation C is a gas phase, non-ionic bond cleavage, as required by the definition of bond dissociation energy.*

Bond energies can be applied to large molecules, if just one bond breaks. For example, the thermochemical equation below has ΔH° equal to the C-C bond dissociation energy of ethane, CH_3-CH_3(g).

$$CH_3\text{-}CH_3(g) \rightarrow CH_3(g) + CH_3(g)$$

If several bonds break, the enthalpy change may represent an *average bond energy*, or a *sum of bond energies*.

***Example 1:** The thermochemical equation below has ΔH° equal to four times the average C-H bond dissociation energy of methane, $CH_4(g)$.*

$$CH_4(g) \rightarrow C(g) + 4\ H(g)$$

***Example 2:** The thermochemical equation below has ΔH° equal to the sum of the N=O bond energy and the N-Cl bond energy.*

$$Cl\text{-}N{=}O(g) \rightarrow Cl(g) + N(g) + O(g)$$

Bond energies are useful in determining the weak point in a molecule. For example, when CFC-11 is exposed to ultraviolet radiation in the stratosphere (Chapter 15), it is a C-Cl bond (BDE = 330 kJ mol^{-1}) rather than the C-F bond (BDE = 485 kJ mol^{-1}) which breaks.

$$CFCl_3\ (g) \rightarrow CFCl_2\ (g) + Cl(g)$$

Bond dissociation energies provide a useful, though approximate, means of calculating reaction enthalpies. Remember that their use is restricted to reactions of covalent compounds in the gas phase. The enthalpy change for the reaction can be estimated by focusing on those bonds which are broken or formed during the reaction. As a simple case, in the $CFCl_3$ example above, the only bond undergoing change is the C-Cl bond, which breaks. Hence ΔH° for the reaction is equal to the C-Cl bond energy, + 330 kJ mol^{-1}; ΔH° is positive, because the bond is broken.

As another example, consider the attack of the hydroxyl radical upon ethane, a reaction which occurs in the lower atmosphere (Chapter 9).

$$CH_3CH_3 + OH \rightarrow H_2O + CH_3CH_2$$

Although there are many bonds in the reactants and products, only two are affected by the reaction. One C-H bond is broken and one O-H bond is formed.

Bonds broken:	C-H	$\Delta H° = + 414$ kJ mol $^{-1}$
Bonds formed:	O-H	$\Delta H° = - 464$ kJ mol $^{-1}$
overall change:		$\Delta H° = - 50$ kJ mol $^{-1}$

Bond dissociation energies can be used to estimate heats of reaction, but the values of $\Delta H°_{rxn}$ thus obtained are approximate because bond energies are **average values**: they do not necessarily apply precisely to the particular bonds being broken and formed in the reaction under consideration. Thus in the example above, the C-H bond energy is a "typical" value for all C-H bonds, rather than the exact value for the C-H bonds in ethane. Likewise the value 464 kJ mol $^{-1}$ for the O-H bond energy is a typical value for O-H bonds in general, rather than applying to the O-H bond in water in particular. This particular example gives an unusually large discrepancy in the estimation of $\Delta H°_{rxn}$ using bond energies, as can be seen by repeating the calculation using standard molar enthalpies of formation.

$$\begin{aligned}\Delta H°(rxn) &= \sum \Delta H°_f(\text{products}) - \sum \Delta H°_f(\text{reactants}) \\ &= \{\Delta H°_f(H_2O,g) + \Delta H°_f(CH_3CH_2,g)\} - \{\Delta H°_f(OH,g) + \Delta H°_f(CH_3CH_3,g)\} \\ &= \{(-241.8) + (+105)\} - \{(+38.9) + (-84.7)\} \\ &= -91 \text{ kJ mol}^{-1}\end{aligned}$$

There is unusually poor agreement between $\Delta H°$ estimated by the use of bond energies, and the more accurate calculation using standard heats of formation. The reason for the poor agreement is that the O-H bond energy in water is very atypical: about 495 kJ mol $^{-1}$, rather than the "average" O-H bond energy quoted as 464 kJ mol $^{-1}$. Usually, the two methods of calculating $\Delta H°_{rxn}$ agree to within 10 kJ mol $^{-1}$ or better.

5.7 Energetics of photochemical reactions

Photochemical reactions are chemical processes that are caused or accelerated by light. Since light (radiation) is a form of energy, under proper conditions it can supply the energy to drive chemical reactions which would otherwise be unfavourable energetically. Photochemical reactions are of great importance environmentally, because sunlight can supply the necessary energy to drive such "uphill" reactions - in fact, the most important reaction for the existence of life on Earth as we know it is a photochemical process: photosynthesis.

Several concepts are needed before we can understand how radiation can drive a chemical reaction, and for simplicity, these are covered below in point form (seven points in all).

1. Visible light is one very small part of the complete spectrum of **electromagnetic radiation**. Figure 4 shows schematically the different kinds of electromagnetic radiation and their wavelengths. Note the logarithmic scale; the radiation which we encounter covers 14 orders of magnitude in its wavelength. Visible light (that is, light which is observable to the human eye) ranges from 400 nm (blue) to 700 nm (red), and is only a tiny part of the whole electromagnetic spectrum. Wavelengths shorter than 400 nm are termed ultraviolet (UV), and those longer than 700 nm are termed infrared (IR). Sunlight comprises wavelengths right across the UV, visible and IR ranges. All these wavelengths are incident on the atmosphere at high altitude, but only those wavelengths > 300 nm penetrate to the Earth's surface; shorter wavelengths are absorbed by the atmosphere. This point is discussed in detail in Chapter 15.

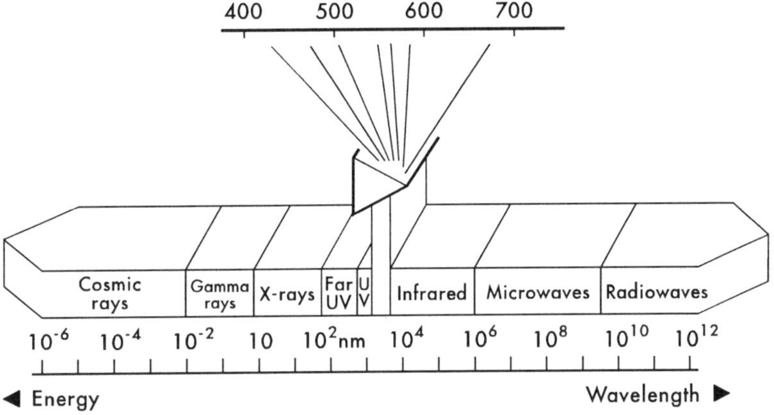

Figure 5.4: Energy and wavelength for different kinds of electromagnetic radiation

2. The nature of light is described in terms of quantum theory. Light is "quantized", meaning that it arrives in discrete packets of energy, known as **photons**. The energy of a single photon is related to its wavelength.

[15] $E_{photon} = hc/\lambda$

h is Planck's constant (6.6×10^{-34} J s), c is the velocity of light (3.0×10^{8} m s^{-1}) and λ is the wavelength of the radiation in meters. E_{photon} is obtained in joules; the energy of a single photon is very small.

Example: *Calculate the energy of a photon of visible (green) light of wavelength 530 nm.*

Answer: *First convert 530 nm to meters (5.30×10^{-7} m), and then substitute into eq. [15].*

$E_{photon} = hc/\lambda = (6.6 \times 10^{-34}\ J\ s)(3.0 \times 10^{8}\ m\ s^{-1})/(5.30 \times 10^{-7}\ m)$

$= 3.7 \times 10^{-19}\ J$

For chemical purposes it is more convenient to consider the energy of a mole of photons, rather than that of a single photon. Photon energy is a source of internal energy, ΔE. To obtain ΔE in J (or kJ) mol $^{-1}$, multiply E_{photon} by Avogadro's constant. When wavelength is in nm and energy is in kJ mol $^{-1}$, the numerical relationship is:

[16] ΔE, kJ mol $^{-1}$ = $(1.2 \times 10^5) / \lambda$, nm

Since energy is inversely related to wavelength, it follows that the energy of a photon increases as the wavelength becomes shorter. In particular, UV radiation is more energetic than visible radiation which, in turn, is more energetic than infrared radiation. UV radiation is the most energetic radiation which reaches the Earth's surface (Table 5.1). Solar radiation having wavelength shorter than about 300 nm is filtered out (absorbed by) the Earth's atmosphere, and hence does not penetrate to ground level.

Energy, kJ mol $^{-1}$	Wavelength, nm	
170	700	
200	600	visible
240	500	
300	400	UV-A
	320	UV-B
400	300	
	290	UV-C

3. The energies in Table 5.1 are of similar magnitude to common bond dissociation energies. Examples: Cl-Cl, 243; C-Cl, 330; C-C, 350 kJ mol $^{-1}$. Consequently, solar radiation has the potential to induce chemical reactions by cleavage of covalent bonds.

4. Radiation can only induce chemical reactions if it is first absorbed by the photochemically active substance. The absorption of light raises the substrate to an "excited", or energized, state, which is generally more reactive chemically than the unenergized or "ground" state.

5. Absorption is a quantum process. This means that 1 photon is absorbed by 1 atom or molecule. It also means that the energy of the photon must be exactly equal to the energy difference between two states of the absorbing molecule. Absorption cannot occur if the energy of the photon is either greater or smaller than the energy gap between the substrate's ground and excited states. If the substrate has no suitably placed energy states, no light can be absorbed and the

substance will be transparent to the radiation in question (see Figure 5.5 in which the matching of energies makes possible absorption of the photon by molecule "A", but not by "B" or "C").

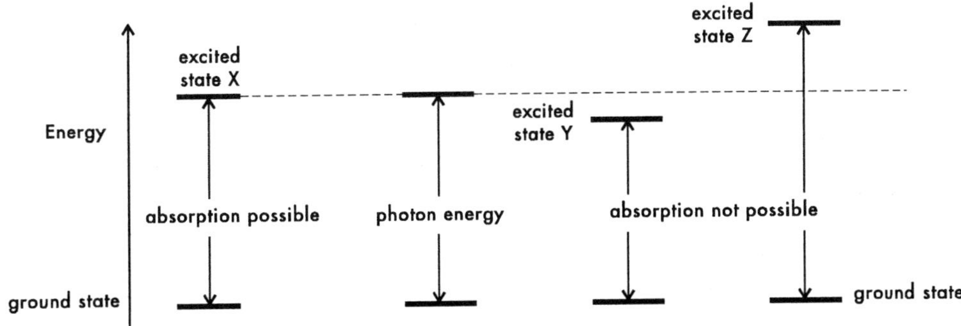

Figure 5.5: Schematic of the matching of photon energy and the gap between energy states of the absorbing substance.

6. The excited state has only a very short lifetime, and its extra energy may:

- be reemitted as light
- be degraded to heat
- be used to drive a chemical reaction, such as cleavage of a covalent bond

The proportion of excited molecules converting their energy into chemical reaction energy varies over the whole range 0-100%, depending on the reaction in question.

7. The energy of the photon is converted to *internal energy* of the substrate ΔE, as noted already. From the First Law of Thermodynamics, $\Delta E = \Delta H - \Delta nRT$; the term ΔnRT is usually much smaller than ΔH, and so $\Delta E \sim \Delta H$. Therefore, it is useful to think of photon energy as essentially the same as a source of enthalpy.[3]

---

[3] In practice, a small proportion of the energy of the photons is "wasted", and so the amount of energy available to the excited state for chemical reactions is usually a little less than the energy of the photons. We will ignore this complication in this book.

Summarizing, two major conditions must both be met before a photochemical reaction can occur:

(i) the substrate must be able to absorb the incident radiation

(ii) the excited state must be sufficiently energetic to drive the reaction *i.e*, $\Delta E_{photon} > \Delta H°_{reaction}$, assuming $\Delta H°_{reaction}$ to be positive.

Example: *The C-Cl bond has bond dissociation energy 330 kJ mol^{-1}, while $CFCl_3$ absorbs radiation having $\lambda < 220$ nm. Will $CFCl_3$ undergo bond cleavage in the lower atmosphere? Explain.*

Answer: *From eq. 16, the wavelength corresponding to 330 kJ mol^{-1} is:*

$$\lambda, \text{nm} = 1.2 \times 10^5/330 = 364 \text{ nm.}$$

Hence any radiation having $\lambda < 364$ nm should be able to cleave this bond. However, $CFCl_3$ does not absorb at 364 nm, only < 220 nm. Radiation reaching the lower atmosphere is exclusively $\lambda > 300$ nm.

Conclusion: $CFCl_3$ is not cleaved by sunlight in the lower atmosphere, because it does not absorb light in the available range. (It is cleaved by UV radiation in the stratosphere, however.)

5.8 Calorimetry

So far we have said nothing about how enthalpy changes are measured. These measurements are referred to as **calorimetry**, and the apparatus used is called a **calorimeter**. A calorimeter is in essence a well-insulated box inside which the chemical reaction occurs. The amounts of substrates are measured carefully and the reaction (usually chosen to be exothermic) is allowed to proceed. The increase in temperature of the calorimeter and its contents is measured, and the heat evolved in the chemical reaction is calculated from a knowledge of how much heat is required to warm up the calorimeter (and contents). The amount of heat required to change the temperature of some object is called its **heat capacity**. Learn these terms.

Definition

Heat capacity: *amount of heat needed to raise the temperature of the object by 1 °C (same as 1 K). Units: J K^{-1}.*

Molar heat capacity: *amount of heat needed to raise the temperature of 1 mol of the substance by 1 °C. Units: J mol^{-1} K^{-1}.*

Specific heat: *amount of heat needed to raise the temperature of 1 gram of the substance by 1 °C. Units: J g^{-1} K^{-1}. Water is often used as the heat absorbing substance in calorimeters; its specific heat is 4.184 J g^{-1} K^{-1}.*

Let's use an example to see how this works in practice.

Example: *A 0.2445 g sample of benzoic acid, $C_7H_6O_2(s)$ is burned in a calorimeter containing 766 g of water. The heat capacity of the calorimeter itself is 431 J K^{-1}. The temperature of the calorimeter and the water inside change from 24.18 °C before combustion to 25.96 °C after combustion. Calculate the standard molar enthalpy of combustion of benzoic acid.*

Answer: *We must determine how much heat was liberated, realizing that for a well-insulated calorimeter, the heat given off in the combustion reaction is equal to the heat absorbed by the calorimeter and its contents. The heat evolved referred to the combustion of 0.2445 g of benzoic acid. The final step in the calculation is to determine the number of moles of benzoic acid in order to report the answer as a molar heat of combustion.*

Temperature rise = 29.96 - 24.18 = 1.78 °C (1.78 K)

Heat absorbed by calorimeter = ΔT x Heat capacity of calorimeter
= 1.78 K x 431 J K^{-1} = 767 J

Heat absorbed by water = ΔT x mass x specific heat
= 1.78 K x 766 g x 4.184 J g^{-1} K^{-1}
= 5705 J

Total heat absorbed = heat evolved in combustion = 5705 + 767
= 6.47×10^3 J

$M(C_7H_6O_2)$ = 122 g mol^{-1}

$n(C_7H_6O_2)$ = 0.2445 g/122 g mol^{-1} = 2.00×10^{-3} mol

$\Delta H°_{comb}$ = - 6.47×10^3 J/2.00×10^{-3} mol
= -3.24×10^6 J mol^{-1} = -3.24 MJ mol^{-1}

In this example, only 3 significant figure precision was obtained; with the best equipment however, very high precision is possible, and heats of combustion are some of the most precisely known thermochemical quantities.

Problems

Sections 5.1 - 5.2

1. Write balanced thermochemical equations conveying the following information:
   a) the enthalpy change when 1 mol NH_3 (g) is formed from its elements is -46.2 kJ
   b) the molar heat of vaporization of liquid mercury is +58.5 kJ mol $^{-1}$
   c) the oxidation of 1 mol CO(g) to 1 mol CO_2 (g) using O_2(g) is exothermic by 283 kJ mol $^{-1}$
   d) the oxidation of benzene (C_6H_6, a liquid) to CO_2 and liquid water is exothermic by 3273 kJ mol^{-1} when O_2 is used as the oxidizing agent.

2. Calculate the change in the internal energy when 1 mole of water is vaporized at 100°C and 1 atm given:

Heat of vaporization of $H_2O(\ell)$	+ 40.5 kJ mol $^{-1}$
Volume of vapour at 100°C	30.14 dm^3
Volume of liquid at 100°C	0.02 dm^3

3. The combustion of propane (C_3H_8) to CO_2(g) and $H_2O(\ell)$ is exothermic by 2220 kJ mol $^{-1}$.
   Calculate:
   a) the heat released by combustion of 15 L of liquid propane (density 0.585 g mL $^{-1}$)
   b) the heat required for a hypothetical reaction in which 6 mol CO_2 (g) and 4 mol $H_2O(\ell)$ are converted into 2 mol C_3H_8 (ℓ) and 8 mol O_2 (g)
   c) the heat change for the partial oxidation of propane into acetone $C_3H_6O(\ell)$, given that the combustion of acetone releases 1790 kJ mol $^{-1}$.

Sections 5.3 - 5.4

1. Write the thermochemical equations having $\Delta H°$ equal to
   a) heat of formation of nitric acid
   b) heat of formation of ammonium carbonate
   c) heat of combustion of pentane (C_5H_{12}, ℓ)
   d) heat of combustion of liquid ethyl alcohol (C_2H_6O)
   e) heat of vaporization of water
   f) heat of sublimation of methane
   g) the heat of atomization of iron

2. Write the thermochemical equations having $\Delta H°$ equal to:
   a) heat of formation of solid copper(I) oxide
   b) heat of formation of solid ammonium sulfate $(NH_4)_2SO_4$
   c) heat of combustion of liquid gasoline (assumed to be C_8H_{18})
   d) heat of combustion of sugar $C_{12}H_{22}O_{11}$ (s)
   e) heat of vaporization of mercury
   f) heat of sublimation of carbon dioxide

3. Write thermochemical equations to define:
   a) the standard molar enthalpy of formation of solid potassium chlorate as -397.7 kJ mol $^{-1}$
   b) the standard molar enthalpy of formation of liquid dichloromethane (CH_2Cl_2) as -121.5 kJ mol $^{-1}$
   c) the standard molar enthalpy of formation of gaseous CFC-11 ($CFCl_3$) as -301.3 kJ mol $^{-1}$
   d) the standard molar enthalpy of formation of gaseous dimethyl mercury (HgC_2H_6) as +94.4 kJ mol $^{-1}$.

4. Use tabulated data to calculate enthalpy changes (298 K, 1 atm) for the following reactions:

a)	$MgSO_4(s)$	→	$MgO(s) + SO_3(g)$
b)	$2Fe_3O_4(s)$	→	$3Fe_2O_3(s) + O_2(g)$
c)	$CH_4(g) + 4Cl_2(g)$	→	$CCl_4(\ell) + 4HCl(g)$
d)	$H_2SO_4(\ell)$	→	$H_2(g) + S(s) + 2O_2(g)$
e)	$HNO_3(\ell)$	→	$H^+(aq) + NO_3^-(aq)$
f)	$KBr(s)$	→	$K^+(aq) + Br^-(aq)$
g)	$2Fe^{2+}(aq) + H_2O_2(\ell) + 2H^+(aq)$	→	$2Fe^{3+}(aq) + 2H_2O(\ell)$

5. Use tabulated data to calculate the heats of reaction for the following processes:
   a) Fe_2O_3 (s) + 3CO (g) → 2Fe (s) + $3CO_2$ (g)
   b) CaO (s) + H_2O (ℓ) → $Ca(OH)_2$ (s)
   c) $6CH_4$ (g) + O_2 (g) → $2C_2H_2$ (g) + 2CO (g) + $10H_2$ (g)
   d) C_2H_2 (g) + $2H_2$ (g) → C_2H_6 (g)

6. Calculate the enthalpy change for the reaction

$$2Hg\ (g) + O_2\ (g) \rightarrow 2HgO\ (s)$$

given (i) heat of vaporization of mercury = 58.46 kJ mol $^{-1}$
(ii) heat of formation of HgO (s) = -90.7 kJ mol $^{-1}$.

7. The heat of combustion of benzene C_6H_6 (ℓ) is 3273 kJ mol $^{-1}$. Use this and the data of Appendix 1 to calculate the heat of formation of benzene.

8. The heat of combustion of H_2S (g) to H_2O (ℓ) and SO_2(g) is 562.6 kJ mol $^{-1}$ and of CS_2 (ℓ) to CO_2 (g) and SO_2 (g) is 1075.2 kJ mol $^{-1}$. Calculate ΔH for the reaction

$$CS_2\ (\ell) + 2H_2O\ (\ell) \quad \rightarrow \quad CO_2\ (g) + 2H_2S\ (g)$$

9. Calculate the change in the internal energy when HCl is oxidized at 25°C according to the equation below:

$$4HCl\ (g) + O_2\ (g) \ \rightarrow \quad 2Cl_2\ (g) + 2H_2O\ (\ell)$$

Section 5.5

1. Write thermochemical equations having $\Delta H°$ equal to the standard molar heats of combustion of:
   a) liquid butane C_4H_{10}
   b) liquid methanol CH_4O
   c) gaseous acetylene C_2H_2
   d) solid naphthalene $C_{10}H_8$

2. a) The standard molar enthalpy of combustion of ethanol (C_2H_6O) is -1369 kJ mol $^{-1}$. Compare the energy content of ethanol (MJ kg $^{-1}$) with that of gasoline (46 MJ kg $^{-1}$).
   b) Assuming ethanol and gasoline to have the same densities, what is the energy content of a "gasohol" of composition 10% ethanol, 90% gasoline?

3. Gasoline has density approximately 0.8 g mL $^{-1}$. If an automobile converts 20% of the energy content of gasoline into mechanical work, calculate the heat released to the atmosphere by driving 25 km in a car which has a fuel efficiency of 100 km per 5.6 L gasoline used.

4. Calculate the mass of SO_2 released to the atmosphere upon burning 5000 t of coal having sulfur content of 2.3%.

5. Calculate the mass of water released into a room upon combustion of 1.0 L of kerosene of average composition $C_{12}H_{24}$, density 0.80 g cm $^{-3}$.

6. Calculate the mass of methane that must be burned to heat a typical house in southern Ontario on a winter day when the total heat requirement is 6.7×10^5 kJ.

Section 5.6

1. Write equations having $\Delta H°$ equal to:
   a) enthalpy of formation of 2H (g)
   b) bond energy of H_2 (g)
   c) bond energy of HI (g)
   d) enthalpy of formation of O (g)

2. Calculate $\Delta H°$ for the following reactions using the bond energies given in Appendix 2. Assume all chemical substances are in the gas phase.
   a) $Cl + CH_4 \rightarrow CH_3 + HCl$
   b) $OH + C_2H_6 \rightarrow C_2H_5 + H_2O$
   c) $CH_3OH + HCl \rightarrow CH_3Cl + H_2O$
   d) $N_2 + 3H_2 \rightarrow 2NH_3$
   e) in part (d), how does the value of $\Delta H°$ calculated by this method compare with the standard molar heat of formation of 2 moles of NH_3?

3. Calculate the strength of the $C \equiv N$ bond given the following information and any needed data from Appendices 1 and 2.

   Heat of formation of HCN (g) + 135 kJ mol $^{-1}$

4. The reaction below is the principal source of ozone in the upper atmosphere.

   $$O\ (g) + O_2\ (g) \rightarrow O_3\ (g)$$

   Calculate ΔH for this reaction given (i) O_2 bond energy = 494 kJ mol $^{-1}$; (ii) $\Delta H_f°\ (O_3,g) = +142.3$ kJ mol $^{-1}$.

5. Calculate the N-O bond energy in NO_2 (structure O-N-O) given the following information:

   $\Delta H_f°\ (NO_2,g) = 33.9$ kJ mol $^{-1}$
   N≡N bond energy = 946 kJ mol $^{-1}$
   O=O bond energy = 494 kJ mol $^{-1}$

6. Calculate the energies of the bonds in CS_2 given the heat of formation of CS_2 (g) is +117 kJ mol $^{-1}$, and the heats of atomization of graphite (+720kJ mol^{-1}) and sulfur (+275 kJ mol $^{-1}$).

Section 5.7

1. Use values of fundamental constants to verify that Einstein's relationship (E_{photon} = hc/λ = (1.2 x 10^5 /λ) kJ nm mol $^{-1}$ when λ is given in nm.

2. Calculate the maximum wavelength of light that would be needed to break each of the following bonds (bond energies are in Appendix 2).
   a) N - H
   b) N ≡ N
   c) C - Cl
   d) O - H
   e) C - H
   f) Cl - Cl

   In this question assume that ΔH° ≈ ΔE.

3. a) Why, throughout Q.2, is it a maximum wavelength that you have calculated?

   b) If light of the wavelengths calculated in Q.2 were incident upon molecules containing the relevant bonds, would bond cleavage automatically occur?

4. a) Calculate ΔH° for the reaction below, and then estimate the maximum wavelength of light capable of causing this reaction.

   $$NO_2\ (g) \rightarrow NO\ (g) + O\ (g)$$

   b) Do you expect this reaction to be possible in the lower atmosphere, given that NO_2 absorbs light throughout the visible and UV solar spectrum?

5. Ozone absorbs radiation having $\lambda < 325$ nm. Is radiation of this wavelength capable of causing the reaction below (all species in the gas phase)?

   $$O_3 \rightarrow O_2 + O$$

Section 5.8

1. A copper coin weighing 38.96 g and initially at 22.4°C is placed in a styrofoam cup containing 50.2 g of water at 46.3°C. When equilibrium is established the temperature is 44.7°C. Calculate the specific heat of copper.

2. A 1.871 g sample of phenacetin $C_{10}H_{13}NO_2$ was burned in excess O_2 in a calorimeter. The calorimeter had a heat capacity of 5.26 kJ K^{-1} and contained 1.503 kg of water. Under these conditions the temperature rose from 22.01°C to 25.34°C. Calculate the heat released on combustion of 1 mol of phenacetin under these conditions.

3. The heat of fusion of water is approximately 6.0 kJ mol^{-1}. Calculate the heat that must be supplied by the oceans to melt completely an iceberg of mass 10,000 t.

4. When 5.00 g of KCl (s) is dissolved in 1.00 L of water in a calorimeter of heat capacity 3.76 kJ per °C, the temperature falls from 24.457°C to 24.244°C. Calculate the molar heat of solution of KCl.

5. Calculate ΔH° for the following reaction from the information provided

   $$Ag^+ (aq) + Cl^- (aq) \rightarrow AgCl (s)$$

   50.00 mL of 0.102 mol L^{-1} $AgNO_3$ is poured into a coffee-cup calorimeter and dilute HCl (in molar excess) is added to give a final mass of 100.00 g. The specific heat of the resulting solution is 4.0 J g^{-1} K^{-1}, and the temperature rises from 22.06°C to 22.81°C.

6 GASES AND THE ATMOSPHERE

6.1 Properties of gases

6.1.1 The ideal gas equation

The three common states of pure substances are solids, liquids, and gases. A solid has a defined size and shape. A given amount of liquid has a defined volume, but no defined shape; it takes on the shape of its container. A gas likewise has no defined shape, but in addition, it has no defined volume; it expands to fill its container completely. Between 1660 and about 1800, researchers discovered a number of relationships between the pressure, the volume, and the temperature of a gas[1]. These were subsequently formulated as the Ideal Gas Equation, eq. [1].

[1] $PV = nRT$

In eq. [1], P is pressure, V is volume, and T is temperature (in kelvins, *not* in °C). R is a constant, known as the Universal Gas Constant.

The important thing to note about the ideal gas equation is that nowhere does it specify the identity of the gas; in other words, all gases are predicted to behave the same as each other. While this is not quite true experimentally, it is nevertheless such a good approximation for ordinary gases at moderate pressure (up to at least 1 atm) that we shall be able to use the Ideal Gas Equation successfully for all the calculations we require in this book.

The units of the universal gas constant R can cause confusion. In SI units, R has the value 8.314 J mol^{-1} K^{-1}, or 8.314 Pa m^3 mol^{-1} K^{-1}, or equivalently 8.314 kPa L mol^{-1} K^{-1}. We shall also have occasion to use the gas constant when the units of pressure are atm (1 atm = 101.325 kPa); under these conditions, R has the value 0.0821 L atm mol^{-1} K^{-1}.

---

1 Boyle's Law: $P \propto 1/V$ at constant T; Charles' Law: $V \propto T$ (temperature in K) at constant P; Avogadro's Law: $V \propto n$ at constant P and T.

Example: *Calculate the pressure, in both kPa and atm, when 0.015 mol of SO_2 (g) is introduced into a previously evacuated 0.62 L vessel at 18°C.*

Answer:

$$P = \frac{nRT}{V}$$

(a) using $R = 8.314\ kPa\ L\ mol^{-1}\ K^{-1}$

$$P = \frac{(0.015\ mol)(8.314\ kPa\ L\ mol^{-1}\ K^{-1})(291\ K)}{0.62\ L}$$

$$= 59\ kPa$$

(b) using $R = 0.0821\ L\ atm\ mol^{-1}\ K^{-1}$

$$P = \frac{(0.015\ mol)(0.0821\ L\ atm\ mol^{-1}\ K^{-1})(291\ K)}{0.62\ L}$$

$$= 0.58\ atm$$

Rearrangement of equation [1] allows us to make useful derived relationships. For example, pressure and concentration can be related [2], and by making use of the definition of definition of molar mass (mass/n), the mass of gas or its density can be related to P or V *e.g*, [3].

[2] $n/V = P/RT$

[3] mass = PV**M**/(RT)

One very familiar derived quantity is the **molar volume** (V/n) of an ideal gas, which has the value 22.4 L mol $^{-1}$ at "STP" (standard temperature and pressure, 273 K and 1 atm). The molar volume is the reciprocal of the concentration at STP.

Example: *Calculate the volume of CO_2 produced, corrected to STP, when 1,000 tonnes of limestone, $CaCO_3$, are roasted to produce quicklime CaO.*

Answer: *The equation is:*

$$CaCO_3\ (s) \rightarrow CaO\ (s) + CO_2\ (g)$$

Step 1: Work out the amount in moles of CO_2 formed.

$\mathbf{M}(CaCO_3) = (40.1 + 12.0 + [3 \times 16.0]) = 100.1\ g\ mol^{-1}$

$n(CaCO_3) = 1{,}000\ Mg/100.1\ g\ mol^{-1}$

$= 10.0\ Mmol.$

Step 2: $n(CO_2) = n(CaCO_3)$ *from equation*

Hence $V(CO_2) = 10.0\ Mmol \times 22.4\ L\ mol^{-1}$

$= 22.4\ ML$ = *same as* $2.24 \times 10^7\ L$ *or* $2.24 \times 10^4\ m^3$

6.1.2 Molecular speeds

A gas fills its container completely because the individual molecules are in constant motion. Most of the container is empty space; the molecules exert pressure by striking the walls of their container. In the course of their random motion, the molecules strike each other, and as a result they can both change their rate and direction of motion, and thereby exchange energy with one another (recall that the kinetic energy of a particle is given by ($\frac{1}{2}mv^2$), where m is the mass and v the velocity of the particle).

If a gas is enclosed in a container of fixed size, its pressure increases as the temperature is raised (Charles' Law). The modern interpretation of this phenomenon is that the average speed of the gas molecules increases with temperature. In pictorial terms, the molecules can be thought to strike the walls of the vessel with greater force because they are travelling at higher speed, thus accounting for the observed increase in pressure.

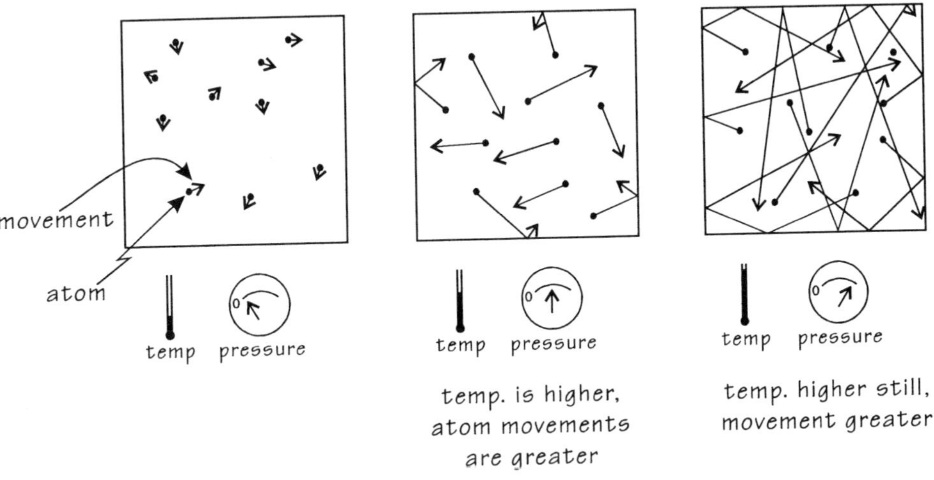

Figure 6.1: Modern interpretation of Charles' Law

In any sample of gas, not all the molecules travel at equal velocities. It is possible to derive a formula for the average velocity for all the molecules in the gas (eq. [4]); as expected, it shows that the velocity increases with temperature - actually as the square root of temperature.

$$[4] \quad v = \left(\frac{3RT}{\mathbf{M}_{kg}}\right)^{\frac{1}{2}}$$

In eq. [4], $\mathbf{M}_{kg}$ is the molar mass in the unusual units of kg mol $^{-1}$ and R has the units J mol $^{-1}$ K $^{-1}$ (same as kg m^2 s $^{-2}$).

Example: *Calculate the average velocities of oxygen molecules at 25 °C and at 300 °C.*
Answer: $M_{kg}(O_2) = 32 \times 10^{-3}$ *kg mol*$^{-1}$

At 298 K:

$$v = \left[\frac{3RT}{M_{kg}}\right]^{1/2} = \{(3 \times 8.314 \text{ kg m}^2 \text{ s}^{-2} \times 298 \text{ K})/(32 \times 10^{-3} \text{ kg mol}^{-1})\}^{1/2}$$

$$= 4.8 \times 10^2 \text{ m s}^{-1}$$

Similarly at 300 °C (573 K):

$$v = \left[\frac{3RT}{M_{kg}}\right]^{1/2} = \{(3 \times 8.314 \text{ kg m}^2 \text{ s}^{-2} \times 573 \text{ K})/(32 \times 10^{-3} \text{ kg mol}^{-1})\}^{1/2}$$

$$= 6.7 \times 10^2 \text{ m s}^{-1}$$

Note the very great speed of typical gas molecules. Even at room temperature, the molecules in air travel about 1 km every 2 s on the average, although of course they change directions and velocities many times during the two seconds - *i.e.*, this is not 1 km in a straight line. Notice also that since the term $\mathbf{M}_{kg}$ occurs in the denominator of eq. [4], it is the lightest molecules that have the greatest average speeds. It is for this reason that the Earth's atmosphere contains very little free hydrogen; H_2 molecules travel so fast that they are able to break free of the Earth's gravitational attraction, and escape into space. By contrast, the heavier planets have atmospheres which are largely hydrogen, because of their greater gravitational fields.

The equation above gives the average speed of molecules of mass $\mathbf{M}_{kg}$ at any temperature, but says nothing about the distribution of speeds. The latter is a complex function, which we will not study in detail. However, we may note that it contains as one term:

[5] $N(E) = N_o \exp(-E/RT)$

This tells us that the fraction of molecules having kinetic energy E is related to the exponent of energy E: the greater the energy, the fewer molecules have this energy or more at a given temperature. We shall use this idea again in Chapter 7. Figure 2 illustrates the distribution of speeds in a gas: taking Curve A as a reference (for example N_2 at 300 K), Curve B could represent N_2 at a higher temperature, or alternatively it could represent a lighter gas (*e.g.* methane) at 300 K. The form of the curves in Figure 1 shows us that the actual function for the distribution of speeds is more complicated than eq. [5] would suggest.

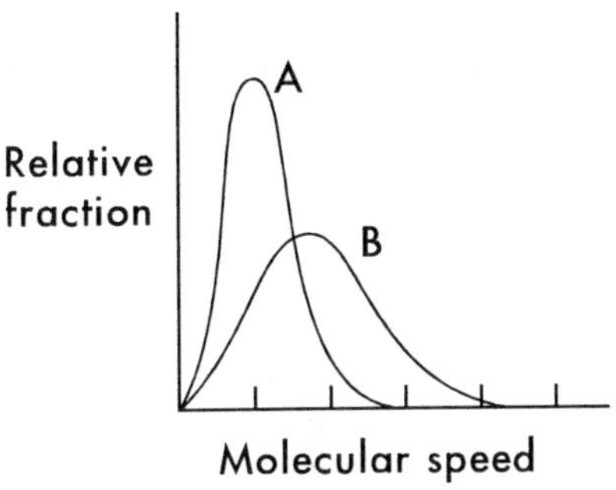

Figure 6.2: Representations of the distribution of speeds in a gas.

6.1.3 Gases and vapours

Oxygen, nitrogen, and methane (natural gas) are examples of "permanent gases" - that is to say, most people only meet these substances in the gaseous form[2]. When a scientific laboratory or a welding shop buys a cylinder of oxygen, the contents are exclusively $O_2(g)$, and the gauge on the cylinder indicates the pressure remaining in the tank at any time. Contrast this with buying a tank of propane; if you move the tank about, you can hear liquid propane "sloshing around" inside. In physico-chemical terms, the contents of the propane tank consist of liquid propane in equilibrium with its vapour, and it is propane vapour that is removed from the tank and burned.

$$\text{propane}(\ell) \quad \rightleftharpoons \quad \text{propane}(g)$$

This equilibrium constant has a fixed value at any temperature (as we shall see in Chapter 8, the equilibrium constant is equal to the **vapour pressure** of propane(g)). Because of this, the pressure of propane in the tank does not fall as the propane is used up: it has a constant value as long as there is any liquid present to maintain equilibrium. This is why propane tanks, unlike oxygen tanks, are not equipped with pressure gauges. After the last of the liquid vaporizes, of course, the pressure falls rapidly as the last of the propane is removed from the tank.

In general, any liquid will tend towards equilibrium with its vapour.

$$X(\ell) \quad \rightleftharpoons \quad X(g)$$

---

2 A more accurate definition of the term "permanent gas" is given in Section 6.1.5.

This is why liquids evaporate if left in open containers; the larger the vapour pressure (that is, the larger the equilibrium constant), the greater the tendency to evaporate. Evaporation is prevented in a closed container, since X(ℓ) and X(g) reach equilibrium.

Example: *Why does water evaporate from puddles on dry days, but not on very humid days?*

Answer: *Whether or not water evaporates depends upon p(H_2O), the pressure of water in the gas phase. The water evaporates if, at the prevailing temperature, p(H_2O) is smaller than the equilibrium value. On very humid days p(H_2O) may approach or even exceed the equilibrium value. Under these conditions evaporation does not occur. In fact, the process occurs in the reverse direction, and rain falls.*

Definition

*The **boiling point** of a liquid is the temperature at which the vapour pressure of the liquid reaches the pressure of the surroundings, so that the vapour can "push back" the surroundings and let the liquid escape into the gas phase.*

Water has the boiling point 100 °C at 1 atm pressure; however, the temperature at which water boils changes with pressure. Above the normal (p = 1 atm) boiling point, the liquid must be pressurized to prevent vaporization because p(X) > 1 atm; this is why propane (normal b.p. -44 °C) is kept in steel tanks in the example above. Conversely, liquids can be made to boil at temperatures well below their normal b.p. if the pressure is reduced; this technique is used on a large scale at oil refineries in order to distil the less volatile fractions of petroleum.

Vapour pressure is of great practical importance for flammable vapours. Many substances commonly used in industry (and in the laboratory) form explosive mixtures when they mix with air in certain proportions. An upper explosive limit occurs because there must be enough oxygen for near-stoichiometric oxidation (combustion) of the vapour; if there is too little oxygen, combustion is inefficient. Conversely, a lower explosive limit occurs because at very low concentrations of the flammable vapour, the vapour is not "rich" enough for a rapid, heat-releasing reaction. Common solvents are labelled with a **flash point**, the minimum temperature at which a spark would cause the contents to ignite in the presence of abundant air. Examples are given below.

Flash points of some common solvents

Solvent	Flash point, °C
pentane	-49
diethyl ether	-45
cyclohexane	-20
acetone	-18
toluene	+4
ethanol	+12

Notice the very low flash points of many common substances. These explain why, for example, smoking is prohibited at gas stations. In industry, where very large amounts of highly flammable substances must be handled safely, great care must be taken to avoid sparks, and all metal equipment is carefully grounded to avoid the build-up of static electricity.

6.1.4 Enthalpy changes associated with vaporization and sublimation

In Chapter 5 we met the terms enthalpy of vaporization and enthalpy of sublimation, which refer, respectively, to the processes below.

$$X(\ell) \rightleftharpoons X(g)$$

$$X(s) \rightleftharpoons X(g)$$

Enthalpies of vaporization and sublimation are both positive quantities: heat is needed to vaporize a liquid or sublime a solid. The amount of heat needed depends upon the strength of the interactions holding the solid or liquid together. Ionic substances such as NaCl have very high enthalpies of vaporization and sublimation by comparison with molecular substances. Vapour formation requires heat to be supplied, a process that is familiar in the doctor's office when the nurse rubs alcohol on your arm prior to an injection; your arm feels cold as your body supplies the heat of vaporization.

Example: *The enthalpy of vaporization of alcohol (C_2H_6O) is 43 kJ mol^{-1}. Calculate the heat supplied by your arm when 100 mg of alcohol evaporates.*
Answer:

$M(C_2H_6O) = (2 \times 12.0) + (6 \times 1.0) + (16.0) = 46.0\ g\ mol^{-1}$
$n(C_2H_6O) = 100\ mg/46.0\ g\ mol^{-1} = 2.2\ mmol$

Heat absorbed $= 2.2\ mmol \times 43\ kJ\ mol^{-1}$
$= 93\ J$

6.1.5 *Supercritical fluids*

Because equilibrium exists between a gas and a liquid, we would guess that applying pressure to a gas should cause it to liquify. As we have seen, propane is liquified to ship it in tanks. The higher the temperature, the greater the pressure that must be applied, because the equilibrium constant for the process $X(\ell) \rightleftharpoons X(g)$ rises with increasing temperature. However, above a certain temperature called the **critical temperature** it is no longer possible to liquify a gas no matter how much pressure is applied. Oxygen and nitrogen at room temperature are way above their critical temperatures, which is why they are called permanent gases; they can only be liquified by lowering the temperature below the critical temperature.

Critical point parameters of some common substances

Substance	**T_c, °C**	**P_c, atm**
O_2	-119	50
N_2	-147	34
methane	-83	46
propane	+97	42
H_2O	+374	220
CO_2	+31	73

When a liquid under pressure is raised above its critical temperature, the meniscus between the liquid and gas phases disappears, and one single phase is formed. This phase, a **supercritical fluid**, has properties somewhere between those of conventional gases and liquids - it behaves like a very dilute and mobile liquid. Extraction using supercritical fluids is an important new technology. In the food industry, for example, decaffeinated coffee is made by removing caffeine from powdered coffee. The traditional method involved extraction with the solvent dichloromethane CH_2Cl_2. Concern was expressed about possible residues of CH_2Cl_2 in the extracted coffee. The newer method uses supercritical CO_2 as the extraction "solvent"; the supercritical fluid is more mobile than a conventional liquid and the extraction is carried out under milder conditions, with less flavour impairment of the coffee. The caffeine is recovered simply by releasing the pressure from the CO_2 which immediately evaporates, leaving the solid caffeine behind. Other promising applications for supercritical fluid extraction are for the extraction of oil from oil-bearing shales, where oil is trapped within strata of rock, and in the remediation (clean-up) of contaminated soils at former industrial sites. Supercritical fluids are also being used in the separation of complex mixtures of substances, in a technique called supercritical fluid chromatography.

6.2 The atmosphere

Atmospheric chemistry is an extremely important part of environmental chemistry. Several of today's societal problems involve atmospheric chemistry: these are global warming, stratospheric ozone depletion, photochemical smog, and acid rain. All these topics will be covered at appropriate places in this book. This section gives an overview of the nature of the Earth's atmosphere.

6.2.1 Composition of the atmosphere

Unpolluted dry air at sea level (total pressure = 1 atm) has the approximate composition 80% nitrogen and 20% oxygen. Water vapour is a variable constituent and ranges from close to zero to about 0.4% (up to 0.004 atm). As we sum the pressures of the various atmospheric components, we make use of the "**Law of partial pressures**", which states that the total pressure in any system is the sum of all the pressures of the individual components.

Major constituents of the atmosphere (atm)

- N_2 0.781
- Ar 0.0093
- O_2 0.209
- CO_2 0.00035

In addition, the atmosphere contains numerous minor components. A common unit of measurement of these substances is "parts per million by volume", ppmv. When the concentration is 1 ppmv, this means that at constant T and P, there is one liter of the component in 1 million liters total of air, or equivalently, 1 mol in 10^6 mol, or 1 molecule in 10^6 molecules. Note the difference between ppmv (for gases, defined by volume, and hence by moles), and ppm for solids and liquids (defined by weight). Parts per billion or trillion by volume, (ppbv, pptv) are defined similarly.

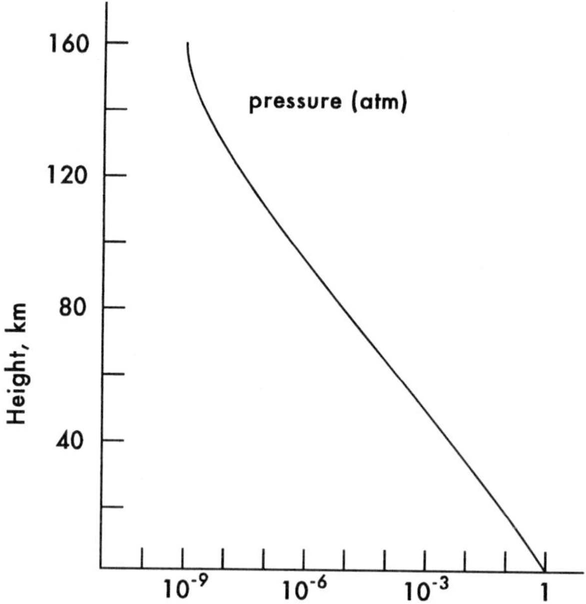

Figure 6.3: Pressure *vs.* altitude in the Earth's atmosphere.

Minor constituents of the atmosphere (ppmv)

- Ne 18
- CH_4 1.7
- H_2 0.5
- He 5.2
- N_2O 0.3
- CO 0.1

Even smaller amounts of NH_3, SO_2, Kr, Xe, O_3 and other gases are present. Atmospheric scientists often use the units molecules per cubic centimeter (molec cm $^{-3}$) for recording the concentrations of these trace components.

Notice in the following example the enormous number of molecules per cubic centimeter even when the partial pressure of the substance is very low. Work out for yourself, the concentration (mol L^{-1} and molec cm^{-3}) corresponding to $p = 1$ atm at 25 °C.[3] Having done so, notice that since 1 atm corresponds to about 0.04 mol L^{-1} at room temperature, the concentration 1 mol L^{-1} in a gas represents a very high pressure - about 25 atm.

[3] (Answers: 0.041 mol L^{-1}; 2.5 x 10^{19} molec cm $^{-3}$).

Example: *On a day of high pollution, ozone levels in the troposphere may reach 120 ppbv. Express this in both atm and molec cm^{-3}. Assume T = 25°C*

Answer: *in atm:*

120 ppbv means 120 L in 10^9 L (constant P,T), or equivalently 120 atm in 10^9 atm (constant V, T). The total atmospheric pressure is taken as 1 atm.

$$\text{Pressure in atm} = 120 \text{ ppbv} \times \frac{1 \text{ atm}}{10^9 \text{ ppbv}} = 1.2 \times 10^{-7} \text{ atm}$$

in molec cm^{-3}:

We already have P in atm, and can use the ideal gas equation to solve for concentration = n/V

$$n/V = P/RT$$

$$= \frac{1.2 \times 10^{-7} \text{ atm}}{0.0821 \text{ L atm mol}^{-1} \text{ K}^{-1} \times 298 \text{ K}}$$

$$= 4.9 \times 10^{-9} \text{ mol L}^{-1}$$

Change to molec cm^{-3}

$$\frac{n}{V} = 4.9 \times 10^{-9} \text{ mol L}^{-1} \times \left[\frac{6.022 \times 10^{23} \text{ molec}}{1 \text{ mol}}\right] \times \left[\frac{1 \text{ L}}{1000 \text{ cm}^3}\right]$$

$$= 3.0 \times 10^{12} \text{ molec cm}^{-3}$$

The total mass of the atmosphere is about 5 x 10^{15} tonnes; gravitational attraction causes it to be concentrated close to the surface; 99% of the total mass lies below 30 km. There is no outer limit to the atmosphere; the concentration of particles simply decreases continuously with altitude until by 150 km altitude the total pressure is about 10^{-9} atm (which would be considered ultra-high vacuum in an Earth-bound laboratory). Although many different regions of the atmosphere may be defined, in this book we shall be principally concerned with the **troposphere** (from the surface to about 15 km) and the **stratosphere** (about 15 to 50 km). As reference points for future discussion, the conditions at the Earth's surface (troposphere) would average out to p = 1.00 atm, T = 288 K, while in the mid-stratosphere the total pressure is around 10^{-2} atm and the temperature approximately 220 K (-50°C).

Atmospheric chemistry is currently a very active area of research, and our understanding of both the polluted and the unpolluted atmosphere has increased tremendously in the last 30 years. Biological processes are extremely important in regulating the chemistry of the Earth's atmosphere. Indeed, the key difference in atmospheric chemistry between the Earth and the other planets is the major role played on Earth by the biological processes through which the major components of the atmosphere are cycled.

6.2.2 Atmospheric constituents

It is impossible to know precisely the amounts of the various constituents of the atmosphere - or any other compartment of the environment. Consequently, the figures quoted in the next few sections should be regarded as estimates only.

6.2.2.1 Oxygen

The total amount of O_2 in the contemporary atmosphere has been quoted as 1.2×10^{18} kg. The principal reaction by which oxygen enters the atmosphere is photosynthesis (4.0×10^{14} kg yr^{-1}). Photosynthesis is almost exactly balanced by processes which consume oxygen: respiration and decay, and the combustion of fossil fuels. Thus the residence time of oxygen in the atmosphere is about 3000 years.

$$\text{residence time} = \frac{\text{total amount in reservoir}}{\text{rate of inflow or outflow}}$$

$$= 1.2 \times 10^{18}\text{kg} / (4.0 \times 10^{14} \text{ kg yr}^{-1}) = 3000 \text{ yr}$$

The long residence time of oxygen in the atmosphere implies the two results below.

- O_2 is well mixed in the atmosphere, *i.e.*, its partial pressure does not vary from place to place. An equivalent statement is that the residence time is much greater than the time needed for the atmosphere to mix.

- There is so much oxygen in the atmosphere that human activities, such as over-use of fossil fuels, are unlikely to deplete the atmosphere of oxygen.

6.2.2.2 Nitrogen

We are tempted to think of the 0.781 atm of nitrogen in the atmosphere as an inert "filler", because in our experience elemental nitrogen is rather unreactive.

The biogeochemistry[4] of nitrogen is extremely complex, but leads us to the conclusion that the atmospheric content of nitrogen, like that of oxygen, is regulated principally by biological processes.

The atmosphere contains some 3.9×10^{18} kg of elemental nitrogen. Major natural sinks are biological nitrogen fixation (2×10^{11} kg yr^{-1}) and the production of NO in thunderstorms and through combustion (7×10^{10} kg of N per year), leading ultimately to deposition of HNO_3 in rainwater.

$$N_2(g) + O_2(g) \xrightarrow{\text{high temp.}} 2\,NO(g) \xrightarrow{\text{oxidize}} NO_2 \longrightarrow HNO_3$$

To this must be added perhaps 5×10^{10} kg fixed industrially by the Haber process:

$$N_2(g) + 3\,H_2(g) \xrightarrow{\text{catalyst, 500°C}} 2\,NH_3(g)$$

Because the rates of transfer of nitrogen in and out of the atmosphere are small in comparison with the size of the reservoir, the calculated residence time of $N_2(g)$ is large and the atmosphere is extremely well mixed with respect to nitrogen.

Example: *Calculate the residence time of nitrogen in the atmosphere from the data given above.*

Answer: *mass in reservoir* $= 3.9 \times 10^{18}$ kg

total rate of outflow, comprising biological nitrogen fixation, combustion, and industrial nitrogen fixation:

$(2 \times 10^{11}\ \text{kg yr}^{-1}) + (7 \times 10^{10}\ \text{kg yr}^{-1}) + (5 \times 10^{10}\ \text{kg yr}^{-1}) = 3 \times 10^{11}\ \text{kg yr}^{-1}$

$$\text{residence time} = \frac{\text{total in reservoir}}{\text{rate of outflow}} = \frac{3.9 \times 10^{18}\ \text{kg}}{3 \times 10^{11}\ \text{kg yr}^{-1}}$$

$$= 1 \times 10^{7}\ \text{yr}$$

Terrestrial and aquatic nitrogen in the forms NO_3^- and NH_4^+ cycles through the biosphere to make proteins and nucleic acids. The processes of decay return the nitrogen to the atmosphere as N_2 (and some nitrous oxide, N_2O) by the action of denitrifying bacteria. Almost all the nitrogen fixed by the Haber process is used as fertilizer, so that increased fertilizer use also increases the rate of return of nitrogen to the atmosphere through biological denitrification. Indeed, as much as half of all nitrogenous fertilizer applied to crops is denitrified by soil microorganisms even before the crop takes it up. Increased use of fertilizer is

[4] Geochemistry = the chemistry of the Earth, including the atmosphere; biogeochemistry = biological aspects of geochemistry

probably the reason for the gradual increase in the atmospheric levels of N_2O, which have risen from 0.29 to 0.31 ppmv over the past two decades. Nitrous oxide is rather unreactive and has a residence time of ~ 20 yr in the atmosphere.

6.2.2.3 Water

Unlike oxygen and nitrogen, water is very unevenly distributed in the atmosphere, consistent with the variation in the weather from place to place, and also from time to time. At any time about 7×10^{14} mol of $H_2O(g)$ are present, a tiny fraction of the 9.5×10^{19} mol present on the surface as $H_2O(\ell)$. Evaporation from the oceans (2.2×10^{16} mol yr^{-1}) and from lakes and rivers (3.5×10^{15} mol yr^{-1}) is balanced by precipitation over the land (5.5×10^{15} mol yr^{-1}) and the oceans (1.9×10^{16} mol yr^{-1}), leading to an average residence time of 3×10^{-2} yr (10 days) for water in the atmosphere. Since complete horizontal and vertical mixing of the troposphere requires several years, water is poorly mixed in the atmosphere, and there is great local variability[5].

The amount of water in the atmosphere depends on the temperature, since the vapour pressure of water increases strongly with temperature. Some representative values are given below.

t, °C	$p(H_2O)$, atm	t, °C	$p(H_2O)$, atm	t, °C	$p(H_2O)$, atm
-20	0.00102	0	0.00603	20	0.02307
-15	0.00163	5	0.00861	25	0.03126
-10	0.00257	10	0.01188	30	0.04187
-5	0.00396	15	0.01683	35	0.05418

These values are saturated, or **equilibrium** vapour pressures. They represent the maximum attainable values of $p(H_2O)$ at the stated temperature. Under most conditions, the atmosphere is sub-saturated with respect to water vapour. The **relative humidity** is the prevailing $p(H_2O)$ as a percentage of the equilibrium value. Note that a given value of the relative humidity does not represent a fixed

[5] What is true for the lifetimes of O_2 and H_2O in the atmosphere and the extent to which they are well mixed is also true for environmental pollutants. For any substance, whether naturally occurring or a pollutant, a long residence time correlates with being well-mixed *i.e.*, widely distributed, in the environment (and *vice versa*). *Example*: chlorofluorocarbons have long residence times and become uniformly mixed in the atmosphere. They represent a global pollution problem (Chapter 15). By contrast, acidic gases survive in the atmosphere only for a few days; acid rain is a much more local or regional phenomenon (Chapter 13).

water vapour content because the equilibrium value of $p(H_2O)$ varies with temperature. Thus a relative humidity of, say, 80% corresponds to $p(H_2O)$ = 0.0048 atm at 0 °C, but to 0.025 atm at 25 °C. Remember also that water vapour can co-exist with ice, so that $p(H_2O)$ is non-zero even below 0 °C; this is why snow can evaporate when the temperature is below freezing.

Example: *On a summer day, the relative humidity is 85% and the temperature is 25 °C. Do you expect dew to fall if the temperature falls to 15 °C during the night?*

Answer: *During the day the value of $p(H_2O)$ is 0.85 x 0.03126 atm (0.02657 atm). At 15 °C, the saturated vapour pressure of water is 0.01683 atm, which is less than 0.02657 atm. Since $p(H_2O)$ is greater than the equilibrium value, the atmosphere is supersaturated with respect to H_2O, and so water must condense in the form of dew.*

6.2.2.4 Carbon dioxide

The amount of carbon dioxide in the atmosphere is 1.4 x 10^{16} mol. In counterpoint to oxygen, a major source of atmospheric CO_2 is respiration, combustion, and decay, and an important sink is photosynthesis (about 1.5 x 10^{15} mol yr^{-1} each). Since CO_2 is somewhat soluble in water, exchange with the oceans must also be taken into account, some 7 x 10^{15} mol yr^{-1} being taken up and 6 x 10^{15} mol yr^{-1} being released by different regions of the oceans. The result is an atmospheric lifetime of about 2 yr, which makes the atmosphere moderately well mixed with respect to carbon dioxide. However, analytical data obtained over many years at the Mauna Loa Observatory in Hawaii show that the atmospheric concentration of CO_2 is not perfectly homogeneous: there is a pronounced one-year cycle in CO_2 concentration, with the peak about April and the trough around October (Figure 4).

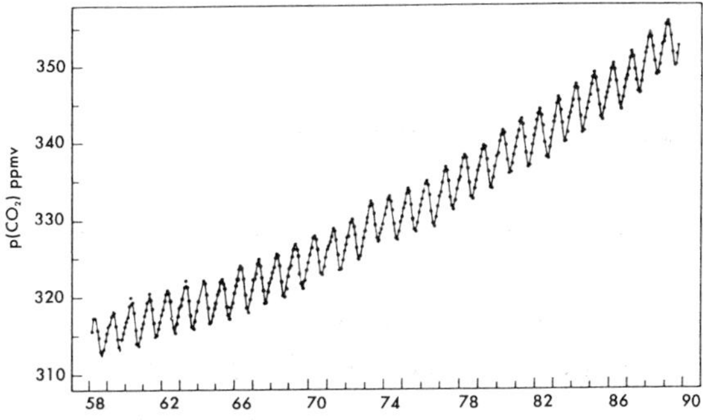

Figure 6.4: Observations of $p(CO_2)$ at the Mauna Loa Observatory for the period 1958 - 1991.

The explanation for the cycle is that photosynthetic activity is highest in the northern hemisphere in the period May to October; during the summer CO_2 is removed from the atmosphere a little faster than it is added, while the reverse situation pertains in the winter months. The peak CO_2 concentration comes at the end of the winter (April) and the trough at the end of the summer (October). Southern hemisphere monitoring stations show the reverse trend.

A second trend in the data of Fig. 4 is the steady increase in the partial pressure of CO_2 over the years, from 315 ppmv in 1958 to about 350 ppmv by 1988. The increase is believed to be due to increasing consumption of fossil fuels coal and oil, which produce CO_2 upon combustion. Predictions are for a doubling of $p(CO_2)$ by the latter part of the twenty-first century.

Coal: $C(s) + O_2(g) \rightarrow CO_2(g)$

Oil: (example) $C_{10}H_{22}(\ell) + 15\frac{1}{2}\ O_2(g) \rightarrow 10\ CO_2(g) + 11\ H_2O(g)$

Example: *Calculate the mass of CO_2 injected into the atmosphere when you burn 1.00 L of gasoline (assume that the approximate simplest formula of gasoline is CH_2, and that its density is 0.80 kg L^{-1}.)*

Answer:

mass of gasoline = 1.00 L x 0.80 kg L^{-1} = 0.80 kg

$n(CH_2)$ = 0.80 kg/(14.0 g mol^{-1}) = 5.7 x 10^{-2} kmol

$n(CO_2)$ = $n(CH_2)$ since carbon is conserved.

$M(CO_2)$ = 44.0 g mol^{-1}

Mass of CO_2 = 5.7 x 10^{-2} kmol x 44.0 g mol^{-1}

= 2.5 kg

Notice in this example that you avoid adding 2.5 kg of CO_2 to the atmosphere if you conserve 1 L of gasoline by walking or cycling.

6.3 Greenhouse gases and global warming

6.3.1 Temperature regulation in the troposphere

Like other stars and planets, the Earth is a "black-body radiator", and emits radiation into space, as well as receiving radiation from the Sun. Because the Earth is relatively cool (average surface temperature *ca.* 288 K), the radiation emitted is of low energy and occurs in the infrared region of the spectrum (unlike the Sun, which is hot, and emits across the ultraviolet, visible, and infrared regions of the spectrum). Some of the infrared radiation is trapped by the gases of the troposphere rather than being lost into space, and so the Earth is thereby kept about 40 °C warmer than it would be if there were no atmosphere. This insulating effect is vital for the existence of life on Earth, for without it, all the water on Earth would be in the form of ice. The insulating effect of the atmosphere is known as the "greenhouse effect", by analogy with the glass of a greenhouse limiting the dissipation of the warmth from inside the greenhouse. (The analogy is really not all that close: in a real greenhouse a major heat-conserving effect is that the glass prevents convectional mixing of the air in the greenhouse with that outside.)

Theoretical prediction and experimental observation agree that not all gases are equally effective at absorbing infrared radiation. The major tropospheric constituents, N_2, O_2, and Ar, are completely ineffective in absorbing infrared radiation,[6] and all the absorption is done by minor atmospheric constituents, including H_2O, CO_2, methane, and other gases as will be discussed below. These latter gases are therefore called "greenhouse gases".

Concentrations of selected trace gases

	1850	1985
carbon dioxide, CO_2	280 ppmv	345 ppmv
methane, CH_4	0.6 - 1.0 ppmv	1.7 ppmv
nitrous oxide, N_2O	280 ppbv	305 ppbv
CFC-11 + CFC-12	0	0.6 ppbv
ozone (troposphere), O_3	probably 10 - 50 ppbv	10 - 50 ppbv

Greenhouse gases play the important role of maintaining the Earth's surface at the +15 °C that is comfortable for life. However, increased concentrations of greenhouse gases would trap infrared radiation more efficiently, and reduce the raction that escaped into space. This could raise the temperature of the troposphere and hence change the climate.

[6] The physical process accompanying infrared absorption is the promotion of a molecule from a lower to a higher vibrational energy state. The requirement for infrared absorption is that the lower and upper vibrational states of the molecule must differ in dipole moment. This condition cannot be fulfilled for atoms such as argon, nor for homonuclear diatomic molecules such as O_2 and N_2, which have zero dipole moment in all their vibrational states.

6.3.2 *Increased levels of greenhouse gases*

The greenhouse gases of principal concern in connection with global warming are carbon dioxide, methane, chlorofluorocarbons, and nitrous oxide. These will be considered in turn. The present contributions of the various greenhouse gases to global warming are estimated to be CO_2, 50% in relative importance; methane and CFCs, 20% each; nitrous oxide, 5%; all other gases 5%. Among these gases, methane, nitrous oxide, and chlorofluorocarbons have atmospheric concentrations which are hundreds or more times less than those of CO_2 and water vapour. This might lead one to expect that their infrared absorbing potential would be insignificant compared with water and CO_2. Such is not the case; these gases are more effective infrared absorbers than CO_2 on a molecule-for-molecule basis.

Carbon dioxide: The increase in the level of CO_2 over 30 years was shown in Figure 4, showing that the sources and sinks of atmospheric CO_2 are not exactly in balance, with the rate of increase in the 1980's a little less than 0.5% annually. If fossil fuel consumption increases as little as 1% annually (it was over 5% from 1945-1975), $p(CO_2)$ is predicted to double, to > 600 ppmv, within a century. Current per-capita releases of CO_2 due to fossil fuel burning range from ~5 tonnes per year for the most highly industrialized countries, through ~¼ tonne per year for developing countries such as India, Brazil, and Mexico, to < 0.1 tonne per year for the least industrialized Third World countries. There is some controversy as to whether all the increase in $p(CO_2)$ should be attributed to increased source strengths due to consumption of fossil fuels; part of the increase may be due to decreased sink strengths, resulting from tropical deforestation and a decreased global rate of photosynthesis. So far, attempts to reach international agreement on limits to CO_2 emissions have not been successful.

Another factor complicating the regulation of CO_2 in the atmosphere is that CO_2 is also taken up by the oceans, where a complex series of equilibria relates $CO_2(g)$ with $CaCO_3(s)$.

$$CO_2(g) \rightleftharpoons H_2CO_3(aq) \overset{-H^+}{\rightleftharpoons} HCO_3^-(aq) \overset{-H^+}{\rightleftharpoons} CO_3^{2-}(aq) \overset{Ca^{2+}}{\rightleftharpoons} CaCO_3(s)$$

The amount of $CO_2(aq)$ in the oceans is sixty times that of $CO_2(g)$ in the atmosphere, suggesting that the oceans can "soak up" most of the additional CO_2 injected into the atmosphere. However, uptake of CO_2 into the surface waters of the oceans is relatively slow (half-life 1.3 years), and in addition, the surface waters of the ocean (zero to ~100 m depth) mix with the deep waters even more slowly (half-life 35 years). Thus in the medium term, the surface waters have the capacity to remove only a fraction of any increase in the load of gaseous CO_2.

Ultimately, carbonate rocks such as $CaCO_3$ constitute an enormous reservoir of CO_2 which is presently locked up. If conditions changed so that some of this

CO_2 began to be released there would be a very strong positive feedback: a "runaway" greenhouse effect in which injection of CO_2 into the atmosphere would raise atmospheric and ocean temperatures, and hence cause more CO_2 to be released into the atmosphere from the oceans. Fortunately, this possibility seems remote[7].

Water: Global warming, a predicted consequence of an "enhanced" greenhouse effect, would increase the average amount of water in the atmosphere through evaporation from the oceans; this is because the equilibrium vapour pressure of water rises with temperature. This situation would represent positive feedback, *i.e.*, increased temperature leading to a rise in $p(H_2O)$, causing in turn a further increase in the efficiency of trapping infrared radiation.

Methane: The concentration of tropospheric methane is currently about 1.7 ppmv but is rising at 1-2% annually. Examination of the methane levels in air bubbles trapped in ice cores indicates that the historical level of this gas was ~ 0.7 ppmv until about two centuries ago. Since then, its rate of increase has been accelerating (Figure 5). The chief sources of atmospheric methane all involve anaerobic decay. As its common name "marsh gas" implies, wetlands are the most important natural source of methane. Human agricultural activity adds greatly to the natural background in the form of anaerobic fermentation in the rumens of ruminant animals, chiefly cattle, and emissions from rice paddies.

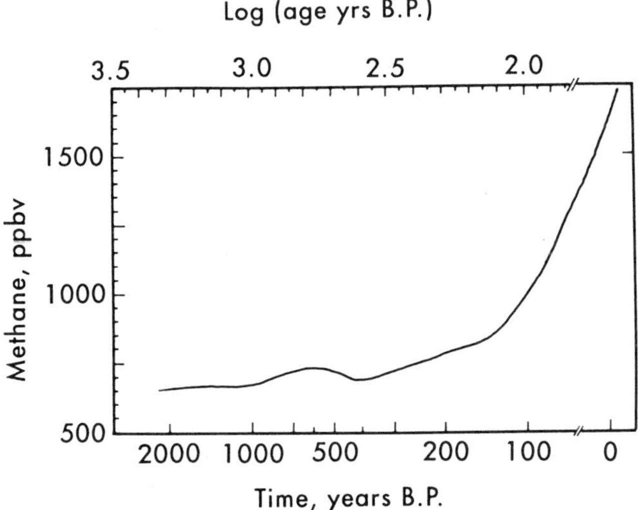

Figure 6.5: Historical trends in the concentration of atmospheric methane. Reproduced from *J. Geophys. Res.* 1984, **89**, 11599.

---

[7] A runaway greenhouse effect does exist on Venus, where the surface temperatures are > 700 K. No liquid water is present, and most of the total CO_2 is in the atmosphere rather than locked away in rocks.

Nitrous oxide, N_2O: The present tropospheric concentration of nitrous oxide, 0.3 ppmv, is increasing at the rate of about 0.2% per year. N_2O has no known sinks in the troposphere; it eventually diffuses into the stratosphere and is destroyed there. Biological denitrification is the chief source of atmospheric N_2O.

Chlorofluorocarbons, CFC's: Like N_2O, these substances have no tropospheric sinks. They are of concern in connection with the destruction of stratospheric ozone (Chapter 15). Up to 1984, the tropospheric concentrations of three of the major commercial CFC's were each growing at an annual rate of ~ 6%. The 1987 Montréal Protocol has set internationally agreed targets for limiting emissions of CFCs in order to protect *stratospheric* ozone. In the context of the *tropospheric* problem of global warming, it is important that the substances chosen to replace the present generation of CFCs are more benign to both stratospheric ozone and to global warming in the troposphere. CFCs exemplify a pollutant which causes two quite different undesirable effects, namely stratospheric ozone depletion and tropospheric warming, and it is important to be clear in any context which of these effects is under discussion.

6.3.3 Climate change

Increased concentrations of greenhouse gases may be anticipated to lead to an increase in the temperature of the troposphere. Data from Antarctic ice cores have shown a close correlation between the local temperature and the atmospheric concentration of CO_2 over the past 160,000 years. Present CO_2 levels are now higher than at any time during that period, with the implied conclusion that climate warming must follow (Figure 6).

News media coverage of the greenhouse effect has, understandably, concentrated on the more sensational of the predictions of a climate change caused by atmospheric warming: melting of the polar ice-caps, and consequent rise in sea level and inundation of major coastal cities, desertification and massive changes in agriculture in the temperate zones, and wholesale extinctions of species. It must be remembered that such predictions are based on very complex computer models which attempt to extrapolate the behaviour of the climate into the future. Assumptions must be made as to the strengths of all sources and sinks for the greenhouse gases, as to the patterns of air circulation and how they might change with the changing climate, and as to possible changes in the proportions of radiation reflected from or trapped by clouds. Such models must account for mixing of the air masses with latitude, with longitude, and with altitude.

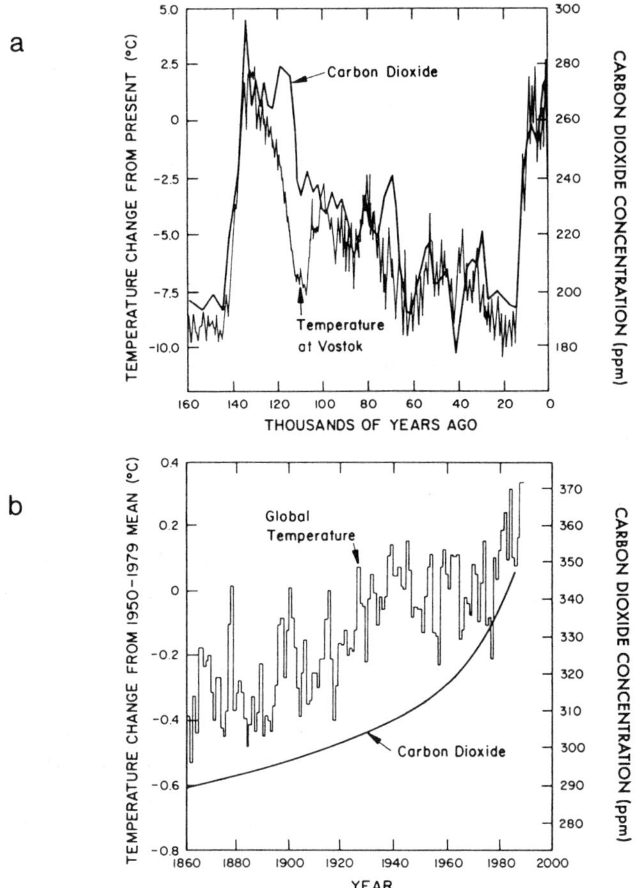

Figure 6.6: Temperature *vs* p(CO_2). a) There has been a close correlation between local temperature and atmospheric concentration of CO_2 over the past 160,000 years. b) During the past 100 years, CO_2 levels have risen markedly, with the implication that global warming must follow.

Current models suggest a possible increase in average global temperatures of 1-3 °C over the next century. This is a very rapid change in climate however; the difference in global temperatures between today and the last ice age is only 4-5 °C, the same order of magnitude as is under discussion. This is why many scientists believe that major climate change is likely to follow the predicted increase in atmospheric temperature. Since this change is predicted to occur over less than a century, severe biological consequences could ensue, since species will have essentially no time to adapt to changes in their habitats.

These are the major changes predicted, and their likely effects upon agriculture:

- Less rainfall in the temperate zones (American Midwest, Canadian Prairies, Russian steppes). These are some of the current "bread-baskets" of the world. A reduction in rainfall could therefore make infeasible the growing of grain in those areas that are currently marginal for this activity.

- Extension of agriculture to higher latitudes. This could offset in part unfavourable consequences of decreased rainfall in the temperate zones.

- Increased rainfall in the drier regions of Africa, northern India, and the U.S. Southwest. This could encourage agriculture in these presently dry regions.

- Development of a tropical climate in the southern U.S.

In southern Canada, predicted effects would include milder winters (bad news for ski operators), and warmer, drier summers. However, it must be stressed that these are predictions and not certainties.

The prediction of climate change due to the release of "radiatively active gases" (greenhouse gases) has led for calls to take measures to limit their emission rates. So far these calls have had only limited success. (This contrasts with the situation with chlorofluorocarbons and stratospheric ozone depletion (Chapter 15), where international agreement has been reached on phasing out the use of the most environmentally-damaging substances.) Success is evident in the regulation of CFCs as just noted (these compounds are environmentally unfriendly in two contexts - greenhouse gases as well as stratospheric ozone depletion).

The major problem however remains CO_2 emissions, and this is bound up with the world's insatiable appetite for energy (the fossil fuels coal and oil). No international agreement has been reached to limit CO_2 emissions. Moreover, the prospects for any such agreement are diminished by the aspirations of developing countries, whose continued industrial development will entail additional energy use, rather than the reverse.

Problems

Section 6.1

1. What mass of $KClO_3$ must be decomposed to KCl and O_2 to afford 638 mL of O_2 measured at 18°C and 92.6 kPa? Assume 100% yield.

2. A gas has density 2.07 g L^{-1} at 22.7 °C and 98 kPa. What is its molar mass?

3. Calculate the densities of the following gases at 25 °C and 1.00 atm.
   a) nitrogen
   b) methane
   c) carbon dioxide
   d) chlorine

a

b

c

d

The atmosphere plays an important role in maintaining the Earth's surface at the +15°C that is comfortable for life. However, increased concentrations of greenhouse gases, smoke, and dust in the atmosphere can be expected to affect the Earth's surface temperature. See Chapter 6, page 140.

a Industrial plants are point sources of nitrous oxides, a greenhouse gas, and sulfur dioxide, a component of acid rain.

b In Cubatao, Brazil, both point and non-point sources emit heavy amounts of smog and smoke.

c Throughout the Third World, home fires contribute smoke and various greenhouse gases. Here, a mountain hut is heated with an open-hearth fire.

d Point and non-point emissions contribute to a "smaze" (smog-haze) above Linden, NJ (USA).

Top, a core taken from glacial ice is held by Dr. Gerald Holdsworth, Environment Canada. Glacial cores reveal snowfall composition and gas constituents from the past much as tree rings record a forest's growth conditions.

Bottom, a slice from a glacial core contains tiny gas bubbles from 150 years ago. These bubbles contain samples of a long-gone atmosphere.

See Chapter 6, page 142.

4. Calculate the average velocities of the following gases at 300 K (27 °C)
   a) hydrogen
   b) oxygen
   c) carbon dioxide
   d) sulfur hexafluoride

5. Which of the following are supercritical fluids under the stated conditions?
   a) CO_2 at 40 °C and 100 atm
   b) H_2O at 250 °C and 300 atm
   c) O_2 at -150 °C and 60 atm
   d) N_2 at -100 °C and 100 atm

6. What is the concentration of a hydrochloric acid solution if 50.0 cm^3 of solution react with excess zinc to form 125.6 cm^3 of hydrogen measured at STP?

   $$Zn\ (s) + 2HCl\ (aq) \rightarrow ZnCl_2\ (aq) + H_2\ (g)$$

7. One of the steps in the production of nickel involves roasting the sulfide ore in air:

   $$2NiS + 3O_2 \rightarrow 2NiO + 2SO_2$$

   If a plant with no pollution control devices roasts 10,000 metric tons of nickel ore containing 10% sulfur by mass, what volume of air would be polluted by SO_2 to the extent of 10 ppmv? Assume the air is at STP.

8. The number of acidic hydrogens in a molecule may be estimated by allowing a sample of the compound to react with CH_3MgI. The number of acidic hydrogen is represented by n in the equation below:

   $$H_nX + nCH_3MgI \rightarrow \uparrow nCH_4 + X(MgIX)_n$$

   When 173.1 mg of resorcinol $C_6H_6O_2$ was allowed to react with excess CH_3MgI, 77.1 cm^3 of methane were produced, measured at 18 °C and 98.7 kPa. How many acidic hydrogens are present in resorcinol?

9. Nitrous oxide N_2O can be prepared as follows:

$$NH_4NO_3 \rightarrow N_2O + 2H_2O$$

In an experiment 0.672 g of ammonium nitrate is decomposed and the N_2O collected over water at 25°C and 1 atm (1.013×10^5 Pa). If the volume collected is 197 cm^3 what is the percent yield of N_2O? (Vapour pressure of water at 25°C is 3.0 kPa).

10. In the Haber process, the temperature is usually about 450°C. If N_2 (75 atm) and H_2 (225 atm) are mixed at this temperature, calculate the concentrations of N_2 and H_2 in moles per liter before any reaction occurs.

Sections 6.2 - 6.3

1. Express the concentrations/pressures of the following atmospheric constituents in each of the units atm, mol L^{-1}, molecules cm^{-3}, and ppmv.
   a) oxygen (p = 0.21 atm)
   b) carbon dioxide (p = 3.5×10^{-4} atm)
   c) methane (p = 1700 ppbv)
   d) hydroxyl radical (c = 5×10^{5} molec cm^{-3})

2. Calculate the relative humidity when $p(H_2O)$ has the following values
   a) 0.00304 atm at -5°C
   b) 3.8 kPa at 30°C
   c) 0.92 kPa at 15°C

3. The total mass of the atmosphere is about 5×10^{15} t. Calculate the approximate masses of the three most abundant components of the atmosphere, assuming negligible masses of the minor components.

4. Suppose SO_2 levels are 0.31 ppmv over a city that is approximately circular of radius 15 km, and the atmosphere is uniformly mixed to a height of 1.0 km. What mass of SO_2 is present in the city's atmosphere?

5. A different set of data from those in the text for the amounts and fluxes of oxygen in the atmosphere are: amount of O_2, 3.8×10^{19} mol; O_2 produced through photosynthesis, 5.0×10^{15} mol yr^{-1}. Estimate the residence time of O_2 in the atmosphere, and compare the value with that given in Section 6.2.2.

6. The mass of hydrogen in the atmosphere is approximately 1.8×10^5 t, and its atmospheric lifetime is about 2 yr. Also, the proportion of H_2 in the atmosphere is currently rising about 0.6% per year. Calculate (a) the rate of removal of H_2 from the atmosphere (b) the excess of addition over removal of H_2 to the atmosphere per year.

7. The heat capacity of an ideal gas at constant pressure is approximately 7R/2, where R is the ideal gas constant. Suggestions are that the Earth's atmosphere might increase in temperature by 2.5°C over the next century. Estimate the average amount of heat that would need to be added to the atmosphere per year to achieve this heating effect (assume no compensating heat losses).

8. Express the heat capacities of the following gases (7R/2) as specific heats ($J\ g^{-1}\ K^{-1}$) in order to compare the specific heats of gases with that of water ($4.18\ J\ g^{-1}\ K^{-1}$).
   a) helium
   b) nitrogen
   c) carbon dioxide

   Which of these gases would be most efficient as a cooling medium to remove heat from a high temperature industrial process?

7 KINETICS

7.1 Rate of reaction

Kinetics is the study of the **rates** of chemical reactions. Different reactions proceed at different speeds; some are over in a fraction of a second, whereas others may require years to complete. Temperature is a factor in determining the rate of a chemical reaction; almost all reactions proceed faster as the temperature is raised.

Chemical rates are an entirely separate matter from chemical energetics (Chapters 5 and 14). Only those reactions which are energetically favoured can proceed at all, but the mere fact that a reaction is thermodynamically favoured does not mean that it will proceed at a measurable rate. As an example, the reaction between $H_2(g)$ and $O_2(g)$ to form water is highly energy-releasing; nevertheless a mixture of $H_2(g)$ and $O_2(g)$ will survive unchanged for an indefinite period at room temperature (unless you strike a match!). Therefore, rate studies complement studies of chemical energetics; energetics tell us whether a reaction is possible in principle, while kinetics tell us whether an energetically favoured process will proceed at a significant rate.

The concept of rate (speed, velocity) of a chemical reaction indicates that **time** is a factor to be considered. The units of a rate of reaction will usually be **moles per liter per second** (mol L^{-1} s^{-1}) or something similar. Atmospheric chemists frequently use the units **molecules per cubic centimeter per second** (molec cm^{-3} s^{-1}).

Example: *The rate of a gas phase reaction is reported as 2.5 x 10^{-6} mol L^{-1} min^{-1}. Express this in molec cm^{-3} s^{-1}.*

Answer: *This is simply an exercise in unit conversion: moles to molecules; liters to cubic centimeters; minutes to seconds.*

$$rate = 2.6 \times 10^{-6} mol\ L^{-1}\ min-1 \times \left[\frac{6.022 \times 10^{23}\ molec}{1\ mol}\right] \times \left[\frac{1L}{1000cm^3}\right] \times \left[\frac{1\ min}{60\ s}\right]$$

$$= 2.6 \times 10^{13}\ molec\ cm^{-3}\ s^{-1}$$

Something which is not intuitively obvious is that the rates of almost all chemical reactions change as the reactions progress; they go more and more slowly as the reactants are used up. We shall see why this is so once we examine the "rate equations" for different chemical processes.

7.2 Order of reaction

7.2.1 Meaning of the term "reaction order"

The "order" of a chemical reaction is the same as the number of chemical concentration terms upon which the rate depends. Speaking generally for the moment in terms of substances A, B, C, etc, we can write examples of possible rate laws. In these rate laws, the square brackets [ ] represent the concentration of the indicated substance at the time the rate is measured - in other words, it is a time-dependent concentration, not the initial or stoichiometric concentration.

[1] rate = $k[A]$

[2] rate = $k[B]^2$

[3] rate = $k[C][D]$

Equation [1] is an example of a **first order** rate expression. The rate of the reaction depends upon only one concentration term - the concentration of substance A. The proportionality constant *k* (lower case, italicized), is called the **rate constant** of the reaction.

Equations [2] and [3] are both examples of **second order** rate expressions. The rate depends upon two concentration terms - the second power of the concentration of substance B in equation [2], and the first power of the concentrations of both C and D in equation [3]. Rate expression [2] is said to be "second order in substance B", while rate expression [3] is "first order in substance C and also first order in substance D".

The equations below *might* be reactions that would follow the rate equations [1], [2], and [3]. However, as we shall see in Section 7.7, it is not always possible to make a direct connection between the rate equation and the chemical equation.

$$A \rightarrow \text{products}$$

$$2\,B \rightarrow \text{products}$$

$$C + D \rightarrow \text{products}$$

Note that the rates of these three reactions depend only on the concentrations of the reactants, not on the concentrations of the products. Exceptions to this rule occur when the reaction is chemically reversible, but then the rate equation is always more complex (see Section 7.7). Sometimes the experimental rate equation involves the concentrations of reactants to fractional powers, and sometimes the rate depends on the concentration of catalysts, which are not even included in the overall chemical equation (reactants → products). We shall concentrate on the simpler cases in this book.

Example 1: *The breakdown of peroxyacetyl nitrate (PAN) is the major sink for this highly toxic and irritating byproduct of photochemical smog (Chapter 9) in the atmosphere. It is a first order reaction.*

$$CH_3CO.OO.NO_2(g) \rightarrow CH_3CO.OO(g) + NO_2(g)$$

$$rate = k[CH_3CO.OO.NO_2]$$

Example 2: *The reaction of 2 moles of NO is the chief route for decomposition of this compound at high temperatures in the laboratory (but not at low temperatures in the environment). This reaction is second order in the concentration of NO.*

$$2\ NO(g) \rightarrow N_2(g) + O_2(g)$$

$$rate = k[NO]^2$$

Example 3: *Reaction with the hydroxyl radical (OH) is the first step in the atmospheric oxidation of SO_2 to SO_3 (Chapter 13). This reaction is first order in the concentration of OH, and also first order in the concentration of SO_2.*

$$SO_2(g) + OH(g) \rightarrow HSO_3(g)$$

$$rate = k[SO_2][OH]$$

The intermediate substance HSO_3 subsequently loses its hydrogen atom to molecular oxygen, in another second order reaction.

Example: *Write the rate law for the second order reaction between HSO_3 and molecular oxygen.*

Answer: *On the basis of the discussion above, the chemical reaction is first order in HSO_3 and first order in O_2. Hence this reaction is like equation [3] and so the rate equation is:*

$$rate = k[HSO_3][O_2]$$

The chemical reaction, including products, is shown below. Note that we did not need to write the complete chemical equation in order to write the rate law.

$$HSO_3(g) + O_2(g) \rightarrow SO_3(g) + HO_2(g)$$

7.2.2 Rate vs. rate constant

The distinction between a rate and a rate constant is a frequent point of confusion. The units are helpful here. A **rate** is always in units of concentration (or sometimes amount) of substance per unit time: *e.g.*, moles per liter per second; grams per liter per hour; molecules per cubic centimeter per second. **Rate constants**, which are the proportionality constants between the rate and the concentration terms in the rate expression, have different units according to the order of the reaction, as will be discussed shortly. A rate constant is a characteristic property of the reaction under discussion; it is called a rate *constant* to remind us that, in contrast with the reaction *rate*, it does not depend on the concentrations of substances reacting. Rate constants do vary with temperature however; they almost always increase with temperature.

7.2.3 Initial rate method for determining reaction order

The rate of a chemical reaction depends upon the instantaneous (or current) concentration(s) of reactants, raised to powers corresponding to their order.

Generally:

$$\text{rate} = k[A]^m[B]^n...$$

where A, B, ... are the chemical substances, and *m*, *n*, are the corresponding kinetic orders.

A useful method for determining the kinetic order of a reaction experimentally is to study how the *rate* changes with different concentrations of the reactants. If the reaction is first order in a particular component, the rate will *double* if the initial concentration of substance doubles.

Examples

(i) $CH_3CO.OO.NO_2(g) \rightarrow CH_3CO.OO(g) + NO_2(g)$

rate $= k[CH_3CO.OO.NO_2]$

The rate of decomposition of PAN doubles if its concentration doubles (or triples if the concentration triples, etc).

(ii) $SO_2(g) + OH(g) \rightarrow HSO_3(g)$

rate $= k[SO_2][OH]$

The rate of reaction doubles if the concentration of either SO_2 or *OH doubles. If both double simultaneously, the rate increases fourfold.*

(iii) $2\ NO(g) \rightarrow N_2(g) + O_2(g)$

rate $= k[NO]^2$

The rate of reaction quadruples (2^2) if the concentration of NO doubles.

A systematic variation of the concentrations of reactants can be used to assess the kinetic order of a reaction. Consider the following example, where the experimental results have been tabulated. They refer to the reaction below in the gas phase at 25 °C (all substances are (g)).

$$2\ NO + O_2 \rightarrow 2\ NO_2$$

	[NO], mmol L^{-1}	$[O_2]$, mmol L^{-1}	rate of reaction of O_2[a]
#1	0.28	1.44	0.69
#2	0.84	1.44	6.2
#3	2.52	1.44	56.
#4	0.84	0.36	1.6

a in μmol L^{-1} s^{-1}

Table entries #1, #2, and #3 are run at constant $[O_2]$, and can be used to determine the order with respect to NO. Conversely, table entries #2 and #4 are run at constant NO, and can be used to determine the order with respect to O_2.

The rate expression is:

$$\text{rate} = k[NO]^m[O_2]^n$$

and we want to find the values of constants m and n. There are two ways to proceed. The "intuitive method" is to compare the ratios of the rates with the ratios of concentrations. Thus comparing table entries #1 and #2 at constant $[O_2]$:

$$[NO](\#2)/[NO](\#1) = 0.84/0.28 = 3.00$$
$$\text{rate}(\#2)/\text{rate}(\#1) = 6.2/0.69 = 8.99 \approx 9$$

Since increasing [NO] 3x increased the rate 9x, it is likely that the reaction is second order (3^2) in NO. (Check that this relationship is also true for table entries #2 and #3).

Now compare table entries #2 and #4, at constant [NO].

$$[O_2](\#2)/[O_2](\#4) = 1.44/0.36 = 4.00$$
$$\text{rate}(\#2)/\text{rate}(\#4) = 6.2/1.6 = 4.2 \approx 4$$

Since increasing $[O_2]$ 4x increased the rate 4x, it is likely that the reaction is first order in O_2.

Thus the rate equation is:

$$\text{rate} = k[NO]^2[O_2]^1$$

The same result can be obtained in a more formal way by writing the rate equation in logarithmic form, and comparing the variation of ln(rate) with ln[reactant].

$$\ln(\text{rate}) = \ln(k) + m\ln([\text{NO}]) + n\ln([\text{O}_2])$$

If numerous data points are available, ln(rate) can be plotted against ln[reactant], and the reaction order for that reactant is the slope of the straight line obtained.

7.3 First order reactions

7.3.1 Time course of a first order reaction

The rate of a first order reaction depends on the concentration of one substance only. In order to describe the kinetic behaviour of a reaction of given kinetic order more completely, it is useful to use the language of calculus, in which the reaction rate is expressed as the rate of change of the concentration of reactant with time.

For the hypothetical reaction:

$$\text{A} \xrightarrow{k_1} \text{B}$$

the rate of reaction is described by

[1] $\text{rate} = -d[\text{A}]/dt = +d[\text{B}]/dt = k_1[\text{A}]$

In eq. [1], t is time, and k_1 is the first order rate constant[1]. [A] is the concentration of substance "A" at the time the measurement is made. The rate of the reaction is the same as the rate of change of the concentration of "A" with time, or -d[A]/dt. (The negative sign indicates that "A" is used up as the reaction progresses.) Equally, we can write that the rate of reaction is equal to the rate of change of the concentration of "B" with time, or +d[B]/dt. (The concentration of "B" increases as the reaction proceeds.)

The reaction rate depends upon the concentration of reactant present at that moment, which is why the reaction rate falls as the reactant is used up in the course of the reaction. Figure 1 illustrates this point; the reaction proceeds more and more slowly with time because the reactant "A" is used up, and its time-dependent concentration becomes smaller and smaller.

---

[1] The subscript 1 on the rate constant is not necessary; it is included to draw attention to the fact that this rate constant is first order.

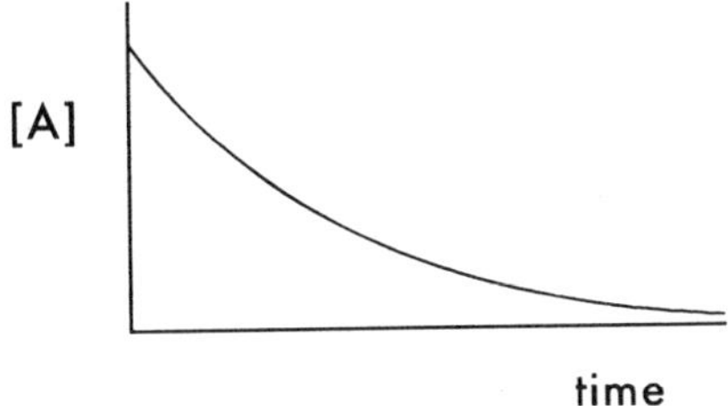

Figure 7.1: Time course of a first order reaction

First order reactions of environmental importance usually involve the breakage of a single covalent bond. They are relatively uncommon. The decomposition of peroxyacetyl nitrate (PAN) has already been mentioned. Another example from tropospheric chemistry is the breakdown of N_2O_5 into NO_2 and NO_3: this reaction is the source of the moderately reactive radical NO_3 in the nighttime troposphere.

$$N_2O_5(g) \rightarrow NO_2(g) + NO_3(g)$$
$$\text{rate} = k[N_2O_5]$$

Radioactive decay is another example of a first order process. Certain atomic isotopes are inherently unstable, and undergo nuclear transformation to more stable nuclei. These processes may vary between very fast (complete in $\ll 1$ s) and very slow (not complete in 1 million years). The spontaneous decay of the rare isotope of carbon, ^{14}C, which is formed in the upper atmosphere by the action of cosmic rays on nitrogen atoms (^{14}N), is used in the technique of radiocarbon dating of historical and archaelogical artefacts. A recent well publicized example of radiocarbon dating was the discovery that the Shroud of Turin could not have been the authentic burial shroud of Christ, because the linen from which it is made dates from the 12th century A.D.

$$^{14}C \rightarrow {}^{14}N + e^-$$
$$\text{rate} = k[^{14}C]$$

The principle of radiocarbon dating is that every living thing is in equilibrium with the tiny amount of radioactive $^{14}CO_2$ present in the atmosphere. Plants take up $^{14}CO_2$ through photosynthesis, and animals eat plants. All of us contain small amounts of such naturally occurring radioisotopes in our bodies. At death, equilibration with the atmospheric reservoir of $^{14}CO_2$ ceases and the organism's own reservoir of ^{14}C begins to decay. If the amount of radioactivity that would have been present at death can be estimated, the present level of ^{14}C allows one to calculate the time elapsed since the object was living. This is done using the rate constant, as described below. Artefacts such as cloth, wood, and bone can be dated directly; objects such as pottery can also be dated by examining the charcoal that was used to fire the clay, making the assumption that the pottery and the wood used to make the charcoal were contemporaneous.

7.3.2 First order rate constants

First order rate constants always have the units (time) $^{-1}$. This is clear if you check eq. [1]. Suppose concentration is in moles per liter, and the rate of the reaction is expressed in moles per liter per second. Then in order for this equation to be dimensionally consistent, k_1 must have the units s $^{-1}$. If you are given a rate constant having the units (time) $^{-1}$, you automatically know that the reaction is first order.

$$\text{rate (mol L}^{-1}\text{ s}^{-1}) = k\ (\text{s}^{-1})\ [\text{A}]\ (\text{mol L}^{-1})$$

Consider again the radioactive decay of ^{14}C. The rate constant has the value 1.2×10^{-4} yr $^{-1}$ (a small rate constant, suggesting a slow reaction).

Example: *When examined in a machine called a scintillation counter, the carbon dioxide obtained from a sample of cloth underwent 525 ^{14}C disintegrations per gram of cloth per minute. What was the concentration of ^{14}C atoms in the cloth?*

Answer: *The rate of reaction was 525 atoms per minute. Since we know the rate constant (1.2×10^{-4} yr $^{-1}$) we can calculate the amount of ^{14}C present in each gram of cloth when the measurement was made.*

$$-d[^{14}C]/dt = k_1[^{14}C]$$

Convert the rate from atoms per minute to atoms per year:

$$\text{rate} = 525 \text{ atoms min}^{-1} \times \frac{(60 \text{ min})}{1 \text{ hr}} \times \frac{(24 \text{ hr})}{1 \text{ day}} \times \frac{(365 \text{ days})}{1 \text{ yr}}$$

$$= 2.78 \times 10^{8} \text{ atoms yr}^{-1}$$

$$2.78 \times 10^{8} \text{ atoms yr}^{-1} = 1.2 \times 10^{-4} \text{ yr}^{-1}\ [^{14}C]$$

$$[^{14}C] = 2.78 \times 10^{8} \text{ atoms yr}^{-1} / 1.2 \times 10^{-4} \text{ yr}^{-1} = 2.32 \times 10^{12} \text{ atoms}$$

Note the exquisite sensitivity of scintillation counting. The total amount of ^{14}C per gram cloth was only 2.32×10^{12} atoms/6.02×10^{23} mol $^{-1}$, or 3.8 pmol, and the amount actually detected by the scintillation counter was 525 disintegrations per minute (8.7×10^{-22} mol min $^{-1}$).

For the first order reaction, the rate of reaction depends on the **first power** of the concentration of the substance "A". If the concentration of "A" doubles, the **rate** of reaction doubles, *i.e.*, twice as many moles react in each liter per second, but the rate constant does not change; as its name implies, it is a constant.

Example: *In a laboratory study of the decomposition of PAN at 300 K, the first order rate constant is $4.9 \times 10^{-4}\ s^{-1}$. What is the initial rate of reaction when [PAN] = (i) 5.8×10^{17} molec cm^{-3}? (ii) 8.7×10^{17} molec cm^{-3}?*

Answer:

rate = k [PAN]

(i) rate = $(4.9 \times 10^{-4}\ s^{-1})(5.8 \times 10^{17}$ molec $cm^{-3}) = 2.8 \times 10^{14}$ molec $cm^{-3}\ s^{-1}$

(ii) rate = $(4.9 \times 10^{-4}\ s^{-1})(8.7 \times 10^{17}$ molec $cm^{-3}) = 4.3 \times 10^{14}$ molec $cm^{-3}\ s^{-1}$

Check for yourself that the rates of reaction are in the same ratio as the initial concentrations for cases (i) and (ii).

7.3.3 Differential and integrated forms of a first order rate equation

The rate of a chemical reaction is, by definition, the change in the concentration of a reactant with time. The following two mathematical statements are identical with reference to the breakdown of N_2O_5.

rate = $k[N_2O_5]$ *or* $-d[N_2O_5]/dt = k[N_2O_5]$

The second version uses the language of calculus, and translates into words as follows: "the change in the concentration of N_2O_5 with time is proportional to the (current) concentration of N_2O_5". The negative sign is used to remind us that as N_2O_5 reacts, its concentration decreases with time, and therefore allows the rate to be a positive, experimental quantity.

The integrated form of the first order rate equation is given by equation [5] (Footnote 2).

[5] $[A] = [A_o] \exp(-k_1 t)$

Other useful, logarithmic forms of eq. [5] are given below.

[5a] $\ln([A_o]/[A]) = k_1 t$ *or* $\ln[A_o] - \ln[A] = k_1 t$

---

2 This is seen as follows. For the general reaction A → B, the rate expression is:

	$-d[A]/dt = k_1.[A]$
Rearranging:	$-d[A]/[A] = k_1.dt$
Integrating:	$-\ln([A]) = k_1.t$ + constant
Evaluate constant:	When t = 0, [A] = $[A_o]$, defining $[A_o]$ as the initial concentration of A.
Therefore:	$\ln([A_o]) - \ln([A]) = k_1 t$
Equivalently:	$\ln([A]/[A_o]) = -k_1 t$
In exponentials:	$[A] = [A_o].\exp(-k_1 t)$

Equation [5a] gives the easiest method of determining k_1: plot the logarithm of the amount of starting material remaining at any time against time; $-k_1$ is the slope of the line, and $\ln[A_0]$ is the intercept on the y axis. (see Figure 7.2)

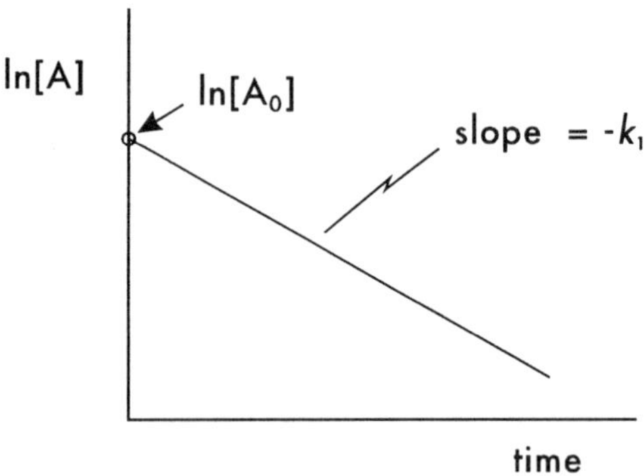

Figure 7.2: Logarithmic plot showing how to obtain a first order rate constant using equation 5a.

The **half-life** of any reaction is the time taken for the concentration of a reactant to decline to half its original. At time $t_{½}$, $[A] = [A_0]/2$. Substituting $[A] = [A_0]/2$ into eq. [5a], we obtain eq. [6]. Because the units of [A] and ½[A] cancel in eq. [5a], any units can be used to calculate the half-life of a first order reaction.

[6] $t_{½} = (\ln 2)/k_1 = 0.693/k_1$

Definition

*The **half-life** of a first order reaction is independent of the initial concentration of the reactant.*

Example: *Suppose the concentration of peroxyacetyl nitrate (PAN) in the troposphere is 0.4 ppbv and $t_{½}$ is 6 h. After 6 h, [PAN] = 0.2 ppbv; after 12 h, [PAN] = 0.1 ppbv; after 18 h, [PAN] = 0.05 ppbv, and so on. The concentration of reactant falls by one half during each half-life.*

The half-life concept allows us to understand a practical limitation of radiocarbon dating. When the sample is very old, the rate of decay of ^{14}C slows down greatly because there is little ^{14}C remaining in the sample. This means that a very large sample of the artefact must be sacrificed in order to make the measurement. If the artefact is very valuable, this may be impossible. This situation existed until recently with the Shroud of Turin, which was too valuable to allow dating by the conventional method. It was dated in 1988, when a new, more sensitive analytical method was developed.

Example: *Suppose the original concentration of ^{14}C in an artefact were 2.4×10^{12} atoms per gram, and it is acceptable to sacrifice 1 gram of sample for the analysis. What is the maximum age of a sample that can be dated if the scintillation counter cannot detect fewer than 80 disintegrations of ^{14}C per minute? The half-life of ^{14}C is 5730 yr.*

Answer: *First determine the rate constant for radioactive decay of ^{14}C.*

$$k = \ln(2)/t_{1/2} = 0.693/5730\,yr = 1.2 \times 10^{-4}\ yr^{-1}$$

From the disintegration rate at time "t" (80 disintegrations per minute), we can work out the concentration of ^{14}C now present in the sample, using the first order rate equation.

$$rate = k[^{14}C_o]$$

$$[^{14}C] = (80\ atoms\ min^{-1})/(1.2 \times 10^{-4}\ yr^{-1})$$

Work out disintegration rate per year:

$$rate = (80\ atoms\ min^{-1})(60\ min/1\ h)(24\ h/1\ day)(365\ days/1\ yr)$$

$$= 4.2 \times 10^{7}\ atoms\ yr^{-1}$$

Now calculate the concentration of ^{14}C:

$$[^{14}C] = rate/k$$

$$= (4.2 \times 10^{7}\ atoms\ yr^{-1})/(1.2 \times 10^{-4}\ yr^{-1})$$

$$= 3.5 \times 10^{11}\ atoms\ (per\ gram\ of\ sample)$$

First order rate equation, in its integrated form:

$$\ln([^{14}C_o]/[^{14}C]) = kt$$

$$\ln((2.4 \times 10^{12})/(3.5 \times 10^{11})) = (1.2 \times 10^{-4}\ yr)t$$

$$t = 6.9/(1.2 \times 10^{-4}\ yr^{-1}) = 5.7 \times 10^{4}\ yr = 57{,}000\ years$$

7.4 Second order reactions

The rate of a second order reaction depends on **two** concentration terms. These may be the same (eq. [2]) or different (eq. [3]). Using the previous generalized examples:

$$2\ B \rightarrow products$$

[2] $rate = k_2[B]^2$ *or* $-d[B]/dt = k_2[B]^2$

$$C + D \rightarrow products$$

[3] $rate = k_2[C][D]$ *or* $-d[C]/dt = -d[D]/dt = k_2[C][D]$

As for first order reactions, the *rate* of reaction is measured in units such as moles per liter per second. However, the second order *rate constant* has the units L mol^{-1} s^{-1} (or something equivalent such as cm^3 $molec^{-1}$ s^{-1}). Again, the units of k allow us to realize what order of reaction is involved, and the subscript

2 has been included only as a reminder that k_2 is a second order rate constant.

Example: *Provide a set of possible units for a third order rate constant.*
Answer: *A third order reaction would have a rate law which depended on three concentration terms: A, B, and C, which might or might not be different chemical species.*

$$rate = -d[A]/dt = k_3[A][B][C]$$

The rate might have the units mol L^{-1} s^{-1}, and this would correspond to concentrations in mol L^{-1}. Thus dimensionally:

$$mol\ L^{-1}\ s^{-1} = k_3\ (mol\ L^{-1})^3$$
$$\text{units of } k_3 = L^2\ mol^{-2}\ s^{-1}$$

The integrated forms of second order rate equations are met less commonly than first order ones. The integrated form of eq. [2] is well known, but occurs infrequently because not very many reactions follow this rate law.

$$2\ B \rightarrow \text{products}$$

[2] rate = $k_2[B]^2$ *or* $-d[B]/dt = k_2[B]^2$

The integrated form[3] is given by equation [7] and the half-life by eq. [8].

[7] $1/[B] - 1/[B_o] = k_2t$

Substitute $[B] = [B_0]/2$ into equation [7].

[8] $t_{½} = 1/(k_2[B_o])$

Unlike a first order reaction, **a second order reaction has a half-life that depends on the initial reactant concentration**; if the initial concentration of reactant is higher, the reaction is not only faster in terms of moles of reactant consumed per liter per second but the half-life is shorter too.

The integrated form of eq. [3] is rarely used, because it is difficult to work with experimentally. Most second order reactions of the type "C + D → products" are treated in practice as pseudo-first order reactions (Section 7.5).

$$C + D \rightarrow \text{products}$$

[3] rate = $k_2[C][D]$ *or* $-d[C]/dt = k_2[C][D]$

---

3 $-d[B]/dt = k_2[B]^2$
Rearranging: $-d[B]/[B]^2 = -[B]^{-2}dB = k_2.dt$
Integrate: $+[B]^{-1} = k_2t$ + Constant
Evaluate constant: When $t = 0$, $[B] = [B_o]$
Therefore: $1/[B] - 1/[B_o] = k_2t$

We will not discuss the integrated form of this equation.

Second order reactions of the type "C + D $\rightarrow$ products" are very numerous. The following examples are drawn from atmospheric chemistry, and involve the hydroxyl radical OH, which is the most important substance for the oxidation of both organic and inorganic compounds in the atmosphere. Previously, we saw its role in the oxidation of the pollutant SO_2; the reactions below show its reactions with NO_2 (to form gaseous nitric acid), and methane (to form the methyl radical, CH_3).

$$OH(g) + NO_2(g) \rightarrow HNO_3(g)$$
$$OH(g) + CH_4(g) \rightarrow H_2O(g) + CH_3(g)$$

The reaction of OH with NO_2 is the chief sink for nitrogen oxides in the atmosphere, because the $HNO_3(g)$ precipitates to earth either by dissolution in raindrops (wet precipitation) or by adsorption on the surface of solid particles (dry deposition). The reaction with CH_4 is different because CH_3 is not the final product. Instead, it is the first **intermediate** product in a long sequence of reactions by which methane is oxidized to CO_2 and H_2O. Details of these two reactions are given in Chapter 9.

The rates of second order reactions depend on **two** concentrations. Thus the rate of reaction of OH with CH_4 is halved if the concentration of *either* CH_4 *or* OH is halved; if both are halved at once, the rate drops to one quarter. Likewise, increasing the concentration of HI(g) in the reaction below by a factor of three raises the rate by 3^2, or 9 times.

$$2\ HI(g) \rightarrow H_2(g) + I_2(g)$$

$$\text{rate} = k_2[HI]^2$$

7.5 Pseudo-first order reactions

The term **pseudo-first order** is used to describe the kinetics of a reaction which is actually of second or higher order, but which follows first order kinetics under particular experimental conditions. For second order reactions, these conditions occur when for some reason the concentration of one of the reactants does not change during the reaction.

$$C + D \rightarrow \text{products}$$

[3] $\text{rate} = k_2[C][D]$ *or* $-d[C]/dt = k_2[C][D]$

If the concentration of D does not change during the reaction (*i.e.*, is constant), then the product k_2[D] will be constant.

[9] rate = k'[C] *or* -d[C]/dt = k'[C], where $k' = k_2$[D]

In equation [8], k' is called a **pseudo-first order rate constant**. Since k' is the product of a second order rate constant (typical units, L mol $^{-1}$ s $^{-1}$) and a concentration (typical units, mol L^{-1}), it has the units of (time) $^{-1}$, just like a regular first order rate constant.

The half-life of a pseudo-first order reaction is given by:

$t_{½} = 0.693/k'$

The half-life of a pseudo-first order reaction is independent of the concentration of C (the reactant whose concentration changes during the reaction). In contrast, it does depend on the concentration of D, the reactant whose concentration stays constant during the reaction. This can be verified experimentally by running the reaction at two different concentrations of D, while keeping the concentration of D fixed during the course of each of these reactions. The half-life $t_{½}$ is then seen to depend on the concentration of a substance which seems not to be part of the rate equation, and this is the difference between the pseudo-first order reaction and a true first order reaction.

Pseudo-first order conditions arise if the concentration of one of the reactants is constant or nearly so. This can happen in one of two ways.

1. The substance whose concentration is constant (D) is regenerated at the same rate as it is consumed. For every molecule that reacts, another is formed, so the concentration does not change. This is the situation with OH in the atmosphere, or any situation where D is a catalyst[4] for the reaction. In this situation the relative magnitudes of the concentrations of reactants C and D are immaterial.

2. D is present in **large excess** compared with C. Even when all of C has reacted, the concentration of D has hardly altered. Consider, for example, the second order reaction between oxygen atoms and oxygen molecules to form ozone, a reaction which is important in the upper atmosphere (Chapter 15).

$O(g) + O_2(g) \rightarrow O_3(g)$

The concentration of oxygen atoms is very low (a few thousand atoms per cm^3), whereas O_2 is present in the upper atmosphere at a pressure of > 0.001 atm

---

[4] Catalysts are discussed in Section 7.9.

(> 3 x 10^{16} molec cm $^{-3}$). Even if all the oxygen atoms react, the concentration of O_2 molecules is not materially changed.

$$\text{rate} = k[O][O_2] = k'[O], \text{ where } k' = k[O_2]$$

Example: *Consider the reaction of OH with methane, for which the second order rate constant at 25 °C is 6.3 x 10 -15 cm^{-3} $molec^{-1}$ s^{-1}.*

(i) *Write the pseudo-first order rate law for this reaction when the concentration of OH is held constant, and evaluate k′ numerically if [OH] = 1.2 x 10^6 molec cm^{-3} (a typical value in the daytime atmosphere).*

Answer: *The second order rate equation is:*

$$\text{rate} = k[OH][CH_4]$$

Since [OH] is constant, it can be included in k′:

$$\text{rate} = k'[CH_4] \quad \text{where } k' = k[OH]$$

Since $k' = k[OH]$, $k' = (6.3 \times 10^{-15}\ cm^{-3}\ molec^{-1}\ s^{-1})(1.2 \times 10^{6}\ molec\ cm^{-3})$

$= 7.6 \times 10^{-9}\ s^{-1}$

(ii) *Calculate the atmospheric half-life of methane if [OH] = 1.2 x 10^6 molec cm $^{-3}$.*

Answer:

$$t_{½} = ln(2)/k' = 0.693/7.6 \times 10^{-9}\ s^{-1} = 9.2 \times 10^{7}\ s = 2.9\ yr$$

7.6 Lifetimes, residence times, and half-lives

The terms lifetime, residence time, and half-life are not all equivalent. **Half-life** has been met in this chapter: it is the time taken for the concentration of reactant to drop to half of its original value. We have learned how to calculate the half-life of first order, second order, and pseudo-first order reactions.

Residence time can be used interchangeably with **lifetime**. These terms were met in Chapter 1, and defined as shown below.

$$\text{Residence time} = \frac{\text{total amount of substance in a reservoir}}{\text{rate of inflow to, or outflow from, a reservoir}}$$

The units of the numerator and denominator must be compatible *e.g.*, moles and moles per year, or grams per liter and grams per liter per second, so that they cancel to give the answer in units of time.

Another, but equivalent definition of a residence time is the reciprocal of the sum of all the first order and pseudo first order rate constants for all the processes by which the substance **disappears** from the reservoir.

$$\text{Residence time (lifetime)} = \frac{1}{(k_1 + k'_2 + k_3 \ldots)}$$

The rate constants **must** all be first order or pseudo first order if the result is to have the units of time. This definition is useful in calculating the lifetime of an excited state in a photochemical process.

The difference between a half-life and a residence time (lifetime) is as follows. The half-life is the time taken for the concentration of the substance to drop to 50% of its initial value. The lifetime (usual symbol τ) is the time taken for the initial concentration to fall to $1/e$ (about 37%) of the original value. This is shown as follows. (See Figure 3.)

$\tau = 1/k_1$ (definition of τ)

For a first order reaction, taking time = τ:

$$\ln([A_0]/[A]) = k_1\tau = k_1/k_1 = 1$$

Therefore $\ln([A_0]/[A]) = 1$, and so $[A_0]/[A] = e$, $[A] = [A_0] \times (1/e)$.

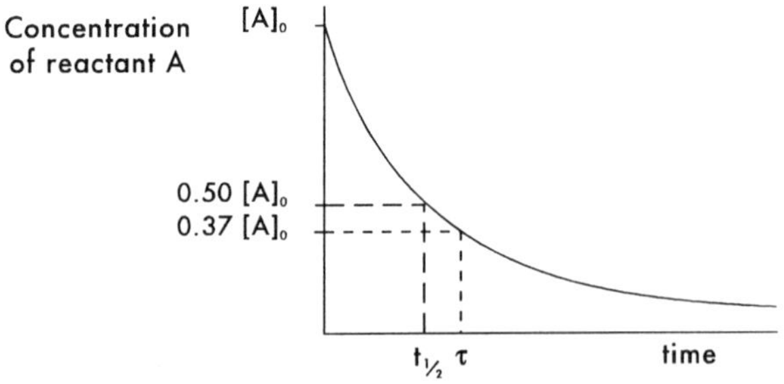

Figure 7.3: The difference between a half-life and a residence time (lifetime)

7.7 Kinetics and mechanism

The reactions discussed so far in this chapter all proceed in one step from reactants to products; that is, there are no substances formed as "intermediates" along the way. However, many reactions are multi-step processes. We have already alluded to one such multi-step reaction: the oxidation of methane to CO_2 and water. Kinetic studies are an important tool in allowing chemists to deduce the mechanism of a reaction. Given an experimental rate law, certain possible mechanisms will be consistent with the rate law, while others are inconsistent, and can be excluded from further consideration.

7.7.1 Elementary reactions

These are reactions which involve only a single step *i.e.*, there are no intermediate substances formed along the reaction pathway. Once you know that a reaction is an elementary reaction (and this must be found out by experiment), the rate law follows from the stoichiometry of the reaction. All the reactions examined so far in this chapter are elementary reactions; their rate equations simply involve the concentrations of all the reactants. Thus for an elementary reaction, but **only** for an elementary reaction, the rate equation can be written down just by looking at the balanced chemical equation.

Definition

*The **mechanism of a multi-step reaction** is the sequence of elementary reactions which, taken together, describes the complete series of chemical events that transform the reactants into the products of the stoichiometric reaction.*

The sum of the elementary steps must give the overall, stoichiometric equation, and also, the rate equation derived from the postulated mechanism must agree with the rate law obtained experimentally.

7.7.2 Order and molecularity

Distinction between these terms is sometimes confusing. Just remember these definitions.

Definition

*The **kinetic order** of a reaction is what is determined experimentally from the rate equation. You can talk about the order of the reaction, or the order with respect to any reactant.*

*The term **molecularity** can only be applied to an elementary reaction. It describes the number of chemical species participating in the elementary reaction. For elementary processes, the molecularity is the same as the overall kinetic order.*

Example: *The oxidation of NO in the laboratory follows the following rate experimental law.*

$$CO(g) + NO_2(g) \rightarrow NO(g) + CO_2(g)$$

$$rate = k[NO_2]^2$$

(i) Is this an elementary reaction? (ii) Does it involve a bimolecular reaction between NO_2 and CO?

Answer: *The answer to each question is "no". This cannot be an elementary reaction, because then the rate law would have been given by:*

$$rate = k[NO_2][CO]$$

Furthermore, only the rate equation just given would be compatible with a bimolecular reaction between NO_2 and CO. The reaction is second order, and we will consider its mechanism shortly.

7.7.3 Rate limiting reactions

The reaction mechanism is the dissection of a complicated reaction sequence into its elementary steps. The overall rate of a complex reaction is the rate of the slowest of these elementary steps, which is called the **rate determining** or the **rate limiting** step. A useful analogy is the presence of an obstacle on a multi-lane freeway; at the obstruction, traffic is restricted to fewer lanes. The rate at which traffic can flow along the freeway is **limited** by the rate at which it can pass the obstacle. Traffic is backed up ahead of the obstacle, and even though it flows freely beyond the obstruction, this does not increase the overall rate of traffic flow along the freeway. Likewise, in a chemical reaction, the reaction rate is determined by the events leading up to and including the rate determining step, but not by the (fast) reactions which occur afterwards. (See Figure 4.)

For example, the following mechanism is suggested for the previously-mentioned reaction of $NO_2(g)$ with CO(g):

Step 1, slow: $2\ NO_2(g) \rightarrow NO(g) + NO_3(g)$

Step 2, fast: $NO_3(g) + CO(g) \rightarrow NO_2(g) + CO_2(g)$

Thus the rate equation includes 2 mol NO_2, which were involved at the slow step of the reaction, but does not include CO, which did not become involved in the reaction until after the slow step. Since the rate of the reaction does not depend on the concentration of CO, the reaction is said to be **zero order** in CO.

Figure 7.4: A traffic obstruction is an analogy for the rate-limiting step of a reaction sequence

Example: *The decomposition of ozone in the laboratory follows the stoichiometry below:*

$$2\ O_3(g) \rightarrow 3\ O_2(g)$$

Suggest a rate equation compatible with the following proposed mechanism.

$O_3(g) \rightarrow O(g) + O_2(g)$ *fast, and proceeds in both directions*

$O(g) + O_3(g) \rightarrow 2\ O_2(g)$ *slow, rate determining*

Answer: *The slow step of the reaction is bimolecular, hence its rate law is:*

$$rate = k[O][O_3]$$

By contrast, the stoichiometric equation would have suggested a bimolecular, second order reaction, with rate = $k[O_3]^2$.

The two examples just covered show that one should never postulate a rate law on the basis of the stoichiometric equation (unless you already know that you are dealing with an elementary reaction). A rate law can be deduced if the mechanism is known, and conversely, a mechanism can be postulated if the experimental rate equation is known.

The rate law deduced in the last example for the decomposition of ozone is incomplete in that it contains the concentration of oxygen atoms, an intermediate in the reaction. Since the concentrations of intermediates are frequently not attainable experimentally, the following is an important convention.

Definition

*A **rate law** is always expressed only in terms of experimentally observable concentrations, rather than those of intermediates.*

In order to tackle this problem we shall need to consider the **steady state approximation**.

7.7.4 Steady states

By definition, a substance is said to be present in a steady state whenever its concentration does not change with time.

Mathematically, d[conc.]/dt = zero.

A trivial reason for having a (nearly) steady state was seen in Section 7.5 (pseudo-first order reactions), when the concentration of one of the reactants was in very large excess over the other. The common reason for the existence of a steady state is that *the rates of formation and destruction of the substance are equal.*

The **steady state approximation** is a useful device in the kinetic analysis of multi-step reactions. The assumption is made that all intermediate species in the reaction rapidly achieve a small and essentially constant concentration. Since these intermediate substances are often not detectable experimentally, the steady state approximation provides a method of eliminating their concentrations from the kinetic rate expression, and hence representing the rate law only in terms of substances whose concentrations are observable experimentally. The previous example of the decomposition of ozone illustrates the concept. There are three reactions, shown with rate constants labelled k_a, k_b, and k_c.

(a)	$O_3 \rightarrow O_2 + O$	k_a
(b)	$O + O_2 \rightarrow O_3$	k_b
(c)	$O + O_3 \rightarrow 2\ O_2$	k_c

All these are elementary reactions, so we can write rate equations for each of them.

$$\text{rate(a)} = k_a[O_3]$$
$$\text{rate(b)} = k_b[O][O_2]$$
$$\text{rate(c)} = k_c[O][O_3]$$

Recall from the previous section that step *(c)* is rate limiting.

In order to express rate(c) in terms of experimental observables, we need to eliminate the concentration [O]. The steady state approximation permits this. If the concentration of [O] reaches a steady state, then the rate of forming [O] (reaction *(a)*) must exactly balance the rate of removal of [O] (reactions *(b)* and *(c)* together).

$$\text{rate(a)} = \text{rate(b)} + \text{rate(c)}$$
$$k_a[O_3] = k_b[O][O_2] + k_c[O][O_3]$$

We can now isolate the concentration [O] in terms of observables.

$$[O] = \frac{k_a[O_3]}{k_b[O_2] + k_c[O_3]}$$

Hence we can eliminate the concentration [O] from rate(c) which, as discussed above, is the rate law for the overall reaction.

$$\text{rate(c)} = k_c[O][O_3] = \frac{k_a k_c[O_3]^2}{k_b[O_2] + k_c[O_3]}$$

Thus eq. [10] represents the rate law for the decomposition of ozone, as derived from the postulated reaction mechanism.

$$[10] \quad \text{rate} = \frac{k_a k_c[O_3]^2}{k_b[O_2] + k_c[O_3]}$$

Notice how complex the rate expression can become, even for a mechanism involving only three elementary steps! Notice also for the first time the presence of a product species ($[O_2]$) in the rate expression. This arises because one of the steps in the reaction is reversible. Reaction *(b)* is the reverse of reaction *(a)*. We can expect to see one or more product concentrations in the rate expression whenever the slow step of the reaction is preceded by a reversible reaction.

Equation [10], as just noted, is the theoretical rate equation for the decomposition of ozone, as derived from the reaction mechanism. In order to know whether this is a possible mechanism for the reaction, we must determine whether this theoretical rate equation is consistent with the experimental rate equation. The experimental rate equation is obtained by studying how the rate of the reaction depends upon the concentrations of reactants (and sometimes, as in this case, products).

The experimental rate law [11] for the decomposition of ozone is consistent with the one deduced from the proposed mechanism.

$$[11] \quad \text{rate} = \frac{k_1[O_3]^2}{k_2[O_2] + k_3[O_3]}$$

Rate equations [10] and [11] are consistent with each other. Rate constant k_1 is the product $k_a k_c$, and k_2 and k_3 replace k_b and k_c respectively.

For practical reasons, the typical experimental conditions for studying the decomposition of ozone involve $[O_2] >> [O_3]$. Under these conditions, a simpler experimental rate equation is obtained.

$$\text{rate} \sim \frac{k[O_3]^2}{[O_2]}$$

This rate law is also compatible with the theoretical rate equation [10]. Since under these experimental conditions $k_2[O_2] >> k_3[O_3]$, then the second term of the denominator of the rate law is insignificant compared with the first and can be omitted. The ratio of rate constants k_1 and k_2 can then be combined into a single constant k. The more complex rate law is predicted to be seen at high $[O_3]$, low $[O_2]$ (conditions not normally used because O_3 is explosive at high concentrations).

The discovery that the theoretical rate equation is consistent with the experimental rate equation is an important step in **establishing** the mechanism of any chemical reaction. Notice the use of the word "establish". We never say that a mechanism is **proved**, because we can never be sure of this. An established mechanism is one that is consistent with all presently known facts. (Tomorrow, someone may discover new facts that require the mechanism to be reconsidered.) Kinetics are an important part of establishing the reaction mechanism; the experimental rate law of the reaction must be consistent with the rate expression which can be deduced on paper from the proposed mechanism. However, **kinetics alone can never be used to establish a mechanism**; there will usually be several possible mechanisms that could be written down, all of which are consistent with the experimental kinetic rate law.

7.8 Activation energies

Virtually all reactions proceed faster as the temperature is raised: rate constants are only constant at constant temperature. At the turn of the century, Arrhenius showed empirically the typical dependence of a rate constant upon temperature, eq. [12], which is now known as the Arrhenius equation.

[12] $k = A \exp(-E_a/RT)$

or, in its logarithmic form,

[12a] $\ln k = \ln(A) - E_a/RT$.

Experimentally the activation parameters E_a and A are obtained from a plot of ln k *vs.* 1/T, where T is in kelvins, not °C.

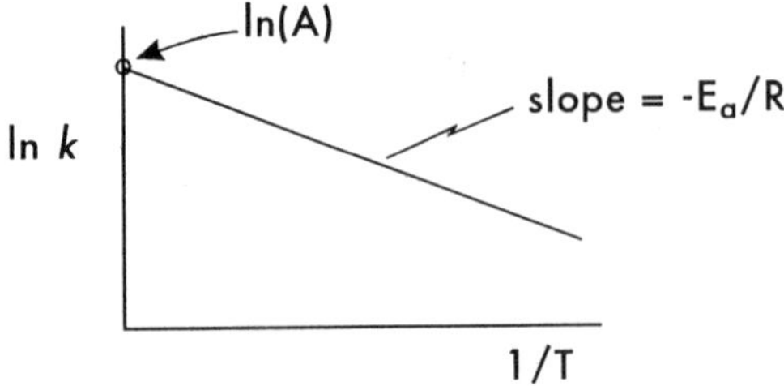

Figure 7.5: Arrhenius plot of the variation of a rate constant with temperature

The physical interpretation is that the activation energy E_a, or E_{act}, is the energy the reactants need to bring to their encounter to make the reaction "go". The significance of temperature is that the average velocities of the molecules and hence their kinetic energies (= $\frac{1}{2}mv^2$) increase with temperature, as was shown for the case of gas phase molecules in Chapter 6. It was also noted in Section 6.1 that the fraction of molecules having energies greater than the threshold value E_a depends on the term $\exp(-E_a/RT)$. Therefore, the exponential part of the Arrhenius equation exactly mirrors the exponent in the dependence of molecular speeds (hence energies), the fraction of the molecules which at any temperature have the specified activation energy. (Figure 7.6.)

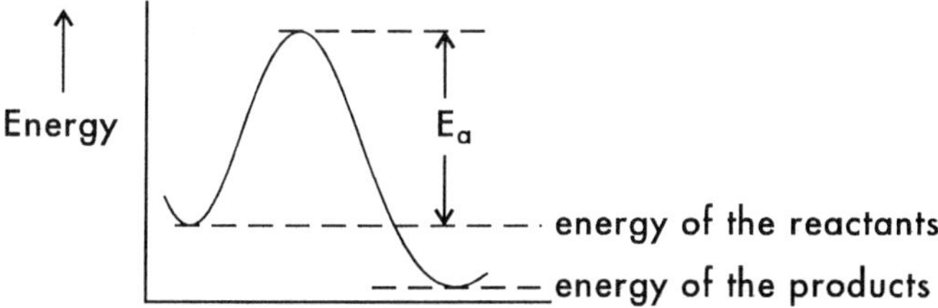

Figure 7.6: The activation energy E_a determines the rate constant of a chemical reaction

The constant "A", called the pre-exponential factor, has the same units as *k*; it can be thought of as what the rate constant would have been without the constraining influence of the activation energy. Another way of putting it is that "A" represents what *k* would be at infinite temperature, when $E_a/RT \rightarrow$ zero. At one time it was thought for gas phase reactions that the Arrhenius "A" could be calculated as the rate of collisions between the molecules. However, in practice A is always smaller than the collision rate. The interpretation is that not all collisions between molecules occur with the correct spatial orientation for reaction to occur, even when the molecules bring sufficient activation energy to their encounter. However, the fastest rate a reaction could possibly have would be if E_a were to be zero, and also if the value of the "A" were to be equal to the rate of collisions between reacting species.

In summary, reactions may be slowed down either because of an unfavourable (large) activation energy, or because of unfavourable spatial requirements when encounter occurs (small pre-exponential factor). A modern interpretation is that E_a can be associated with enthalpic constraints of the reaction in overcoming its "activation barrier". The pre-exponential factor is related to entropic factors. Since entropy will not be covered until Chapter 14, we will not consider it further here.

The basis of the Arrhenius equation is that E_a is independent of temperature. This is not always so, especially when very large temperature intervals are involved - a situation which is relevant to atmospheric chemistry, where the temperature ranges from about 300 K at the Earth's surface down to 200 K in the stratosphere[5]. Under these conditions, curved plots are obtained instead of straight lines, and modified Arrhenius equations must then be used.

A convenient method of writing the rate constant as a function of temperature is given below, eq. [13]. This equation allows one to calculate the rate constant for a reaction at any temperature, given only the constants A and B.

[13] $k = A \exp(-B/T)$

A is the pre-exponential factor; -B incorporates $-E_a/R$, and has the units K^{-1}.

Example: *The rate constant for the oxidation of NO by O_3 in the gas phase ($NO + O_3 \rightarrow NO_2 + O_2$) is given by:*

$$k = 2.3 \times 10^{-12} \exp(-1450/T)\ cm^3\ molec^{-1}\ s^{-1}$$

Calculate E_a, and evaluate k at 22 °C.

Answer:

$$-1450\ K^{-1} = -E_a/R = -E_a/8.314\ J\ mol^{-1}\ K^{-1}$$

$$E_a = 1.2 \times 10^4\ J\ mol^{-1} = 12\ kJ\ mol^{-1}$$

$$k_{295} = 2.3 \times 10^{-12} \exp(-1450/295) = 1.7 \times 10^{-14}\ cm^3\ molec^{-1}\ s^{-1}$$

7.9 Catalysis

A catalyst is a substance which alters (almost always increases) the rate of a chemical process but is unchanged as a result. Since it is present in both the reactants and the products, the catalyst does not appear in the stoichiometric equation for the reaction.

The concentration of the catalyst affects the reaction rate, and normally appears in the rate equation even though it does not appear in the overall stoichiometric chemical equation. The dehydration of ethyl alcohol to ethylene is catalysed by acid, and illustrates this point.

$$CH_3CH_2OH(g) \rightarrow CH_2 = CH_2(g) + H_2O(g)$$

The mechanism of this reaction is given below.

1) $CH_3CH_2OH + H^+ \rightleftarrows CH_3CH_2OH_2^+$ fast and reversible

2) $CH_3CH_2OH_2^+ \rightarrow CH_3CH_2^+ + H_2O$ slow

3) $CH_3CH_2^+ \rightarrow CH_2 = CH_2 + H^+$ fast

The rate constants are defined as follows: k_1 is the rate constant for forming

[5] Even greater temperature ranges are found if one includes the extreme upper reaches of the atmosphere, where T rises to about 1500 K.

$CH_3CH_2OH_2^+$, k_{-1} is the rate constant for reversion of $CH_3CH_2OH_2^+$ back to CH_3CH_2OH, and k_2 is the rate constant for the rate-limiting step in the reaction.

The rate of the slow, elementary step of the reaction is:

$$\text{rate} = k_2[CH_3CH_2OH_2^+]$$

After eliminating the concentration of the intermediate species $CH_3CH_2OH_2^+$ by the use of the steady state approximation, the theoretical rate equation is consistent with the experimental one.

$$\text{rate} = k[H^+][CH_3CH_2OH]$$

You can show by application of the steady state approximation that in this equation k is a composite rate constant $(k_1k_2/(k_{-1} + k_2))$. Notice what happens to the term $(k_1k_2/(k_1 + k_2))$ as the relative magnitudes of $k_1$and k_2 change. If $k_{-1} >> k_2$, the rate expression simplifies as follows:

$$\text{rate} = \{k_1.k_2/k_{-1}\}[H^+][CH_3CH_2OH]$$

Conversely, if $k_{-1} << k_2$, step 1 has by definition become rate limiting (make sure you agree with this statement!) and the rate equation simplifies so as to reflect this.

$$\text{rate} = k_1[H^+][CH_3CH_2OH]$$

Acid catalyses this reaction in the following way. The first step in the reaction is the addition of H^+ to the OH group of ethyl alcohol. This makes it easier for water to be split off. Consequently, the catalyst H^+ provided a path to allow the reaction to proceed more easily - in other words a pathway of lower activation energy. Figurc 7.7 illustrates the energy profile for the reaction just described. Notice the following points about this diagram.

1. The activation energy for the uncatalysed reaction (dotted line) is lower than that of the catalysed reaction (solid line). Catalysts almost always act by providing a new reaction pathway of lower activation energy, so that the specific case given in Figure 7.7 is an example of a general phenomenon. Very occasionally, the catalysed reaction has larger E_a than the uncatalysed process, but in these cases the larger E_a is more than offset by a more favourable (larger) A factor.

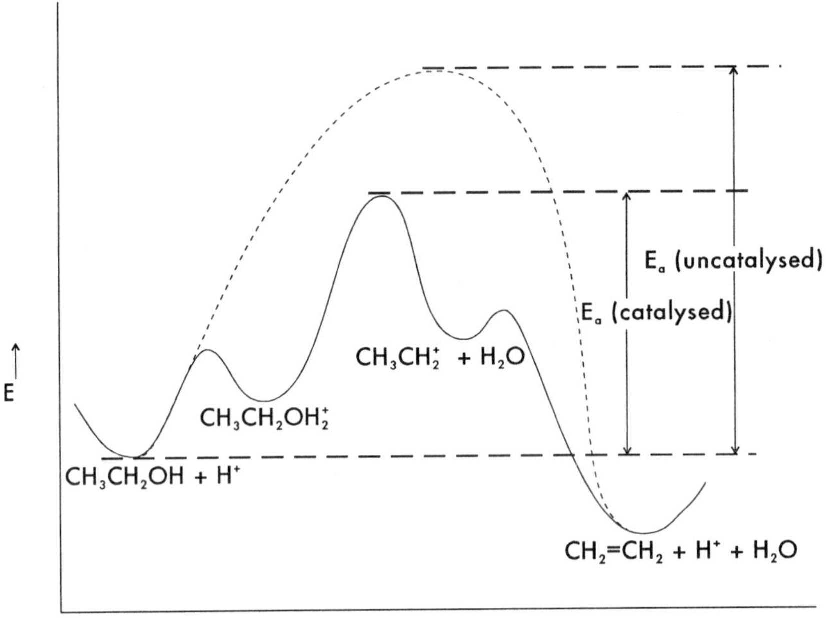

Figure 7.7: Effect of catalyst on the rate of conversion of CH_3CH_2OH to $CH_2 = CH_2 + H_2O$

2. The step preceding the slow step requires less activation than the overall process. Again, this is a common situation[6], and under these conditions the steps preceding the slow step will not only be fast but also reversible. We shall return to this point in Section 8.12.

---

[6] The alternative condition is that the slow step of the reaction is bimolecular and its rate is limited by low concentration of one of its reactants. This condition also leads to reversibility of the reactions preceding the rate limiting step.

Problems

Sections 7.1 - 7.2

1. Are the following quantities rates or rate constants for a reaction?
   a) 1.2×10^{-5} mol L^{-1} s^{-1}
   b) 6,300 s^{-1}
   c) 1.3×10^{6} L mol^{-1} s^{-1}
   d) 4.5×10^{13} molec cm^{-3} s^{-1}
   e) 85 g L^{-1} h^{-1}
   f) 8.3×10^{-33} cm^{6} $molec^{-2}$ s^{-1}

2. Give the order of reaction with respect to each component and the overall order of reaction in the following rate laws.
   a) rate = $k[N_2O_5]$
   b) rate = $k[OH][CH_4]$
   c) rate = $k[NO]^2$
   d) rate = $k[HNO_3][H^+][\text{benzene}]$
   e) rate = $k[CH_3I][OH^-]$

3. The reaction below has $k = 1.7 \times 10^{-14}$ cm^3 $molec^{-1}$ s^{-1} at 295K

   $$O_3 + NO \xrightarrow{k} O_2 + NO_2$$

   a) Calculate the rate constant in the units L $mol^{-1}s^{-1}$
   b) What is the kinetic order of the reaction?
   c) Show that the units of the rate constant are compatible with the following rate law

   $$\text{rate} = k[O_3][NO]$$

   d) Calculate the rate of reaction in acceptable units for $p(O_3)$ = 120 ppbv and $p(NO)$ = 40 ppbv in the troposphere (p_{total} = 1.00 atm) and 295 K.

4. At -10°C the reaction below is first order in Cl_2 and second order in NO and has $k = 1.8 \times 10^2$ L^2 mol^{-2} s^{-1}

   $2NO\ (g) + Cl_2\ (g) \rightarrow 2NOCl$

   a) Calculate the initial rates of reaction for:

      (i) [NO] = 0.010 mol L^{-1}, $[Cl_2]$ = 0.020 mol L^{-1}
      (ii) [NO] = 0.010 mol L^{-1}, $[Cl_2]$ = 0.040 mol L^{-1}
      (iii) [NO] = 0.030 mol L^{-1}, $[Cl_2]$ = 0.020 mol L^{-1}

   b) Would the rate of reaction stay the same through the complete course of the reaction (complete consumption of reactants)?

   c) Would the rate law for the reaction remain constant through the complete course of the reaction?

5. The decomposition of di-tert butyl peroxide is first order, and has rate constant 3.3×10^{-4} s^{-1} at 150 °C. What is the initial rate of reaction for initial concentrations of reactant:

   a) 9.1×10^{-3} mol L^{-1}

   b) 2.6×10^{-4} mol L^{-1}

   c) 7.2×10^{-5} mol L^{-1}

Section 7.3

1. The gas phase decomposition of di-(1,1-dimethylpropyl) peroxide is first order. At 132°C the rate constant is 7.2×10^{-5} s^{-1} and at 150°C it is 4.8×10^{-4} s^{-1}. Calculate the half-life and the lifetime at each of these temperatures.

2. The following radioisotopes have half-lives of decay as given below. Calculate the first order rate constant in each case.

   a) ^{3}H 12.26 years

   b) ^{235}U 7.1×10^8 years

   c) ^{18}F 109 minutes

   d) ^{13}O 0.0087 seconds

3. At 45°C the gas phase decomposition of N_2O_5 is first order, $k = 5.0 \times 10^{-4}$ s^{-1}.

   a) Calculate the fraction of N_2O_5 remaining if a sample is heated to 45°C for 2.6 h.

   b) If the initial $p(N_2O_5)$ is 36 kPa, what is the value of $p(N_2O_5)$ after 35 min?

4. The decomposition of PAN in the atmosphere is a first order process, and has $k = 4.9 \times 10^{-4}\ s^{-1}$ at 27°C, but $k = 5.6 \times 10^{-6}\ s^{-1}$ at 0°C.

   a) If we take these temperatures as representative of summer and winter respectively, calculate the half-life of PAN under summer and winter conditions.

   b) If the initial concentration of PAN is 15 ppbv, calculate the initial rate of decomposition under summer and winter conditions.

   c) For the conditions of part (b) calculate the amount of PAN remaining after 1 day, if there are no new emissions of PAN during that time.

5. The thermal isomerization of cis-azobenzene $C_{12}H_{10}N_2$ in solution follows first order kinetics.

   a) Write the first order rate equation for this reaction in both the differential and integrated forms.

   b) At 50°C, the rate constant has the value $4.0 \times 10^{-5}\ s^{-1}$. Calculate:
   (i) the initial rate of reaction when a $4.7 \times 10^{-3}\ mol\ L^{-1}$ solution of cis-azobenzene is heated to 50°C;
   (ii) the half-life of the reaction at 50°C;
   (iii) the fraction of cis-azobenzene that has reacted after 45 minutes at 50°C.

6. Predict the effect of the following on a first order reaction.

   a) What happens to the rate of reaction when the concentration of reactant is doubled?

   b) What happens to the half-life of the reaction when the concentration of reactant is doubled?

   c) What happens to the rate constant for the reaction when the concentration of reactant is doubled?

   d) What happens to the rate of reaction when the temperature rises by 20°C?

7. Express the following first order rate constants in different units

   a) $6.5 \times 10^{-4}\ s^{-1}$ in h^{-1}

   b) $3.7 \times 10^{5}\ yr^{-1}$ in s^{-1}

Section 7.4

1. The gas phase decomposition of HI follows a rate law which is second order in HI.
   a) What happens to the rate of reaction when the concentration of HI is increased 3 times?
   b) What happens to the rate constant for the reaction when the concentration of HI increases 3 times?
   c) What happens to the half-life of the reaction when the concentration of HI increases 3 times?

2. The following reactions are both second order, and also bimolecular, in the gas phase.

$$2NO \rightarrow N_2 + O_2$$
$$NO + O_3 \rightarrow NO_2 + O_2$$

   a) Write a rate equation for each of these reactions.
   b) What is the effect on the rate of each reaction of increasing the concentration of NO by a factor of 2.5 times (but in the second reaction leaving the initial concentration of O_3 unchanged)?
   c) The rate constant for the reaction between NO and O_3 has the value 3.0 x 10^6 L mol $^{-1}$s $^{-1}$ at 37°C. Calculate the initial rate of this reaction when p(NO) = 50 ppbv, p(O_3) = 120 ppbv and the total pressure is 1.0 atm.

3. At 900 K, N_2O decomposes into N_2 and O_2.

$$N_2O\ (g) \rightarrow N_2\ (g) + \tfrac{1}{2}O_2\ (g)$$

For an initial concentration of N_2O of 0.0050 mol L^{-1} the half-life of the reaction was 297 min, while when the initial concentration of N_2O was 0.0200 mol L^{-1} the half-life of the reaction was 75 min.
   a) What was the kinetic order of the reaction?
   b) Write the rate law in differential and integrated forms.
   c) Calculate the rate constant.
   d) Calculate the amount of N_2O that would be left after 1.00 h if an initial concentration of 0.0250 mol L^{-1} N_2O were maintained at a constant temperature of 900 K.

4. At 400°C the rate constant for the reaction below is 2.3×10^{-2} L mol^{-1} s^{-1}.

   $$H_2\,(g) + I_2\,(g) \rightarrow \quad 2HI\,(g)$$

   a) What is the reaction order?
   b) The rate of the reaction changes if either the concentration of H_2 or the concentration of I_2 is changed. Write the kinetic rate equation.
   c) Calculate the rate of reaction under conditions where $[H_2] = 0.0226$ mol L^{-1} and $[I_2] = 0.0438$ mol L^{-1}.

5. The hydrolysis of acetamide (A) has been studied under conditions where [A] = $[OH^-] = 0.120$ mol L^{-1}. At 63.2°C the second order rate constant has the value 6.5×10^{-4} L mol^{-1} s^{-1}, and the rate depends upon the concentrations of both reactants.

   $$CH_3CONH_2 + OH^- \rightarrow \quad CH_3CO_2^- + NH_3$$

   a) Write the rate equation.
   b) Calculate the half-life for the reaction at 63.2°C
   c) Calculate the amount of acetamide remaining after 6.5 h.

6. Express the following rate constants in different units.
   a) 6.2×10^9 L mol^{-1} s^{-1} in cm^3 molecule^{-1} s^{-1}
   b) 3.8×10^{-15} cm^3 molecule^{-1} s^{-1} in L mol^{-1} s^{-1}
   c) 2.5×10^5 L mol^{-1} s^{-1} in L mol^{-1} min^{-1}

Section 7.5

1. The gas phase reaction between NO and O_2 is a third order reaction

   $$\text{rate} = k[NO]^2[O_2]$$

   How would the reaction be described in kinetic terms if
   (i) [NO] were held constant during the course of the reaction
   (ii) $[O_2]$ were held constant during the course of the reaction.
   (iii) Initial $p(O_2) = 0.21$ atm, and initial p(NO) = 100 ppmv.

2. The hydrolysis of ethyl acetate by hydroxide ion is a second order reaction (first order in each reactant), having $k = 8.4 \times 10^{-1}$ L mol $^{-1}$ s $^{-1}$ at 54°C.
   a) Calculate the pseudo first order rate constant if $[OH^-]$ is held constant at 0.015 mol L^{-1}.
   b) Calculate the half-life of ethyl acetate under the conditions of part a).
   c) Calculate the amount of ethyl acetate remaining unreacted after 10 minutes at 54°C with $[OH^-]$ = 0.015 mol L^{-1} (constant), and initial [ethyl acetate] = 4.4×10^{-3} mol L^{-1}.

3. The oxidation of CO by the hydroxyl radical is a first order in each reactant, with $k = 2.7 \times 10^{-13}$ cm^3 molec $^{-1}$ s $^{-1}$ at 300 K.
   a) With [OH] constant at 4.5×10^6 molec cm^{-3} write the pseudo-first order rate expression, and evaluate the pseudo-first order rate constant numerically.
   b) What is the rate of the reaction for [CO] = 45 ppmv and [OH] = (i) 4.5×10^6 mole cm $^{-3}$ (ii) 1.5×10^6 mole cm $^{-3}$?
   c) What is the half-life of CO under each of the conditions described in part (b)?
   d) A laboratory experiment is carried out in which CO is oxidized by OH at 300 K. A flow system is used to keep the concentrations of CO and OH in the reaction chamber both at constant values:

      [CO] = 8.7×10^{10} molec cm $^{-3}$. [OH] = 2.1×10^7 molec cm $^{-3}$.

      What is the rate of the reaction, and how does it change with time?

4. The oxidation of Fe^{2+} in acidic solution at constant pH by dissolved O_2 has been found to follow the following rate law:

   $$-d[Fe^{2+}]/dt = k[Fe^{2+}]^2\, p(O_2)$$

   At 35°C and high $[H^+]$, k has the numerical value 3.7×10^{-3}
   a) What are the units of k? (Be careful: check the rate law.)
   b) For $p(O_2)$ held constant at 0.21 atm, what is the apparent kinetic order of the reaction? Give the numerical value and the units of the rate constant under these conditions.
   c) Calculate the half-life of Fe^{2+} (aq) under each of the following conditions:
      (i) $[Fe^{2+}]_o = 1.2 \times 10^{-3}$ mol L^{-1}; $p(O_2)$ = 0.21 atm
      (ii) $[Fe^{2+}]_o = 3.6 \times 10^{-3}$ mol L^{-1}; $p(O_2)$ = 0.21 atm
      (iii) $[Fe^{2+}]_o = 3.6 \times 10^{-3}$ mol L^{-1}; $p(O_2)$ = 0.040 atm

5. The hydrolysis of cane sugar in aqueous solution is catalysed by $[H^+]$ and follows the rate equation:

   $$-d[\text{sugar}]/dt = k[H^+][\text{sugar}]$$

   a) What is the apparent rate equation when $[H^+]$ is held constant at 0.010 mol L^{-1} HCl?

   b) Calculate the length of time needed for 85% of the initial amount of sugar to hydrolyse at 30°C, given that the half-life for hydrolysis under these conditions is 78 h.

Section 7.6

1. Calculate residence times corresponding to the following half-lives:
   a) 6.5×10^3 s
   b) 28 days

2. Calculate half-lives corresponding to the following residence times:
   a) 86 years
   b) 2.3×10^{-2} s
   c) Calculate the rate constants appropriate to parts (a) and (b).

3. Calculate, where possible, residence times appropriate to rate constants of the following values:
   a) 6.1×10^{-4} s^{-1}
   b) 2.6×10^5 L mol^{-1} s^{-1}
   c) 4.7×10^{-14} cm^3 $molec^{-1}$ s^{-1}

4. The hydroxyl radical reacts in the unpolluted atmosphere principally with the natural amounts of CO and CH_4. The reactions are given below, with rate constants at 300 K.

   $$OH + CO \rightarrow CO_2 + H \qquad k = 2.7 \times 10^{-13}\ cm^3\ molec^{-1}\ s^{-1}$$

   $$OH + CH_4 \rightarrow CH_3 + H_2O \qquad k = 6.3 \times 10^{-15}\ cm^3\ molec^{-1}\ s^{-1}$$

   Calculate the residence time of OH under conditions where [CO] = 0.12 ppmv and $[CH_4]$ = 1.7 ppmv.

5. Calculate the residence time of sulfur dioxide in the atmosphere if the following are its only sinks

$$SO_2 + O_2 \xrightarrow{k_1} \text{oxidation to } SO_3$$

$$SO_2 \xrightarrow{k_2} \text{deposition}$$

Take $p(O_2) = 0.21$ atm, $k_1 = 0.065$ atm $^{-1}$ h $^{-1}$ and $k_2 = 0.030$ h $^{-1}$.

Section 7.7

1. Decide whether the following reactions could be elementary reactions based on the kinetic information provided.

   a) $CH_3I + OH^- \rightarrow CH_3OH + I^-$
   rate = $k[CH_3I][OH^-]$

   b) $I^- + OCl^- \rightarrow Cl^- + OI^-$
   rate = $k[I^-][OCl^-][H^+]$

   c) $Cl_2 + CO \rightarrow COCl_2$
   rate = $k[Cl_2]^{3/2}[CO]$

   d) $N_2O_5 \rightarrow 2NO_2 + \frac{1}{2}O_2$
   rate = $k[N_2O_5]$

   e) $2NOCl \rightarrow 2NO + Cl_2$
   rate = $k[NOCl]^2$

   f) $2NO_2Cl \rightarrow 2NO_2 + Cl_2$
   rate = $k[NO_2Cl]$

2. For problem 1(b) above, indicate which of the following reactions could be compatible with the kinetic rate equation:

A: $I^- + OCl^- \rightarrow (IOCl)^{2-}$ slow
$(IOCl)^{2-} + H^+ \rightarrow HOI + Cl^-$ fast
$HOI \rightarrow H^+ + OI^-$ fast

B: $OCl^- + H^+ \rightleftharpoons HOCl$ fast, reversible
$HOCl + I^- \rightarrow (HOClI)^-$ slow
$(HOClI)^- \rightarrow HOI + Cl^-$ fast
$HOI \rightarrow H^+ + OI^-$ fast

C: $OCl^- + H^+ \rightleftharpoons HOCl$ fast, reversible
$HOCl + I^- \rightarrow HOI + Cl^-$ slow
$HOI \rightarrow H^+ + OI^-$ fast

D: $OCl^- + H^+ \rightarrow HOCl$ slow
$HOCl + I^- \rightarrow H^+ + OI^- + Cl^-$ fast

E: $H^+ + OCl^- \rightleftharpoons HOCl$ fast, reversible
$HOCl + I^- \rightarrow ICl + OH^-$ slow
$ICl + OH^- \rightarrow HOI + Cl^-$ fast
$HOI \rightarrow H^+ + OI^-$ fast

3. The reaction of an enzyme (E) with its substrate(s) generally follows the following mechanism:

$E + S \xrightarrow{k_1} ES$ fast

$ES \xrightarrow{k_2} E + S$ fast

$ES \xrightarrow{k_3}$ Products slow

Use the steady state approximation to determine the kinetic rate expression for the loss of the substrate - *i.e.*, -d[S]/dt.

4. The decomposition of H_2O_2 is believed to occur by the following mechanism:

$$H_2O_2 \rightarrow 2OH$$
$$OH + H_2O_2 \rightarrow H_2O + HO_2$$
$$HO_2 + OH \rightarrow H_2O + O_2$$

The rate law is: rate = $k[H_2O_2]$.

Specify the rate limiting step of the reaction.

5. The iodination of acetone C_2H_6O to C_3H_5OI in the presence of base is believed to follow the mechanism below.

$$C_3H_6O + OH^- \rightarrow C_3H_5O^- + H_2O \quad \text{slow}$$
$$C_3H_5O^- + I_2 \rightarrow C_3H_5OI + I^- \quad \text{fast}$$

Give the rate law that is consistent with this mechanism.

6. The reaction below is found to be kinetically of the first order in each of H^+, BrO_3^-, and Br^-.

$$BrO_3^- + 5Br^- + 6H^+ \rightarrow Br_2 + 3H_2O.$$

Could the following mechanism be consistent with the facts above?

$$BrO_3^- + H^+ \rightarrow H_2BrO_3 \quad \text{fast}$$
$$HBrO_3 + H^+ \rightarrow H_2BrO_3^+ \quad \text{fast}$$
$$H_2BrO_3^+ \rightarrow BrO_2^+ + H_2O \quad \text{slow}$$
$$BrO_2^+ + Br^- \rightarrow Br_2O_2 \quad \text{fast}$$
$$Br_2O_2 + Br^- \rightarrow Br_2 + BrO_2^- \quad \text{fast}$$

followed by further (rapid) reactions of BrO_2

Section 7.8

1. The hydrolysis of ethyl acetate using NaOH is a second order reaction having E_{act} = 44 kJ mol $^{-1}$ and A = 6.3 x 10^6 s $^{-1}$. Calculate the value of the rate constant at 30°C, 45°C, and 60°C.

2. The second order reaction between O_3 and NO in the gas phase has been studied at several temperatures. Deduce E_{act} and A from the following data:

Temp, °C	-78	-43	-13	25	96
k_1, L mol $^{-1}$ s $^{-1}$	1.1 x 10^9	3.0 x 10^9	5.4 x 10^9	1.2 x 10^{10}	3.6 x 10^{10}

3. The first order decomposition of N_2O_5 has been studied at several temperatures

Temp, °C	0	25	45
k, s^{-1}	7.9×10^{-7}	3.5×10^{-5}	5.0×10^{-4}

Calculate the expected value of k, and also the reaction half-life, at 75°C.

4. Radioactive decay of radionuclides is independent of temperature. What can you say about the activation energy for this process?

5. Oxygen atoms are very efficient at destroying ozone in the gas phase

$$O + O_3 \rightarrow 2O_2 \quad k = 1.5 \times 10^{-11} \exp(-2218/T) \text{ cm}^3 \text{ molec}^{-1} \text{ s}^{-1}$$

Calculate:

a) the activation energy for this reaction
b) the rate constant for this reaction at 300 K
c) the half life of ozone under conditions where [O] remains constant at 6.4×10^5 molec cm^{-3}.

Section 7.9

1. State the effect of a catalyst on each of the following characteristics of a reaction (all at a fixed temperature)

a) its rate
b) its rate constant
c) its half life
d) its activation energy
e) the enthalpy difference between reactants and products.

2. The reaction $I^- + ClO^- \rightarrow IO^- + Cl^-$ is catalysed by H^+

a) Is H^+ used up in the reaction?
b) Is $[H^+]$ likely to appear in the rate expression?

3. Hydrogen peroxide is decomposed in the presence of H^+ and I^-. The stoichiometric equation and the rate equation are given below.

$$2H_2O_2 \rightarrow 2H_2O + O_2$$

$$\text{rate} = k[H_2O_2][H^+][I^-]$$

Which substances, if any, are catalysts for this reaction?

8 GAS PHASE EQUILIBRIA

8.1 Concept of equilibrium

So far, we have assumed that chemical reactions proceed from reactants to products until all the limiting reactant has been used up. We describe such reactions as going "to completion". In practice, not all reactions proceed to completion; some appear to stop short of complete conversion of reactants into products. One such system which was studied over 80 years ago involves gaseous H_2, I_2, and HI. When H_2 and I_2 are heated to about 400 °C, they combine to form HI.

$$H_2(g) + I_2(g) \rightarrow 2\ HI(g)$$

However, the reaction stops short of completion. Conversely, when HI(g) is heated to the same temperature, there is partial dissociation to H_2 and I_2.

$$2\ HI(g) \rightarrow H_2(g) + I_2(g)$$

Again, this reaction stops short of completion.

When the progress of these reactions is followed as a function of time, graphs similar to Figure 1 are obtained.

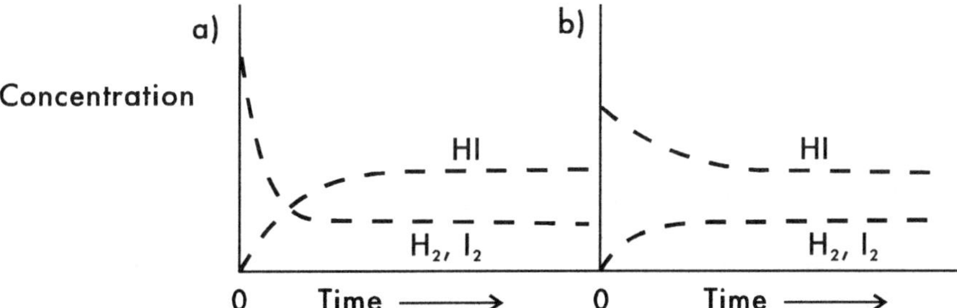

Figure 8.1: Time course of the reactions a) $H_2(g) + I_2(g) \rightarrow 2\ HI(g)$ and b) $2\ HI(g) \rightarrow H_2(g) + I_2(g)$.

Notice that the progress of the reaction reaches an asymptote, after which there is no further change, and also that the same asymptote is reached whether the initial reactants are $H_2 + I_2$, or HI. At the asymptote, **equilibrium** is said to have been reached. Equilibrium is not a static state, however; it is now known that at equilibrium both "forward" and "reverse" reactions are proceeding at equal rates. Thus in the example taken, at equilibrium, HI is being formed from H_2 and I_2 at exactly the same rate as HI is reacting to form H_2 and I_2. This is why the concentrations of substances show no further change once equilibrium has been reached. Because equilibrium is a dynamic process, we use the special equilibrium arrow ($\rightleftharpoons$) to remind us that the reaction proceeds in both directions.

$$H_2(g) + I_2(g) \rightleftharpoons 2\ HI(g)$$

For convenience (only) of discussion in this chapter, we will refer to "reactants" as the substances on the left hand side of an equilibrium as written, and "products" as those substances on the right hand side. We must remember, however, that it would be perfectly possible to reverse the reaction, which would interchange the identities of "reactants" and "products".

At equilibrium, not only are the rates of the "forward" and "reverse" reactions equal, but also the concentrations of reactants and products no longer change with time, as is evident from Figure 1. It would therefore be reasonable to describe the concentrations of reactants and products as having reached a steady state (*cf.* Section 7.7). However, it is incorrect to use the terms "equilibrium" and "steady state" interchangeably. It is quite possible to achieve a steady state different from equilibrium, **provided that energy is continuously supplied from outside in order to maintain the (non-equilibrium) steady state**. Many photochemical reactions in the environment fall into the latter category, with solar radiation acting as the outside energy source. An example is the production of ozone in the stratosphere (Chapter 15). A useful concept is that equilibrium is a special case of a steady state. All reactions at equilibrium are in a steady state, but not all steady states are at equilibrium. For further discussion, see Chapter 15.

8.2 The equilibrium constant

The degree to which a given reaction proceeds to completion is not arbitrary, but can be predicted for a given reaction at a given temperature. The equilibrium state can be defined in terms of an **equilibrium constant**, that is, in terms of the amounts of the original reactants and the products that are present at equilibrium. The equilibrium constant is an experimental quantity, whose units depend on the units chosen to describe the amounts of the reactants and products. The equilibrium constant is a characteristic property of the reaction system, and is therefore constant: that is, independent of the amounts of reactants used. Equilibrium constants vary with temperature however, and so the numerical value of an equilibrium constant must always be accompanied by the temperature at which it is applicable.

K_c is the symbol for the equilibrium constant when the concentrations of reactants and products are expressed in moles per liter (the subscript "c" indicates an equilibrium constant K defined in terms of concentration). Note the use of the capital K for an equilibrium constant, in order to avoid confusion with the lower case italic *k* to denote a rate constant. In dealing with equilibria, we take care to distinguish molar concentrations *at equilibrium* by the use of the square brackets [ ]; stoichiometric concentrations will be denoted by c(X).

Examples:

1. $2\,SO_2(g) + O_2(g) \rightleftharpoons 2\,SO_3(g)$

$$K_c = \frac{[SO_3]^2}{[SO_2]^2[O_2]}$$ *units of* K_c, $(mol\ L^{-1})^{-1}$, *or* $L\ mol^{-1}$

2. $N_2(g) + 3\,H_2(g) \rightleftharpoons 2\,NH_3(g)$

$$K_c = \frac{[NH_3]^2}{[N_2][H_2]^3}$$ *units of* K_c, $(mol\ L^{-1})^{-2}$, *or* $L^2\ mol^{-2}$

3. $H_2(g) + I_2(g) \rightleftharpoons 2\,HI(g)$

$$K_c = \frac{[HI]^2}{[H_2][I_2]}$$ K_c *is dimensionless*

In the general case:

$$a\,A + b\,B \rightleftharpoons c\,C + d\,D$$

$$K_c = \frac{[C]^c[D]^d}{[A]^a[B]^b}$$

K_c is written as the (mathematical) product of all the concentrations of the reaction products, each raised to the power of its coefficient in the chemical equation, divided by the product of all the concentrations of the reactants, each raised to the power of its coefficient in the chemical equation. *It is very important to note that K_c is defined in terms of the chemical equation for the equilibrium reaction.* The consequence of this last statement is that the equilibrium constant changes if the chemical equation is written a different way.

Examples:

$$2\ SO_2(g) + O_2(g) \rightleftharpoons 2\ SO_3(g)$$

$$K_c = \frac{[SO_3]^2}{[SO_2]^2[O_2]} \qquad \text{units of } K_c\text{, } (mol\ L^{-1})^{-1}\text{, or } L\ mol^{-1}$$

(a) *Divide the chemical equation through by 2:*

$$SO_2(g) + \tfrac{1}{2}\ O_2(g) \rightleftharpoons SO_3(g)$$

$$K_c = \frac{[SO_3]}{[SO_2][O_2]^{1/2}} \qquad \text{units of } K_c\text{, } (mol\ L^{-1})^{-1/2}\text{, or } L^{1/2}\ mol^{-1/2}$$

The new value of K_c is the square root of the original K_c .

(b) *Reverse the equation:*

$$2\ SO_3(g) \rightleftharpoons 2\ SO_2(g) + O_2(g)$$

$$K_c = \frac{[SO_2]^2[O_2]}{[SO_3]^2} \qquad \text{units of } K_c\text{, } mol\ L^{-1}$$

The new value of K_c is the reciprocal of the original K_c.

To repeat: K_c is defined in terms of the chemical equation for the equilibrium reaction.

Question: *Can you write an equilibrium constant K_c for the equilibrium reaction between $SO_2(g)$, $O_2(g)$, and $SO_3(g)$?*

Answer: *No, not without information as to how the equation is written. As we saw in the last example, three different forms of K_c were written for this system, depending on how the chemical reaction was formulated.*

So far, we have discussed only K_c , the equilibrium constant defined in terms of molar concentration. Other equilibrium constants would be obtained if the amounts of substances were written in units other than mol L^{-1}, and these alternative equilibrium constants would therefore have different units. For gas phase reactions K_p , the equilibrium constant defined in terms of pressure in atmospheres, is a very useful equilibrium constant because pressure is a natural method of measuring gas phase "concentrations". K_p is defined analogously to K_c , except that every concentration term is replaced by a pressure term. Taking the previous examples:

1. $2\ SO_2(g) + O_2(g) \rightleftharpoons 2\ SO_3(g)$

$$K_p = \frac{\{p(SO_3)\}^2}{\{p(SO_2)\}^2 p(O_2)}$$ units of K_p, atm $^{-1}$

2. $N_2(g) + 3\ H_2(g) \rightleftharpoons 2\ NH_3(g)$

$$K_p = \frac{\{p(NH_3)\}^2}{p(N_2)\{p(H_2)\}^3}$$ units of K_p, atm $^{-2}$

3. $H_2(g) + I_2(g) \rightleftharpoons 2\ HI(g)$

$$K_p = \frac{\{p(HI)\}^2}{p(H_2)p(I_2)}$$ K_p is dimensionless

The relationship between K_p and K_c may be deduced from the ideal gas equation. Recall the correlation between pressure and molar concentration.

$n/V = P/RT$ *or* $P = (n/V)RT$

Therefore, to change K_c into K_p, multiply every concentration term by (RT), remembering to raise (RT) to the same power as the concentration term appears in the K_c expression. If concentration is in mol L^{-1}, and pressure is in atm, remember to use R = 0.0821 L atm mol $^{-1}$ K $^{-1}$, and T in kelvins.

Example: *The equilibrium below has $K_c = 6.4 \times 10^5$ mol L^{-1} at 500 K. Calculate K_p for the same reaction at 500 K.*

$$2\ NO(g) + O_2(g) \rightleftharpoons 2\ NO_2(g)$$

Answer:

$$K_c = \frac{[NO_2]^2}{[NO]^2[O_2]} \text{ mol L}^{-1}$$

$$K_p = \frac{p(NO_2)^2}{p(NO)^2 p(O_2)} \text{ atm}^{-1}$$

$$= \frac{[NO_2]^2(RT)^2}{[NO]^2(RT)^2[O_2](RT)} \text{ atm}^{-1}$$

$$= K_c/RT$$

$$= \frac{6.4 \times 10^5 \text{ mol L}^{-1}}{0.0821 \text{ L atm mol}^{-1} \text{ K}^{-1} \times 500 \text{ K}}$$

$$= 1.6 \times 10^4 \text{ atm}^{-1}$$

8.3 Equilibria involving pure solids and liquids in addition to gases

In all the examples discussed so far, all the reactants and products have been gases. However, many important equilibria involve substances in other states of matter. In this section we consider equilibria in which one or more of the reactants and products is a pure solid or a pure liquid. We defer to Chapters 11 and 12 consideration of substances present in solution.

The evaporation of water is a simple example of an equilibrium involving a pure liquid.

$$H_2O(\ell) \rightleftharpoons H_2O(g)$$

K_p for this process is given as $K_p = p(H_2O)$, and no mention is made about $H_2O(\ell)$ - it is omitted from the equilibrium constant expression. We may formulate the following general rule.

Rule: *When an equilibrium involves one or more pure solids or liquids in addition to gaseous species, the concentration (or pressure) of any solid or liquid is omitted from the equilibrium constant expression.*

The decision to omit the concentrations/pressures of pure substances when defining K_p or K_c may be justified on the grounds that the pure solid or liquid is not the substance which is present in the gas phase. A more formal discussion of this issue will be presented in Chapter 14.

Example: *Write equilibrium constants K_p for the following physical processes, assuming each time that 1 mol of substance is involved in the reaction. (i) sublimation of "dry ice" (CO_2); (ii) vaporization of liquid nitrogen; (iii) condensation of steam to liquid water.*

Answers: *(i)* *Equation:* $CO_2(s) \rightleftharpoons CO_2(g)$

$$K_p = p(CO_2,g)$$

(ii) *Equation:* $N_2(\ell) \rightleftharpoons N_2(g)$

$$K_p = p(N_2,g)$$

(iii) *Equation:* $H_2O(g) \rightleftharpoons H_2O(\ell)$

$$K_p = 1/p(H_2O(g)$$

We return to the example previously discussed.

$$H_2O(\ell) \rightleftharpoons H_2O(g)$$

As stated already, $K_p = p(H_2O)$; in other words, K_p is equal to the vapour pressure of water. This value of $p(H_2O)$ is also known as the equilibrium vapour pressure of water, or the saturated vapour pressure of water (Chapter 6). It corresponds to a relative humidity of 100% at the temperature of interest. At a lower relative humidity, $p(H_2O)$ is less than the equilibrium value.

The following reactions are examples of commercially important gas phase equilibria involving a pure solid or liquid.

Decomposition of limestone

$$CaCO_3(s) \rightleftharpoons CaO(s) + CO_2(g) \qquad K_p = p(CO_2)$$

This reaction was encountered in Chapter 4. It is involved in the production of lime, and also in the manufacture of glass and cement. The reaction as practised commercially is made to go virtually to completion, by roasting the limestone ($CaCO_3$) in open kilns. This allows the escape of the CO_2 as it forms, and so its concentration (pressure) does not build up inside the kiln. If instead the reaction were done in a closed container, the reaction would stop (*i.e.*, come to equilibrium) as soon as $p(CO_2)$ inside the container became equal to K_p.

Purification of nickel

Newly manufactured nickel contains both metallic and non-metallic impurities. One method of purification involves heating the crude product with carbon monoxide, whereupon the volatile nickel tetracarbonyl is formed.

$$Ni(s) + 4\ CO(g) \rightleftharpoons Ni(CO)_4(g)$$

The $Ni(CO)_4(g)$ is pumped away, leaving solid impurities behind. Under conditions of low temperature and high pressures of CO, excellent conversion of Ni to $Ni(CO)_4$ can be achieved. However, the product is easily decomposed back to nickel at higher temperatures. Since the impurities were left behind at the first stage of the process, the redeposited nickel is in the purified form. The CO is then recycled.

$$Ni(CO)_4(g) \rightleftharpoons Ni(s) + 4\ CO(g)$$

Gaseous fuels from coal

Coal is the chief energy resource of many industrializing countries, as indeed it was in Europe and North America before oil and natural gas became readily available in the 20th century. One reaction by which coal may be converted to a gaseous material is the "producer gas reaction", in which the carbon of coal is partially oxidized to carbon monoxide by reaction with a limited supply of air.

$$2\ C(s) + O_2(g) \rightleftharpoons 2\ CO(g)$$

This affords a relatively low grade fuel, because of the presence of 4 mol N_2 for every 1 mol of O_2 used. Consequently, producer gas has a relatively low energy output per m^3 of fuel. The production of CO from carbon is exothermic, and so no heat needs to be applied to sustain the reaction.

Another coal-derived product is "water gas", also called "synthesis gas", which is formed by passing steam over hot coal. It provides the simplest method of obtaining hydrogen from coal[1].

$$C(s) + H_2O(g) \rightleftharpoons H_2(g) + CO(g)$$

This reaction is endothermic, and so it is not self-sustaining in the absence of heat. One approach is to alternate passage of air (exothermic) and steam (endothermic) over coal, affording a mixture of CO, H_2 and nitrogen.

Synthesis gas is the starting point for the production of synthetic gasoline from coal. Over certain metallic oxide catalysts, the H_2 reduces the CO to methane and also to mixtures of more complex hydrocarbons, which can be used as liquid fuels such as synthetic gasoline.

$$m\ CO + n\ H_2 \rightarrow CH_4 + C_2H_6 + C_3H_8 + \ldots\ldots + H_2O$$

This area of chemistry was developed in Germany in the 1930s, and successfully powered the German military in World War II. The subsequent availability of cheap oil led to a loss of interest in coal-to-gasoline chemistry in most countries (apart from a brief period in the West at the time of the 1970s oil crisis). In South Africa however, this technology is used on a huge scale, since South Africa has abundant coal reserves but no oil.

---

[1] In a hydrocarbon-based economy (oil and natural gas), hydrogen is usually obtained from natural gas at elevated temperature, rather than from synthesis gas. (Chapter 4)

$$CH_4(g) + 2\ H_2O(g) \rightleftharpoons 4\ H_2(g) + CO_2(g)$$

8.4 Calculations involving equilibrium constants

The equilibrium constant is a useful quantity, in that it allows us to calculate how much of a certain product will have formed when a reaction proceeds to equilibrium. Conversely, the equilibrium constant is measured experimentally by determining the amounts of products and unreacted reactants that are present when a reaction is allowed to proceed to equilibrium.

Example: *In a determination of the equilibrium constant for the Haber process at 400 °C, the following gas phase concentrations were found:* N_2, *0.10 mol* L^{-1}; H_2, *0.10 mol* L^{-1}; NH_3, 7.1×10^{-3} *mol* L^{-1}. *Calculate* K_c *for the reaction:*

$$N_2(g) + 3\ H_2(g) \rightleftharpoons 2\ NH_3(g)$$

Answer: *Write out the expression defining* K_c*:*

$$K_c = \frac{[NH_3]^2}{[N_2][H_2]^3}\ (\text{mol L}^{-1})^{-2}$$

Then substitute the appropriate values into the equilibrium constant expression.

$$K_c = \frac{(7.1 \times 10^{-3}\ \text{mol L}^{-1})^2}{(0.10\ \text{mol L}^{-1})(0.10\ \text{mol L}^{-1})^3} = 0.50\ (\text{mol L}^{-1})^2$$

The calculation of how much product (for example) is formed at equilibrium, given the starting concentrations of reactants and the equilibrium constant, can be more complicated. Fortunately, there is a step-by-step method for approaching such problems.

Steps for solving equilibrium calculations:

1. Write out the chemical equation which defines K_p or K_c, whichever is appropriate.

2. Set up a table of <u>initial</u> and <u>equilibrium</u> concentrations (or pressures) for all reactants and products. It is generally necessary to define one of the concentrations or pressures as *x*, in which case, all other concentrations or pressures should be defined in terms of the same *x*.

3. Write out the equilibrium constant expression K_p or K_c.

4. Substitute the equilibrium concentrations or pressures from your table into the K_p or K_c expression, and solve for *x*.

Example: *At 500 K the equilibrium below has $K_p = 3 \times 10^{-18}$ (no units).*

$$N_2(g) + O_2(g) \rightleftharpoons 2\ NO(g)$$

Calculate the equilibrium pressure of NO in a container containing initially 1 atm each of N_2 and O_2 at 500 K. Express your answer in atm and in ppbv.

Answer: *We already have the chemical equation. Set up a table of pressures (all values in atm). Define the equilibrium pressure of NO (which is required as the answer) as 2x. Note: it is very helpful to make the coefficients of x match those of the chemical equation. Hence define final NO as 2x not x.*

	N_2	O_2	NO
initial p	1.0	1.0	zero
change in p	- x	- x	+ 2x
equilibrium p	(1.0-x)	(1.0-x)	2x

The points below explain how the table was filled out.

1. *Initial NO = 0; final NO = 2x.*
2. *Change in NO = + 2x (goes from zero to 2x).*
3. *From the stoichiometry of the reaction, the changes in N_2 and O_2 (both negative) are each - x. This is why it is better to call final NO "2x"; if you called it x, the changes in N_2 and O_2 would each be -½x.*
4. *The values of $p(N_2)$ and $p(O_2)$ at equilibrium are obtained by adding together the "initial" and "change" values.*

Now write out the expression for K_p and substitute in the values.

$$K_p = \frac{p(NO)^2}{p(N_2)(pO_2)}$$

$$3 \times 10^{-18} = x^2/(1.0\text{-}x)^2$$

This equation is particularly easy to solve, because the right hand side is a perfect square.

$$1.7 \times 10^{-9} = x/(1.0 - x)$$

Hence x = 1.7×10^{-9}

Looking at the table of values, we note that p(NO) = 2x , hence the desired answer is p(NO) = 3.4×10^{-9} atm (or, 3×10^{-9} atm to one significant figure).

To convert to ppbv, note that p(total) = 2.0 atm

$$p(NO) = 3.4 \times 10^{-9} \text{ atm} \times (10^9 \text{ ppbv}/2 \text{ atm}) = 2 \text{ ppbv}.$$

As already noted, this example was chosen to be easy with respect to solving the final equation for *x*. More complex examples are considered in section 8.6. In terms of the result obtained from the example just given, note that even at 500 K (227 °C), the equilibrium concentration of NO is very small. This is relevant to the issue of nitrogen oxide emissions from automobiles, see section 8.8 and also Chapter 9.

8.5 The method of successive approximations

In Section 8.4 we discussed how to do equilibrium calculations. The example chosen, however, was a particularly easy one in terms of the final calculation. Much more often, the algebraic equation to be solved involves a quadratic, cubic, or higher order equation. Not only are these equations tedious to solve, but they may not have an analytical solution at all.

The method of approximations offers a way out of this difficulty - in fact, it is also one of the most powerful methods for solving equations using a computer. Let's first take examples in which we restrict the discussion to solving the equation, without considering any chemistry.

Example 1: *Solve the equation*

$$2.7 \times 10^{-5} = (x)^2/(0.13 - x)$$

This particular type of example will be met repeatedly in Chapter 11.

Answer: *Make the* ***approximation*** *that x << 0.13. At this stage, we do not know whether x actually is << 0.13; we are just "floating a trial balloon". Then solve the equation on the basis of the assumption, and then check to see whether the assumption was correct. If it was, then the calculated value of x is a correct solution to the equation.*

$$2.7 \times 10^{-5} \approx (x)^2/(0.13)$$

$$x^2 = (0.13)(2.7 \times 10^{-5}) = 3.5 \times 10^{-6}$$

$$x = 1.9 \times 10^{-3}$$

Check the assumption: 1.9×10^{-3} is only about 0.01 (1%) as large as 0.13, therefore it was correct to say that x << 0.13. Therefore this value of x is a valid solution to the equation.

The general rule is that it is usually justified to neglect x when x is no more than 5% of the value with which it is being compared. This is the same in practice as neglecting x when the magnitude of x is so small that it does not change the last significant digit in the quantity with which x is being compared. The justification for the "5% rule" is that equilibrium constants are rarely known to better than 2 significant figure precision, and under these circumstances the 5% rule and the significant digit concept amount to the same thing.

You may wonder why it was acceptable to neglect x in the expression $(0.13 - x)$, but not acceptable to neglect the x^2 term. The answer is that since the x^2 was not added to or subtracted from another quantity, to neglect it would be the same as saying $x \ll$ zero, which does not make sense.

Rule: *Only make as your first approximation $(A \pm x) \approx (A)$.*

What if x were greater than 5% of the quantity with which it was being compared? In that case, one can make use of **successive approximations**. Consider this example, which is very similar to the previous one.

Example: *Solve the equation*

$$2.7 \times 10^{-3} = (x)^2/(0.13 - x)$$

Answer: *Proceed as before, assuming* $x << 0.13$

$$2.7 \times 10^{-3} \approx (x)^2/(0.13)$$

$$x^2 = (0.13)(2.7 \times 10^{-3}) = 3.5 \times 10^{-4}$$

$$x = 1.9 \times 10^{-2}$$

This time, x is found to be 1.9×10^{-2}*, which is 15% of 0.13* $\{100 \times (1.9 \times 10^{-2})/0.13\}$*. This is not negligible. It is both > 5% of 0.13, and changes the last significant digit of the number with which x is compared:* $(0.13 - x) \rightarrow 0.11$.

Call x_1 *the new value of x. Now approximate again, but instead of saying* $x << 0.13$ *(which we now know is not true), approximate by writing:*

$$(0.13 - x) \approx (0.13 - x_1)$$

$$2.7 \times 10^{-3} \approx (x)^2/(0.13 - 0.019) = x^2/0.11$$

$$x^2 = (0.11)(2.7 \times 10^{-3}) = 3.0 \times 10^{-4}$$

$$x = 1.7 \times 10^{-2}$$

If necessary, call this new value x_2 *and approximate again.*

$$(0.13 - x) \approx (0.13 - x_2)$$

$$2.7 \times 10^{-3} \approx (x)^2/(0.13 - 0.017) = x^2/0.11$$

$$x^2 = (0.11)(2.7 \times 10^{-3}) = 3.0 \times 10^{-4}$$

$$x = 1.7 \times 10^{-2}$$

Since x did not change upon a further iteration, this must be the correct solution.

At first sight, this looks like a lot of work. In practice, several iterations can be carried out much more quickly using a pocket calculator than a complete solution of even a quadratic (let alone a higher order) equation.

8.6 Some examples of equilibrium problems involving the approximation method

***Example 1**: At 250 °C, the reaction*

$$Ni(s) + 4\,CO(g) \rightleftharpoons Ni(CO)_4(g) \text{ has } K_p = 1.7 \times 10^{-4}\ atm^{-3}.$$

Calculate the partial pressure of nickel tetracarbonyl when Ni(s) and 20 atm CO(g) are allowed to come to equilibrium in a closed container at 250 °C.

***Answer**: Set up a table of pressures of all substances, remembering that p(Ni) is irrelevant to working the problem, since Ni is a solid.*

	p(Ni)	p(CO)	$Ni(CO)_4$
initial pressure	n.a.	20	zero
change in pressure	n.a.	-4x	+x
equilibrium pressure	n.a.	20-4x	x

Write out the equilibrium constant expression, and substitute the values.

$$K_p = \frac{p(Ni(CO)_4)}{\{p(CO)\}^4}$$

$$1.7 \times 10^{-4} = x/(20-4x)^4$$

Approximate: $(20 - 4x) \sim 20$

$$1.7 \times 10^{-4} \sim x/20$$

$$x = 3.4 \times 10^{-2}$$

Check assumption: $4x = 0.14$, *which is* $<< 20$

Hence $x = p(Ni(CO)_4) = 3.4 \times 10^{-2}$ *atm*

Example 2: *Nitrosyl chloride (NOCl) is formed from nitric oxide and chlorine.*

$$2\ NO(g) + Cl_2(g) \rightleftharpoons 2\ NOCl(g) \qquad K_p = 0.080\ atm^{-1} \text{ at } 25\ °C$$

Calculate p(NOCl) when 1.0 atm NO and 0.50 atm Cl_2 are react together until equilibrium is reached.

Answer: *We already have the chemical equation. Set up a table of pressures (all values in atm). Define the equilibrium pressure of NOCl (which is required as the answer) in terms of x.*

	NO	Cl_2	NOCl
initial p	1.0	0.50	zero
change in p	- 2x	-x	+ 2x
equilibrium p	(1.0 - 2x)	(0.50 - x)	2x

Let us review the order in which the table was filled out.

1. *Initial NOCl = 0; final NOCl = 2x, making the coefficients of x match those of the chemical equation.*
2. *Change in NOCl = + 2x (goes from zero to 2x).*
3. *From the stoichiometry of the reaction, the changes in NO and Cl_2 (both negative) are - 2x and - x respectively.*
4. *The values of p(NO) and p(Cl_2) at equilibrium are obtained by adding together the "initial" and "change" values.*

Now write out the expression for K_p and substitute in the values.

$$K_p = \frac{p(NOCl)^2}{p(NO)^2 p(Cl_2)}$$

$$= 4x^2/(1.0 - 2x)^2(0.5-x)$$

Approximate: $x << 0.5;\ 2x << 1.0$

$$K_p \sim 4x^2/(1.0)^2(0.5)$$

$$0.080 = 8x^2$$

$$x = 0.10$$

In this case, x is 20% of the value 0.5, which is not negligible, and a second approximation must be made.

$$K_p = 4x^2/(1.0 - 2x)^2(0.5 - x)$$

Approximate:

$$K_p = 4x^2/(1.0 - 2x_1)^2(0.5 - x_1)$$

$$K_p \sim 4x^2/(1.0 - 0.2)^2(0.5 - 0.1) = 4x^2/(0.8)^2(0.4)$$

$$x^2 = (0.080)(0.8)^2(0.4)/4 = 5.1 \times 10^{-3}$$

$$x = 0.072$$

Since p(NOCl) = 2x (see table of values), the required answer is p(NOCl) = 0.14 atm at equilibrium.

8.7 Systems not at equilibrium

The work of the French chemist Le Chatelier guides our qualitative understanding in this area.

Definition

Le Chatelier's Principle: *When a system, initially at equilibrium, is subjected to a change which displaces it from equilibrium, the reaction proceeds back towards equilibrium by reacting in such a way as to oppose the initial change.*

xample: *At 400 °C, the gases N_2 (0.10 mol L^{-1}), H_2 (0.10 mol L^{-1}), and NH_3 (7.1 x 10^{-3} mol L^{-1}) re at equilibrium. What happens (qualitatively) when $p(H_2)$ is suddenly raised to 0.5 mol L^{-1}?*

nswer: *The chemical equation is:*

$$N_2(g) + 3\,H_2(g) \rightleftharpoons 2\,NH_3(g)$$

he system was initially at equilibrium, then more reactant was added to the N_2/H_2 side of the eaction. The "stress" on the reaction is relieved by formation of more NH_3.

A similar result (more NH_3 formed) would have been obtained if some of the ammonia were removed from the reaction at equilibrium. Again, more H_2 would have combined with N_2 to relieve the "stress" on the system caused by removing NH_3.

The following possibilities summarize the way in which Le Chatelier's principle operates.

1. Add or remove one of the reactive substances. This is the situation covered in the last example. The equilibrium shifts so as to reduce the concentration of the substance added, or increase the concentration of the substance that was removed.

2. Change the total pressure at which the process operates. Again, let us use the $N_2/H_2/NH_3$ reaction as an example. As the reaction was written in the last example, 4 mol of gas (1 N_2 + 3 H_2) react to form 2 mol NH_3. Therefore, running the reaction in the forward direction would reduce the pressure, and so the forward reaction is favoured by raising the total pressure in the reaction vessel.

The following commercial reactions all involve fewer moles of product gases than reactant gases, and so all operate most economically at high pressure.

$$N_2(g) + 3\ H_2(g) \rightleftharpoons 2\ NH_3(g)$$

$$2\ SO_2(g) + O_2(g) \rightleftharpoons 2\ SO_3(g)$$

$$Ni(s) + 4\ CO(g) \rightleftharpoons Ni(CO)_4(g)$$

Conversely, the following reactions all have more moles of gases on the "products" side of the reaction as written, and operate best at low pressures (usually 1 atm).

$$CaCO_3(s) \rightleftharpoons CaO(s) + CO_2(g)$$

$$2\ C(s) + O_2(g) \rightleftharpoons 2\ CO(g)$$

$$C(s) + H_2O(g) \rightleftharpoons H_2(g) + CO(g)$$

3. **Change the temperature**. So far, we have not considered the influence of energetics on equilibria, apart from noting that an equilibrium constant will only remain constant provided the temperature does not change. However, in the Le Chatelier's sense, temperature change is a stress on an equilibrium. Raising the temperature disfavours the products when the reaction is exothermic (produces heat, Chapter 5), while it favours the products when the reaction is endothermic.

Rule: *In order to predict the effect of temperature change upon an equilibrium it is first necessary to know* $\Delta H°$ *reaction.*

For the commercially important reactions just discussed, the effects of temperature are given below. The first group of reactions is endothermic and hence favoured in the forward direction at high temperatures.

Reaction	$\Delta H°$
$CaCO_3(s) \rightleftharpoons CaO(s) + CO_2(g)$	$\Delta H° = 178\ kJ\ mol^{-1}$
$C(s) + H_2O(g) \rightleftharpoons H_2(g) + CO(g)$	$\Delta H° = 131\ kJ\ mol^{-1}$

These reactions should be run at as high a temperature as consistent with economic considerations *i.e.*, fuel usage.

The second group of reactions is shown below.

$N_2(g) + 3\ H_2(g) \rightleftharpoons 2\ NH_3(g)$	$\Delta H° = -92$ kJ mol $^{-1}$ (Note a)
$2\ SO_2(g) + O_2(g) \rightleftharpoons 2\ SO_3(g)$	$\Delta H° = -199$ kJ mol $^{-1}$
$Ni(s) + 4\ CO(g) \rightleftharpoons Ni(CO)_4(g)$	$\Delta H° = -161$ kJ mol $^{-1}$
$2\ C(s) + O_2(g) \rightleftharpoons 2\ CO(g)$	$\Delta H° = -221$ kJ mol $^{-1}$

a In thermochemical equations, remember that "per mole" means per mole of equation as written. In the case of ammonia formation, for example, the heat evolution is 92 kJ per **2** mol of NH_3 formed.

All the reactions in the second group are exothermic, and therefore the "products" side of the equilibrium is favoured by running the reaction at as low a temperature as possible. In practice, the requirement for low temperature conflicts with the rate at which equilibrium can be attained (recall that reactions proceed faster as the temperature is raised: Chapter 7). Practical operating conditions for exothermic equilibria are therefore a compromise between low temperature (slow reaction, but highly favoured equilibrium), and high temperature (fast attainment of equilibrium, but less product present in the equilibrium mixture).

The following changes to the reaction conditions are not relevant to Le Chatelier's Principle.

- **Increase the pressure by the addition of an inert gas**. By "inert" is meant a gas which is neither a reactant nor a product of the reaction; nor does it react itself with any of the reactants or products. As an example, consider operating the Haber process at a total pressure of reactants and products of 300 atm, and then introducing 100 atm of argon as an inert gas, for a total pressure of 400 atm. *There is no effect upon the position of equilibrium.* The reason is seen most clearly by reference to the definition of K_p (or K_c).

$$K_p = \frac{\{p(NH_3)\}^2}{p(N_2)\{p(H_2)\}^3}$$

Since Ar does not appear in the equilibrium constant expression, its addition has no effect upon the position of equilibrium.

Note by contrast that the same result would not be obtained if the gas introduced reacted with one of the components of the equilibrium. For example, if oxygen were added to the $N_2/H_2/NH_3$ system, it would react with hydrogen.

This would change the position of equilibrium indirectly, by removing one of the reactive components (H_2) from the system.

- **Addition of a catalyst.** The key property of a catalyst is that it changes the rate of a chemical reaction, but is chemically unchanged as a result of the reaction. Almost all catalysts increase reaction rates. As we saw in Section 7.9, the catalyst provides a new reaction pathway, having a lower activation energy than the uncatalysed reaction. The catalyst increases the rate of both forward and reverse reactions. *Since the catalyst does not appear in the equilibrium constant expression, the position of equilibrium is unchanged.* The effect of catalyst on the Haber process is shown in Figure 2. Strictly, Figure 2 is an oversimplification, because it implies that the catalysed and uncatalysed reactions are both elementary processes. This could certainly not be possible, at least for the catalysed reaction, where there has to be a mechanistic role for the catalyst (compare Figure 7.7).

We shall see in Chapter 14 that the position of equilibrium is determined by the "free energy" difference between reactants and products; the presence of a catalyst affects the speed at which equilibrium is reached, but not the composition of the equilibrium mixture.

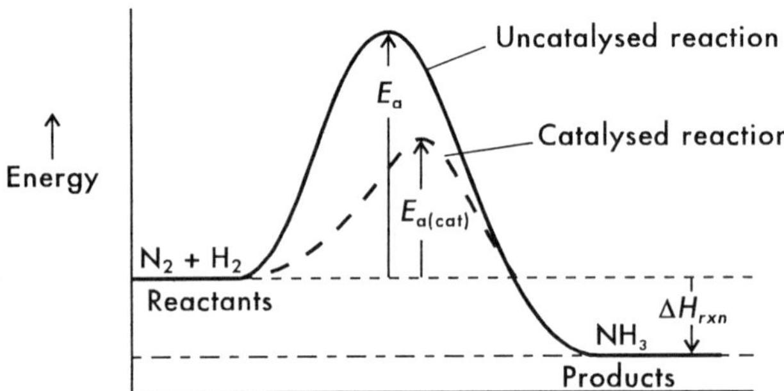

Figure 8.2: Schematic of the effect of catalyst on the energetics of the Haber process.

8.8 Practical applications of Le Chatelier's Principle

We now summarize the previous section by reviewing the conditions under which some commercially important reactions are carried out.

Haber Process

$$N_2(g) + 3\,H_2(g) \rightleftharpoons 2\,NH_3(g) \qquad \Delta H^\circ = -92 \text{ kJ mol}^{-1}$$

The reaction is exothermic and involves a change in the number of moles of gas from 4 to 2, so ammonia formation is favoured by low temperature and high pressure. This is shown by the data in Table 8.1. In order to run the reaction economically, the formation of ammonia must be rapid. The compromise conditions used commercially are: very high pressure (300 atm); moderately high temperature (400-500 °C); and use of a Co_2O_3 + Fe_2O_3 catalyst to speed the attainment of equilibrium. As noted in Chapter 4, the U.S. produces 15 million tonnes of ammonia annually by the Haber process. Canadian production is about 3.5 million tonnes. Most of the Canadian ammonia plants are situated in Alberta, where natural gas is cheap (natural gas is the major raw material, needed for the production of H_2).

Table 8.1: Percentage conversion of a 1: 3 molar ratio of N_2: H_2 to ammonia under different conditions.

	% NH_3 at equilibrium at a total pressure of:			
Temp., °C	10 atm	50 atm	300 atm	1000 atm
200	51	74	90	98
300	15	39	71	93
400	3.9	15	47	79
500	1.2	5.6	26	57
600	0.5	2.3	14	31
700	0.2	1.1	7.3	13

Oxidation of sulfur dioxide

$$2\ SO_2(g) + O_2(g) \rightleftharpoons 2\ SO_3(g) \qquad \Delta H^\circ = -199\ \text{kJ mol}^{-1}$$

The situation is analogous to the Haber process: exothermic reaction; fewer moles of product than reactants. The reaction is carried out at moderate temperature (450-550 °C, a compromise between optimum K_p and convenient reaction rate) and moderate pressure (~ 20 atm), catalysed by $V_2O_5(s)$. Although the conversion of SO_2 into SO_3 is highly favoured energetically and hence has a very favourable equilibrium constant, SO_2 is indefinitely stable at room temperature in the presence of atmospheric oxygen. Oxidation of SO_2 occurs only at elevated temperature and in the presence of catalyst. It does not take place at room temperature because the activation energy is too high.

Purification of nickel

$$Ni(s) + 4\ CO(g) \rightleftharpoons Ni(CO)_4(g) \qquad \Delta H° = -161\ kJ\ mol^{-1}$$

Again, this is an exothermic reaction involving reduction in the number of moles of gases. The original technology operated at 50 °C, taking advantage of the favourable equilibrium at low temperature. Low pressure operation was practised in the early technology, because of the difficulty of preventing leakage of the highly toxic CO and $Ni(CO)_4$. Improvements in chemical engineering now permit Inco Ltd. to operate at high pressure (70 atm) and higher temperature (170 °C). The higher temperature allows more rapid equilibration and hence greater throughput in the plant. In order to protect the safety of workers at the facility, the plant is equipped with automatic sensors which sound alarm bells for evacuation of the plant should the concentration of $Ni(CO)_4$ in the plant ever exceed 5 ppbv.

The purified nickel is recovered at high temperature (≈ 300 °C) and low pressure (1 atm), to favour the endothermic decomposition of 1 mol $Ni(CO)_4$ into 4 mol CO.

Decomposition of limestone

$$CaCO_3(s) \rightleftharpoons CaO(s) + CO_2(g) \qquad \Delta H° = 178\ kJ\ mol^{-1}$$

This endothermic reaction is operated at temperatures in excess of 800 °C and at atmospheric pressure. In the atmosphere $p(CO_2)$ is only ~ 3×10^{-4} atm, and so the escape of CO_2 as it forms continuously removes this product from the reaction, thereby driving the process to completion. This is important, because at 800 °C, K_p for this reaction is only 0.5 atm. If the CO_2 could not escape, the reaction would come to equilibrium before any significant fraction of the limestone had decomposed.

Combination of nitrogen and oxygen

$$N_2(g) + O_2(g) \rightleftharpoons 2\ NO(g) \qquad \Delta H° = 181\ kJ\ mol^{-1}$$

This reaction was practised commercially in the early years of this century as a means of fixing nitrogen (Chapter 4), until it was rendered obsolete by the Haber process. Since the numbers of moles of reactant and product gases are equal, the composition at equilibrium is indifferent to pressure. The reaction is highly endothermic; K_p increases with temperature, but only reaches the rather small value of 0.1 even at 2000 °C, the temperature of the former commercial process.

The importance of this reaction today is that small amounts of NO are formed whenever air is heated. It was mentioned in Chapter 6 that this reaction is responsible for natural nitrogen fixation during thunderstorm activity. Nitrogen is also fixed, unintentionally, whenever combustion occurs - for example, in automobile exhaust, and in the emissions of blast furnaces and coal burning electricity generating stations. Even when these gases cool down, the nitrogen remains fixed. This is because at ambient temperature the rate of reversion of NO into N_2 and O_2 is imperceptibly slow, even though the equilibrium constant for the reverse reaction is highly favourable. The NO formed at high temperature is thus said to be "frozen in". As a result, there is frequently much more NO in urban air than in unpolluted rural air. This creates a problem, because nitrogen oxides are the precursors of "photochemical smog" (Chapter 9).

8.9 Calculations on systems not at equilibrium: the reaction quotient

Suppose at 400 °C, a vessel contained 0.88 mol L^{-1} $N_2(g)$, 0.40 mol L^{-1} $H_2(g)$ and 0.47 mol L^{-1} $NH_3(g)$. How could we determine whether this mixture was at equilibrium, and if it were not, in which direction would the reaction proceed to reach equilibrium?

$$N_2(g) + 3\ H_2(g) \rightleftharpoons 2\ NH_3(g) \qquad K_c = 0.50\ (\text{mol L}^{-1})^{-2} \text{ at } 400\ °\text{C}$$

The way to approach this question is to set up a **reaction quotient**. The reaction quotient has the same form as the equilibrium constant, but the values inserted into the expression are the actual concentrations (Q_c) or pressures (Q_p), which are not necessarily the same as the equilibrium values. In this instance we are interested in Q_c, to make the comparison with K_c.

$$Q_c = \frac{c(NH_3)^2}{c(N_2)c(H_2)^3}$$

Note the use of $c(NH_3)$ etc rather than the square brackets [ ]. The latter are used only for equilibrium concentrations.

We now substitute the experimental values into the Q_c expression.

$$Q_c = \frac{c(NH_3)^2}{c(N_2)c(H_2)^3} = \frac{(0.47)^2}{(0.88)(0.40)^3} = 3.9$$

The value of Q_c was 3.9 $(\text{mol L}^{-1})^{-2}$, compared with $K_c = 0.50\ (\text{mol L}^{-1})^{-2}$. The following conclusions may be drawn.

1. Since $Q_c \neq K_c$, the reaction was not at equilibrium.

2. Since $Q_c > K_c$, we conclude that equilibrium would be approached if Q_c became smaller. This would require the numerator of the Q_c expression to decrease and/or the denominator to increase. These conditions are met by the reaction proceeding from right to left as written, reducing $c(NH_3)$ and increasing both $c(N_2)$ and $c(H_2)$.

This example allows us to draw general conclusions about the consequence of the various possible ratios Q_c:K_c (or equivalently, Q_p:K_p).

- $Q_c > K_c$ (the example just described). Equilibrium is approached if the reaction as written proceeds in the reverse (right to left) direction.
- $Q_c < K_c$ Equilibrium is approached if the reaction as written proceeds in the forward (left to right) direction.
- $Q_c = K_c$ The system is already at equilibrium, so there is no net change.

The Q_c /Q_p concept is completely consistent with the predictions of Le Chatelier's Principle.

Example: *A mixture of SO_2 (0.10 atm), O_2 (0.25 atm) and SO_3 (1.2 atm) is at equilibrium at 450 °C. Predict the effect of adding 0.20 atm SO_2.*

Answer:

Le Chatelier approach:

$$SO_2(g) + \tfrac{1}{2}\, O_2(g) \rightleftharpoons SO_3(g)$$

Addition of SO_2, a reactant, to a mixture already at equilibrium drives the reaction to the right.

Q_p approach:

Work out K_p:

$$K_p = \frac{p(SO_3)}{p(SO_2)p(O_2)^{1/2}} = \frac{(1.2)}{(0.10)(0.25)^{1/2}} = 24 \text{ atm}^{-1/2}$$

$$Q_p = \frac{p^\circ(SO_3)}{p^\circ(SO_2)p^\circ(O_2)^{1/2}} = \frac{(1.2)}{(0.30)(0.25)^{1/2}} = 8.0 \text{ atm}^{-1/2}$$

Since $Q_p < K_p$ the reaction proceeds in the forward direction.

8.10 Multiple equilibria

In Chapter 5, we encountered Hess' Law, which was useful in allowing the calculation of $\Delta H°$ for a certain reaction, given enthalpy data for other reactions involving combinations of the same reactants and products. A similar rule applies to equilibria, only this time addition of two chemical equations requires the corresponding equilibrium constants to be multiplied together, not added together.

Example: *Calculate K_c at 500 K for the reaction below.*

$$MnO_2(s) + 2\,CO(g) \rightleftharpoons Mn(s) + 2\,CO_2(g)$$

Use the following data, relevant to 500 K.

(1) $MnO_2(s) + 2\,H_2(g) \rightleftharpoons Mn(s) + 2\,H_2O(g) \qquad K_c = 180$

(2) $CO(g) + H_2O(g) \rightleftharpoons CO_2(g) + H_2(g) \qquad K_c = 0.052$

Answer: *Double equation (2) - which will square K_c - and add the result to equation (1). This affords the desired equation.*

$MnO_2(s) + 2\,H_2(g) \rightleftharpoons Mn(s) + 2\,H_2O(g) \qquad K_c = 180$

$2\,CO(g) + 2\,H_2O(g) \rightleftharpoons 2\,CO_2(g) + 2\,H_2(g) \qquad K_c = (0.052)^2 = 2.7 \times 10^{-3}$

Hence required $K_c = K_c(1) \times K_c(2)^2 = (180)(2.7 \times 10^{-3}) = 0.49$ (all K_cs in this problem are dimensionless).

8.11 Kinetics *vs*. equilibrium

From Figure 2 (see Section 8.7), it was seen that the *rate* of attaining equilibrium gives no guide to the *position* of equilibrium. Thus, the presence of a catalyst did not affect the *composition* of an equilibrium mixture. However, there is a link between reaction rate and the equilibrium constant. The relationship depends on the concept that *at equilibrium, the rates of the forward and reverse reactions are equal.* This, we recall, was why equilibrium is regarded as a dynamic process; the reaction does not simply "stop" once equilibrium is reached.

The relationship is seen most easily when the rates of reaction are expressed in mol L^{-1} time^{-1} and K_c is the equilibrium constant. The reaction between N_2, O_2, and NO will be used to illustrate the argument.

$$N_2(g) + O_2(g) \rightleftharpoons 2\ NO(g)$$

$$K_c = \frac{[NO]^2}{[N_2][O_2]}$$

Both the forward and the reverse reactions are elementary processes. Thus we can write rate expressions for each of them.

$$\text{rate}(f) = k_f[N_2][O_2]$$
$$\text{rate}(r) = k_r[NO]^2$$

At equilibrium, rate(f) = rate(r)

Hence $k_f[N_2][O_2] = k_r[NO]^2$

Rearranging:

$$\frac{[NO]^2}{[N_2][O_2]} = \frac{k_f}{k_r} = K_c$$

Rule: *In general, the equilibrium constant may be expressed as the ratio of rate constants for the forward and reverse reactions.*

This relationship is useful if for some reason the rate of a desired reaction cannot be measured experimentally, because it is too fast or too slow. The unknown rate constant may be calculated provided that the equilibrium constant and the rate of the reverse reaction are available.

8.12 Kinetic rate laws with pre-equilibrium steps

The relationship between kinetics and reaction mechanism was discussed in Section 7.7. In Section 7.9 it was noted that the elementary reactions preceding the slow step of a reaction are always reversible. This means that the concentrations of substances in these fast, equilibrium processes can be described in terms of equilibrium constants. Consider again the example from Section 7.9.

1) $CH_3CH_2OH + H^+ \rightleftharpoons CH_3CH_2OH_2^+$		fast, reversible
2) $CH_3CH_2OH_2^+ \rightarrow CH_3CH_2^+ + H_2O$		rate limiting
3) $CH_3CH_2^+ \rightarrow CH_2{=}CH_2 + H^+$		fast

In Section 7.9, we noted that the rate equation could be written as shown below. Rate equation (a) is based on the limiting elementary reaction, and rate equation (b) was obtained by applying the steady state approximation under conditions that reaction (1) is fast in both directions.

a) rate $= k_2[CH_3CH_2OH_2^+]$

b) rate $= \{k_1k_2/k_{-1}\}[H^+][CH_3CH_2OH]$

However, the equilibrium constant for reaction (1) is defined as follows.

$$K_c = \frac{[CH_3CH_2OH_2^+]}{[CH_3CH_2OH][H^+]}$$

From Section 8.10, we can write $K_c = k_1/k_{-1}$, and so another way of transforming rate equation (a) into rate equation (b) is to write:

b′) rate $= k_2K_c[H^+][CH_3CH_2OH]$ $\qquad k(\text{observed}) = k_2K_c$

It is almost always possible to say that a reversible reaction preceding the rate limiting step of a reaction can be treated as an equilibrium. (The few exceptions occur when k_2 and k_{-1} are of similar magnitude: refer back to Section 7.9). The value of the equilibrium constant for the reversible reaction can be used to calculate the rate constant for the rate limiting elementary reaction. In the present example k(observed) $= k_2K_c$. Therefore if k(observed) is measured and K_c is known, k_2 can be obtained.

Here is another example of a reaction whose slow step is preceded by rapid equilibria: the nitration of benzene, using a mixture of nitric acid and a large excess of sulfuric acid. The currently accepted mechanism is as follows.

$HNO_3 + H_2SO_4 \rightleftharpoons H_2NO_3^+ + HSO_4^-$	fast
$H_2NO_3^+ + H_2SO_4 \rightleftharpoons NO_2^+ + H_3O^+ + HSO_4^-$	fast
$NO_2^+ + C_6H_6 \rightarrow C_6H_6NO_2^+$	slow
$C_6H_6NO_2^+ + HSO_4^- \rightarrow C_6H_5NO_2 + H_2SO_4$	fast

The rate law corresponding to the rate limiting elementary reaction is:

$$\text{rate} = k[NO_2^+][C_6H_6]$$

However, the rate law must be expressed in terms of experimental observables. Adding together the first two equilibria, we obtain:

$$K_c = \frac{[NO_2^+][H_3O^+][HSO_4^-]^2}{[HNO_3][H_2SO_4]^2}$$

This expression can be substituted into the elementary rate law.

$$\text{rate} = \frac{kK_c[C_6H_6][HNO_3][H_2SO_4]^2}{[H_3O^+][HSO_4^-]^2}$$

The reaction turns out to be first order in each of the reactants C_6H_6 and HNO_3, in accord with experimental observation at constant $[H_2SO_4]$. It is accelerated by high concentrations of sulfuric acid, and retarded by both water (which is converted to H_3O^+ in sulfuric acid solution) and HSO_4^-. In such a case, the observed rate constant is a composite of the rate constant for the elementary rate limiting step, and one or more equilibrium constants.

Problems

Sections 8.1 - 8.3

1. Write equilibrium constants K_p for the equilibria below:

    a) $CO\ (g) + H_2O\ (g) \rightleftharpoons CO_2\ (g) + H_2\ (g)$

    b) $NO\ (g) + \frac{1}{2}O_2\ (g) \rightleftharpoons NO_2\ (g)$

    c) $C\ (s) + H_2O\ (g) \rightleftharpoons CO\ (g) + H_2\ (g)$

    d) $NiO\ (s) + CO\ (g) \rightleftharpoons CO_2\ (g) + Ni\ (s)$

    Include the units of K_p in each case.

2. Write expressions for K_c for these reactions:

   a) $3O_2\ (g) \rightleftharpoons 2O_3\ (g)$

   b) $CaCO_3\ (s) + SO_2\ (g) + \frac{1}{2}O_2\ (g) \rightleftharpoons CaSO_4\ (s) + CO_2\ (g)$

   c) $Br_2\ (g) + 3F_2\ (g) \rightleftharpoons 2BrF_3\ (\ell)$

   d) $MnO_2\ (s) + 2H_2\ (g) \rightleftharpoons Mn\ (s) + 2H_2O\ (g)$

   Include the units of K_c in each case.

3. The following data relate to equilibria at 700°C

Reaction	K_c
$C(s) + CO_2(g) \rightleftharpoons 2CO\ (g)$	0.025 mol L^{-1}
$SO_2\ (g) + NO_2\ (g) \rightleftharpoons SO_3\ (g) + NO$	9.0
$2SO_2\ (g) + O_2\ (g) \rightleftharpoons 2SO_3\ (g)$	420 L mol^{-1}
$CO_2\ (g) + H_2\ (g) \rightleftharpoons CO\ (g) + H_2O\ (g)$	0.65
$CaCO_3 \rightleftharpoons CaO\ (s) + CO_2\ (g)$	5.7×10^{-4} mol L^{-1}

   Calculate equilibrium constants for the following processes, all at 700°C.

   a) $SO_3\ (g) \rightleftharpoons SO_2\ (g) + \frac{1}{2}O_2\ (g)$

   b) $C\ (s) + H_2O\ (g) \rightleftharpoons CO\ (g) + H_2\ (g)$

   c) $CaCO_3\ (s) + C\ (s) \rightleftharpoons CaO\ (s) + 2CO\ (g)$

   d) $NO\ (g) + \frac{1}{2}O_2\ (g) \rightleftharpoons NO_2\ (g)$

Sections 8.4 - 8.6

1. At 500°C iodine pentafluoride decomposes by the reaction shown:

   $7IF_5\ (g) \rightleftharpoons I_2\ (g) + 5IF_7\ (g)$

   In such an experiment carried out in a 1.00 dm^3 vessel at 500°C, the pressure dropped from 62.6 kPa initially to 59.8 kPa. At this point in the reaction, what proportion of the IF_5 had decomposed?

2. At 400 K, the equilibrium below has $K_c = 0.23\ (\text{mol L}^{-1})^3$.

   $Ni(CO)_4\ (g) \rightleftharpoons Ni\ (s) + 4CO\ (g)$

   If the equilibrium concentration of CO (g) is 0.0447 mol dm^{-3} at 400 K, what is the equilibrium concentration of $Ni(CO)_4$ (g)?

3. At 1200 K $K_c = 0.2$ for the reaction

   $C\ (s) + H_2O\ (g) \rightleftharpoons CO\ (g) + H_2\ (g)$

   Steam at a concentration 0.088 mol dm^{-3} is introduced into a 10.0 liter vessel containing 25 g of solid carbon at 1200 K. How much carbon remains at equilibrium?

4. At 600°C the reaction below has $K_c = 0.014$.

   $N_2\ (g) + 3H_2\ (g) \rightleftharpoons 2NH_3\ (g)$

   Calculate the percentage of the N_2 which is converted to NH_3 at equilibrium when the starting concentrations are $[N_2] = 0.100$ mol dm^{-3}; $[H_2] = 0.300$ mol dm^{-3}; $[NH_3] = 0$.

5. Consider the reaction $\frac{1}{2}N_2O_4\ (g) \rightleftharpoons NO_2\ (g)$ at 100°C. Pure N_2O_4 is introduced into a reaction vessel at a pressure of 310 kPa. At equilibrium, the total pressure is 496 kPa. Calculate K_p.

6. The vapour pressure of mercury at 25°C is 245 Pa.

   a) Calculate K_p and K_c for the reaction: $Hg\ (\ell) \rightleftharpoons Hg\ (g)$ at 25°C

   b) Mercury is spilled on the floor of a laboratory measuring 5.0 m long x 4.0 m wide x 3.0 m high. What mass of mercury will be vaporized at equilibrium?

7. At 700°C the reaction below has $K_c = 20.5\ (\text{mol L}^{-1})^{-1/2}$.

   $SO_2\ (g) + \frac{1}{2}O_2\ (g) \rightleftharpoons SO_3\ (g)$.

   Calculate the percent conversion of SO_2 to SO_3 at equilibrium, if the starting concentrations are SO_2, 2.6 mol L^{-1}; O_2, 2.4 mol L^{-1}.

Sections 8.7 - 8.11

1. Predict (qualitatively) the effect of making the following changes to the reaction below, initially at equilibrium.

   $C\ (s) + H_2O\ (g) \rightleftharpoons CO\ (g) + H_2\ (g)$

   (i) increase the size of the vessel from 10 to 20 liters
   (ii) add extra H_2 (g)
   (iii) add Ne (g)
   (iv) add more C (s)
   (v) remove CO (g)

2. At 520°C $K_c = 1.5 \times 10^{-2}$ for the reaction

   $2HI_{(g)} \rightleftharpoons H_2\ (g) + I_2\ (g)$

   a) A concentration of 0.042 mol dm^{-3} HI is introduced into a flask at 520°C. Calculate the concentrations of all species at equilibrium.

   b) To the equilibrium mixture in (a) 0.007 mol dm^{-3} H_2 (g) is added. What are the new concentrations of all species?

3. Consider the reaction $CaCO_3\ (s) \rightleftharpoons CaO\ (s) + CO_2\ (g)$ for which $K_c = 5.7 \times 10^{-4}$ mol L^{-1} at 700°C.

   a) Calculate the equilibrium concentration of CO_2 (g) at 700°C
   b) Calculate ΔH° for the reaction as written
   c) Use your answer in (b) to predict the effect on the equilibrium if the temperature is raised to 1000°C.

4. A mixture of N_2, H_2, and NH_3 is at equilibrium in a 2.00 dm^3 vessel when the vessel contains 0.600 mol N_2, 0.800 mol H_2, and 0.200 mol NH_3. How many moles of N_2 must be added in order to double the equilibrium concentration of NH_3?

5. N_2 (75 atm) and H_2 (225 atm) are allowed to equilibrate at 500°C, which leads to the percent conversion of N_2 to NH_3 being 25%. Calculate K_p at 500°C.

6. For the reaction conditions given in Q.5, predict the effects of the following changes:
   a) raise the temperature to 550°C
   b) add more H_2
   c) add a catalyst
   d) add 50 atm of argon.

7. Determine the direction in which each for the following reaction mixtures must proceed in order to reach equilibrium. The reaction is:

   $CO\ (g) + Cl_2\ (g) \rightleftharpoons COCl_2\ (g)$ $K_c = 3.0\ L\ mol^{-1}$ at 1000°C

   a) $[CO] = 0.040\ mol\ L^{-1}$; $[Cl_2] = 1.5\ mol\ L^{-1}$; $[COCl_2] = 0.71\ mol\ L^{-1}$
   b) $[CO] = 0.72\ mol\ L^{-1}$; $[Cl_2] = 0.46\ mol\ L^{-1}$; $[COCl_2] = 1.4\ mol\ L^{-1}$
   c) $[CO] = 1.3\ mol\ L^{-1}$; $[Cl_2] = 0.21\ mol\ L^{-1}$; $[COCl_2] = 1.1\ mol\ L^{-1}$

8. The data below refer to 1000 K.

   $CaCO_3\ (s) \rightleftharpoons CaO\ (s) + CO_2\ (g)$ $K_p = 0.039$ atm

   $C\ (s) + CO_2\ (g) \rightleftharpoons 2CO\ (g)$ $K_p = 1.9$ atm

   a) Calculate K_p for the reaction below at 1000 K.

      $CaCO_3\ (s) + C\ (s) \rightleftharpoons CaO\ (s) + 2CO\ (g)$

   b) Calculate the direction in which the reaction will proceed in a closed vessel in which p(CO) = 0.05 atm.

9. The following data refer to 1000°C.

   $CH_4\ (g) + 2H_2O\ (g) \rightleftharpoons CO_2\ (g) + 4H_2\ (g)$ $K_p = 37\ atm^2$

   $CO\ (g) + H_2O\ (g) \rightleftharpoons CO_2\ (g) + H_2\ (g)$ $K_p = 1.4$

   a) Calculate K_p for the reaction below at 1000°C.

      $CH_4\ (g) + H_2O\ (g) \rightleftharpoons CO\ (g) + 3H_2\ (g)$

   b) Determine ΔH° for the reaction given in part (a) and hence determine whether K_p will increase or decrease if the temperature is raised.

10. The reaction below has $K_c = 0.23$.

$$Ni(CO)_4(g) \rightleftharpoons Ni(s) + 4CO(g)$$

The initial concentration of $Ni(CO)_4$ is 0.011 mol L^{-1}. Calculate the concentration of CO(g) which must be present in order to ensure that no more than 1% of the $Ni(CO)_4$ decomposes.

11. Repeat Problems 2 and 3 of Section 7.7 assuming that steps labelled "fast, reversible" are equilibria.

9 PHOTOCHEMICAL SMOG AND GROUND LEVEL OZONE

The **troposphere** is the technical name for the lower atmosphere: from ground level to approximately 15-20 km altitude (Chapter 6). Tropospheric chemistry is dominated by the oxidation of trace atmospheric components. As a result of these oxidations, organic compounds such as methane and other hydrocarbons are converted into carbon dioxide and water. Under conditions where the atmosphere contains excessive amounts of oxidizable material, some of the intermediate compounds formed during oxidation can build up to cause pollution problems. One such pollution issue is **photochemical smog**, which will be a major focus of this chapter.

9.1 The hydroxyl radical as a tropospheric oxidant

The hydroxyl radical is central to the chemistry of the troposphere, even though it is present in only very tiny amounts. In this section we examine its formation and its typical reactions. The hydroxyl radical OH carries no charge, and is therefore chemically distinct from the hydroxide ion, OH^-. It is extremely reactive, and exists in the troposphere for less than 1 second before undergoing chemical reaction. It must therefore be continually regenerated, and this reaction requires sunlight. Because of the requirement for sunlight, the concentration of the hydroxyl radical falls to zero at night. The hydroxyl radical has two characteristic reactions.

1. Abstraction of a hydrogen atom from a suitable substrate.
2. Addition to an "unsaturated centre".

Examples of these reactions are:[1]

[1] $OH(g) + CH_4(g) \rightarrow CH_3(g) + H_2O(g)$

[2] $OH(g) + NO_2(g) \rightarrow HNO_3(g)$

---

[1] For the rest of this chapter, all species in chemical equations will be assumed to be in the gas phase unless noted otherwise.

Hydrogen abstraction (eq. [1]) is the preferred reaction for most substrates which contain hydrogen. This is explained by the very high O-H bond energy in water; the H-OH bond strength is much greater than those of C-H bonds. Among the few organic substances which are not attacked by hydroxyl radicals are chlorofluorocarbons; this is the reason the latter are so inert in the troposphere (see Chapter 15).

Example: *Calculate the enthalpy change for reaction [1], given standard molar enthalpies of formation:*

$OH(g)$, +38.9; $CH_4(g)$, -74.8; $CH_3(g)$, +138.9; $H_2O(g)$, -241.8 kJ mol^{-1}.

Answer:

$$\Delta H^\circ_{rxn} = \sum \Delta H^\circ_f(\text{products}) - \sum \Delta H^\circ_f(\text{reactants})$$
$$= \Delta H^\circ_f(H_2O,g) + \Delta H^\circ_f(CH_3,g) - \Delta H^\circ_f(OH,g) - \Delta H^\circ_f(CH_4,g)$$
$$= (-241.8) + (+138.9) - (-74.8) - (+38.9) = -67.0 \text{ kJ mol}^{-1}$$

Addition to an unsaturated centre means either addition to a multiple bond (double bond or triple bond) or to a site of electron deficiency. The second situation pertains in the addition of OH to NO_2 (eq. [2]), in which the total valence electron count (Chapter 1) is 17 (5 for N, and 6 each for O), which is an odd number. Substances with an odd electron count are called **free radicals.** Both NO_2 and OH (valence electron count 6 + 1 = 7) are free radicals; when they combine they yield the even-electron, stable molecule HNO_3.

When a free radical such as hydroxyl reacts with an "even electron" substance such as methane, one of the products is inevitably also a free radical (since odd number + even number = odd number). Most free radicals are very reactive towards molecular oxygen, which is a major constituent of the atmosphere. Thus the free radical CH_3 derived from methane adds to molecular oxygen, eq. [3].

[3] $CH_3 + O_2 \rightarrow CH_3OO$

The reaction sequence [1], [3] leads to the replacement of one of the C-H bonds in methane by a C-O bond. Although the details of the chemistry are complex, we can imagine that successive reactions of this type will eventually oxidize the organic substrate completely, to CO_2 and H_2O. Details are given in Section 9.2.2.

The hydroxyl radical is formed in the troposphere by a complicated sequence of reactions, driven by sunlight (eq. [4]-[9]). These reactions will be considered in turn.

[4] $N_2 + O_2 \rightarrow 2\ NO$

[5] $NO + O_3 \rightarrow NO_2 + O_2$

[6] $NO_2 \xrightarrow{h\nu,\ \lambda < \approx 400\ nm} NO + O$

[7] $O + O_2 \rightarrow O_3$

[8] $O_3 \xrightarrow{h\nu,\ \lambda < 320\ nm} O_2^* + O^*$

[9] $O^* + H_2O \rightarrow 2\ OH$

Reaction [4]: $N_2 + O_2 \rightarrow 2\ NO$

Nitric oxide is formed whenever air is heated. The reaction is endothermic, and so the equilibrium concentration of NO increases with temperature (Le Chatelier's Principle). Lightning and combustion both cause NO to be formed, and because the reverse of reaction [4] is slow, any NO formed at the elevated temperature is "frozen in" when the air cools (Section 8.7).

An example of the phenomenon of freezing in occurs when NO forms in an automobile engine. The exhaust gases mix with the cool ambient air, and at this temperature most of the NO should revert to N_2 and O_2 if equilibrium were to be reached. However, the reverse of reaction [4] is slow at room temperature, which is why automobile exhaust can lead to a "higher than equilibrium" concentration of NO in urban areas.

Example: *At 298 K, the equilibrium constant for reaction [4] has the value 5×10^{-31}.* (i) *Calculate the partial pressure of NO in equilibrium with air at 298 K;* (ii) *if the experimental concentration of NO in the troposphere is 0.01 ppmv, is the system $N_2/O_2/NO$ at equilibrium?*

Answer: (i) *For reaction [4],*

$$K_p = \frac{\{p(NO)\}^2}{p(N_2)p(O_2)}$$

K_p is dimensionless, so $K = K_c = K_p$

In air, $p(O_2) = 0.21$ atm and $p(N_2) = 0.78$ atm (Chapter 6)

Substituting,

$$5 \times 10^{-31} = \frac{\{p(NO)\}^2}{(0.21)(0.78)}$$

Solving, $p(NO) = 3 \times 10^{-16}$ atm

(ii) *0.01 ppmv $\equiv$ 0.01 ppmv x (1 atm/10^6 ppmv) = 1×10^{-8} atm*

Since p(experimental) is not the same (actually, it is much larger) than p(equilibrium), the system is not at equilibrium.

Reaction [5]: $NO + O_3 \rightarrow NO_2 + O_2$

Nitric oxide is oxidized in the troposphere to nitrogen dioxide by reaction [5], and nitrogen dioxide is photolyzed to nitric oxide by reaction [6]. Both NO and NO_2 have an "odd" valence electron count, so that the "odd nitrogen" species cycle back and forth between NO and NO_2. At the concentrations at which it is found in the troposphere (<< 1 ppmv) NO is not oxidized by molecular oxygen at an appreciable rate, since the reaction is second order in [NO], and third order overall.

$$2\ NO + O_2 \rightarrow 2\ NO_2$$

$$\text{rate} = k[NO]^2[O_2]$$

Example: *The third order rate constant for the reaction just given is 2.8 x 10^9 L^2 mol^{-2} h^{-1} at 25 °C. Calculate the half-life of NO if p(NO) = 0.1 ppmv, and assuming that this reaction is the only sink for NO.*

Answer: *rate =* $k[NO]^2[O_2]$

First obtain the concentrations of NO and O_2 in mol L^{-1}, using the ideal gas equation:

$$c = n/V = P/RT$$

$$p(NO) = 0.1\ ppmv \equiv 0.1\ ppmv \times (1\ atm/10^6\ ppmv) = 1 \times 10^{-7}\ atm$$

$$c(NO) = \frac{1 \times 10^{-7}\ atm}{0.0821\ L\ atm\ mol^{-1}\ K^{-1} \times 298\ K}$$

$$= 4.1 \times 10^{-9}\ mol\ L^{-1}$$

$$p(O_2) = 0.21\ atm$$

$$c(O_2) = \frac{0.21\ atm}{0.0821\ L\ atm\ mol^{-1}\ K^{-1} \times 298\ K}$$

$$= 8.6 \times 10^{-3}\ mol\ L^{-1}$$

Because $c(O_2) >> c(NO)$ the reaction is pseudo-second order

$$\text{rate} = k'[NO]^2 \quad \text{where } k' = k[O_2] = 2.4 \times 10^7\ L\ mol^{-1}\ h^{-1}$$

For a second order reaction $t_{1/2} = 1/(k'[NO]_o)$

$$t_{1/2} = \frac{1}{(2.4 \times 10^7\ L\ mol^{-1}\ h^{-1})(4.1 \times 10^{-9}\ mol\ L^{-1})} = 1 \times 10^2\ h$$

In the troposphere, the direct reaction of NO with tropospheric O_2 is not a major route for the oxidation of NO to NO_2 because it is too slow, as seen in the half-life just calculated. Oxidation therefore takes other routes, involving more powerful oxidants. Ozone (O_3) is one of the most important environmental oxidants for NO, as shown in reaction [5].

Reaction [6]: $NO_2 \xrightarrow{h\nu,\ \lambda < \approx 400\ nm} NO + O$

Nitrogen dioxide is a brown gas, which absorbs in both the visible and the ultraviolet regions of the spectrum. However, visible light is not sufficiently energetic to cause N-O bond cleavage, which takes place only with radiation having wavelength less than about 400 nm.

Example: *Calculate the maximum wavelength of sunlight capable of dissociating NO_2 according to reaction [6].*

Answer: *We need to calculate the enthalpy change for reaction [6], and then relate the enthalpy change to photon wavelength.*

Step 1: *Look up standard enthalpies of formation of NO_2, g (+33.2), NO,g (+90.3), and O,g (+249.2 kJ mol^{-1}).*

Step 2: *Calculate $\Delta H°_{rxn}$*

$$\begin{aligned}\Delta H°_{rxn} &= \sum \Delta H°_f(\text{products}) - \sum \Delta H°_f(\text{reactants}) \\ &= \Delta H°_f(NO,g) + \Delta H°_f(O,g) - \Delta H°_f(NO_2,g) \\ &= (+90.3) + (+249.2) - (+33.2) \\ &= +306.3\ kJ\ mol^{-1}\end{aligned}$$

At this point, recall that photon energy is a source of internal energy (Chapter 5).

$$\Delta E = \Delta H° - \Delta nRT$$

In this reaction, Δn = +1 mol because 1 mol NO_2 was cleaved to 1 mol each of NO and O.

$$\begin{aligned}\Delta E &= +306.3\ kJ\ mol^{-1} - (8.314\ J\ mol^{-1}\ K^{-1})(298\ K)(1\ kJ/1000\ J) \\ &= +303.8\ kJ\ mol^{-1}\end{aligned}$$

Step 3: *Relate photon energy to wavelength*

$$\begin{aligned}\Delta E &= (1.19 \times 10^5/\lambda) \\ &= (1.19 \times 10^5\ kJ\ mol^{-1}\ nm)/(303.8\ kJ\ mol^{-1}) \\ &= 392\ nm\end{aligned}$$

Because NO and NO_2 are continuously cycled back and forth, it is useful to judge urban air quality in terms of the sum of their concentrations. This sum is commonly designated NO_x. Although the concentrations of NO and NO_2 are usually small (< 10 ppbv), substantial amounts of NO_x may be present in the urban troposphere: *i.e.*, a low concentration but a large total number of moles.

Reaction [7]: $O + O_2 \rightarrow O_3$

The formation of ozone is extremely fast; the highly reactive oxygen atom formed in reaction [6] is immediately consumed by the large excess of O_2 molecules (p = 0.21 atm).

Example: *The rate constant for the elementary reaction [7] at 298 K is* 1.5×10^{-14} cm^3 $molec^{-1}$ s^{-1}. *Calculate the half-life of the oxygen atom under these conditions.*

Answer: *Since* O_2 *is present in large excess, the reaction is pseudo-first order.*

$$rate = k'[O], \text{ where } k' = k[O_2]$$

Calculate $c(O_2)$ *in mol* L^{-1}:

$$p(O_2) = 0.21 \text{ atm}$$

$$c(O_2) = \frac{0.21 \text{ atm}}{0.0821 \text{ L atm mol}^{-1} \text{ K}^{-1} \times 298 \text{ K}}$$

$$= 8.6 \times 10^{-3} \text{ mol L}^{-1}$$

Change units to molec cm^{-3}:

$$c(O_2) = (8.6 \times 10^{-3} \text{ mol L}^{-1})(1 \text{ L}/1000 \text{ cm}^{-3})(6.022 \times 10^{23} \text{ molec}/1 \text{ mol})$$

$$= 5.2 \times 10^{18} \text{ molec cm}^{-3}$$

Calculate k′ *and hence* $t_{1/2}$:

$$k' = k[O_2] = (1.5 \times 10^{-14} \text{ cm}^3 \text{ molec}^{-1} \text{ s}^{-1})(5.2 \times 10^{18} \text{ molec cm}^{-3})$$

$$= 7.8 \times 10^{-4} \text{ s}^{-1}$$

$$t_{1/2} = \ln(2)/k' = 0.693/(7.8 \times 10^{-4} \text{ s}^{-1}) = 8.9 \times 10^{-6} \text{ s} = 8.9 \ \mu s$$

Reaction [8]: $O_3 \xrightarrow{h\nu,\ \lambda < 320 \text{ nm}} O_2^* + O^*$

The photochemical cleavage of ozone yields O_2 and O. The reaction has an interesting twist to it, in that the oxygen atom may be formed either in its ground state or in an excited state (designated O*). The ground state oxygen atoms are identical to those formed in reaction [6], and immediately combine with O_2, regenerating O_3 (eq. [7]).

Reaction [9]: $O^* + H_2O \rightarrow 2\ OH$

Reaction [9] is only possible when the oxygen atoms formed in Reaction 8 are in their excited states. Ground state oxygen atoms, which have less energy than excited state oxygen atoms, are insufficiently energetic to cleave the O-H bond of water.

Example: *The excitation energy of the oxygen atom is 188 kJ mol^{-1}. Calculate the enthalpy change for reaction [9] for the case of (i) a ground state oxygen atom; (ii) an excited state oxygen atom. Standard molar enthalpies of formation: $H_2O(g)$, -241.8; OH(g), +38.9, O(g, ground state), +249.2 kJ mol^{-1}.*

Answer:

(i) *Calculate $\Delta H°_{rxn}$*

$$\Delta H°_{rxn} = \sum \Delta H°_f(products) - \sum \Delta H°_f(reactants)$$
$$= 2\ \Delta H°_f(OH,g) - \Delta H°_f(O,g) - \Delta H°_f(H_2O,g)$$
$$= (2 \times +38.9) - (+249.2) - (-241.8) = +70.4\ kJ\ mol^{-1}$$

The reaction is endothermic.

(ii) *Repeat the calculation for the excited oxygen atom. First calculate $\Delta H°_f(O^*,g)$, remembering that the excitation energy converts ground state oxygen atoms to excited state oxygen atoms.*

$$O(g) \rightarrow O^*(g)$$
$$\Delta H°_{rxn} = \sum \Delta H°_f(products) - \sum \Delta H°_f(reactants)$$

For this reaction $\Delta H°_{rxn} = \Delta H°_f(O^*,g) - \Delta H°_f(O,g)$

$$+188 = \Delta H°_f(O^*,g) - (+249.2)$$
$$\Delta H°_f(O^*,g) = 188 + 249 = 437\ kJ\ mol^{-1}$$

Now calculate the energetics of reaction [9] for an excited oxygen atom.

$$\Delta H°_{rxn} = \sum \Delta H°_f(products) - \sum \Delta H°_f(reactants)$$
$$= 2\ \Delta H°_f(OH,g) - \Delta H°_f(O^*,g) - \Delta H°_f(H_2O,g)$$
$$= (2 \times +38.9) - (+437) - (-241.8) = -117\ kJ\ mol^{-1}$$

The excitation (extra) energy of $O \rightarrow O^*$ compared with O(ground state) is sufficient to make the reaction exothermic.

Although reaction [9] has a large rate constant at temperatures near 25 °C, only a small fraction of all the excited oxygen atoms formed in reaction [8] actually go on to afford hydroxyl radicals. This is because O^* is rapidly deactivated through collisions with any molecule (M) in the air, the excitation energy of O^* being converted into kinetic energy in the process, eq. [10].

[10] $O^* + M \rightarrow O + M +$ kinetic energy

Reactions [9] and [10] are thus in competition.

The hydroxyl radical is the most important substance for initiating the oxidation of tropospheric contaminants, both natural and anthropogenic. It can be regarded as the chief clean-up agent in the troposphere. However, its concentration is exceedingly tiny: on a globally- and seasonally-averaged basis its concentration is between 500,000 and 1 million radicals per cm^3, or about 10^{-15} moles per liter. However, since OH is chemically very reactive, its atmospheric lifetime is very short (<< 1 s). Since in addition sunlight is required for the production of OH, its concentration varies widely: from essentially zero during the hours of darkness to about 10^7 molec cm $^{-3}$ under conditions of high urban pollution (Section 9.4).

Example: *The rate constant for the bimolecular reaction of OH(g) with naphthalene ($C_{10}H_8$, a polycyclic aromatic hydrocarbon which is formed in automobile exhaust) is 2.3×10^{-11} cm^3 $molec^{-1}$ s^{-1} at 298K. What is the lifetime (not half-life) of naphthalene in polluted urban air if the steady state concentration of OH is 8.0×10^6 molec cm^{-3}?*

Answer: *The bimolecular reaction is:*

$$OH(g) + C_{10}H_8(g) \rightarrow \text{products (identities immaterial)}$$

Since OH is present at a steady state the reaction is pseudo-first order, and k′ = k[OH]

$$k' = (2.3 \times 10^{-11}\ cm^3\ molec^{-1}\ s^{-1})(8.0 \times 10^6\ molec\ cm^{-3})$$
$$= 1.8 \times 10^{-4}\ s^{-1}$$

Since lifetime $(\tau) = 1/k'$, $\tau = 5.4 \times 10^3\ s = 1.5\ h$

9.2 Oxidations involving the hydroxyl radical

9.2.1 Oxidation of carbon monoxide

Carbon monoxide is a natural tropospheric constituent (~ 0.1 ppmv). It is formed biologically, and also by incomplete combustion (*e.g.*, forest fires and human activities). It is also produced in the troposphere as an intermediate in the oxidation of substances such as methane. Incomplete combustion can lead to local concentrations of CO which are greatly in excess of 0.1 ppmv. For example, city air may often contain 2 - 20 ppmv of carbon monoxide, mainly due to pollution by automobiles, and peak concentrations up to 100 ppm have been recorded in vehicular subways and tunnels. This is on the threshold of representing a health hazard, since carbon monoxide combines irreversibly with the body's hemoglobin, making it unavailable for transporting oxygen. (At 100 ppm, up to 15% of a person's hemoglobin would be converted to carboxyhemoglobin at equilibrium.) Frequent motor vehicle testing programs and switching to oxygenated fuels have been proposed as ways to reduce urban CO levels, see also Section 9.7.

The sinks for atmospheric CO are oxidation by OH (eq. [11]) and uptake in soil, followed by microbial oxidation to CO_2. In the unpolluted troposphere, 70%

of all hydroxyl radicals disappear through reaction with CO, eq.[11][2]; most of the remainder react with methane, eq.[1].

[11] $OH + CO \rightarrow H + CO_2$ $k = 2.7 \times 10^{-13}$ cm^3 molec^{-1} s^{-1} at 300 K

The byproduct of reaction [11] is a hydrogen atom, which rapidly forms the hydroperoxy radical HO_2. HO_2 is another powerful oxidant, one of whose reactions is the oxidation of NO to NO_2, thereby regenerating another OH radical. Under conditions of high NO_x in the polluted troposphere, reaction with NO may be a major sink for HO_2.

[12] $H + O_2 \rightarrow HO_2$

[13] $HO_2 + NO \rightarrow NO_2 + OH$

The reaction sequence [11] - [13] is an example of how reactive species such as OH can be destroyed and re-formed in the troposphere, allowing them to maintain a steady state. This sequence can be considered as a reaction cycle for OH. Another way of thinking about this is to consider OH as a catalyst for the oxidation of CO.

Example: *Write the overall reaction corresponding to [11]-[13]. Why does this sequence suggest that OH can be a catalyst for the oxidation of CO?*

Answer: $CO + NO + O_2 \rightarrow CO_2 + NO_2$

The role of OH is catalytic, because it appears in the reaction sequence ([11] - [13]) yet is not consumed in the overall reaction.

9.2.2 *Oxidation of methane*

Methane is the simplest hydrocarbon. As noted in connection with the global warming problem (Chapter 6), methane is a natural constituent of the atmosphere whose chief source is anaerobic decay. The mechanism for its complete oxidation to CO_2 and H_2O is very complex. These are some (but by no means all) of the reactions which occur. As you study these reactions, check which bonds are being made and broken.

---

[2] Reaction [11] does not seem to fit into the generalization that OH reacts either by hydrogen abstraction or by addition; however, [11] has been suggested to be a two-step process rather than an elementary reaction.

[11a] $OH + CO \rightarrow HOCO$

[11b] $HOCO \rightarrow H + CO_2$

Scheme 1

[1] $OH + CH_4 \rightarrow CH_3 + H_2O$

[3] $CH_3 + O_2 \rightarrow CH_3\text{-O-O}$

[14] $CH_3\text{-O-O} + HO_2 \rightarrow CH_3\text{-O-O-H} + O_2$

[15] $CH_3\text{-O-O-H} + OH \rightarrow H_2O + H_2C{=}O + OH$

[16] $H_2C{=}O + OH \rightarrow H_2O + H\text{-}C{=}O$

[17] $H\text{-}C{=}O + O_2 \rightarrow CO + HO_2$

[11] $OH + CO \rightarrow H + CO_2$

The overall picture that emerges is one of successive attack on the C-H bonds of methane and their replacement by CO bonds, or the elimination of H_2O. Free radicals (H, OH and HO_2) are consumed and regenerated in these reactions. Higher hydrocarbons react similarly, but the reaction sequence is even more complicated. Ultimately, the free radical species are removed from the system by reactions such as [2].

[2] $NO_2 + OH \rightarrow HNO_3$

The HNO_3 formed in reaction [2] returns to earth either adsorbed on particles, or dissolved in raindrops, where it is one of the sources of **acid rain** (Chapter 13). Reaction [2] is fast and irreversible, and constitutes the chief sink for the removal of nitrogen oxides from the troposphere.

9.3 Photochemical smog and ground level ozone

At the outset, it must be stressed that the chemistry of photochemical smog is the same as that described already in this chapter: the formation of OH, and the oxidation of hydrocarbons. Oxidation in the unpolluted and the polluted troposphere differ only in detail, not in principle. As its name implies, photochemical smog is initiated by sunlight, with the photochemical cleavage of NO_2 to NO and O being the most important initiating step.

Physical characteristics of photochemical smog include a yellow-brown haze, which reduces visibility, and the presence of substances which both irritate the respiratory tract and cause eye-watering. The yellowish colour is due to NO_2, while the irritant substances include **ozone**, **aldehydes** (of which $CH_2{=}O$ is the simplest), and **organic nitrates**, of which the most notorious is PAN (**p**eroxy**a**cetyl **n**itrate, $CH_3CO\text{-OO-}NO_2$). Photochemical smog was first recognized as a problem in Los Angeles, California in the 1940's, but has since been documented in many other locations.

Four conditions are necessary before photochemical smog can develop.

Nitrogen oxides (NO_x)	**Hydrocarbons**
Sunlight	**Temperatures above ~18 °C**

These factors can be understood in terms of the reactions which have already been encountered. **Nitrogen dioxide** absorbs **sunlight** (eq. [6]) as the first step in the production of first ozone and ultimately the hydroxyl radical. Just as in the unpolluted troposphere, the chemical reactions occurring in photochemical smog involve the attack of hydroxyl radicals on **hydrocarbons**, which are present in abundance in automobile exhaust. The **temperature** requirement arises because many of the reactions involved in oxidation have moderately large activation energies; 18 °C is not an absolute cut-off, but gives an idea of the temperature needed for these atmospheric processes to proceed fast enough for the obnoxious byproducts to build up to the levels that we associate with air pollution. Thus very hot sunny weather provides the conditions when photochemical smog is most likely. Table 1 gives the concentrations of some important pollutants under conditions of photochemical smog.

Table 9.1: Typical pollutant concentrations (ppbv) in unpolluted air and under conditions of photochemical smog

Pollutant	Unpolluted (remote) area	Heavily polluted area
CO	< 200	10,000-50,000
NO_x	< 1	1,000-3,000
O_3	< 50	100-500
Hydrocarbons[a]	< 65	> 1,500
PAN	< 0.05	20-70
$CH_2{=}O$	< 2	20-75

[a] Excluding methane (*ca.* 1700 ppbv), most of which is of natural origin

The terms **photochemical smog** and **ground level ozone** refer to essentially the same phenomenon. Ozone is one of the obnoxious byproducts of photochemical smog, and will tend to reach higher-than-normal levels in the troposphere whenever the conditions mentioned above (air pollution, sunlight,

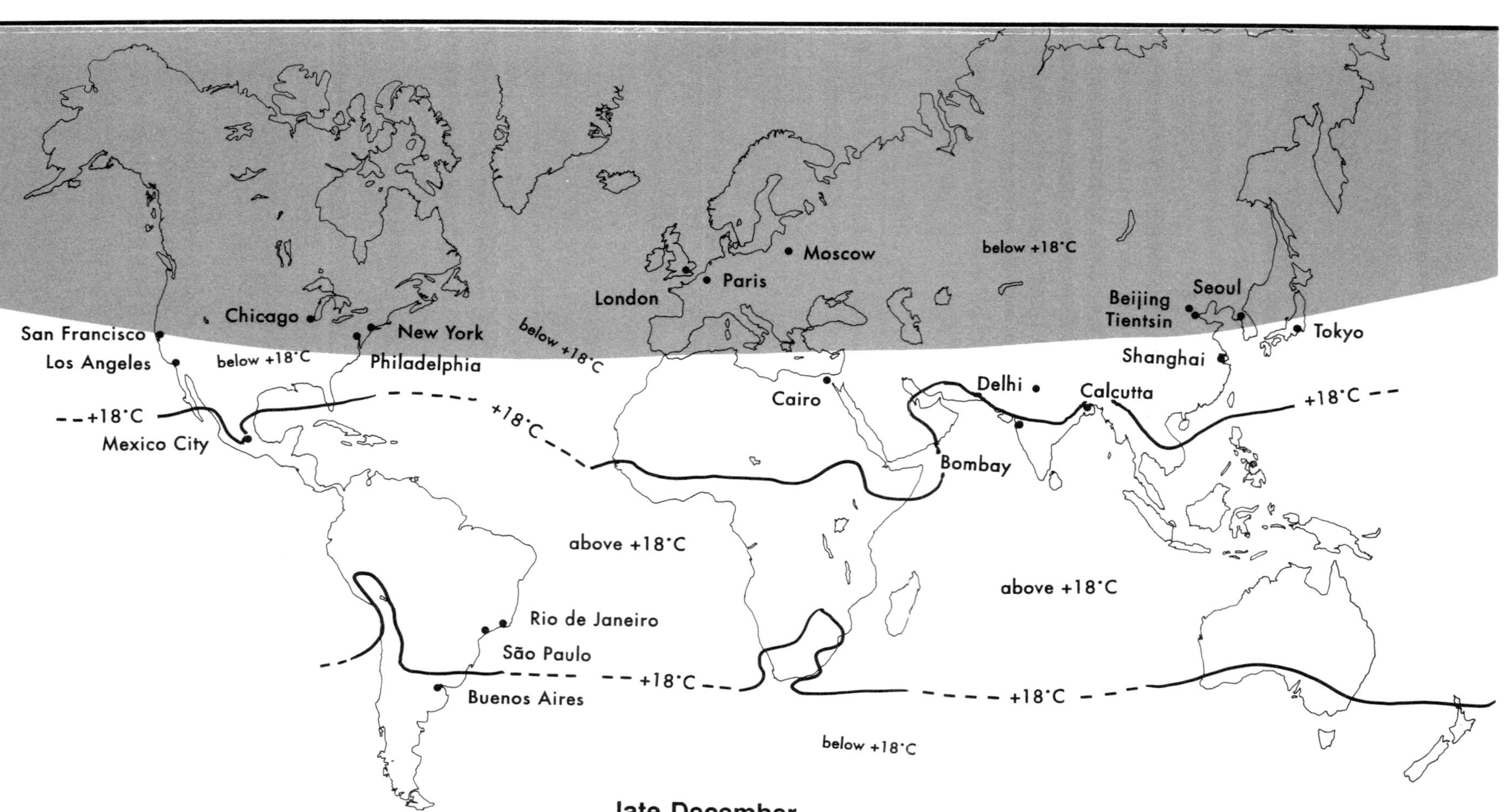

late December

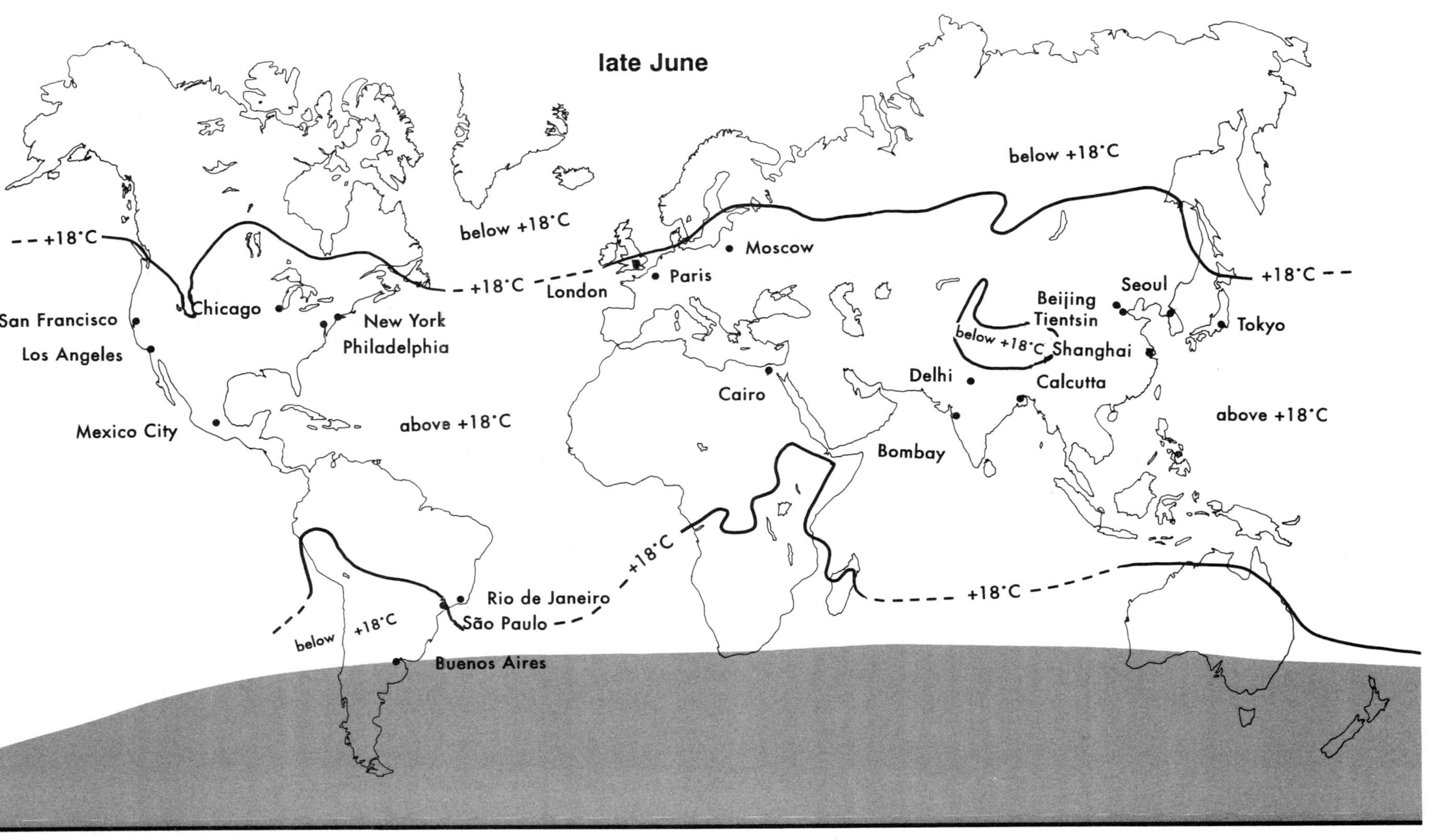

Two of the four "smog criteria" are represented in these maps. During the rush-hour period, sunlight (light areas) and warmth (above +18°C) are crucial; they are found in more cities during June than in December.

high ambient temperature) are met[3]. Ozone is a normal trace component of the troposphere (20-50 ppbv), but its concentration may rise to 100-200 ppbv under conditions of severe air pollution. Ontario's air quality guidelines suggest < 80 ppbv as an acceptable concentration for tropospheric ozone. This value is typically exceeded about 20 days per year in southern Ontario (and also in the Fraser Valley, in British Columbia, the only other region of Canada where ground level ozone is considered a serious problem). By contrast, this same guideline is exceeded more than 200 days each year in southern California. Serious problems exist in other fast-growing tropical and subtropical cities such as Rio de Janeiro and Mexico City. In Mexico City, air pollution reaches three times the Ontario guideline at times, and on occasion the government has attempted to limit air pollution by restricting the use of motor vehicles - half the population being allowed to drive one day, the other half the next.

Since automobiles generate NO_x and hydrocarbons, photochemical smog is a big city phenomenon; however drift of the urban plume can affect neighbouring rural areas. In southern Ontario in summer, much of the tropospheric ozone is the result of transboundary pollution from the U.S. Midwest.

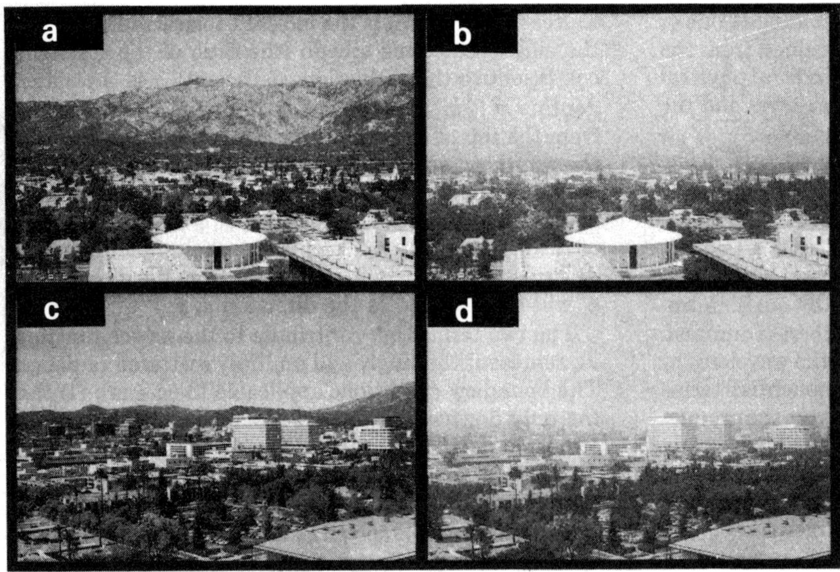

Photographs taken near Los Angeles:

a) clear day, b) smoggy day ... San Gabriel Mountains
c) clear day, d) smoggy day ... Pasadena

[3] In contrast to the troposphere, where abnormally high levels of ozone are considered a pollution problem, there is currently concern that ozone levels maybe decreasing in the stratosphere, where ozone plays the important role of providing shielding from high energy ultraviolet radiation. See Chapter 15.

Example: *In which of the following locations would you expect to encounter abnormally high levels of tropospheric ozone?*

(i) *Houston, Texas, in May*
(ii) *Canadian Prairies in August*
(iii) *New York, NY in February*
(iv) *Southern Ontario in July*

Answers:

(i) *Likely; Houston is a large urban area with considerable traffic and a hot sunny climate*
(ii) *Unlikely; the population density is too small to experience substantial air pollution*
(iii) *Most unlikely; New York in February is too cold, despite its high auto population*
(iv) *Possible, but only as a result of transboundary pollution*

9.4 Mechanisms for forming ozone and PAN

Ozone forms by reactions [6] and [7], shown earlier in this chapter.

[6] $NO_2 \xrightarrow{h\nu,\ \lambda < \approx 400\text{ nm}} NO + O$

[7] $O + O_2 \longrightarrow O_3$

Reaction [18], which is the sum of the reaction sequence [6], [7], is the reverse of reaction [5].

[18] $NO_2 + O_2 + \text{sunlight} \rightarrow NO + O_3$

[5] $NO + O_3 \rightarrow NO_2 + O_2$

Hence reactions [5] - [7] produce a steady state of ozone.

Question: *Does the pair of reactions [5], [18] constitute an equilibrium?*

Answer: *No; there is the continual input of sunlight, an outside energy source. These reactions represent a steady state but not an equilibrium.*

As we see from reaction [18], ozone production is favoured by high $[NO_2]$ and strong sunlight, in accord with the conditions typical of photochemical smog.

PAN and related organic nitrates form as byproducts (minor reaction pathway) of the oxidation of hydrocarbons. Specifically, the **aldehydes** (R-CHO) are direct precursors of organic nitrates. In Scheme 1, reactions [16] and [17] showed the oxidation of the simplest aldehyde $CH_2{=}O$. Let's look at what occurs for the next simplest aldehyde, CH_3-CH=O (acetaldehyde).

Scheme 2

$$CH_3\text{-}CH{=}O + OH \rightarrow CH_3\text{-}C{=}O + H_2O \quad \text{(analogous to [16])}$$
$$CH_3\text{-}C{=}O + O_2 \rightarrow CH_3\text{-}(CO)\text{-}OO$$
$$CH_3\text{-}(CO)\text{-}OO + NO \rightarrow CH_3 + CO_2 + NO_2 \quad \text{(major route)}$$
$$CH_3\text{-}(CO)\text{-}OO + NO_2 \rightarrow CH_3\text{-}(CO)\text{-}OO\text{-}NO_2 \text{ (PAN)} \quad \text{(minor route)}$$

The formation of an organic nitrate is a minor (but important) side reaction in the oxidation of hydrocarbons - or, more specifically, of the aldehydes formed as intermediates during hydrocarbon oxidation.

9.5 Effects of photochemical smog

The consequences of photochemical smog are mainly due to the toxicity of intermediates such as ozone and PAN. Respiratory impairment results from breathing air containing > 0.1 ppmv of ozone, especially by persons suffering from respiratory complaints such as asthma and emphysema.

Although ozone and PAN are toxic to animals and to people, they are much more toxic to plants. At 0.1 ppmv of ozone, most people show little adverse reaction; in plants however, the rate of photosynthesis may be reduced by a factor of more than two. This is a very serious issue in California, where agriculture is such an important industry and must coexist with urban pollution. The annual losses in agricultural production, increased need for health care, and property damage due to ozone and PAN in the atmosphere have been estimated at > $10 billion in southern California. Even in Ontario, where ground level ozone is a much less serious problem, losses through reduced yields of a single crop, white beans, have been estimated at $20 million annually.

The toxicity of ozone can be understood in broad terms because it is such a powerful oxidant. With ΔH°_f for $O_3(g)$ = +142 kJ mol^{-1}, we can appreciate that ozone is a much more powerful oxidant even than elemental oxygen. Cell membranes are particularly susceptible to attack because they contain unsaturated fatty acid residues; unsaturated compounds (alkenes) are well known to suffer bond cleavage (ozonolysis) by gaseous ozone.

9.6 Emission controls and catalytic converters

As noted already, automobile exhaust is a major contributor to photochemical smog through emissions of NO_x and incompletely oxidized hydrocarbons.

Emission control devices are designed to reduce these emissions. Not surprisingly, California was the first jurisdiction to institute emission controls on automobiles, beginning in the 1975 model year. Unburned hydrocarbons and nitrogen oxides are the emissions which are regulated. Since that time, emissions have been regulated by a large number of governments, exemplified by these U.S. data which show the remarkable improvements which have been made in emission standards over the past two decades.

USA auto emissions

	RH[a]	CO[a]	NO_x[a]
pre 1968	5.5	54	2.3
1975	0.9	9.4	1.9
1983 +	0.3	2.1	0.6

[a] all data in g km^{-1}

Emission control devices are used to reduce emissions of hydrocarbons and nitrogen oxides in automobile exhaust to acceptable levels. The functioning of the internal combustion engine allows us to see some of the difficulties which ensue. In the engine, hydrocarbons (gasoline) are burned. There is a stoichiometric relationship between the amount of fuel consumed and the amount of air needed to burn it. For thermodynamic reasons (discussed in Chapter 14), the operating temperature should be as high as possible in order to maximize the fuel efficiency of the engine. However, the higher the temperature, the greater the potential for forming NO; recall that reaction [4] is endothermic; Le Chatelier's Principle reminds us that the equilibrium will increasingly favour NO as the temperature rises.

[4] $N_2 + O_2 \rightleftharpoons 2\ NO \qquad \Delta H° = +180\ kJ\ mol^{-1}$

This presents a dilemma. It is difficult to set the fuel:air mixture in an automobile exactly to the stoichiometric ratio, and so engines must operate with the mixture either "lean" (excess air in the mixture) or "rich" (excess fuel). If the mixture is lean, little unburned hydrocarbon should escape in the exhaust gases, but some of the excess oxygen will react with some of the nitrogen in the air used for combustion, forming nitric oxide. As already stated, this is "frozen in" when the hot gases are exhausted to the atmosphere. Alternatively, a "rich" mixture has more than the stoichiometric amount of hydrocarbons, so the excess must inevitably pass out in the exhaust.

Catalytic converters operate by creating a second zone for combustion outside the engine itself. The canister contains a catalyst which permits oxidation to be completed at a relatively low temperature, so that less NO will form. The catalysts are finely divided noble metals such as platinum, supported on a matrix of inert oxides. Because the active surfaces of the catalyst are inactivated (poisoned) by lead compounds, leaded gasoline is incompatible with the operation of a catalytic converter, and lead-free gasoline is required.

Cars equipped with the early models of catalytic converter had carburation adjusted so that the air:fuel mixture was deliberately made rich, with the aim of limiting the formation of NO. Catalytic oxidation of the excess hydrocarbon was effected in a secondary combustion zone, after admitting additional air. Because the temperature of the secondary combustion was lower, very little NO_x was formed at this stage. The disadvantages of this system were first, that the rich mixture gave poorer fuel economy and tended to foul the spark plugs and valves with soot; second, when the catalytic converter lost its efficiency due to age, even larger amounts of hydrocarbons were emitted than in the absence of any emission control device.

A second generation of catalytic converter, introduced with the 1981 model year, operates in two stages. The mixture entering the engine is approximately in stoichiometric balance between fuel and air. Small amounts of both NO and unburned hydrocarbon therefore leave the primary chamber (the engine). In the first catalytic stage, a ruthenium-based catalyst is used to reduce any NO_x to elemental nitrogen. The reactions involve the production of hydrogen by the reaction of water vapour (a combustion product) with any unburned hydrocarbon, shown here with approximate stoichiometry for methane as an example.

$$H_2O(g) + CH_4 \rightarrow CO + 3\ H_2$$

The hydrogen thus formed reduces the NO_x over the ruthenium catalyst.

$$H_2 + NO_x \rightarrow N_2 + H_2O \qquad \text{(unbalanced)}$$

The second stage of the converter employs oxidation chemistry; air is admitted, and oxidation proceeds at a sufficiently low temperature that NO_x formation is minimized. The oxidation step also serves to convert any NO_x which was over-reduced to ammonia (in the first stage) back to molecular nitrogen. The present catalytic converters are capable of eliminating 96% of unburned hydrocarbons and CO, and 76% of NO_x emissions from exhaust gases. It is expected that by the year 2000, total emissions of RH, CO, and NO_x in North America will decline 57, 73, and 46% respectively from 1989 levels, despite the anticipated increase of 1.8% annually in the size of the fleet.

Automobile emissions comprise two problems: NO_x and unburned hydrocarbons. Together, these lead to photochemical smog under conditions of warmth and sunlight. Recently, it has been realized that efforts to limit hydrocarbon emissions seem not to have the desired effect on air quality. One reason for this appears to be the high natural background levels of hydrocarbons in the air, mostly from pine trees (the fresh scent of the pine forest so beloved of the advertisers of air fresheners). In southern California, as much as 30% of all hydrocarbon emissions are now thought to be of natural origin. This means that efforts to improve air quality should probably focus upon NO_x emissions, because there is less of a natural background of NO_x (remember, *both* hydrocarbons and NO_x are required for photochemical smog to develop).

9.7 Lead in gasoline

Gasoline is one of the fractions obtained by distilling crude oil. Crude petroleum is a mixture of many thousands of components, encompassing a full spectrum of molar masses. Petroleum refining converts this complex, highly viscous mixture into the familiar range of useful petroleum products seen below (Table 2).

Table 9.2: Fractions of crude oil

Number of Carbon Atoms	Name and/or use
1	methane, natural gas
2	ethane, used to make ethylene
3,4	propane and butane, compressed gases
5-7	naphtha, solvent
7-12	gasoline
10-15	kerosene, aviation fuel, home heating
15-30	lubricating oils, heavy fuel oil
30-40	paraffin waxes
> 40	asphalt

Simple distillation of crude petroleum oil affords a "straight run" gasoline, which is a poor fuel for the modern automobile because under compression it burns erratically in a series of small explosions. These explosions are known as "knocking" or "pinging". They represent premature detonation of the gasoline-air mixture before the piston is ready to move upwards in the cylinder, thus greatly reducing the efficiency of the engine. In the 1920's Thomas Midgeley of General Motors' research laboratories discovered that the addition of small amounts of organolead compounds (substances with at least one Pb-C bond) to gasoline greatly reduced the tendency to knocking, and hence improved fuel efficiency.

Tetraethyllead (TEL) has been the most favoured gasoline additive. It is prepared from sodium-lead alloy.

$$4\ Na/Pb + 4\ C_2H_5Cl \xrightarrow{75°\ C,\ 4\ atm,\ acetone} Pb(C_2H_5)_4 + 4\ NaCl + 3\ Pb$$

TEL promotes smooth combustion of gasoline by decomposing evenly into ethyl free radicals, which act as free radical initiators.

$$Pb(C_2H_5)_4 \xrightarrow{heat} Pb(g) + 4\ C_2H_5$$

This reaction serves to introduce a steady source of free radicals into the combustion mixture, promoting smoother combustion through free radical chain oxidation. However, the lead is emitted from the automobile exhaust in the form of semivolatile inorganic lead compounds.

Lead emissions from vehicles were formerly a major source of environmental contamination by lead, and the nature of the source is such that the wide dispersion of the pollutant was inevitable. In Canada in the mid-1970s, leaded gasoline was estimated to be responsible for 70% of all emissions of lead into the environment. Numerous studies, in North America and in Europe, have shown that the soil close to freeways contains high levels of lead, and that the concentration of lead increases both with proximity to the roadway and with traffic volume. Unleaded gasoline was introduced into Canada in the late 1970s, and ultimately the leaded product was removed from the marketplace by December 1, 1990.

In North America the move to remove lead from gasoline originated in the 1970's with the thrust to reduce air pollution from unburned hydrocarbons and nitrogen oxides. The tie-in to lead is that in order to limit hydrocarbon and NO_x emissions, it is necessary to use emission control devices (catalytic converters), whose catalysts used in these converters are rendered inactive ("poisoned") by lead. Consequently the use of leaded gasoline is incompatible with controlling emissions by means of catalytic converters. In Europe, the move against lead in gasoline has been directed largely towards eliminating contamination by lead for its own sake, rather than as a prerequisite for limiting air pollution by automobile combustion products.

9.8 Alternatives to leaded fuels

Given that organolead compounds offer such improvements in combustion efficiency, how can they be replaced? Research has focused on changes in engine design, changes in petroleum refining technology, and developing different additives[4].

Several fractions are obtained when crude oil is fractionally distilled (Table 2), but the composition of a barrel of oil matches the demand for the fractions very poorly; in particular, the demand for the gasoline fraction far exceeds the supply of straight run gasoline. In addition, straight run gasoline is a rather poor fuel, as previously mentioned; it is composed principally of linear alkanes, which have a strong tendency to knock. Branched chain alkanes are superior fuels from this point of view. Modern refinery technology includes two important processes: isomerization, by which linear alkanes are isomerized into branched chain ones; and cracking, by which the more abundant longer chain alkanes are broken catalytically into smaller fragments.

Octane ratings are a means of classifying the knocking tendencies of a given gasoline. The branched alkane 2,2,4-trimethylpentane (incorrectly known in commerce as isooctane) is assigned an octane rating of 100 and heptane, a poor fuel, is given an octane rating of zero. A fuel with an octane rating of 90 (for example) burns under test conditions exactly as efficiently as a mixture of 90 parts of isooctane and 10 parts of heptane. The "regular" grade of unleaded gasoline currently available in North America has an octane rating of about 87, while the "super" grades are usually about 93.

Octane ratings can be improved by the use of additives such as TEL and by increasing the amounts of branched, aromatic, or oxygenated components in the gasoline. Therefore, in order to manufacture a lead-free gasoline, changes in refinery operation are needed so as to change the composition of the blend. Besides optimizing the proportion of branched and aromatic hydrocarbons, the octane rating can be enhanced by addition of small quantities of "oxygenates" (oxygen-containing organic compounds) to the mixture. Suitable oxygenates include short chain alcohols and ethers. One of these substances, methyl tert-butyl ether (MTBE) had become No. 19 in the ranking of Top 50 chemicals in the U.S. by 1993.

Oxygenates have the reputation of burning more cleanly than hydrocarbons, particularly with respect to hydrocarbon and CO emissions. The U.S. Clean Air Act of 1990 requires a minimum of 2.7% by weight oxygen in automotive fuel sold in wintertime in the 44 areas of the U.S. which have the worst record of ambient CO levels. For those regions of the U.S. having the poorest compliance with ground level ozone standards, a 2.0% minimum oxygen by weight will be required year round from 1995.

Recent information suggests that not all oxygenates are equally effective at curbing air pollution however. Ethers (R-O-R) appear to be more effective than

---

[4] The search for free radical initiators other than TEL has not been successful, with fouling of the engine parts and/or excessive engine wear being unwelcome side effects. Methylcyclopentadienylmanganese tricarbonyl (MMT) has been used in Canada as an octane enhancer, but it is not used in the U.S. on account of toxicological concerns.

alcohols (R-O-H). The reason is probably that alcohols are rather easily oxidized to aldehydes, which as we saw in Section 9.4, are the immediate precursors of organic nitrates such as PAN. Evaporation of gasolines which contain alcohols can therefore afford organic nitrates in few chemical steps. This is of practical importance because a powerful agricultural lobby in the U.S. Midwest has been advocating the use of ethanol (derived by fermentation of surplus grains) as an oxygenated gasoline additive: so-called "gasohol". Although gasohol has been promoted as a renewable, environmentally friendly fuel resource, more research is clearly needed to determine its effect as an air pollutant.

Along with changes in gasoline manufacture, car engines have been redesigned to operate at lower compression ratios since about 1970. This change means that fuels of lower octane number can be used, because premature ignition (knocking) is more prevalent when the fuel/air mixture is at high pressure.

9.9 Other air pollutants

In this chapter we have concentrated on the emissions from automobiles - nitrogen oxides and hydrocarbons. Nitrogen oxide emissions are not unique to automobiles; they form whenever air is heated, and so are also formed during electricity generation, steel production, and so on.

The term "Volatile organic compounds" (VOCs) has come into use to describe emissions of organic compounds generally, especially those emanating from industrial production. To date, regulations on air emissions are less stringent than those pertaining to water quality. A likely trend through the 1990s is increased attention to these other compounds in the atmosphere.

Problems

Section 9.1

1. a) Calculate $\Delta H°$ for each of reactions [6], [7], [8], and [9] without considering the energy provided by solar radiation. Use the information to explain why light is required to drive reactions [6] and [8]. In addition to standard molar enthalpies of formation in Appendix 1, take $\Delta H_f°$ (O*, g) = 437 kJ mol $^{-1}$ and $\Delta H_f°$ (O_2*, g) = 90 kJ mol $^{-1}$.

   b) Now repeat the calculation for reactions [6] and [8] with the energy of the photon included.

a

b

c

d

Non-point sources of pollution contribute to photochemical (Los Angeles type) smog in various cities. In order for this type of smog to form, four key ingredients must be present, including daylight, and temperatures about 18°C or above. See Section 9.3, page 228.

a New Delhi, India
b Manila, Philippines
c Mexico City, Mexico
d Los Angeles, USA

The Great Smoky Mountains, Tennessee, USA. Their attractive bluish haze is caused by oxidation of hydrocarbons emitted by pine trees, and is chemically closely related to photochemical (Los Angeles type) smog. It differs in that nitrogen oxides, a key ingredient of photochemical smog, are not present at elevated levels. See Section 9.3, page 228.

2. a) Calculate the amount of NO formed at equilibrium in an automobile engine at 800 K assuming that inside the engine $p(N_2)$ = 3.0 atm and $p(O_2)$ = 0.01 atm. Take $K_p = 3 \times 10^{-11}$ for the reaction:

$$N_2\ (g) + O_2\ (g) \rightleftharpoons 2NO\ (g)$$

   b) Repeat the calculation for 288 K (15°C) at which temperature $K_p = 3 \times 10^{-32}$, assuming initial $p(N_2)$ = 0.78 atm, and $p(O_2)$ = 0.21 atm. If automobile exhaust at 800 K is emitted to the atmosphere, is p(NO) at equilibrium at 288 K?

3. The numerical value of the rate constant for the elementary reaction below is given by the expression: $k = 2.6 \times 10^{12} \exp(-32{,}000/T)$ L mol^{-1} s^{-1}

$$2NO\ (g) \longrightarrow N_2\ (g) + O_2\ (g)$$

   a) Calculate the activation energy for this reaction
   b) Calculate the rate constant for the reaction at 800 K and at 288 K.
   c) Calculate the half-life of the reaction at 800 K and at 288 K when p(NO) is initially 45 ppbv.

4. Compare the results you obtained in Problems 2 and 3.
   a) Why do we say that NO formed in an automobile engine is "frozen in" when the exhaust gases are released to the environment?
   b) Calculate the rate constant for the reaction

$$N_2(g) + O_2\ (g) \longrightarrow 2NO\ (g)$$

   at 288 K and at 800 K
   c) Calculate the initial rate of formation of NO under the conditions given in the previous problems for 288 K and 800 K.
   d) Comment on the result.

5. Excited oxygen atoms suffer two fates in the atmosphere

(i) $O^* \longrightarrow O$ (ground state, reaction [10]) $k_1 = 7.2 \times 10^8$ s^{-1}

(ii) $O^* + H_2O \longrightarrow 2OH$ (reaction [9]) $k_2 = 2.2 \times 10^{-10}$ cm^3 molec^{-1} s^{-1}

Both rate constants are given for 25°C, at which temperature the saturated vapour pressure of water is 3.2 kPa. Carry out the following calculations for conditions of 25°C, 75% relative humidity:
   a) Calculate the lifetime of the excited oxygen atom
   b) Calculate the fraction of excited oxygen atoms which react with water vapour to form OH.

Section 9.2

1. The principal reactions of OH in the unpolluted troposphere are with CO and CH_4 for which the rate constants at 300 K are 2.7 x 10^{-13} cm^3 $molec^{-1}$ s^{-1} and 6.3 x 10^{-15} cm^3 $molec^{-1}$ s^{-1} respectively. In clean air, the concentrations of CO and CH_4 are 100 ppbv and 1700 ppbv, respectively. Calculate the lifetime of the OH radical under these conditions.

2. The globally averaged concentration of OH is 5 x 10^5 molec cm^{-3} and the total mass of the atmosphere is 5 x 10^{15} t.
   a) Calculate the concentration of OH in pptv.
   b) Estimate the number of moles of gases in the atmosphere, assuming an average **M**(air) = 30 g mol^{-1}.
   c) Estimate the mass of OH in the atmosphere, assuming that its concentration in pptv is constant throughout the atmosphere.
   d) Use the result of Problem 1 to estimate the global rate of formation of OH in the atmosphere in tonnes per hour.

3. The rate constant for reaction [12] is 1.7 x 10^{-13} cm^3 $molec^{-1}$ s^{-1} at 290 K

   [12] $H + O_2 \longrightarrow HO_2$

   a) Estimate the lifetime of a free hydrogen atom in the atmosphere (290 K, 1.00 atm).
   b) Work out the enthalpy change for reaction [12]. Would you expect that this reaction requires sunlight to drive it in the forward direction?

4. The following are believed to be the chief sources of methane into the atmosphere:

wetlands, marshes	150 x 10^6 t yr^{-1}
cattle and other ruminants	120 x 10^6 t yr^{-1}
rice paddies	35 x 10^6 t yr^{-1}
oceans	35 x 10^6 t yr^{-1}
other sources	150 x 10^6 t yr^{-1}

   a) The concentration of methane in the atmosphere is currently 1.7 ppmv. Estimate the residence time. Hint: See Problem 2c to determine how to find the mass of CH_4 in the atmosphere.
   b) The concentration of methane in the atmosphere is currently increasing by 1.5% per year. What is the net annual increase in the atmospheric load of methane?

Sections 9.3 - 9.4

1. The equilibrium NO (g) + O_3 (g) $\rightleftharpoons$ NO_2 (g) + O_2 (g) has $K_p = 3.6 \times 10^{34}$ in the absence of sunlight at 300 K.
   a) Is the system p(NO) = 40 ppbv, p(O_3) = 120 ppbv, p(NO_2) = 86 ppbv, p(O_2) = 0.21 atm at equilibrium.
   b) Would it be possible for the conditions given in part (a) to represent a steady state?

2. Under conditions of heavy air pollution p(O_3) may reach 150 ppbv.
   a) Express this in the units moles per liter at 300K.
   b) Would the injection of NO (g) in to the atmosphere be a practical strategy for the removal of O_3 by the reaction:

$$O_3 + NO \longrightarrow NO_2 + O_2$$

3. Some scientists favour describing the oxidation of tropospheric traces gases in terms of "cycles" of reactions, in which the formation and destruction of all free radical intermediate species are made to cancel out. One such cycle is given below for the oxidation of CO.

$$CO + OH \longrightarrow CO_2 + H$$
$$H + O_2 \longrightarrow HO_2$$
$$HO_2 + NO \longrightarrow NO_2 + OH$$
$$NO_2 + h\nu \longrightarrow NO + O$$
$$O + O_2 \longrightarrow O_3$$

   a) What is the overall reaction that corresponds to this "cycle"?
   b) According to this mechanism, what is the net effect of oxidizing CO in the atmosphere?
   c) Which substances act as catalysts in the above cycle?
   d) Quantitatively, what is the effect on the atmosphere in a large city of oxidizing 100 ppbv of CO?

4. The rate of the reaction O_3 (g) + NO (g) $\longrightarrow$ O_2 (g) + NO_2 (g) has been measured at several temperatures. Calculate the activation energy from the data below, and also calculate the rate of reaction at -20°C.

t, °C	-30	-10	+10	+30
k, cm^3 $molec^{-1}$ s^{-1}	5.9×10^{-15}	9.3×10^{-15}	1.4×10^{-14}	1.9×10^{-14}

5. Consider the reaction sequence:

   (i) NO_2 (g) + OH (g) $\longrightarrow$ HNO_3 (g) $k = 2.0 \times 10^{-11}$ cm^3 molec^{-1} s^{-1}

   (ii) HNO_3 (g) + H_2O $\longrightarrow$ HNO_3 (aq)

   a) Explain how this could be considered a sink for tropospheric NO_x

   b) Work out the enthalpy change for reaction (i). Is the enthalpy change compatible with the large rate constant for this reaction?

   c) The following rate constants (300 K) are for reactions of the substances stated with OH:

   NO_2: 2.0 x 10^{-11} cm^3 molec^{-1} s^{-1}

   CO: 2.7 x 10^{-13} cm^3 molec^{-1} s^{-1}

   Hydrocarbons: ~ 5 x 10^{-14} cm^3 molec^{-1} s^{-1}

   Under the following conditions, which is the predominant sink reaction for OH: [NO_2] = 80 ppbv, [CO] = 10 ppm, [Hydrocarbon] = 2.0 ppm?

   d) Calculate the lifetime of OH under these conditions. How does it compare with the lifetime of OH in the unpolluted troposphere: Section 2, Problems 1 and 2?

Sections 9.6 - 9.9

1. Estimate the strength of the Pb-C bonds in tetraethyllead from the following data:

Enthalpy of atomization of lead (s)	195.0 kJ mol^{-1}
Enthalpy of formation of the ethyl radical (C_2H_5,g)	105 kJ mol^{-1}
Enthalpy of formation of tetraethyllead (ℓ)	52.7 kJ mol^{-1}
Enthalpy of vaporization of tetraethyllead (ℓ)	57 kJ mol^{-1}

2. The destruction of tetraethyllead which escapes into the atmosphere is initiated by reaction with the hydroxyl radical

   $$Pb(C_2H_5)_4 + OH \longrightarrow Pb(C_2H_5)_3(C_2H_4) + H_2O$$

   The rate constant for this reaction is 2.0 x 10^{-11} cm^3 molec^{-1} s^{-1} at 20°C.

   a) Write the rate expression for this reaction

   b) Does the rate equation simplify if the concentration of OH is constant?

   c) Calculate the half-life of tetraethyllead in the atmosphere under conditions where [OH] has the constant value 3.8 x 10^6 molec cm^{-3}.

3. Methyl tert-butyl ether (MTBE) is synthesized by the reaction below from methanol and isobutylene

$$CH_3OH + (CH_3)_2C{=}CH_2 \longrightarrow (CH_3)_3\ C\text{-}OCH_3$$

In 1991, the production of MTBE was 4.4×10^6 t

a) What mass of isobutylene was used to synthesize MTBE?

b) What is the percent by weight oxygen in MTBE?

c) What mass of MTBE would need to be added to 1.0 t of gasoline (assumed composition CH_2 as its empirical formula) in order to make an oxygenated fuel that was 2.0% oxygen by weight?

4. Trichloroethylene (TCE) is an example of a volatile organic compound. Its bimolecular reaction with OH has $k = 2.3 \times 10^{-12}$ cm^3 molec^{-1} s^{-1} at 300 K. Estimate the lifetime of TCE in the atmosphere on a day when the average concentration of OH is 2.0×10^6 molec cm^{-3}.

10 WATER

10.1 Physical properties of water

Water is a remarkable substance. It is an excellent solvent, and as a result many chemical reactions take place in aqueous solution. Besides this, water is extremely common in the environment; the oceans cover 70 % of the Earth's surface, and few locations exist on Earth where substances which react chemically with water can survive unchanged. In addition, a vast array of environmentally important reactions takes place in aqueous solution - in raindrops, in rivers and lakes, and in the oceans. Almost all biochemical reactions take place in water, and water is the major constituent of all living matter (about 62% in the case of humans).

Some of the properties of water are familiar to everyone: its melting point (0 °C) and its boiling point at 1 atm (100 °C). As a practical point, water is the only substance which most people have ever encountered in all three states of matter: solid (ice), liquid (water), and gas (steam). The physical parameters of water are quite remarkable, as is evident by comparing it with neighbouring hydrides in the Periodic Table (Table 1).

Table 10.1: Physical properties of simple hydrides

Substance	CH_4	NH_3	H_2O	HF	H_2S
Melting point, °C	-182	-78	+0	-83	-86
Boiling point, °C	-164	-33	+100	+19	-61

The high m.p. and b.p. of water are important for the existence of life on this planet, because water is a liquid at most places on the Earth. We saw in Chapter

6 that the insulating property of the troposphere warms the Earth's surface to an average +15 °C, where water is a liquid, instead of the -30 °C that would otherwise be expected. At -30 °C, all the water on earth would be frozen, and life as we know it would not be possible.

Another unusual property of water which is important for the existence of life is that the liquid has a greater density than the solid (density of $H_2O(\ell)$ = 1.0 g cm $^{-3}$ *vs.* 0.9 g cm $^{-3}$ for $H_2O(s)$). Very few substances are more dense in the liquid state than in the solid state. But for this property, ice would fall to the bottom of lakes in winter, and would not melt completely in the summer. The turnover of lakes in fall and spring, which also has the effect of recycling nutrients, occurs because water has a maximum density at 4 °C. In the fall, as the temperature of many lake surfaces drops to 4 °C, the surface layers become denser than the layers underneath, and therefore fall, mixing the lake vertically.

As noted in Chapter 6, the vapour pressure of water rises sharply with increasing temperature, and so water, like other liquids, can evaporate below its normal boiling point. Evaporation recycles water from the oceans into the atmosphere, and back to earth as rain and snow, without the liquid actually having to boil.

Water is classified as a polar liquid because the hydrogen and oxygen atoms share the electrons of the covalent O-H bond unequally. The hydrogen atoms thus acquire a partial positive character, and the oxygen atoms a partial negative character. The partial positive charge on hydrogen causes attraction with non-bonding electrons on the oxygen atoms, a phenomenon known as **hydrogen bonding**, because it is unique to the chemistry of very polar X-H bonds. The unusually high m.p. and b.p. of water are attributed to the capacity of individual water molecules to hydrogen bond to their neighbours; water is thus an "associated liquid". In a sense, water has high m.p. and b.p. because it is a "bigger" molecule than the simple formula H_2O would indicate.

The polar nature of water makes it a good solvent for other polar substances. These may be covalent compounds, especially compounds containing nitrogen or oxygen atoms (with which water can form hydrogen bonds), or ionic compounds. Non-polar substances (hydrocarbons, oils, greases, gasoline etc) are not appreciably soluble in water. Water is a quite reactive substance, and some substances react with water rather than simply dissolving. For example, CCl_4 is a non-polar liquid which is only slightly soluble in water, while $SiCl_4$ (same group of the Periodic Table) reacts vigorously with water.

$$SiCl_4 + 2\ H_2O \rightarrow SiO_2 + 4\ HCl$$

Unfortunately, there are no simple rules for predicting which compounds will be decomposed by water in this way.

When an ionic compound dissolves in water, the substance is transformed from an ordered crystal in which the cations and anions are held together by electrostatic attraction into a solution in which each cation and anion is **hydrated** (surrounded by water molecules)[1]. Not all ionic compounds dissolve in water to an appreciable extent; we shall learn **solubility rules** to predict which ionic compounds are water-soluble in Chapter 12. However, we can state here that in order for a compound to have appreciable solubility, the process X(s) → X(aq) must be thermodynamically favourable. In the case of ionic compounds, dissolution will occur only if the energy of hydration of the ions is sufficient to overcome the electrostatic attraction between the oppositely charged ions in the crystal. Many insoluble salts are composed of cations and anions with charges > 1. In such cases, the energy of hydration may not be sufficient to compensate for the loss of the strong electrostatic attraction of the ions in the crystal.

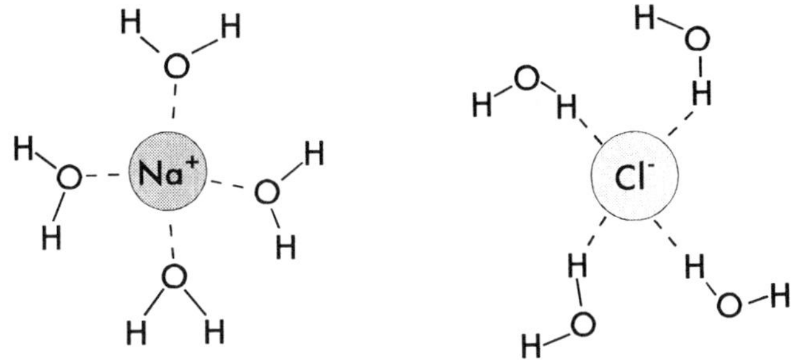

Cations and anions may be hydrated, or surrounded by water molecules.

> ***Example:*** *NaCl is soluble in water; the stabilization of the* Na^+ *and* Cl^- *ions by hydration more than compensates for the loss of the electrostatic attraction of these ions for each other in the crystal.*
>
> $$NaCl(s) \rightarrow Na^+(aq) + Cl^-(aq)$$
>
> $CaCO_3$ *is insoluble in water; the attraction of the doubly charged calcium ions and carbonate ions is so strong that the solvation energy does not compensate.*
>
> $$CaCO_3(s) \not\rightarrow Ca^{2+}(aq) + CO_3^{2-}(aq)$$

The energetics of dissolution in water are rather more complex than the foregoing argument suggests. We shall return to this question in Section 14.6, after we have discussed entropy changes in physical and chemical reactions. At this point, we note simply that the reason that non-polar substances such as

[1] Hydrated is the word used specifically when water is the solvent. The corresponding general term for solvent stabilization of a solute is **solvated**.

hydrocarbons fail to dissolve in water is **not** because the reaction is enthalpically unfavourable, but because it is unfavourable in the entropic sense. In particular, many substances have low solubility because they cause substantial loss of entropy of the solvent water. In other cases, high solubility can result even when ΔH° is positive, because the entropy change is favourable.

The previous paragraph alerts us to an important property of solvents, and especially water. Solvents are not continuous media which are indifferent to the presence of the solute. Frequently, the solvent plays a key role in directing chemical and biochemical processes. For example, when the protein hemoglobin binds molecular oxygen in the lungs, as many as 70 additional moles of water of hydration are taken up per mole of hemoglobin. The energy of hydration of the oxygenated compared with the deoxygenated form of the protein is important in holding the protein in the correct molecular shape to accept the oxygen molecules.

10.2 Occurrence of water

Two thirds of the Earth's surface is covered with water. Most of this is ocean, as seen in the figures below.

Distribution of the Earth's surface water	
Oceans	9.5×10^{19} mol (> 99%)
Lakes and Rivers	1.7×10^{15} mol
Atmosphere	7.2×10^{14} mol

As well, there is water deep underground, called "ground water".

Water is a vital commodity, with agriculture and industry making enormous demands on this resource. In many parts of the world, water use is greater than can be sustained without importation.

10.2.1 Dissolved solids in natural waters

The amount of solid dissolved in natural waters varies widely. The values in Table 2 are typical of river and ocean water, although river water is quite variable in its mineral content. Ground water is at least as high in dissolved solids as lake and river water; sometimes it is much higher, with total dissolved solids exceeding 1000 ppm.

Table 10.2: Typical concentrations of ions in river and sea water

Ion	c(river) mmol L^{-1}	c(ocean) mmol L^{-1}	Annual input to oceans, Tmol yr^{-1}
Cl^-	0.22	550	7.2
Na^+	0.27	460	9.0
Mg^{2+}	0.17	54	5.5
SO_4^{2-}	0.12	28	3.8
K^+	0.059	10	1.9
Ca^{2+}	0.038	10	12
HCO_3^-	0.095	2.3	32

Example: *Use the data of Table 10.2 to calculate (a) the residence time of Ca^{2+} in the oceans (b) the total dissolved solids in the oceans in grams per liter.*

Answer:

$$\textit{residence time} = \frac{\textit{amount in reservoir}}{\textit{rate of input}}$$

Volume of ocean $= (9.5 \times 10^{19}\ mol) \times 18\ g\ mol^{-1} \times (1\ mL/1\ g)$
$= 1.7 \times 10^{21}\ mL = 1.7 \times 10^{18}\ L$

amount in reservoir $= 10\ mmol\ L^{-1} \times 1.7 \times 10^{18}\ L$
$= 1.7 \times 10^{16}\ mol$

rate of input $= 12 \times 10^{12}\ mol\ yr^{-1}$

residence time $= 1.7 \times 10^{16}\ mol/1.2 \times 10^{13}\ mol\ yr^{-1}$
$= 1 \times 10^{3}\ yr$

(b) Sum all components in Table 10.2:

conc $(g\ L^{-1}) = $ *conc* $(mmol\ L^{-1}) \times \mathbf{M}(g\ mol^{-1}) \times (1\ mol/1000\ mmol)$

Cl, 19.5; Na, 10.6; Mg, 1.3; SO_4, 2.7; K, 0.4; Ca, 0.4; HCO_3, 0.1

Total $= 35\ g\ L^{-1}$

Seawater is a much more concentrated salt solution than almost any other natural water, with the exception of inland lakes in areas of high evaporation, such as the Dead Sea in Israel and the Great Salt Lake in Utah. Over the millennia, rivers have carried dissolved minerals from the land, but the oceans can lose only the water and not the salts by evaporation. As a result, the oceans are the major reservoir of "soluble" ions such as Na^+ and Cl^-. Naturally occurring deposits of substances such as NaCl (rock salt) are believed to have formed through the evaporation of ancient seas.

With the exceptions of Ca^{2+} and HCO_3^-, there is a parallel between the average concentrations of ions in fresh and in ocean water. There is relatively more Ca^{2+} and HCO_3^- in river water because rivers **dissolve** ancient rocks containing $CaCO_3$, whereas the oceans **precipitate** $CaCO_3$ in the form of marine organisms' exoskeletons.

10.2.2 Dissolved solids and the freezing point of water

As already noted, the freezing point of pure water at 1 atm pressure is 0 °C (273.15 K). However, the freezing point of water changes when solutes are dissolved in it — as does the freezing point of any other pure substance when it contains added solutes. The direction of the change in the freezing point is always to lower the freezing temperature below that of the pure liquid. Thus, blood, sea water, beer, wine and maple syrup all freeze at a temperature lower than 0 °C.

The extent to which a solute lowers the freezing point of water, or another solvent, can be described quantitatively.

$$\Delta T = k_f m$$

ΔT is the difference in temperature between the freezing point of the pure solvent and that of the solution, k_f is a constant called the **molal freezing point depression constant**, and m is the concentration of the solution, expressed in the units mol solute per kg solvent (molality). The constant k_f has the units K kg mol $^{-1}$; what is especially noteworthy about k_f is that it is a property of the *solvent*, not a property of the solute. In other words, for a given mass of a particular solvent, equal moles of all solutes produce the same depression of the freezing point. This is true of all solutes, not just solid solutes. Table 10.3 gives some k_f values for different solvents.

Table 10.3: Molal freezing point depressions of different pure liquids

Substance	k_f, K kg mol $^{-1}$	Substance	k_f, K kg mol $^{-1}$
water	1.86	acetic acid	3.90
acetone	2.40	carbon tetrachloride	29.8
cyclohexane	20.2	benzene	4.90

The freezing point depression of a solution depends on the molal concentration of solute particles, not on their identity. When the solvent is water, and the solute is an electrolyte, the number of moles of solute particles is the total number of moles of ions. For example, if 1 mol of a non-ionic solute such as ethyl alcohol dissolves in 1 kg of water, the freezing point is depressed by 1.86 K (from 0.00 °C to -1.86 °C). However, if 1 mol of $CaCl_2$ dissolves in 1 kg of water, the freezing point is calculated to fall to -5.58 °C (*i.e.* 3 x 1.86 °C) because $CaCl_2$ dissolves to gives 3 moles of ions[2].

Example: *The normal freezing point of benzene is 5.5 °C. What mass of toluene (C_7H_8) must be dissolved in 10.0 g of benzene to lower its freezing point to 4.9 °C?*

Answer: *Benzene has k_f = 4.90 K kg mol^{-1}*

m = $\Delta T/k_f$ = (0.6 K)/(4.90 K kg mol^{-1}) = 0.12 mol kg^{-1}

c = 0.12 mol kg^{-1} x **M**(C_7H_8) = 0.12 mol kg^{-1} x 92 g mol^{-1} = 11 g kg^{-1}

In 10.0 g benzene:

mass C_7H_8 = (11 g toluene/kg benzene) x 0.0100 kg benzene = 0.1 g

We are now able to explain several observations of environmental interest. Salt (as NaCl or $CaCl_2$) is spread on roads and sidewalks in the winter to melt snow. The melting point (freezing point) of the snow is lowered when the salt dissolves. Urea (H_2N-CO-NH_2) is also used as a sidewalk deicer, but it is less effective mole-for-mole, because it is non-ionic[3]. A bottle of liquor, usually 40% ethyl alcohol by volume, does not freeze even in the freezer section of your refrigerator.

Sea water does not freeze even at temperatures well below 0 °C, because it is a rather concentrated electrolyte solution. Certain fish which live in Antarctic waters produce their own antifreeze substances which are secreted into the bloodstream in order to prevent them from freezing in the frigid water. This is necessary, even though blood contains dissolved solutes; the molal concentration of solutes in blood is considerably lower than the total concentration of solutes in sea water. Many trees produce antifreeze substances to prevent their cells from freezing, and hence bursting, during a harsh winter. These substances are broken down with the approach of warmer spring weather. This explains why a tree may

---

[2] This is something of an oversimplification. In practice, the electrostatic attraction of oppositely charged ions means that at any moment, some of the ions will be travelling through the solution as **ion pairs** (A^+B^-), rather than as independent ions A^+ and B^-. The extent of ion pairing can be measured experimentally. The neglect of ion pairing causes the freezing point depression of an electrolyte solution to be overestimated.

[3] The disadvantage of using NaCl or $CaCl_2$ is that their solutions are electrolytes, and promote corrosion of metals such as steel (Chapter 17). The use of salt as a deicer causes rusting of automobiles and other metal structures. Urea is less prone to promote rusting because it is a non-electrolyte.

survive the winter, yet suffer severe frost damage during a late cold snap in the spring: by this time the antifreeze substances have been degraded, leaving the tree unprotected.

10.2.3 Dissolved solids and irrigation

Many parts of the world have a climate which is well suited to agriculture, but lack sufficient rainfall; these include parts of the southern Canadian Prairies and many parts of the U.S. Great Plains and Southwest. Irrigation has brought successful agriculture to areas that were formerly desert or near-desert; when you fly over Nevada or southern Alberta, it is easy to spot the green irrigated areas among the brown natural background. The irrigated areas of the U.S. Southwest can produce a crop of alfalfa hay every 30 days through the growing season. Besides agricultural irrigation, of course, an enormous amount of water is needed to supply the fast-growing cities of the U.S. Southwest (and to create green suburban lawns in the desert).

The price of irrigation technology is paid in the form of increased salinity and aquifer depletion. Increased salinity arises because irrigation in a dry climate is inevitably accompanied by high rates of evaporation. When the water evaporates, any dissolved solids remain in the soil. Continued irrigation therefore leads to a buildup of salts in the soil, until plant growth becomes impossible because of the salinity of the soil. One theory for the decline of the ancient civilizations in the Tigris and Euphrates valleys is that increasing salinity may have reduced the fertility of the soil. In modern times, inorganic residues from fertilizers compound the problem.

Salinity was not recognized as a problem when irrigation was first introduced in North America, and some farms went through a complete cycle of desert - high productivity - reduced productivity - abandonment in as little as twenty years. Increasing salinity is being recognized as a serious problem in Pakistan, where loss of land to salinity threatens to wipe out the increases in agricultural production that have been made in the past generation through the application of intensive farming methods. In North America, modern irrigation technology extends the working life of the land by "back-flushing": after irrigation for a certain time, a heavy application of water is made to the land in order to wash out the accumulated salts with the run-off, which is then returned to a convenient river.

The example of the degradation of the quality of the Colorado River due to irrigation has been extensively discussed. The use of its waters for irrigation has substantially reduced its flow, and in addition, its waters are used and reused for irrigation so many times that the water delivered to farms way down-stream is already rather highly saline. Several tributaries of the Colorado have concentrations of dissolved solids in excess of 1 gram per liter (the average dissolved salt concentration in river water as calculated from the data of Table 10.2 is 0.04 g L^{-1}). A huge desalination plant has been built in Arizona to remove dissolved salts from the Rio Grande in order to permit its continued reuse for irrigation downstream, but in practice the technology has been expensive and of only limited success. This technology involves **reverse osmosis** (Section 10.7).

Irrigation water comes either from rivers or from underground aquifers. Much of the southern U.S. is irrigated from the immense Ogallala Aquifer, which underlies parts of eight states. In some desert and semi-desert areas, the rates of water use from the aquifer are one to three orders of magnitude greater than the rates of recharge. Put another way, the consumption of this resource is ten to one thousand times the annual rainfall. By 1980, Texas alone had already consumed (mined) over 20% of its share of this water. Furthermore, in some parts of the aquifer the water contains high concentrations of dissolved salts even before it is used for irrigation. These issues suggest that intensive agricultural production on the more arid parts of the U.S. Plains may not be sustainable in the long term.

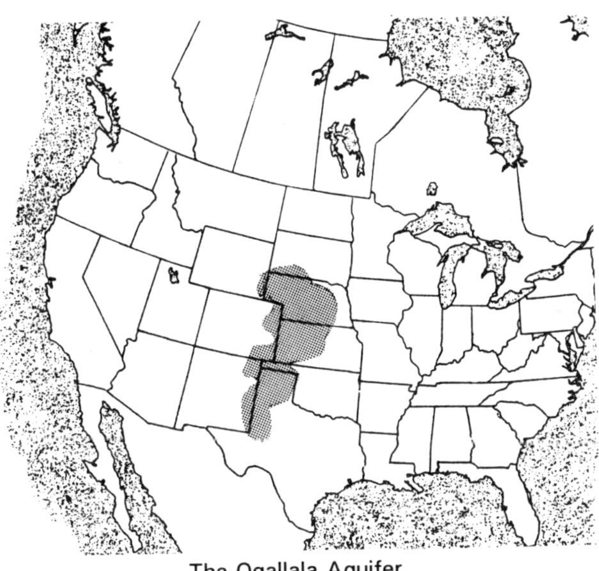
The Ogallala Aquifer

Depletion of natural waters as result of irrigation is not confined to North America. The huge aquifer beneath the Sahara Desert is being consumed rapidly for agricultural irrigation in Libya; its rate of recharge is essentially zero, so that its use amounts to mining a non-renewable resource. Likewise, the shores of the Dead Sea in Israel, and the Aral Sea in Russia have each retreated many kilometers in the last few decades. Agricultural irrigation has reduced the inflow from their rivers to the point that it is quite possible that these inland seas may cease to exist in the foreseeable future.

10.3 Water quality and water pollution

The issues raised in the preceding section are one aspect of water quality - in that case, the degradation of water quality when "used" irrigation water is returned to a river. One of the greatest difficulties in discussing water quality is that there is no absolute standard. For instance in my home town of Guelph, Ontario, the municipal water is of excellent quality for drinking, yet it is so laden with minerals that it would be completely useless for raising steam in an industrial process (the boiler and pipes would rapidly accrete a thick layer of scale due to the "hardness" of the water: Section 10.4).

Water quality depends upon the use to which the water is to be put. Water for drinking must not be contaminated with microorganisms or with toxic trace metals such as lead, cadmium, and mercury. However, its hardness is immaterial. The opposite criteria apply to industrial water for raising steam. Industrial cooling water - for example, for power stations - is subject to few restrictions, provided that it is sufficiently cool, and not corrosive to the heat exchanging equipment.

Water pollution is usually discussed in the context of substances of industrial origin: both inorganic (metallic) and organic (*e.g.*, pesticides). For example, concern has been raised about the water quality in the Great Lakes, the Rhine and the Danube River, all of which contain traces of industrial compounds such as polychlorinated biphenyls (PCBs), dioxins, and pesticides. One source of such contamination is the discharge of water containing traces of industrial products from manufacturing facilities into rivers and streams. Similarly, the leaching (dissolution) of toxic substances from contaminated soil at industrial plant sites, and from municipal and chemical waste dumps, can cause contamination of underlying aquifers.

Another definition of poor water quality is that the water is unfit for its intended use. This approach to pollution is becoming increasingly relevant because of advances in analytical chemistry, which have made it possible to detect almost any contaminant of interest in almost any environmental matrix. If, as in the Great Lakes, the water contains several parts per trillion of polychlorinated biphenyls, should the water be considered polluted? Our answer would be "No" if we want to bathe in it, or use it for boating, but may well be "Yes" if we are concerned about some of the trends which seem to be taking place in wildlife populations (unusually large numbers of fish cancers, and birth deformities in aquatic birds). Ironically, these waters still appear to be safe for drinking; the amounts ingested by someone drinking an average 2 L of water per day have not been shown to cause harm, whereas the fish which lives its whole life in the water has a much higher exposure to contaminants (see further discussion in Section 12.4).

On the basis of fitness for an intended purpose, the city of Regina, Saskatchewan has had a longstanding problem with **geosmin**, which is a naturally occurring organic compound produced by fungi during hot summer weather. Although geosmin is not highly toxic, it imparts a strong earthy flavour to the water, which residents find objectionable. Geosmin is not caused by industrial pollution, but nevertheless makes the water unfit for its intended purpose. As another example, the hot springs in both Jasper and Banff National Parks in the Canadian Rockies contain relatively large amounts of sulfides and toxic metals. In this context, "relatively large" means that the water would be unfit as a drinking water source, but is perfectly acceptable for bathing.

Specific examples of water pollution will be taken up in other chapters. These include biomagnification (Chapter 12), acid rain (Chapter 13), mercury pollution from the chlor-alkali industry (Chapter 16), and acid mine drainage (Chapter 17).

Other important environmental issues include oxygen depletion of natural waters due to thermal pollution or due to the presence of excessive quantities of dissolved material which can be oxidized biologically (Section 12.2). This can greatly compromise the survival of aquatic life. Aquatic life is also adversely affected by **eutrophication** (accelerated aging of lakes) through excessive inputs of phosphates into waterways. Phosphates enter aquatic ecosystems through run-off of phosphate fertilizers from farms and through the use of detergents containing phosphates. One manifestation of eutrophication is the growth of large mats ("blooms") of algae, which deplete the water of oxygen when they die and decay. Lack of oxygen in the water causes suffocation of aquatic animals such as fish, and also promotes anaerobic decomposition of the algae, with the formation of unpleasant-smelling byproducts of decay. Eutrophication has declined sharply in North America during the past 20 years, as a result of legislation to limit the phosphate content of detergents and through advanced methods of sewage treatment which remove phosphate at the sewage works (Section 12.3).

Phosphorus is a major "villain" in the context of eutrophication because it is the "limiting nutrient" in the growth of algae. Microorganisms require carbon, nitrogen, and phosphorus as major nutrients. Just as chemical reactions stop once the limiting reactant has been consumed (Section 3.4), so the growth of algae is limited by the availability of nutrients in the water. The uptake of nutrient elements into biomass takes place in the ratio C: N: P = 100: 15: 1. Carbon is almost never limiting in water because it can be resupplied from the atmosphere as CO_2 (Section 12.2), and blue-green algae can supplement the dissolved nitrogen compounds by fixing atmospheric nitrogen. Thus phosphorus is usually the limiting nutrient, even though it is needed in the smallest amount.

In the 1950s and 1960s, certain phosphate salts such as sodium tripolyphosphate ($Na_3H_2P_3O_{10}$) were added in large amounts to detergent formulations in order to regulate the pH of the wash and to keep ions such as Ca^{2+} (next section) in solution. A high proportion of the phosphate was passed through sewage treatment facilities into streams and lakes, causing annual problems of algal blooms. In Canada, the amount of phosphate which can be used in detergents was limited to 20% by weight in 1970 and 5% in 1973. As a result, the average phosphorus content of raw Ontario sewage dropped from 10 ppm in 1969 to 5 ppm in 1974, and has remained below 5 ppm since then.

Additionally, advanced methods of sewage treatment have been introduced in many communities to remove phosphate from sewage (Section 12.3). Today the chief source of phosphates in natural waters in North America is agriculture, in the form of run-off from fertilizer application to the land, and run-off of manure from feedlots.

The improvement in water quality in the Great Lakes basin has been most evident in Lake Erie, which was generally regarded as "dead" in the 1970s — stinking algal blooms were an annual occurrence, and the commercial fishery was no longer viable. With the reduction in phosphate loadings, Lake Erie is essentially free of algal blooms and once again supports commercial fishing. Lake Erie's health was restored relatively quickly, because the residence time of the water in the lake is short — about 2.7 years. Pollutants in the lake can thus be flushed out relatively quickly.

The Great Lakes Basin

The speed with which a water body can become polluted depends in part upon the residence time of the water. Table 10.4 shows the residence times of the water in the five Great Lakes.

Table 10.4: Residence times of water in the Great Lakes

Lake	**Volume, km^3**	**Residence time, yr**
Superior	1.2×10^4	190
Michigan	4.9×10^3	100
Huron	3.5×10^3	22
Erie	4.8×10^2	2.7
Ontario	1.6×10^3	6.0

In the 1960s, Lake Erie was most susceptible to pollution because of two factors: high population density, and small volume of water to receive pollutants. Fortunately, it was able to recover fastest because of the short residence time. At the other end of the scale, Lake Superior is resistant to contamination because of its large volume, but would take an extremely long time to recover should it become seriously polluted.

Example: *If the concentration of a pollutant in Lake Superior were 1.0 ppm today, how long would it take for its concentration to fall to 0.1 ppm if all input of the pollutant into the lake ceased immediately?*

Answer: *Calculate the first order rate constant for loss of pollutant from the residence time of the water in the lake.*

$$k = (\text{residence time})^{-1} = (190\ yr)^{-1} = 5.3 \times 10^{-3}\ yr^{-1}$$

For a first order process, $\ln(c_o/c) = kt$

$$t = \ln(c_o/c)/k = \ln(1.0/0.1)/(5.3 \times 10^{-3}\ yr^{-1})$$
$$= 4.3 \times 10^2\ yr = \text{roughly the year 2400}$$

10.4 Hard and soft water

"Hard" is the term used to describe water which contains appreciable (> several ppm) amounts of the "hardness cations", calcium and magnesium. These are the cations which cause precipitation of scum with soaps, and deposits of scale in boilers, water heaters, and hot water pipes. They originate when natural water dissolves them (usually as carbonates) from underlying rocks. In limestone areas the concentration of Ca^{2+} that of Mg^{2+}. An exception occurs where the underlying rock is dolomite (dolomitic limestone, $CaCO_3.MgCO_3$) and significant concentrations of both Ca^{2+} and Mg^{2+} are present in the water.

From the foregoing discussion, it may be guessed that soft water contains only low concentrations of calcium and magnesium. Depending on its source, it may contain alkali metal cations, particularly sodium, or it may contain very little dissolved solid at all. Soft water is encountered commonly in regions where the underlying rock is granite, which is very insoluble. New England, much of eastern Canada, Scotland, and Scandinavia are all areas where soft water is common. Rainwater contains extremely low concentrations of dissolved solids, which is why it was formerly collected in cisterns and used for laundry (no scum formation with soap). This is no longer necessary, because modern detergents are formulated to maintain their effectiveness even in hard water.

The chemistry associated with water hardness is as follows. Hardness is introduced into the water when rainwater (soft) comes in contact with the underlying rocks of streams. The commonest mechanism for acquiring hardness is when CO_2-laden water comes in contact with chalk or limestone (both $CaCO_3$).

$$CaCO_3(s) + H_2O(\ell) + CO_2(aq) \rightarrow Ca(HCO_3)_2(aq)$$

Scale is deposited when the water is heated, expelling CO_2 from the solution.

$$Ca(HCO_3)_2(aq) \rightarrow CaCO_3(s) + H_2O(\ell) + CO_2(g)$$

More details on these reactions are given in Chapter 12.

The analysis for calcium is carried out by titrating the water sample against a standard solution of ethylenediaminetetraacetic acid (EDTA), which is supplied as its tetrasodium salt, $EDTA^{4-}$.

$$Ca^{2+} + EDTA^{4-} \rightarrow (CaEDTA)^{2-} \qquad K_c = 5 \times 10^{10}\ L\ mol^{-1}$$

All these substances are colourless, so an indicator, Eriochrome Black T, is used to signal the end-point. Free Eriochrome Black T is dark blue in solution. Also added to the solution at the start of the titration is Mg^{2+}, which forms a red complex with Eriochrome Black T. The approach of the endpoint is signalled by the colour of the solution changing from red (Mg^{2+}/Eriochrome) to blue (free Eriochrome) because the Mg^{2+} detaches from the indicator and associates with the EDTA, once the last of the calcium ions have become bound to EDTA, and free EDTA remains in solution.

Example: *A 0.100 L sample of water is titrated against 0.00334 mol L^{-1} Na_4EDTA solution, and requires 8.84 mL to reach the endpoint. Calculate the hardness of the water in ppm of calcium.*

Answer: *From the equation (given above) the stoichiometry of the reaction is 1:1. First calculate the concentration of Ca^{2+} in mol L^{-1} or equivalent units.*

$n(EDTA) = c \times V = 3.34\ mmol\ L^{-1} \times 8.84\ mL = 29.5\ \mu mol$

$n(Ca^{2+}) = n(EDTA) = 29.5\ \mu mol$

$c(Ca^{2+}) = n/V = 29.5\ \mu mol/0.100\ L = 295\ \mu mol\ L^{-1}\ (0.295\ mmol\ L^{-1})$

Now convert the concentration to ppm, same as mg/L in aqueous solution.

$c(Ca^{2+}) = 0.295\ mmol\ L^{-1} \times M(Ca^{2+})$

$= 0.295\ mmol\ L^{-1} \times 40.1\ mg\ mmol^{-1}$

$= 11.8\ mg\ L^{-1} = 11.8\ ppm$

10.5 Water softening

Tap water, or water drawn from a lake or river, is unsuitable for many industrial purposes. Most importantly, large quantities of industrial water are used to make steam, for large central heating installations, and to drive turbines in electricity generating plants. Any calcium or magnesium salts in the water would be left behind as **scale** when the water vaporized; this would greatly retard heat transfer to the water, and would eventually fill the boiler. In hot water systems in the home, scale has the same effect on water heaters and hot water pipes when the water is hard. In addition, calcium and magnesium salts give scummy, or curdy, precipitates with soap[4]. Precipitation as scum renders the soap ineffective for cleaning, as well as leaving an unsightly mess in the bath-tub.

Water softening removes the hardness ions from water, and thus protects water installations from scaly deposits in steam and hot water systems. There are several methods of softening water.

10.5.1 Lime softening

Treatment with lime ($Ca(OH)_2$) is used to soften water on a large scale, as for example at an electric power station. The raw water is treated with the stoichiometric amount of $Ca(OH)_2$.

$$Ca(OH)_2(aq) + Ca(HCO_3)_2(aq) \rightarrow 2\ CaCO_3(s) + 2\ H_2O(\ell)$$

Because lime softening involves addition of one calcium compound (lime) to precipitate another, the exact stoichiometry must be maintained, otherwise the treated water will still contain Ca^{2+}. Lime softening is only practical industrially, where the composition of the raw water can be closely monitored. Its benefit is the low cost of lime as a softening agent.

10.5.2 Ion exchange

Ion exchangers are available for both home and industrial use. A cation exchanger is an insoluble inorganic or organic polymer, which carries multiple negative charges on its backbone. Its multiple negative charges are balanced by the positive charges of cations. However, the cations are not an integral part of the structure, but exist in the aqueous phase, loosely bound to the anionic sites on the solid polymer. The cations are free to move within the structure of the polymer, and one cation can be replaced (exchanged) by another. Many

---

[4] This is because soaps are the sodium salts of organic acids. The corresponding calcium salts are insoluble (scum), and precipitation therefore removes the soap from the water before it can exert any cleaning action.

$$2\ Na^+(soap)^-(aq) + Ca^{2+}(aq) \rightarrow Ca(soap)_2(s) + 2\ Na^+(aq)$$

substances have cation exchange properties, but all have in common that the polymer is a solid polyanion, which is insoluble in water. The anionic sites on the polymer backbone attract cations electrostatically, and one can write an equilibrium reaction for this association. This can be written as a chemical equation by designating the anionic sites on the polymer as (A^-).

$$X^+(\text{aq, a cation}) + (A^-) \rightleftharpoons (X^+A^-)$$

When the ion exchanger is ready for use the associated cations are normally Na^+. When hard water passes over a bed of the ion exchange resin, the hardness cations in the water can exchange with the sodium ions associated with the resin. That is, calcium (or magnesium) from the water becomes associated with the ion exchanger, and sodium leaves the resin and enters the water.

(1) $Ca^{2+}(aq) + 2\ Na^+(A^-) \rightleftharpoons 2\ Na^+(aq) + Ca^{2+}(A^-)_2$

When most of the sodium ions have been exchanged for calcium or magnesium, the ion exchanger loses its effectiveness. It can be regenerated by passing a concentrated salt solution through the resin, whereupon the reverse reaction occurs.

(2) $Ca^{2+}(A^-)_2 + 2\ Na^+(aq) \rightleftharpoons Ca^{2+}(aq) + 2\ Na^+(A^-)$

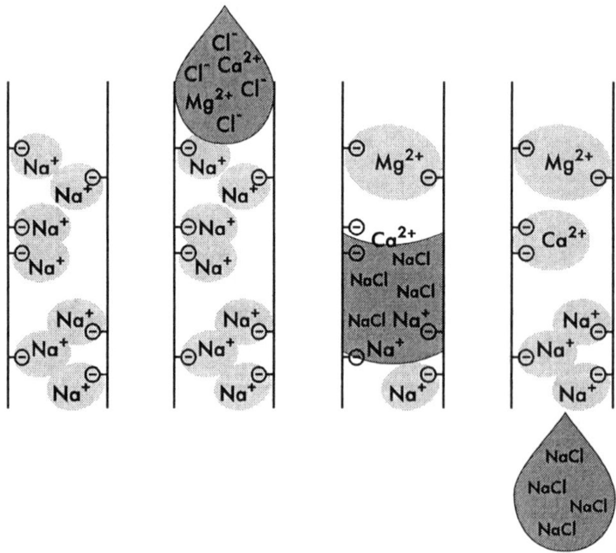

Ion exchange. The anionic sites on the polymer backbone attract cations electrostatically.

The calcium released is discharged to the drain, together with the excess NaCl. The success of the method depends upon the magnitudes of the equilibrium constants for reactions (1) and (2). Reaction (1) is clearly favoured by a large equilibrium constant, which allows efficient removal of even low concentrations of hardness cations from the water; however, if this equilibrium constant is *too* large, the reverse reaction will fail, and it will not be possible to regenerate the resin.

Cation exchangers replace Ca^{2+} in the water with Na^{+}. The ingestion of large amounts of sodium is believed to increase one's risk of heart disease. For this reason, many people who use water softeners in their homes prefer to keep one faucet supplied with unsoftened water for drinking and cooking.

10.6 Distilled and deionized water

These methods are used when water of very high purity is needed. Deionized water is free of ionic solutes, but may still contain dissolved gases and non-ionic solutes. The quality of deionized and distilled water is determined by measuring its electrical resistance; any residual electrolytes in the water will lower its electrical resistance.

10.6.1 Distillation

Distillation is a method for producing extremely pure water. The battery in your car requires distilled water rather than tap water for "topping up", because hardness salts in tap water would precipitate on the lead electrodes and reduce the efficiency of the battery. Distillation involves heating water above the boiling point so that it is converted to steam, and allowing the steam to condense back to pure water in a separate chamber. The impurities (solutes) are left behind. Distillation is not practical on a large scale because of the high energy requirements ($\Delta H°$ of vaporization of water = 41 kJ mol $^{-1}$ at 100 °C).

The production of distilled water of the very highest purity is very laborious. The water must first be distilled in the presence of a substance which can oxidize any non-ionic impurities: $KMnO_4$ is often used for this purpose. Distillation also removes dissolved gases from the water, since all gases become less soluble at higher temperatures (Section 12.2). The distillate is then re-distilled at least once, and must be stored without access to the air if it is to be maintained in an ultra-pure state.

10.6.2 Deionized water

Deionized water is prepared by the use of two ion exchangers in series. The first unit is a cation exchanger, which is similar to that described in Section 10.5, except that H^+ is the counter ion in the active form of the resin, instead of Na^+.

$$Ca^{2+}(aq) + 2\ H^+(A^-) \rightarrow 2\ H^+(aq) + Ca^{2+}(A^-)_2$$

After use, the resin is regenerated with HCl rather than NaCl:

$$Ca^{2+}(A^-)_2 + 2\ H^+(aq) \rightarrow Ca^{2+}(aq) + 2\ H^+(A^-)$$

The second exchanger is an anion exchanger. This time the polymer backbone carries positively charged (cationic) substituents, and the counter ions are OH^-. Writing (C^+) as a cationic site on the resin, and Cl^- as an example of an anion to be exchanged, the reaction for anion exchange is:

$$Cl^-(aq) + (C^+)OH^- \rightarrow OH^-(aq) + (C^+)Cl^-$$

This resin is regenerated with NaOH solution.

The cation exchanger replaced unwanted cations by $H^+(aq)$. Charge balance requires that the anion exchanger replaces unwanted anions by an equal number of OH^- ions. These react with the $H^+(aq)$ to form water, thus removing all the ions from the solution.

$$H^+(aq) + OH^-(aq) \rightarrow H_2O(\ell)$$

Deionized water is much less expensive than distilled water, which has high energy costs, but distilled water is of higher purity than deionized water. Deionized water may still contain small concentrations of non-ionic solutes and dissolved gases, notably CO_2. As we shall see in Chapter 12, the presence of CO_2 makes deionized water slightly acidic.

10.7 Osmosis and reverse osmosis

10.7.1 Osmosis

The concept of osmosis is probably familiar from biology. If a dilute solution and a concentrated solution are separated by a semipermeable membrane, water will flow from the dilute solution into the concentrated one. A semipermeable membrane is one which allows solvent molecules to pass through it, but excludes most types of solute molecules. Normally, this discrimination occurs on the basis of size: the solvent molecules are smaller than the solute molecules. Under these conditions, a substantial difference in height between the two water columns may be established, Figure 1. The difference in pressure exerted on the semipermeable membrane by the two columns is called the **osmotic pressure**. The difference in pressure arises because of the natural tendency of two solutions to equalize their concentrations.

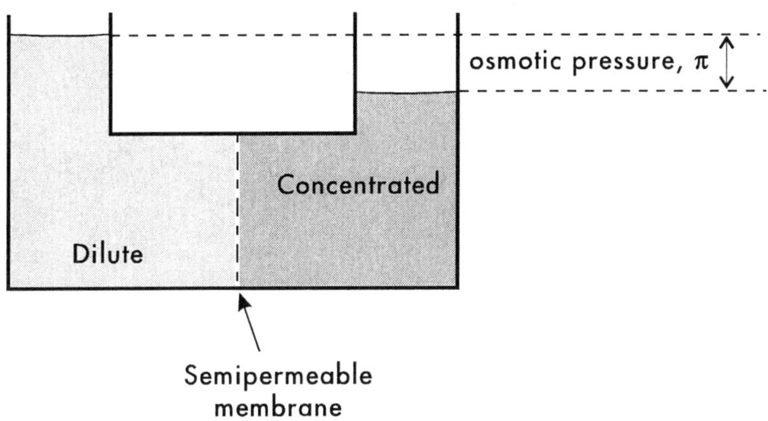

Figure 10.1: Osmotic pressure across a semipermeable membrane

The difference in height of the columns of water on each side of a semipermeable membrane can be related quantitatively to the concentrations of the solutions on each side of the membrane. For the simplest case, where pure water is on one side of the membrane, the relationship is given by eq. [3], whose form is reminiscent of the ideal gas equation (in fact, the Greek letter π is the equivalent of the roman "p").

[3] $\pi = cRT$

When R is expressed in L atm mol^{-1} K^{-1} and c is in mol L^{-1}, π (the osmotic pressure) has the units of atm.

In Section 6.1 we learned that the pressure of a fixed number of moles of any ideal gas at a specified volume and temperature is independent of the identity of the gas: all gases were predicted to behave the same. Likewise, the osmotic pressure of a specified concentration of any solute at a defined temperature is independent of the identity of both the solvent and the solute. Like the freezing point depression of liquids (Section 10.2), the osmotic pressure depends only upon the total number of moles of solute particles present in a fixed volume of solvent. For electrolytes in water, this means that once again we must consider the number of ions produced when the substance dissolves.

Example: *Compare the osmotic pressure produced by each of the following solutions at 25 °C: 0.10 mol L^{-1} sucrose, 0.10 mol L^{-1} NaCl, 0.10 mol L^{-1} $Fe_2(SO_4)_3$.*

Answer: $\pi = cRT$

sucrose (non-electrolyte):

$$\pi = (0.10 \text{ mol L}^{-1})(0.0821 \text{ L atm mol}^{-1} \text{ K}^{-1})(298 \text{ K})$$
$$= 2.4 \text{ atm}$$

NaCl (1:1 electrolyte):

$$\pi = (0.2 \text{ mol L}^{-1})(0.0821 \text{ L atm mol}^{-1} \text{ K}^{-1})(298 \text{ K})$$
$$= 4.9 \text{ atm}$$

$Fe_2(SO_4)_3$ (2:3 electrolyte):

$$\pi = (0.5 \text{ mol L}^{-1})(0.0821 \text{ L atm mol}^{-1} \text{ K}^{-1})(298 \text{ K})$$
$$= 12 \text{ atm}$$

The calculated values for the ionic solutions are only approximate, because of charge interactions between the ions, except at very low concentrations (refer back to footnote 2 in this chapter).

What is noteworthy about the values just calculated is their magnitude: 2.4 atm for a 0.1 mol L^{-1} solution.

Question: *What is 2.4 atm in terms of the height of a column of water?*

Answer: *Water has density 1 kg L^{-1}. A pressure of 1 atm corresponds to 1.01×10^5 Pa, where 1 Pa ≡ 1 N m^{-2} and 1 N is the force exerted by 1 kg of water (mass x gravitational attraction).*

Hence:

$$\pi = 2.4 \text{ atm} = 2.4 \times 10^5 \text{ N m}^{-2} = 2.4 \times 10^5 \text{ (kg m s}^{-2}\text{) m}^{-2}/(9.8 \text{ m s}^{-2})$$
$$= 2.5 \times 10^4 \text{ kg m}^{-2}$$

Since density of water = 1 kg L^{-1} ≡ 1000 kg m^{-3}:

$$h = \text{column height} = 2.5 \times 10^4 \text{ kg m}^{-2}/1000 \text{ kg m}^{-3} = 25 \text{ m}$$

10.7.2 Applications of osmosis

A chemical application of osmosis is the determination of the molar masses of very large molecules -- synthetic polymers and biological macromolecules. The pressure difference exerted by a solution of known concentration (grams per liter) is measured, allowing the molar mass of the solute to be determined. In the case of a biological molecule, water would normally be the solvent, but any appropriate solvent can be used, provided a membrane can be found which is permeable to the solvent but not to the solute.

Example: *A solution of polystyrene in toluene had concentration 5.78 g L^{-1}, and produced an osmotic pressure of 1.03 x 10^{-2} atm at 23 °C. Estimate the molar mass of the polystyrene sample.*
Answer: *From eq. [3]:*

$$c = \pi/RT = \frac{1.03 \times 10^{-2}\ atm}{0.0821\ L\ atm\ mol^{-1}\ K^{-1} \times 296\ K}$$

$$= 4.24 \times 10^{-4}\ mol\ L^{-1}$$

$$M = 5.78\ g\ L^{-1}/4.24 \times 10^{-4}\ mol\ L^{-1}$$

$$= 1.36 \times 10^{4}\ g\ mol^{-1}$$

"Osmometry" is a useful method of determining the molar mass of a large molecule because of the large osmotic pressure developed by a small concentration of analyte.

Appreciation of the importance of osmotic pressure is crucial to understanding many biological phenomena. In this respect, there is an important difference between plant cells and the cells of animals and bacteria. Plant cell walls are constructed so as to withstand a difference in osmotic pressure; in fact, they maintain their rigidity by taking water in from outside. This is done by maintaining a higher total concentration of solutes inside the cell than outside. In the absence of sufficient water outside, as in time of drought, the plant wilts because there is insufficient osmotic pressure to maintain rigidity. For the same reason, plants cannot grow in soil that is highly saline, as for example on land that has been improperly irrigated (p.254).

Animal cell walls are less strong. Any substantial difference in osmotic pressure causes them to rupture, killing the cell. Scientists who work with cell cultures and living tissue specimens must maintain the solution in which the cells are kept at the same osmotic pressure as that inside the cells. Such a solution is called an **isotonic** solution, and has "osmolar" concentration (that is, total dissolved solute concentration) *ca.* 0.3 mol L^{-1}. A 0.9% by weight solution of sodium chloride is a commonly used isotonic solution.

If living animal or bacterial cells are placed in a **hypotonic** solution (less concentrated than isotonic), water rushes through the cell membrane into the cell, causing it to swell, and eventually burst. Conversely, if a **hypertonic** (more concentrated) solution is used, water is lost from the cell, which visibly shrinks and becomes dehydrated. Cell death may follow if the loss of water is sufficient to dehydrate proteins and causes them to change their shape from the one that is biologically active. (Such changes of shape are usually irreversible.)

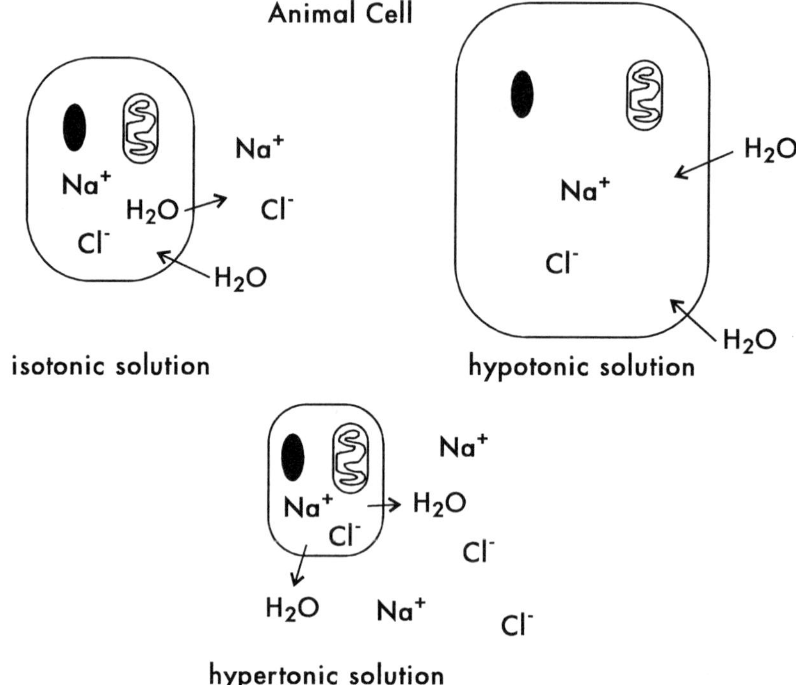

Animal cells are very susceptible to changes in osmotic pressure.

Osmotic pressure is therefore an important consideration when live cells are to be frozen but yet kept viable, as for example in the cryopreservation of living embryos for embryo transplant. Cryopreservation means preserving the embryo in a cryogen such as liquid nitrogen (Section 4.2). The technology involves freezing the fertilized embryo in an aqueous solution containing other non-polar solutes; details of the osmotic pressure changes that occur upon cooling are still not completely understood.

10.7.3 Reverse osmosis

Reverse osmosis is a technology for producing pure water from a solution that contains significant concentrations of salts. One example already mentioned is the use of reverse osmosis to desalinate water that has been used and reused for irrigation. The technology has been extensively used in the Middle East for the production of drinking water from sea water or from wells that are "brackish" -- that is, the water is somewhat salty, and hence unfit for drinking. Reverse osmosis is also used on a large scale to purify industrial waste water before returning it to the environment.

In reverse osmosis, pressure is applied externally to one side of a semipermeable membrane (Figure 10.2). Under these conditions, the solutions on either side of the membrane can be maintained at different concentrations. This is the reverse of osmosis. Water flows through the membrane from the high concentration, high pressure side to the low pressure side, leaving the salts behind on the high pressure side.

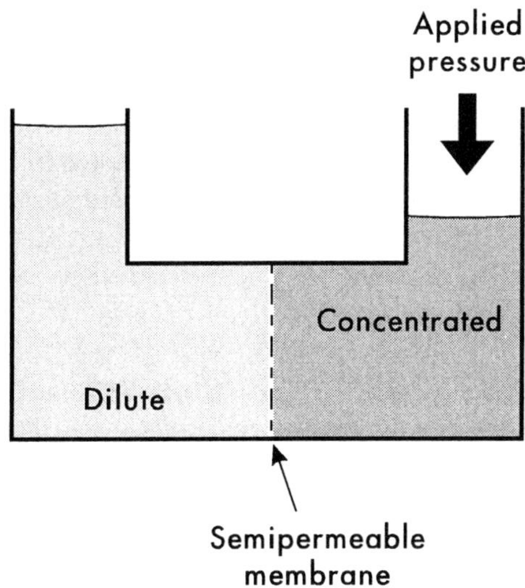

Figure 10.2: Reverse osmosis across a semipermeable membrane

Consider Figure 2 as applied to the reverse osmosis of sea water. Sea water is subjected to high pressure in a chamber containing a semipermeable membrane. Water is able to pass through the membrane, but the dissolved salts cannot. Hence purified water passes through the membrane, leaving a more concentrated solution behind, to be discarded.

10.8 Drinking water

No other public health or medical innovation comes close in importance to access to a safe clean supply of drinking water. Those of us who live in the developed parts of the world can turn a tap and have safe, fresh drinking water without a second thought. World-wide, the story is very different; millions of people, mostly women, must spend many hours every day in carrying water, often of dubious quality, from a distant well to their homes. Between 15 and 20 million young children die every year as result of water-borne diarrheal diseases such as typhoid fever, ameboid dysentery, and cholera. But for municipal water purification, Western cities too would be rife with disease, as indeed they were as recently as the last century. In the context of drinking water, biological purity is a more important issue than chemical contamination.

10.8.1 Sources of drinking water

Potential drinking water sources are classified either as **ground water** or **surface water**. Ground water is recovered from deep in the ground by means of wells, which may be from tens to hundreds of meters deep. Sometimes the water in these aquifers may be replaced only very slowly. As noted in Section 10.3, the "fossil water" in some of these aquifers is estimated to be thousands of years old, and constitutes a resource which is non-renewable in our lifetime. One such aquifer, under Mexico City, is being consumed so rapidly for drinking and industrial water that the city itself may become unsustainable during the next century.

Surface water is drawn from a lake or a river; it almost always has a higher content of suspended materials and a higher microbial count than ground water, and consequently requires more processing to make it safe to drink. Many major waterways, such as the Great Lakes in North America and the Rhine and Danube Rivers in Europe are used for drinking water and for other purposes by a large number of communities. Communities lying down-river draw water which has been contaminated by sewage outfall and industrial use upstream. For communities near the mouth of the Rhine, as little as 40% of the water withdrawn is "new" water *i.e.*, has not previously been discharged by another city. The task of the municipal, or usually in Europe the regional, water engineer is to make this rather unpromising raw material fit to drink.

Ground water tends to be less contaminated than surface water because organic matter in the water has had time to be decomposed by soil bacteria. The ground itself acts as a filtering device so that less suspended matter (including microorganisms) is present. Indeed, filtration of river water through sand was the first successful method of municipal water treatment. Its introduction in London, England in the middle 1800's led to an immediate decline in the incidence of water-borne disease.

Before discussing municipal water treatment in detail, we should remember that even in the developed world by no means everyone has access to municipal water. Millions of people in rural areas depend on individual wells, or even springs or streams, for their drinking water. Surface water and shallow ground water are particularly vulnerable to pollution, and the users normally consume this water without any treatment or disinfection. In developed countries, public health departments usually offer free analysis of individual water, especially for bacterial content, but the citizens of poorer countries are less fortunate in this regard.

10.8.2 Water treatment

There are four steps in the treatment of surface water. Ground water, as already noted, tends to require less treatment. For example, the water in my home town is fit to drink as it is drawn from the ground, and requires only a mild disinfection to ensure that it does not become biologically contaminated in the distribution system.

1. Primary settling. The water is brought into a large holding basin to allow particulate matter to settle. Lime may be added at this stage if the pH of the water is too low (< 6.5).

2. Aeration. The clarified water is agitated with air. This promotes the oxidation of easily oxidizable substances in the water. These would otherwise consume the chlorine or other disinfecting material to be added later in the treatment process.

3. Coagulation and secondary settling. Primary settling is not sufficient to remove the finest particles from water. These fine particles include colloidal mineral particles, bacteria, pollen, spores, etc. They are removed in order to give the finished water a clear, sparkling appearance. The commonest filter aid is **filter alum** $Al_2(SO_4)_3.18H_2O$. Upon dissolution in water at pH 6-8, $Al(OH)_3(s)$ is formed.

$$Al^{3+}(aq) + 3\ HCO_3^-(aq) \rightarrow Al(OH)_3(s) + 3\ CO_2(aq)$$

Aluminum hydroxide is produced in this reaction as a gelatinous precipitate which settles very slowly, over the course of several hours; as it does so, it carries down with it all the fine particles in the water. "Secondary settling" is thus an integral part of the coagulation treatment.

4. Disinfection. **Chlorine** is the most commonly used agent for disinfecting water. Other disinfectants are **ozone**, **chlorine dioxide**, and **ultraviolet radiation**. Ozone is a very effective disinfectant; it is used mostly in very large installations because of the high capital cost of the ozone generator[5]. Regional waterworks in Europe commonly use ozonation. In Canada, the only municipality to use ozonation is Montreal. Chlorine dioxide is usually used on a small scale, and may be used as an alternative to chlorine at times of pollution of the water supply (see below).

---

[5] Ozone is made by passing an electric discharge at 15,000 V through dry air. This converts some of the O_2 in the air into O_3.

Chlorine, chlorine dioxide, and ozone all act as disinfectants by reacting chemically with the cell walls of microorganisms. Ultraviolet disinfection is slightly different in that biomolecules such as proteins and DNA absorb ultraviolet radiation, and this intake of energy causes destructive photochemical reactions. The radiation source used for disinfection is the low pressure mercury vapour lamp, which emits radiation principally at 254 nm. This corresponds to an energy of ≈ 470 kJ mol^{-1}, higher than the bond dissociation energy of most covalent bonds. Ultraviolet disinfection was developed in Switzerland, and is now used by a number of European water authorities. Its advantage is that it is quick acting, and the water can be treated on a flow-through basis; in contrast, the chemical agents all require a large disinfection tank to be built so that the residence time of the water in the tank, in contact with the disinfectant, will be 30-60 minutes.

Disinfection is the most essential part of water treatment. Filtration and coagulation afford water that is pleasant to look at, but it is disinfection that makes it safe to drink. Disinfection kills those microorganisms which have escaped being filtered away but, at least as important, it **prevents recontamination during the time the water is in the distribution system**. If you live in the suburbs of a large city, the water may have been in the distribution system for five days or more before you drink it. Five days is enough time for any "missed" microorganisms to start to multiply, besides which, leaks and breaks in the water mains allow plenty of opportunities for recontamination, especially at the extremities of the distribution system where the water pressure is lower.

Chlorine is unique in that it is the only disinfectant possessing residual disinfectant activity; in other words, it maintains its protection of the drinking water throughout the distribution system. All the other disinfectants mentioned must be followed with a low dose of chlorine in order to keep up the protection. The concentration of chlorine in the treated water is called the **chlorine residual**; it is usually expressed in ppm of Cl_2, even though the chlorine actually reacts chemically with the water as shown in the equations below.

$$Cl_2 + H_2O(\ell) \rightleftharpoons HCl(aq) + HOCl(aq)$$

$$HOCl(aq) \rightleftharpoons H^+(aq) + OCl^-(aq)$$

The second equilibrium becomes important only at pH greater than about 7 (see Chapter 11). For typical water samples at pH < 7, HOCl is the predominant form of "chlorine"; at pH > 8, OCl^- predominates. Between pH 7 and 8, both HOCl and OCl^- are present.

The chlorine residual is often lower than the amount of chlorine supplied to the water: the **chlorine dose**. This occurs whenever the water contains substances which react chemically with chlorine: the so-called **chlorine demand**.

chlorine residual = chlorine dose - chlorine demand

b

a

c

d

Many parts of the world have a climate which is well suited to agriculture, but lack sufficient rainfall. An enormous amount of water is needed. The increasing scarcity of ground water has led to diverse, even unusual, effects. See Chapter 10, page 254.

a The Nicosio Reservoir (Marin County, California) bottomed out in 1977-78, and has remained much below normal for most of the years since.

b Wells in the San Joaquin (Central Valley, California) provided irrigation water for many years, but the progressive subsidence of the water level also caused the land to subside. The ground surface sank 8 meters in 52 years.

c Pecos County (Texas) was once prime cotton-growing land, due to the Ogallala Aquifer. By 1973, the water table had dropped as much as 100 meters locally, and these machines for cotton-picking now stand idle.

d In Baytown, Texas (near Houston), ground water subsidence has caused the land to drop by 3 meters or more. As a result, some 600 homes have sunk into Houston Bay. Here, a homeowner holds a snapshot of what, until recently, was the family home.

Water is a remarkable substance. Surface waters, for instance, are highly regarded for their recreational value. However, they are often more contaminated than ground waters because the organic matter in the water decomposes. These waterfalls are justly famous for their marvellous scenery, yet the water is unfit for drinking without purification. See Chapter 10, page 270.

Niagara Falls, North America
Thomson's Falls, Africa

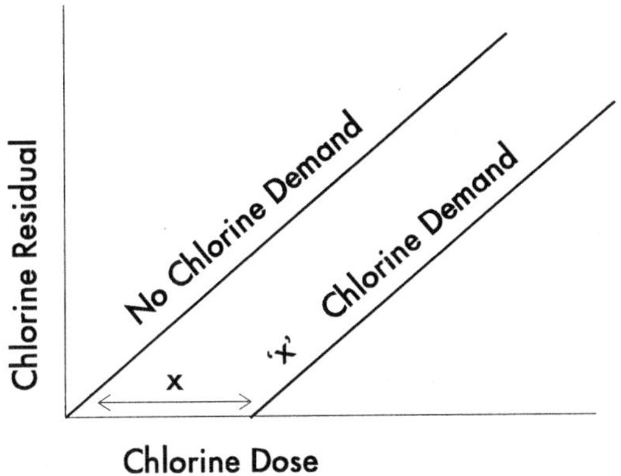

The concentration of chlorine which remains in treated water is the chlorine residual.

Recontamination in the distribution system is a serious problem in the urban slums which are found at the fringes of the rapidly growing cities of the "Third World", in India, Asia, and South America. Often these develop faster than the city can extend its distribution system to them, and the populace must depend on contaminated ground or surface waters. Even if these communities are provided with municipally treated water, it may not be of the highest quality. Reasons include:

- The low pressure at the outer edges of the distribution system and many breaks because of inadequate installation allow contamination from the ground. The high pressure in a properly installed system means that the flow through any leak is always from inside to outside the pipe. Low pressure at the fringe of the system occurs when the city grows so rapidly that the demand on the system exceeds its capacity.

- The absence of outlets in each individual home. Water is drawn from a common, often dirty, communal outlet.

- The illegal, and usually unsanitary, tapping into the city water supply at a point close to the newly developing "barrio" when the latter has not been supplied by the city.

In developed countries, the issue has been raised as to whether all domestic water should be of a quality fit for drinking. The average person drinks only 2 L of water daily, but uses many times that for bathing, laundry, flushing toilets, cleaning cars, watering lawns, etc. None of these other uses require water of potable quality. However, it is probably not practical to provide every household with two different water supplies.

10.8.3 Drinking water guidelines

Each jurisdiction has its own standards. Those quoted here are the recommendations of the Canadian federal government but they are in line with those approved elsewhere.

Table 10.5: Canadian drinking water guidelines

Substance	M.A.C., ppm[a]	Substance	M.A.C.,ppm
Inorganics			
aluminum	0.2 (EEC standard)	arsenic	0.05
cadmium	0.005	chloride	250
copper	1.0	cyanide	0.2
fluoride	1.5	iron	0.3
lead	0.005	manganese	0.05
mercury	0.001	nitrate	45
selenium	0.01	silver	0.05
sodium	lowest practical	sulfate	500
uranium	0.02	zinc	5.0
pH	6.5 - 8.5		
Organics[b]			
Aldrin/Dieldrin	0.0007	DDT	0.03
Lindane	0.004	Parathion	0.035
phenols	0.002	trihalomethanes	0.35

a M.A.C. = maximum acceptable concentration

b Aldrin, Dieldrin, DDT, Lindane, and Parathion are pesticides; phenol is usually of industrial origin; trihalomethanes are formed by the action of chlorine disinfectant on natural substances in the water

10.8.4 Problem substances in drinking water

Iron is a potential nuisance in water, because it stains bathroom fixtures and gives an unpleasant "metallic" taste to the water, but it is rarely present in amounts that might be toxic. Iron contamination can be a serious problem with untreated well water, usually due to dissolution of the slightly soluble $FeCO_3$, which contains iron as Fe(II). Upon exposure to air, Fe(II) is oxidized to Fe(III), which precipitates as Fe_2O_3 (rust). By contrast, most municipal water supplies have little trouble with iron, because it is oxidized during aeration and then precipitates during secondary settling.

Lead, cadmium, and **mercury** are all of concern in water since they are cumulative poisons. This means that they have long residence times in the body, so that their concentrations tend to increase upon continued exposure. All three elements are similar in that they have no known biological function as trace minerals. They may be present in water as a result of industrial activities, for example ground water contamination from municipal and chemical waste dumps, and in the vicinity of mines (Chapter 17).

Lead in drinking water is also a problem in homes containing lead plumbing. The word plumbing comes from the Latin word for lead; it was the Romans who first introduced lead piping as a means of distributing water. Roman houses had indoor plumbing and hot water systems as long as two thousand years ago, and some of these still survive in Europe.

In hard water areas lead plumbing causes few problems, since a deposit of scale ($CaCO_3$) quickly covers the surface of the lead piping so that the water does not actually come in contact with the metal. Soft waters, which are low in minerals, do not provide this protection. Soft waters are often somewhat acidic, which promotes the dissolution of lead. A small degree of exposure to lead occurs even with copper piping because lead solder is used to join the pipes. In addition, water from the "hot water" side of the system is likely to have a higher lead content than the "cold side" because (i) more rapid dissolution will occur at higher temperature and (ii) scale tends to deposit in the hot water heater, leaving the hot water pipes with less protection by scale. Householders with lead piping should not use the hot water tap for drinking or cooking, and should make a particular point of not using, even from the cold tap, the "first draw" water (the water that has been standing in the pipes overnight). Instead they should run the water away for one or two minutes in the morning in order to flush this more contaminated water out of the system.

The toxicity of lead is a particular problem for children; whereas adults typically retain <10% of the lead they ingest, the unborn fetus absorbs 50%, and the proportion drops as a child grows older. Lead can cross the blood-brain barrier, again in the order fetus > child > adult, leading to mental retardation in severe cases, and IQ deficiency or behavioural problems at lower levels. Clinical signs of retardation have been observed in children born to women who used first draw water having [Pb] > 0.8 ppm during pregnancy. These potential health problems have caused the allowable concentration of lead in drinking water to be reduced from 50 to 5 ppb in many jurisdictions.

Aluminum has generally been considered to be innocuous in drinking water and in the diet until recently. However, the current situation regarding the toxicology of aluminum is extremely confused. Although aluminum is very abundant in the Earth's crust, it appears to have no natural biochemical function. A link between aluminum and Alzheimer's Disease (senile dementia) has been proposed, with the brains of dementia patients showing characteristic "tangles" of neurons and "hot-spots" of unusually high aluminum concentration; whether the aluminum is a cause of the disease or simply a side effect is not known. Several studies have been reported in recent years in an attempt to determine whether there is a predisposition towards Alzheimer's disease among populations whose diets are high in aluminum, but the results are not clear cut. The typical dietary intake of aluminum in North America has been estimated at 22 mg/day, whereas the highest concentration of aluminum measured in a survey of 200 drinking water systems was 2.7 ppm. At an estimated 2 L of water drunk per day, this amounts to 5.4 mg of aluminum, or < 25% of the average daily intake. One buffered aspirin tablet may contain as much as 35 - 200 mg of aluminum; no evidence has yet been presented that long term use of this medication is a health hazard. Most likely, healthy individuals eliminate dietary aluminum rather than absorb it.

Aluminum cookware is another potential source of aluminum in the diet. However, the concentrations of aluminum in foods cooked in aluminum cookware are insignificant, with one exception. Aluminum forms strong complexes with citrate ion, which is present in fruits, and it has recently been found that up to 100 ppm of aluminum can be detected in foods such as stewed rhubarb and tomato sauce cooked in aluminum cookware. The combination of high $[H^+]$ and high [citrate] is responsible.

High levels of **nitrate ion** (up to 100 ppm) are sometimes found in drinking water on farms with shallow wells. The source of this contaminant is run-off of fertilizer from fields or of manure from barnyards, feedlots, and the holding tanks used to contain liquid manure. Nitrate contamination of ground water is a particular problem in the Netherlands, where a high population density is combined with intensive hog production.

Even at the maximum recommended concentration of 50 ppm in drinking water, most people take in 3 or 4 times as much nitrate from food (mainly vegetables) as from drinking water. A specific toxic effect of nitrate ion is a condition of infants called **methemoglobinemia**, in which hemoglobin is converted to a form which cannot carry oxygen to the tissues. Severe cases of methemoglobinemia can result in mental retardation.

Organics in drinking water sources arise mostly from agricultural run-off (pesticides), from industrial operations, and by the action of chlorine on natural organic compounds present in the water (trihalomethanes, of which chloroform, $CHCl_3$, is the most important). Phenols from industrial operations are a particular nuisance, because they react with chlorine at the disinfection stage to produce **chlorinated phenols**, whose antiseptic odour is so powerful as to make the water undrinkable, even when they are present at only ppb concentrations. Municipalities experiencing "taste and odour" problems caused by phenols frequently switch from chlorine as disinfectant to chlorine dioxide for the duration of the problem. Public concern over **chloroform** in drinking water stems from the discovery that chloroform may be carcinogenic at high doses in laboratory animals. However, the concentrations reached in drinking water (typically 10-20 ppb) are many orders of magnitude lower than those which have been reported to show toxic effects. In addition, the risk associated with ingestion of tiny amounts of chloroform is orders of magnitude less than the very real risk of dying from cholera or typhoid when untreated water is drunk.

The issues just mentioned involve the chemical reactivity of chlorine with substances which may be present in raw drinking water. As a result, some people advocate switching from chlorine to other disinfectants. In doing so, it is important to keep in mind that chlorination chemistry has been well documented, and any risks may in fact be better understood than the (less documented) risks from the use of ozone or chlorine dioxide. Furthermore, even if alternative methods of disinfection are employed, it is still necessary to use a low concentration of chlorine to retain residual disinfecting power in the water distribution system.

10.9 Fluoridation of drinking water

Fluoridation is a public health issue which raises passionate arguments as to whether fluoride ion should be added to municipal water supplies. In favour of fluoridation are those who see a medical advantage in the prevention of tooth decay, while those opposed claim that fluoride is a poison and people should not be medicated against their will. The chemistry of fluoridation is relatively simple. Tooth enamel is composed of a mineral called **hydroxylapatite**, $Ca_5(PO_4)_3OH$. Tooth decay occurs when the tooth enamel is damaged, allowing the entry of bacteria which destroy the tooth material underneath. It was noted in the 1930's that certain areas of the United States where the water was naturally high in fluoride ion (> 1 ppm) had an unusually low incidence of tooth decay. Fluoride ion can replace the hydroxide ion in hydroxylapatite, yielding a new mineral, fluorapatite, which is less soluble.

$$F^-(aq) + Ca_5(PO_4)_3OH(s) \rightarrow Ca_5(PO_4)_3F(s) + OH^-(aq)$$

The suggestion was therefore made that it would be a good idea to add fluoride ion artificially to the water supply in areas where natural fluoride was low or absent, in order to reduce the incidence of dental caries, especially among children. Today about half of all North Americans drink artificially fluoridated water (1 ppm of F^-), and in addition, pastes and solutions containing fluoride ion are routinely applied to patients' teeth during visits to the dentist's office. A level of 1 ppm in drinking water is usually recommended; higher concentrations (3-5 ppm) lead to mottling of the teeth (dental fluorosis). In Ontario, drinking water may be supplemented with up to 1.2 ppm of fluoride, but must be rejected if the F^- concentration exceeds 2.4 ppm.

Until recently, the statistical (but not necessarily the ethical) case in favour of fluoridation seemed very convincing. The incidence of dental caries, especially among children has declined dramatically during the 45 years since fluoridation was first introduced. It seemed that a cause and effect relationship existed between fluoridation and healthier teeth. However, the incidence of dental caries seems to have declined to a comparable extent both in communities which practise fluoridation and those which do not. This may be a situation where a statistical correlation exists between two events (here, fluoridation and tooth decay), but one is not a consequence of the other. At present, we cannot answer definitively the question of whether fluoridation is valuable as a public health measure.

Problems

Sections 10.1 - 10.3

1. Use the data of Table 10.2 to estimate the lifetime of (a) Na^+ (b) K^+ in the oceans.

2. The average lifetime of water vapour in the atmosphere is about 11 days. Estimate the annual rate of vaporization of water from lakes and oceans, using the data in Section 10.2.

3. Estimate the total concentrations of dissolved solids in river water, using the data of Table 10.2, and compare this with the value quoted for sea water of 35 g L^{-1}.

4. Estimate the freezing point of a solution of benzene produced by dissolving 0.1239 g of benzoic acid in 8.244 g of pure benzene. The m.p. of pure benzene is 5.5°C.

5. What is the "colligative molality" (the total number of moles of solutes per kg solvent of the following solutions?
   a) 2.5 g of glucose $C_6H_{12}O_6$ dissolved in 25 g water
   b) 8.3 g K_2SO_4 dissolved in 146 g water
   c) 2.7 g hexane (C_6H_{14}) dissolved in 57 g cyclohexane (C_6H_{12})
   d) 2.5 g NaCl plus 1.6 g alcohol C_2H_6O dissolved in 27.3 g water.

6. Estimate the freezing point of the following aqueous solutions
   a) 3.8 g glucose $C_6H_{12}O_6$ in 94.3 g water
   b) 3.1 g $CaCl_2$ in 67 g water
   c) 0.0100 mol $Na_2SO_4.10H_2O$ dissolved in 28.6 g water
   d) 10% by mass ethylene glycol (antifreeze, $C_2H_6O_2$) in water.

7. Calculate the anticipated molal freezing point depressions and use them to explain the following facts:
   a) A 0.100 mol kg^{-1} solution of K_2SO_4 in water has freezing point -0.38°C.
   b) A 0.200 mol kg^{-1} solution of acetic acid in cyclohexane has freezing point 4.5°C (the pure solvent has freezing point 6.5°C).
   c) A 12.00% by mass solution of NaCl has freezing point -8.176°C.

8. Camphor is an organic compound, m.p. 178.8°C whose molal freezing point depression constant has the unusually large value 39.7 kg K mol^{-1}. It is frequently used to determine the molar masses of unknown compounds, as in the experiment below.

   A sample of cholesterol (86.3 mg) is dissolved in 1.0115 g of melted camphor. The solution freezes at 170.0°C. Estimate the molar mass of cholesterol.

9. The concentration of phosphorus in Lake Huron has been steady for 20 years at 5 ppm. Estimate the annual rate of deposition of phosphorus into Lake Huron.

10. A dog goes for a swim, and after shaking itself, retains 150 g of water in its coat. Calculate a) the amount of heat needed to evaporate all this water at 25°C; b) the amount of glucose that the dog must metabolise to provide the necessary heat. Glucose has heat of combustion 2803.0 kJ mol^{-1}.

Sections 10.4 - 10.6

1. A 0.100 L sample of water is titrated against 0.01208 mol L^{-1} EDTA solution to the Eriochrome Black T endpoint. This requires 13.86 mL of the EDTA solution. Calculate the hardness of the water in both mmol L^{-1} of Ca^{2+} and ppm of Ca^{2+}.

2. A soap has the chemical composition $C_{18}H_{35}O_2Na$. What mass of scum is formed if 100 g of soap is used in a washing machine of capacity 60 L, if the water used has hardness 45 ppm of Ca^{2+} and the scum is assumed to be completely insoluble in water?

3. A water supply contains 38 ppm of Ca^{2+}. What mass of lime should be used to soften 2.2×10^4 m^3 of this water?

4. An ion exchange resin has exchange capacity 1.1 meq cm^{-3} and density 0.81 g cm^{-3}.

   [1 meq ≡ milliequivalent ≡ the number of mmol of cations of charge +1 that can be absorbed by the resin]. What mass of resin would you need in your water softener to soften the 1.0 m^3 of water used daily in a typical household, if the resin is to be regenerated no more often than once a week? Assume that the raw water supply contains 50 ppm of Ca^{2+} and that the ion exchange reaction proceeds "to completion".

5. Calculate the cost of providing distilled water to a typical home in Canada on the assumption that electricity costs 6.8 ¢ per kWh. Calculate the cost on an annual basis per household; assume that the molar heat of vaporization of water is 41 kJ mol^{-1} at 100°C, and that the energy required to heat water from 15°C to 100°C can be recovered in the heat released when H_2O (g) condenses back to H_2O (ℓ).

Section 10.7

1. Calculate the osmotic pressures of the following solutions at 20°C
   a) 20 mg sucrose ($C_{12}H_{22}O_{11}$) in 1.73 g of water
   b) 72.6 mg KCl in 100.0 g of water assuming complete dissociation
   c) 6.3 mg of vitamin D (M = 385 g mol^{-1}) in 8.26 g of toluene.

2. Calculate the percent association or dissociation of the following solutions, given the osmotic pressures at 25°C.
   a) 1.0×10^{-4} mol L^{-1} acetic acid in water, $\pi = 3.5 \times 10^{-3}$ atm
   b) A 2.0 g L^{-1} solution of NaCl in water has $\pi = 1.67$ atm
   c) A 1.00% solution of $FeCl_3$ in water has $\pi = 5.06$ atm.

3. Calculate the molar masses of the following substances from osmotic pressure data
   a) 0.110 g of hemoglobin in 100 mL water has osmotic pressure 5.8×10^{-3} atm
   b) A 0.63% by weight solution of the soluble fraction of starch in water has osmotic pressure 1.7×10^{-3} atm.
   c) A solution of polypropylene 426 mg in 100 mL toluene has osmotic pressure 1.13 kPa.

4. Calculate the freezing point and the osmotic pressure of an isotonic solution of NaCl.

5. Estimate, on the basis of Table 10.2, the pressure that would need to be applied to a solution of sea water to produce pure water by reverse osmosis.

6. An industrial waste water contains total dissolved solids of 2.3 g L^{-1}, principally $CaCl_2$. Estimate the pressure that would need to be applied to produce pure water by reverse osmosis.

Section 10.8

1. A 1.00 L sample of chlorinated water is treated with acidified KI, which converts all the Cl_2 into I_2. The I_2 is then titrated agains 0.01038 mol L^{-1} $Na_2S_2O_3$, of which 7.46 mL are needed to react with all the I_2. Calculate the concentration of Cl_2 in the water in ppm.

2. A chlorination facility is constructed to the specification that the water will have residence time 25 min in the tank, and produce 2.0×10^3 m^3 of water per hour.
   a) Calculate the rate at which Cl_2 should be injected into the tank in order to achieve a chlorine concentration of 2.0 ppm.
   b) Calculate the size of the tank.
   c) Calculate the rate of Cl_2 injection if the water contains oxidizable substances that will consume 0.44 ppm of Cl_2, if the final Cl_2 concentration is to remain at 2.0 ppm.

3. The solubility of O_3 in water at 20°C is given by the following expression:

   $[O_3](aq, mg L^{-1}) = 0.41 [O_3](g, mg L^{-1})$

   a) A water sample at 20°C is equilibrated with O_3 at 2.6×10^{-3} atm partial pressure. Calculate the concentration of ozone in the water, in ppm.

   b) The rate of decomposition of ozone in water follows the rate law below

   $$\text{rate} = (2.2 \times 10^5)[O_3]^2[OH^-]^{0.55} \text{ mol } L^{-1} s^{-1}.$$

   Calculate the rate of decomposition of ozone at pH 7.55 under the conditions of part (a).

   c) Calculate how long it would take for $[O_3, aq]$ to fall to 1.0 ppm at 20°C, pH 7.55.

   d) Based on your answer to (c), why can ozonation not be used as the sole disinfection agent for drinking water?

4. Calculate the mass of NaF that must be added to 5,000 m^3 of raw water whose F^- concentration is 0.09 ppm, so that the final fluoride concentration will be 1.05 ppm.

5. Chlorine dioxide is prepared for water treatment by the reaction below.

   $$10NaClO_2 + 5H_2SO_4 \rightarrow 8ClO_2 + 5Na_2SO_4 + 2HCl + 4H_2O$$

   a) Calculate the mass of $NaClO_2$ needed to generate enough ClO_2 to treat 5.0×10^5 m^3 of water with 2.3 ppm of ClO_2 if the percent yield of the reaction is 83%.

   b) ClO_2 levels in water can be determined by iodometric titration

   $$ClO_2 + 4H^+ + 5I^- \rightarrow \tfrac{5}{2}I_2 + Cl^- + 2H_2O$$

   A 200.0 mL sample of water is treated with acidified KI and the I_2 liberated is titrated with 4.26×10^{-3} mol L^{-1} $Na_2S_2O_3$, of which 9.66 mL are needed to react with all the I_2. Calculate the concentration of ClO_2 in the water.

11 Acids and Bases

11.1 Definition of an acid and a base

Of the many definitions of acidity that have been proposed over the years, the one suggested by Brønsted and Lowry (1923) is especially useful for aqueous solutions. According to this definition:

Definition

*An **acid** is a substance which can donate a proton (H^+) to a base, and a **base** is a substance which can accept a proton from an acid.*

Note that a proton is a positively charged hydrogen atom. In terms of bonding theory, a base has a pair of unshared electrons which are used to make the new bond to H^+.

Thus: **Acid: proton donor** **Base: proton acceptor**

According to the Brønsted-Lowry definition, a substance cannot function as an acid unless there is a base available to accept the proton; neither can another substance function as a base unless there is an acid available to donate the proton. This concept is illustrated by the following examples of acid-base reactions; the acidic and basic substances are identified.

[1] $HNO_3(\ell) + H_2O(\ell) \rightleftharpoons H_3O^+(aq) + NO_3^-(aq)$

acid base acid base

[2] $HF(aq) + H_2O(\ell) \rightleftharpoons H_3O^+(aq) + F^-(aq)$

acid base acid base

[3] $NH_3(aq) + H_2O(\ell) \rightleftharpoons NH_4^+(aq) + OH^-(aq)$

base acid acid base

[4] $CN^-(aq) + H_2O(\ell) \rightleftharpoons HCN(aq) + OH^-(aq)$

base acid acid base

Several points can be noted from these equations.

- Acid-base reactions are **equilibria**. The equilibrium constant for a particular acid-base reaction may be much greater than 1 or much less than 1. When the equilibrium constant for the reaction between an acid and water is >> 1, the acid is said to be **completely dissociated** in water. Another term for expressing the same concept is that an acid which is completely dissociated in water is said to be a **strong acid**. Conversely, a **weak acid** is incompletely dissociated in water, and its equilibrium constant for dissociation is usually << 1. The terms "strong" and "weak" acid are therefore used in the same sense as strong electrolyte and weak electrolyte in Chapter 1.
- Since acid-base reactions involve proton transfers, it follows that an acid is converted (at least in principle) into a base when it loses its proton, and the base is converted into an acid upon gaining the proton. This becomes more obvious if we compare eq. [1] with its reverse (remember that any equilibrium can be written as being approached from either side).

$$[1] \quad \underset{\text{acid}}{HNO_3(\ell)} + \underset{\text{base}}{H_2O(\ell)} \rightleftharpoons \underset{\text{acid}}{H_3O^+(aq)} + \underset{\text{base}}{NO_3^-(aq)}$$

$$[1a] \quad \underset{\text{acid}}{H_3O^+(aq)} + \underset{\text{base}}{NO_3^-(aq)} \rightleftharpoons \underset{\text{acid}}{HNO_3(\ell)} + \underset{\text{base}}{H_2O(\ell)}$$

We use the term **conjugates** to describe substances such as HNO_3 and NO_3^-, which differ in chemical formula by one proton. In equations [1] and [1a], HNO_3 would be called the conjugate acid of NO_3^-, or equivalently NO_3^- would be called the conjugate base of HNO_3. Likewise, H_2O and H_3O^+ are conjugates. Note that conjugates differ in both proton count and in charge.

Definition

Conjugates *are chemical species which differ in chemical formula by one proton (H^+).*

Example: *Identify which of the following pairs of species are conjugates.*

A: $HClO_2$ and ClO_2^-
B: HNO_2 and NO_2^+
C: H_2O and OH
D: $H_2PO_4^-$ and HPO_4^{2-}

Answers: *A, yes; B, no (NO_2^- is the conjugate base of HNO_2); C, no (OH^- is the conjugate base of H_2O); D, yes*

- Proton transfers are extremely fast processes, so fast that almost all acid-base reactions reach equilibrium as quickly as the reactants can be mixed together. The few exceptions will not be encountered in this text, hence kinetic considerations can be ignored in dealing with acid-base chemistry.
- Water plays a central role in acid-base equilibria, because it can function both as an acid and as a base. In eq. [1] and [2], water behaves as a base, accepting a proton from HNO_3 (eq. [1]) or from HF (eq. [2]), whereas in eq. [3] and [4], water acts as an acid, donating a proton to NH_3 (eq. [3]) or to CN^- (eq. [4]). Besides this, water is an excellent solvent for solvating the ions which are present in acid-base reactions. Water is the only common solvent which combines excellent solvation of ions with the ability to function as both a proton acceptor and a proton donor.

11.2 Self-ionization of water and the pH scale

11.2.1 Self-ionization of water

Although liquid water consists almost entirely of covalent molecules, at any temperature a tiny proportion of all water molecules undergo an acid-base reaction with each other, a process known as **self-ionization** (or alternatively, auto-ionization).

$$2\ H_2O(\ell) \rightleftharpoons H_3O^+(aq) + OH^-(aq)$$

The self-ionization equilibrium constant for water has the value 1.0×10^{-14} $(mol\ L^{-1})^2$ at 298 K, and is given the special symbol K_w, because of its importance in acid-base chemistry.

$$K_w = [H_3O^+][OH^-] = 1.0 \times 10^{-14}\ (mol\ L^{-1})^2 \text{ at } 298\text{ K}$$

K_w, and all the other equilibrium constants met in this chapter are examples of K_c, the equilibrium constant defined in terms of concentration in moles per liter. Recall that the concentration of $H_2O(\ell)$ does not appear in the equilibrium constant expression, because it is a pure liquid. Throughout this chapter, all concentration terms will be assumed to be (aq), unless stated otherwise.

The importance of the auto-ionization of water is that it automatically regulates the relative concentrations of $H_3O^+(aq)$ and $OH^-(aq)$ in any aqueous solution. For example, suppose we dissolve 0.0100 mol NaOH in enough water to make 1.00 L of solution. Since NaOH is a strong electrolyte, the solution contains 0.0100 mol L^{-1} of both $Na^+(aq)$ and $OH^-(aq)$. Therefore we can calculate the concentration of $H_3O^+(aq)$ in this solution.

$$K_w = [H_3O^+][OH^-]$$
$$[H_3O^+] = K_w/[OH^-] = \{1.0 \times 10^{-14} (\text{mol L}^{-1})^2\}/(0.0100 \text{ mol L}^{-1})$$
$$= 1.0 \times 10^{-12} \text{ mol L}^{-1}$$

Since acid-base reactions are very fast, the concentrations of H_3O^+(aq) and OH^- (aq) in any aqueous solution always adjust themselves immediately to satisfy the K_w expression.

A **neutral** solution contains equal concentrations of H_3O^+(aq) and OH^-(aq). If we set $[H_3O^+] = [OH^-] = x$, we can solve $K_w = x^2$, and obtain the result that in pure water $[H_3O^+] = [OH^-] = 1.0 \times 10^{-7}$ mol L^{-1}. An aqueous solution containing more H_3O^+ than OH^- is termed **acidic**, while one containing an excess of OH^- over H_3O^+ is termed **alkaline** (also called "basic" by other authors).

11.2.2 Lack of distinction between H^+ (aq) and H_3O^+ (aq)

So far, we have been careful to write H_3O^+(aq) to describe the species that is formed when solvent water accepts a proton from an acid. However, the bracketed (aq) reminds us that the proton is not associated with just *one* water molecule, but that numerous water molecules surround the proton in a solvent sheath. Rather than single out one particular water molecule as the partner of the proton in an aqueous solution, some chemists prefer to write H^+ (aq). This formulation emphasizes that when water accepts a proton from another acid, the proton is solvated by a large number of individual water molecules. Opinions vary as to which is the better chemical formula to write: H^+ (aq) or H_3O^+(aq). Whichever one is chosen, it is important to remember that they refer to the **same** chemical species. However, it is clearly preferable to use "H_3O^+" in a reaction like eq. [1], where one wants to draw attention to the role of water in *accepting* the proton from another acid: the acid does not simply expel a proton into the solution, which would be implied by the alternative formulation of eq. [1b].

[1] $HNO_3(\ell) + H_2O(\ell) \rightleftharpoons H_3O^+(aq) + NO_3^-(aq)$

[1b] $HNO_3(\ell) \rightleftharpoons H^+(aq) + NO_3^-(aq)$

Example: *Write a net ionic equation to show that the hydrogen sulfate ion acts as an acid in water.*

Answer: *If HSO_4^- acts as an acid, it must donate a proton to water:*

$HSO_4^-(aq) + H_2O(\ell) \rightleftharpoons SO_4^{2-}(aq) + H_3O^+(aq)$

11.2.3 pH

From Section 11.2.1, we can deduce that the range of values of $[H_3O^+]$ and $[OH^-]$ in commonly encountered aqueous solutions is very wide. For example, a 1.0 mol L^{-1} solution of HCl (a strong acid) would be calculated[1] to have $[H_3O^+]$ = 1.0 mol L^{-1} and $[OH^-]$ = 1.0 x 10^{-14} mol L^{-1}. Conversely, a 1.0 mol L^{-1} solution of NaOH would be calculated to have $[OH^-]$ = 1.0 mol L^{-1} and $[H_3O^+]$ = 1.0 x 10^{-14} mol L^{-1}. The pH scale has been devised in order to avoid the inconvenience of writing exponentials to describe the acidity of every aqueous solution. pH may be operationally defined as follows.

Definition

$$pH = -\log_{10} [H^+] \text{ or, equivalently } pH = -\log_{10} [H_3O^+]$$

Notice the following points about the pH scale.

- The scale is logarithmic; a change of 1 pH unit corresponds to a **tenfold** change in the hydrogen ion concentration.

- The scale is a **negative** logarithm; high pH values correspond to low acidity, and low pH values correspond to high acidity. The pH of the 1.0 mol L^{-1} solution of HCl mentioned above would be 0.00, because pH = $-\log_{10}[H_3O^+]$ = $-\log_{10}(1.0)$ = 0.00. Likewise, the pH of the corresponding 1.0 mol L^{-1} solution of NaOH would be 14.00, because pH = $-\log_{10}[H_3O^+]$ = $-\log_{10}(1.0 \times 10^{-14})$ = 14.00. Commonly encountered values of pH lie between 0 (very acidic) and 14 (very alkaline). Extremely pure water has pH 7.00 ($-\log_{10}[H_3O^+] = -\log_{10}(1.0 \times 10^{-7}) = 7.00$).

  Although ultra-pure water does indeed have pH 7.00, a natural water, which can equilibrate with atmospheric gases and underlying rocks, is usually slightly acidic or slightly alkaline depending on its source. This deviation from neutrality is due to the presence of the substances that are dissolved in the water (see Sections 11.11 and 12.2).

- From the value of K_w at 298 K, one can deduce the following useful relationship.

  pH + pOH = 14.00

---

[1] In practice, the experimental values of $[H_3O^+]$ and $[OH^-]$ would not quite match the calculated ones, but we shall not consider this complication in the present chapter. The reason will be discussed briefly in Chapter 14.

This is obtained by taking the negative logarithm of both sides of the equation defining K_w, and defining pOH as $-\log_{10}[OH^-]$.

$$K_w = [H_3O^+][OH^-] = 1.0 \times 10^{-14} \text{ (mol L}^{-1})^2$$

$$-\log_{10}K_w = -\log_{10}[H_3O^+] + -\log_{10}[OH^-]$$

$$-\log_{10}(1.0 \times 10^{-14}) = 14.00 = \text{pH} + \text{pOH}$$

- Like other equilibrium constants, K_w varies with temperature. The reaction $H_2O(\ell) \rightarrow H^+(aq) + OH^-(aq)$ is endothermic, and so K_w increases at higher temperatures.

Example: *At 40 °C,* $K_w = 2.9 \times 10^{-14}$ *(mol L*$^{-1}$*)*^{2}*. What would be the reading of a pH meter in pure deoxygenated water at this temperature?*

Answer: $H_2O(\ell) \rightarrow H^+(aq) + OH^-(aq)$

In pure water, $[H^+] = [OH^-] = x$

$K_w = [H^+][OH^-]$

$2.9 \times 10^{-14} = x^2$; $x = 1.7 \times 10^{-7}$; pH = 6.77

Note: *The water is not acidic, it still contains equal concentrations of* $H^+(aq)$ *and* $OH^-(aq)$.

Table 11.1 gives approximate pH values of some commonly encountered substances. They range from very acidic to very alkaline.

Table 11.1: Approximate pH values for familiar substances

Substance	pH	
Concentrated (37%) HCl	-1	strongly acidic
1 mol L^{-1} HCl	0	
Stomach acid	1.5	
Lemon juice, vinegar	2	
Wine	3.5	
"Acid rain"	<4.5	
Unpolluted rain	5.6	
Milk	7	
Blood	7.4	
Borax solution	9	
Household ammonia	12	
1 mol L^{-1} NaOH	14	
Drain cleaner (not diluted)	15	strongly alkaline

The measurement of pH is carried out experimentally with an instrument called a "pH meter". The heart of this instrument is a very thin membrane, constructed from a special glass which is somewhat permeable to H^+ (aq). On the inside of the membrane is sealed a solution whose H^+ concentration is precisely known; the outside of the membrane is dipped into the solution whose pH is to be measured. An electrical potential is generated whenever the concentration of H^+ (aq) in the test solution is different from that sealed inside the instrument.

The electrical potential is amplified and multiplied by a calibration constant, so that the meter reads directly in pH units rather than in volts. As we shall see in Chapter 16, electrical potential is always proportional to log(concentration) rather than directly to concentration, which is a second reason to define pH as a logarithmic function (the first was the avoidance of using scientific notation to describe acidity).

11.3 The equilibrium constant K_a and the pH of a solution of an acid

11.3.1 Definition of K_a

Let us put the definition in terms of a hypothetical acid HA, which donates a proton to solvent water.

$$\underset{\text{acid}}{HA(aq)} + \underset{\text{base}}{H_2O(\ell)} \rightleftharpoons \underset{\text{acid}}{H_3O^+(aq)} + \underset{\text{base}}{A^-(aq)}$$

The equilibrium constant K_c for this reaction is given the special symbol K_a, and is called the **acid dissociation constant**. By convention, K_a is always taken to refer to the dissociation of 1 mol of acid.

$$K_a = \frac{[H_3O^+][A^-]}{[HA]} \qquad \textit{or equivalently} \qquad K_a = \frac{[H^+][A^-]}{[HA]}$$

Again, the concentration of the pure liquid H_2O is omitted from the equilibrium constant expression, and so K_a always has the units mol L^{-1}.

11.3.2 Types of acid

Strong acids: There are six common strong acids - those which dissociate completely in water, and hence have $K_a >> 1$. They are H_2SO_4, HNO_3, HCl,

HBr, HI, and $HClO_4$. For strong acids, the concentration of undissociated acid present at equilibrium in water is too small to measure. Therefore, we write their dissociations as occurring completely.

e.g. $HClO_4(aq) + H_2O(\ell) \rightarrow H_3O^+(aq) + ClO_4^-(aq)$

Weak acids: A large variety of substances can function as weak acids (in the presence of a suitable base). They include neutral molecules (examples: HF, HNO_2, acetic acid), a few anions (HSO_4^-, HCO_3^-, $H_2PO_4^-$, HPO_4^{2-}), and numerous cations (NH_4^+, related nitrogen-containing cations, and many aqueous metal cations). Most of these acids lose a proton which was formerly bonded to oxygen or nitrogen; hydrogen atoms bonded to carbon are only rarely acidic.

In the case of the metals, the proton is actually lost from one of the water molecules hydrating the metal ion, rather than from the metal itself. Examples of the chemical equations and the corresponding K_a expressions follow. A tabulation of K_a values is found in Appendix 3.

- $HNO_2(aq) + H_2O(\ell) \rightleftharpoons H_3O^+(aq) + NO_2^-(aq)$

$$K_a = \frac{[H_3O^+][NO_2^-]}{[HNO_2]} \quad \textit{or equivalently} \quad K_a = \frac{[H^+][NO_2^-]}{[HNO_2]}$$

- $HCO_3^-(aq) + H_2O(\ell) \rightleftharpoons H_3O^+(aq) + CO_3^{2-}(aq)$

$$K_a = \frac{[H_3O^+][CO_3^{2-}]}{[HCO_3^-]} \quad \textit{or equivalently} \quad K_a = \frac{[H^+][CO_3^{2-}]}{[HCO_3^-]}$$

- $NH_4^+(aq) + H_2O(\ell) \rightleftharpoons H_3O^+(aq) + NH_3(aq)$

$$K_a = \frac{[H_3O^+][NH_3]}{[NH_4^+]} \quad \textit{or equivalently} \quad K_a = \frac{[H^+][NH_3]}{[NH_4^+]}$$

- $Ag(H_2O)_2^+(aq) + H_2O(\ell) \rightleftharpoons H_3O^+(aq) + Ag(H_2O)(OH)(aq)$

$$K_a = \frac{[H_3O^+][Ag(H_2O)(OH)]}{[Ag(H_2O)_2^+]} \quad \textit{or equivalently} \quad K_a = \frac{[H^+][Ag(H_2O)(OH)]}{[Ag(H_2O)_2^+]}$$

11.3.3 Calculations involving K_a

Before reading this section, review Sections 8.4 and 8.5 on how to do equilibrium calculations.

One common calculation is to determine the pH of a weak acid of a given concentration, given the value of K_a for the acid.

Example: *Calculate the pH and the percent dissociation of a 0.010 mol L^{-1} solution of acetic acid, for which $K_a = 1.8 \times 10^{-5}$ mol L^{-1}.*

Answer: *Write out the net ionic equation and set up a table of initial and equilibrium concentrations.*

	HAc	⇌	H^+	+	Ac^-
initial conc	0.010		-		-
change in conc	-x		+x		+x
equilib. conc	0.010-x		x		x

Write out the equilibrium constant expression

$$K_a = [H^+][Ac^-]/[HAc]$$

Substitute in the values from the table

$$1.8 \times 10^{-5} = x^2/(0.010-x)$$

Make the approximation that x << 0.010 mol L^{-1}. Remember that at this stage we are only testing the possibility that x may be < 0.010 mol L^{-1}.

$$1.8 \times 10^{-5} \approx x^2/(0.010)$$

$$x^2 \approx 1.8 \times 10^{-7};\ x = 4.2 \times 10^{-4}$$

Check approximation: 4.2×10^{-4} << 5% of 0.010, so approximation was justified.
Since x was defined as $[H^+]$ at equilibrium,
pH = -log(x) = 3.37 (two significant figures).
Percent dissociation = 100 x (conc dissociated/initial conc)
= 100 x (x/0.010) = 4.2 %

As a point of generalization, you will find that the approximation $([A]-x) \approx [A]$ is usually justified whenever $[A] > 1000\ K_a$.

In carrying out K_a calculations there are essentially three quantities to consider: (1) the equilibrium constant; (2) the initial, or stoichiometric, concentration of the acid; (3) the equilibrium concentration of H^+ (or the pH, or the percent dissociation). Any one of these can be calculated provided the other two are already known. The next example shows how to calculate the initial (stoichiometric) concentration, given the pH and K_a for the acid.

Example: *Calculate the concentration of ammonium chloride in a solution whose pH is measured as 5.13, given* K_a *for the ammonium ion as* 5.6×10^{-10} *mol* L^{-1}*.*

Answer: *From the pH, calculate* $[H^+] = 7.4 \times 10^{-6}$ *mol* L^{-1}*.*

Write out the net ionic equation and set up a table of initial and equilibrium concentrations. This time, define x as the stoichiometric concentration of NH_4^+*. Fill in the table in the following order: (i) initial* $c(NH_4^+)$, $c(H^+)$*, and* $c(NH_3)$*; (ii) equilibrium* $[H^+]$*; (iii) equilibrium* $[NH_3]$*; (iv) equilibrium* $[NH_4^+]$*.*

	NH_4^+	$\rightleftharpoons$	H^+	+	NH_3
initial conc	x		-		-
equilib. conc	$x-7.4\times10^{-6}$		7.4×10^{-6}		7.4×10^{-6}

Write out the equilibrium constant expression

$$K_a = [H^+][NH_3]/[NH_4^+]$$

Substitute in the values from the table

$$5.6 \times 10^{-10} = (7.4\times10^{-6})^2/(x-7.4\times10^{-6})$$

This time, no approximation is needed, and x = stoichiometric concentration = 9.8×10^{-2} *mol* L^{-1}*.*

11.4 Bases and the equilibrium constant K_b

11.4.1 The equilibrium constant K_b

By analogy with our discussion of K_a, let us put the definition of K_b in terms of a hypothetical base B, which accepts a proton from solvent water.

$B(aq) + H_2O(\ell)$	$\rightleftharpoons$	$BH^+(aq) + OH^-(aq)$
base acid		acid base

The equilibrium constant for this reaction is given the special symbol K_b. By convention, K_b always refers to the protonation of 1 mol of base.

$$K_b = \frac{[BH^+][OH^-]}{[B]}$$

Again, the concentration of the pure liquid H_2O is omitted from the equilibrium constant expression, and K_b has the units mol L^{-1}. Some example reactions follow.

[3] $NH_3(aq) + H_2O(\ell)$	$\rightleftharpoons$	$NH_4^+(aq) + OH^-(aq)$
base acid		acid base

$$K_b = \frac{[NH_4^+][OH^-]}{[NH_3]}$$

The value of K_b for this reaction has the value 1.8 x 10^{-5} mol L^{-1} at 298 K. In other words, a solution of ammonia in water has only a slight tendency to form NH_4^+(aq) and OH^-(aq); it consists mainly of undissociated NH_3 molecules. For this reason it is inaccurate to call such a solution "ammonium hydroxide"; it is more properly described as a solution of ammonia in water.

[4] $CN^-(aq) + H_2O(\ell) \rightleftharpoons HCN(aq) + OH^-(aq)$
base acid acid base

$$K_b = \frac{[HCN][OH^-]}{[CN^-]}$$

11.4.2 *Strong and weak bases*

The common **strong bases** are the soluble hydroxides of the periodic groups IA and IIA metals. The commonly encountered strong bases are NaOH, KOH, $Ca(OH)_2$ and $Ba(OH)_2$, all of which dissociate completely in water to give stoichiometric quantities of OH^-. The group IA and IIA hydroxides not mentioned are either less common (but still strong bases, *e.g.*, LiOH, RbOH, CsOH, $Sr(OH)_2$), or are insoluble ($Mg(OH)_2$); beryllium does not form a simple hydroxide. The hydroxides of all other metals are insoluble. "Ammonium hydroxide" is not a true hydroxide, as discussed above.

Besides metal hydroxides there are many substances which react with water stoichiometrically to form OH^-. One example is sodamide $NaNH_2$, which contains the amide ion NH_2^-.

$$NaNH_2(s) + H_2O(\ell) \rightleftharpoons Na^+(aq) + NH_3(aq) + OH^-(aq) \qquad K_b >> 1$$

Sodamide is a strong base in the sense that it furnishes OH^- stoichiometrically, but it would be equally correct to describe sodamide as being decomposed by water.

Weak bases react with water to form OH^-, but with equilibrium constants << 1. Ammonia is one example.

[3] $NH_3(aq) + H_2O(\ell) \rightleftharpoons NH_4^+(aq) + OH^-(aq)$
base acid acid base

Weak bases can be neutral molecules (many are nitrogenous bases) or anions (*e.g.*, NO_2^-, CO_3^{2-}, CN^-). In the case of the nitrogenous bases, the proton is accepted at the nitrogen atom through the formation of an N-H bond. There are no common cationic bases. Weak bases are generally the conjugates of weak acids; this point is discussed below in Section 11.4.4.

11.4.3 Calculations involving K_b

These are very much analogous to those involving K_a. There are essentially three quantities to consider: (1) the equilibrium constant; (2) the initial, or stoichiometric, concentration of the base; (3) the equilibrium concentration of OH^- (or the pOH or the pH or the percent dissociation). Any one of these can be calculated provided the other two are already known, as the following two examplcs show.

Example: *Calculate K_b for the cyanide ion, given that a 1.6×10^{-2} mol L^{-1} solution of potassium cyanide has pH 10.76.*

Answer: *The solution contains the weak base CN^-. Write the net ionic equation showing CN^- acting as a base, and set up a table of initial and equilibrium concentrations. Remember that since pH = 10.76, pOH = 3.24, and therefore the equilibrium concentration $[OH^-] = 5.8 \times 10^{-4}$ mol L^{-1}.*

	CN^-	+	H_2O	$\rightleftharpoons$	HCN	+	OH^-
initial conc	1.6×10^{-2}		-		-		-
equilib. conc	$1.6\times10^{-2}-5.8\times10^{-4}$		-		5.8×10^{-4}		5.8×10^{-4}

Write out the K_b expression and substitute.

Note that $[CN^-]$ at equilibrium is approximately 1.6×10^{-2} mol L^{-1}.

$$K_b = [HCN][OH^-]/[CN^-]$$
$$= (5.8\times10^{-4})^2/(1.6\times10^{-2}) = 2.1 \times 10^{-5} \text{ mol } L^{-1}$$

To summarize the ideas of Sections 11.4.1 to 11.4.3, compare the addition of 0.01 mol of NaOH to 1 L of water with the addition of 0.01 mol NH_3 to 1 L of water. The strong base NaOH gives a concentration of 0.01 mol L^{-1} OH^- (pH 12). The 0.01 mol NH_3 ($K_b = 1.8 \times 10^{-5}$ mol L^{-1}) yields only 4×10^{-4} mol L^{-1} of OH^- (pH $\approx$ 10.6), and the remaining NH_3 ($\approx$ 0.01 mol L^{-1}) does not affect the pH.

Rule: *Only the H^+ and the OH^- present in any solution determine the pH; the pH is unaffected by other (weak) acids and bases, except insofar as they dissociate to give H^+ or OH^-.*

Example: *Calculate the pH of a 0.0025 mol L^{-1} solution of the weak base methylamine (CH_3NH_2), given that K_b is 4.5×10^{-4} mol L^{-1}.*
Answer: *Write out the net ionic equation and set up a table of initial and equilibrium concentrations.*

	CH_3NH_2	+ H_2O	⇌ $CH_3NH_3^+$ +	OH^-
initial conc	0.0025	-	-	-
change in conc	- x	-	+x	+x
equilib. conc	0.0025-x	-	x	x

Write out the equilibrium constant expression and substitute in the values from the table

$$K_b = [CH_3NH_3^+][OH^-]/[CH_3NH_2]$$

$$4.5 \times 10^{-4} = x^2/(0.0025-x)$$

Make the initial approximation that x << 0.0025 mol L^{-1}.
Remember that at this stage we are only testing the possibility *that x may be < 0.010 mol L^{-1}.*

$$4.5 \times 10^{-4} \approx x^2/(0.0025)$$

$$x^2 \approx 1.1 \times 10^{-6}; \quad x = 1.1 \times 10^{-3}$$

Check approximation: 1.1×10^{-3} = 42% of 0.0025, so approximation was not justified. In this problem we must use the method of successive approximations to complete the calculation.
Approximate again: (0.0025-x) ≈ (0.0025-0.0011).

$$4.5 \times 10^{-4} \approx x^2/(0.0014)$$

$$x^2 \approx 6.4 \times 10^{-7}; \quad x = 8.0 \times 10^{-4}$$

Repeat approximation: : (0.0025-x) ≈ (0.0025-0.0008).

$$4.5 \times 10^{-4} \approx x^2/(0.0017)$$

$$x^2 \approx 7.8 \times 10^{-7}; \quad x = 8.8 \times 10^{-4}$$

Repeat approximation: : (0.0025-x) ≈ (0.0025-0.0009).

$$4.5 \times 10^{-4} \approx x^2/(0.0016)$$

$$x^2 \approx 7.4 \times 10^{-7}; \quad x = 8.6 \times 10^{-4}$$

Further approximation gives no change in the value of x. Since x was defined as $[OH^-]$ at equilibrium, pOH = -log(x) = 3.07, and pH = 14.00 - pOH = 10.93 (two significant figures).

11.4.4 Relationship between K_a and K_b

There is a numerical relationship between K_a and K_b *for the special case where the acid and the base are conjugates of each other.* We can deduce this relationship using the "rule of multiple equilibria" (Chapter 8). The example selected involves HCN and CN^-, but the result is generally applicable.

Dissociation of HCN

(i) $HCN(aq) + H_2O(\ell) \rightleftharpoons H_3O^+(aq) + CN^-(aq)$ $\quad K_c = K_a$ for HCN

Dissociation of CN^-

(ii) $CN^-(aq) + H_2O(\ell) \rightleftharpoons HCN(aq) + OH^-(aq)$ $\quad K_c = K_b$ for CN^-

Add equations (i) and (ii); multiply the equilibrium constants

(iii) $2\,H_2O(\ell) \rightleftharpoons H_3O^+(aq) + OH^-(aq)$ $\quad K_c = ?$

Thus $K_c = K_a(HCN) \times K_b(CN^-)$

However, equation (iii) has already been met as the reaction which defines K_w. Therefore, we obtain the following result (which is generally applicable when the base is the conjugate base of the acid).

Rule: *For the special case of K_a for a weak acid and K_b for its conjugate base: $K_w = K_a \times K_b$*

A very practical consequence of this result is that most compilations of data do not include tables of K_b values. Instead, what is listed is the value of K_a for the conjugate acid of the base in question. This practice is followed in this book; Appendix 3 contains values of K_a for acids, but not values of K_b for their conjugate bases.

A second consequence of the result $K_a \times K_b$ = constant is that *the larger is K_a for a certain acid, the smaller will be K_b for its conjugate base.* Thus if K_a had the value 10^{-2} mol L^{-1} (a moderately strong acid), its conjugate would have $K_b = 10^{-12}$ mol L^{-1} (a very weak base). Table 11.2 summarizes the situation.

Table 11.2: Relationship between K_a and K_b values

Type of acid	**(example)**	K_a	K_b	**Type of base**	**(example)**
Strong	(HNO_3)	> 1	$< 10^{-14}$	Extremely weak	(NO_3^-)
Weak	(NH_4^+)	$10^{-1} - 10^{-14}$	$10^{-14} - 10^{-1}$	Weak	(NH_3)
Extremely weak	(CH_3OH)	$< 10^{-14}$	> 1	Strong	(CH_3O^-)

Rule: *The stronger an acid is, the weaker is its conjugate base, and vice versa.*

Nitric acid is a strong acid (fully dissociated in water). Its conjugate base, the nitrate ion, is such an extremely weak base that its reaction with water does not produce sufficient OH^- ions to change the pH of water. The nitrate ion is therefore considered "neutral" in water.

At the other extreme, methanol (CH_3OH) is such a feeble acid that its reaction with water does not produce sufficient H_3O^+ ions to change the pH of water ($K_a < 10^{-14}$ mol L^{-1}). Methanol is therefore considered "neutral" in water. However, its conjugate base (the methoxide ion, CH_3O^-) is such a powerful base that it reacts completely with water to give OH^- ($K_b > 1$ mol L^{-1}).

$$CH_3O^-(aq) + H_2O(\ell) \rightarrow CH_3OH(aq) + OH^-(aq)$$

In the middle are the conventional weak acids and their conjugate weak bases. The *product* K_aK_b, which must be numerically equal to K_w, 10^{-14}, is composed of K_a and K_b terms which are each < 1. There is thus no contradiction that the conjugate of a weak acid is itself a weak base. Compare these statements:

- *The weaker the acid, the stronger is its conjugate base* (**True**).
- *The conjugate of a weak acid is a strong base* (**False**).

11.5 Polyprotic acids

A polyprotic acid has more than one ionizable hydrogen atom. HNO_3 and HCl each have one ionizable hydrogen atom; H_2SO_4, H_2SO_3, and H_2CO_3 are examples of acids having two ionizable hydrogen atoms, while H_3PO_4 is an example of an acid with three ionizable hydrogen atoms[2]. Let's examine the ionization of three important polyprotic acids, H_2SO_4, H_2CO_3, and H_3PO_4.

Sulfuric acid

$$H_2SO_4 + H_2O \rightarrow H_3O^+ + HSO_4^- \qquad K_a \gg 1$$

$$HSO_4^- + H_2O \rightleftharpoons H_3O^+ + SO_4^{2-} \qquad K_a = 1.7 \times 10^{-2} \text{ mol L}^{-1}$$

The first dissociation of sulfuric acid is complete in water, and affords HSO_4^-. HSO_4^- can function both as the conjugate base of H_2SO_4, and as a moderately strong acid in its own right. $NaHSO_4$, which dissolves in water as Na^+ and

---

[2] Not all the hydrogens in a given molecule are necessarily ionizable. For example, acetic acid $C_2H_4O_2$ has only one ionizable hydrogen atom. The other three hydrogen atoms, which are bonded to carbon, are not ionizable and have no acidic properties. You cannot determine the number of ionizable hydrogen atoms just from the molecular formula. As a generalization, hydrogen atoms bonded to oxygen, nitrogen, sulfur, and the halogens are somewhat acidic, while most hydrogen atoms bonded to carbon are not appreciably acidic.

HSO_4^-, is used in some brands of toilet bowl cleaner.

Carbonic acid

$H_2CO_3 + H_2O$	$\rightleftharpoons$	$H_3O^+ + HCO_3^-$	$K_a = 4.2 \times 10^{-7}$ mol L^{-1}
$HCO_3^- + H_2O$	$\rightleftharpoons$	$H_3O^+ + CO_3^{2-}$	$K_a = 4.8 \times 10^{-11}$ mol L^{-1}

Carbonic acid (H_2CO_3) is formed when CO_2 dissolves in water; pure carbonic acid cannot be isolated; attempts to do so lead to decomposition to CO_2(g). H_2CO_3 is a rather weak acid, unlike H_2SO_4, and its conjugate base HCO_3^- is weaker still when it functions as an acid. H_2CO_3 and HCO_3^- are extremely important in natural systems; they regulate the pH of blood (Section 11.11) and natural waters such as lakes (Chapters 12 and 13).

Phosphoric acid

$H_3PO_4 + H_2O$	$\rightleftharpoons$	$H_3O^+ + H_2PO_4^-$	$K_a = 1.4 \times 10^{-3}$ mol L^{-1}
$H_2PO_4^- + H_2O$	$\rightleftharpoons$	$H_3O^+ + HPO_4^{2-}$	$K_a = 6.3 \times 10^{-8}$ mol L^{-1}
$HPO_4^{2-} + H_2O$	$\rightleftharpoons$	$H_3O^+ + PO_4^{3-}$	$K_a = 4.8 \times 10^{-13}$ mol L^{-1}

Phosphate species participate along with carbonate species in regulating the pH of blood and other biological fluids. The three K_a values for phosphoric acid show clearly that each loss of H^+ affords a successively weaker acid.

Species such as HSO_4^-, HCO_3^-, $H_2PO_4^-$, and HPO_4^{2-} can act either as acids or bases (the technical term is "amphoteric"). How should we describe the solutions of their sodium salts: acidic or alkaline? To answer this question, we have to compare K_a and K_b for the ion in question. Remember that to get K_b for the hydrogen carbonate ion (for example) we calculate K_w/K_a for the conjugate acid): H_2CO_3 in this case. We then compare the magnitudes of K_a (tendency to donate H^+ to water) and K_b (tendency to convert water into OH^-): Table 11.3.

For example; for HCO_3^-: $K_a(HCO_3^-) > K_b(HCO_3^-)$; HCO_3^- has a greater tendency to gain a proton from water than to donate a proton to water. Consequently, a solution of $NaHCO_3$ would react alkaline in water. Notice that in this discussion the identity of water as the proton donor or proton acceptor has been explicitly included. If the solution contained a base stronger than water, such as OH^-, HCO_3^- would act as an acid. If the solution contained an acid stronger than water, such as H^+, HCO_3^- would act as a base.

Table 11.3: Acid-base properties of amphoteric anions[a]

Anion	K_a	K_b	Acidic or basic?
HSO_4^-	1.7×10^{-2}	$<< 10^{-14}$	acidic
HCO_3^-	4.8×10^{-11}	2.4×10^{-8}	basic
$H_2PO_4^-$	6.3×10^{-8}	1.7×10^{-12}	acidic
HPO_4^{2-}	4.8×10^{-13}	1.6×10^{-7}	basic

a equilibrium constants in units of mol L^{-1}

11.6 Acid-base properties of salt solutions

The concepts already discussed in this chapter allow us to predict whether a particular solution will be acidic (pH < 7), alkaline (pH > 7), or neutral (pH ≈ 7). We simply ask ourselves what species the solution contains ... are they acidic, basic, or neutral? Table 11.4 summarizes the information.

Table 11.4: Acids, bases, and neutral ions

Neutral ions:	All group and cations (*e.g.*, Na^+, Ca^{2+}) Anions which are conjugates of the strong acids (*e.g.*, Cl^-, ClO_4^-)[a]
Acids:	Uncharged strong or weak acids (*e.g.*, HCl, HClO, H_2S) Metal cations other than those of Groups IA and IIA Cationic conjugates of nitrogen bases (*e.g.*, NH_4^+, $C_5H_5NH^+$) The two anions HSO_4^- and $H_2PO_4^-$
Bases:	All anions not already listed Uncharged nitrogen bases such as NH_3, C_5H_5N

a SO_4^{2-}, which has K_b = 8.3×10^{-13} mol L^{-1}, is almost neutral

Example: *State whether the following aqueous solutions have pH < 7; pH = 7; or pH > 7: (a)* $CuCl_2$*; (b)* Na_3PO_4*; (c)* C_5H_5NHBr*; (d) MgS*

Answer: *In each case, examine the acid-base properties of the constituent ions*

(a) Cu^{2+}: *acidic;* Cl^-: *neutral* → *pH < 7*

(b) Na^+: *neutral;* PO_4^{3-}: *basic* → *pH > 7*

(c) $C_5H_5NH^+$: *acidic;* Br^-: *neutral* → *pH < 7*

(d) Mg^{2+}: *neutral;* S^{2-}: *basic* → *pH > 7*

11.7 The "common ion" effect

In an earlier problem (Section 11.3) we calculated the percent dissociation and the pH of a 0.010 mol L^{-1} solution of acetic acid. We now address the question of what would happen if acetic acid were allowed to dissociate in a solution which already contained acetate ion. The question could also be posed in terms of what would happen if acetate ion were added to a solution of acetic acid.

The chemical reaction is still the same.

$$HAc(aq) + H_2O(\ell) \rightleftharpoons H_3O^+(aq) + Ac^-(aq)$$

However, Le Chatelier's Principle indicates that addition of acetate ion to a solution of $HAc/H^+/Ac^-$ at equilibrium will shift the reaction towards the left - in other words, *suppress the ionization of acetic acid.* This is called the "common ion effect"; acetate, the "common ion", affects the equilibrium in the manner predicted by Le Chatelier's Principle. We calculate this effect quantitatively in the next two examples. In the first example, we see that the percent dissociation of a weak acid (or equivalently a weak base) is greatly reduced in the presence of its conjugate. In the second example, the presence of the common ion OH^- for a weak base (or equivalently H^+ for a weak acid) so reduces the dissociation of the weak acid or base that the pH of the solution is almost entirely determined by the concentration of excess H^+ or OH^- that may be present.

Example: *In Section 11.3, we calculated the percent ionization of a 0.010 mol L^{-1} solution of acetic acid. Repeat this calculation for the dissociation of 0.010 mol L^{-1} acetic acid in a solution which already contains 0.22 mol L^{-1} sodium acetate.*

Answer: *Write out the net ionic equation and set up a table of initial and equilibrium concentrations. Remember that the "initial" solution already contains 0.22 mol L^{-1} Ac^-.*

	HAc	$\rightleftharpoons$	H^+	+	Ac^-
initial conc	0.010		-		0.22
change in conc	-x		+x		+x
equilib. conc	0.010-x		x		0.22+x

Write out the equilibrium constant expression

$$K_a = [H^+][Ac^-]/[HAc]$$

Substitute in the values from the table

$$1.8 \times 10^{-5} = (x)(0.22+x)/(0.010-x)$$

Make the approximation that $x << 0.010 \text{ mol } L^{-1}$, and therefore also that $x << 0.22 \text{ mol } L^{-1}$.

$$1.8 \times 10^{-5} \approx x(0.22)/(0.010)$$

$x = 8.2 \times 10^{-7}$; *approximation was justified*

percent ionization $= 100 \times (8.2 \times 10^{-7})/(0.010) = 0.008\%$

The percent ionization was reduced from 4% in the absence of Ac^- to 0.008% in the presence of Ac^-.

Example: *Calculate the pH of a solution containing both ammonia (0.050 mol L^{-1}) and NaOH (0.010 mol L^{-1}). Take K_b for ammonia as 1.8×10^{-5} mol L^{-1}.*

Answer: *Write out the net ionic equation and set up a table of initial and equilibrium concentrations. Remember that the "initial" solution already contains 0.010 mol L^{-1} OH^-.*

	NH_3	+ H_2O	$\rightleftharpoons$ NH_4^+	+ OH^-
initial conc	0.050	-	-	0.010
equilib. conc	0.050-x	-	4x	0.010+x

Write out the equilibrium constant expression

$K_a = [NH_4^+][OH^-]/[NH_3]$

Substitute in the values from the table

$1.8 \times 10^{-5} = (x)(0.010+x)/(0.050-x)$

Make the approximation that $x << 0.010$ mol L^{-1}, and therefore also that $x << 0.050$ mol L^{-1}

$1.8 \times 10^{-5} \approx x(0.010)/(0.050)$

$x = 9.0 \times 10^{-5}$; *approximation was justified*

Total $[OH^-] = 0.010+x \approx 0.010$ (Remember that x is not the answer to the problem!)

pOH = 2.00; pH = 14.00 - pOH = 12.00

Almost all the OH^- came from the NaOH, even though NaOH was present in much smaller stoichiometric concentration than NH_3.

A useful application of the common ion effect is a simple method of determining K_a for a weak acid experimentally. Recall the definition of K_a for a weak acid HA.

$$K_a = \frac{[H_3O^+][A^-]}{[HA]} \quad \textit{or equivalently} \quad K_a = \frac{[H^+][A^-]}{[HA]}$$

Suppose we were to mix exactly equal amounts of HA and A^- in the same solution; then the concentrations [HA] and $[A^-]$ would be equal. Under these conditions:

$K_a = [H_3O^+][A^-]/[HA]$ becomes $K_a = [H^+]$, or $\mathbf{pH = pK_a}$

pK_a is defined as $-\log_{10} K_a$, just as $pH = -\log_{10} [H^+]$.

Rule: *The pK_a of a weak acid can be determined experimentally by measuring the pH of a solution which contains equal concentrations of a weak acid and its conjugate weak base.*

Example: *A solution is prepared by dissolving 0.10 mol of solid sodium acetate in 100 mL water. A 10.00 mL aliquot of this solution is half neutralized by adding 0.0050 mol HCl, when the pH = 4.74. What is K_a for acetic acid?*

Answer: *The chemical reaction (as a net ionic equation) is:*

$$Ac^-(aq) + H^+(aq) \rightarrow HAc(aq)$$

When half the Ac^- has been converted to HAc, the concentrations of Ac^- and HAc are equal, and so pH = pK_a = 4.74.

$$K_a(HAc) = \text{antilog}(-4.74) = 1.8 \times 10^{-5}\ mol\ L^{-1}$$

11.8 Acid-base titrations

11.8.1 Titration curves

It will be recalled from Chapter 3 that titration is a method of carrying out chemical analysis in solution. Titration is very well suited to the analysis of either acidic or alkaline solutions, and the procedure is then called an acid-base titration. Frequently, a coloured indicator is used to signal the "end point" in the titration (Section 11.8.5). However, it is possible to follow the progress of an acid-base titration in a different manner, by using a pH meter to follow the pH of the solution being titrated. Suppose, for example, that we wish to monitor the progress of the titration between acetic acid (in the flask) and sodium hydroxide (added from a burette). We would place the pH electrode in the flask along with the acetic acid solution, and record the pH after the addition of each increment of sodium hydroxide solution. The resulting graph of "pH *vs*. Volume of NaOH" is called a **titration curve**. Such a graph is presented in Figure 1.

The stoichiometric and net ionic equations for this reaction are given below (all species are (aq)); it is easier to discuss acid-base chemistry in terms of the net ionic equation.

Stoichiometric equation:

$$HAc + NaOH \rightarrow NaAc + H_2O$$

Net ionic equation

$$HAc + OH^- \rightarrow Ac^- + H_2O$$

Note the following points about Figure 1.

• The pH increases during the course of the titration, as would be expected since we are adding a base to an acid.

• The *x* axis has two labels; the lower one reads "Volume of added NaOH", the upper one reads "moles OH^- per mole HAc". The first of these labels is clearly experimental - the actual volume of NaOH solution added.

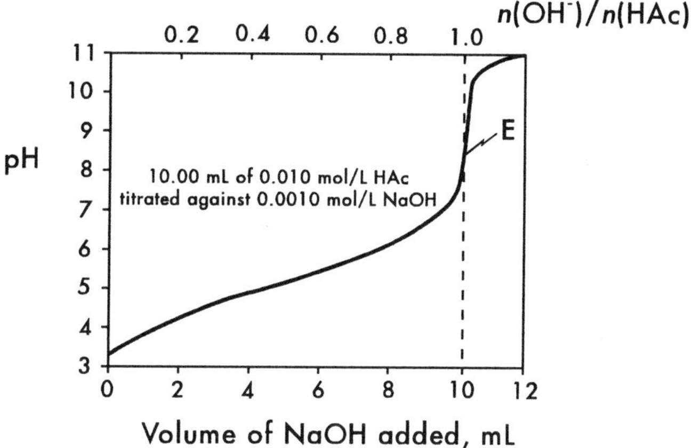

Figure 11.1: pH change as a function of volume of NaOH added in the titration between acetic acid and sodium hydroxide.

The upper axis label interprets this as a molar ratio, moles NaOH added per mole of HAc initially present. The stoichiometry of the reaction is that 1 mol HAc reacts with 1 mol NaOH. Hence HAc is in excess at all points on the graph where $n(OH^-)/n(HAc) < 1$, while NaOH is in excess whenever $n(OH^-)/n(HAc) > 1$.

• The **equivalence point** of the reaction should occur when $n(OH^-)/n(HAc) = 1$. At this point, all the HAc initially present has been converted into Ac^-. The equivalence point is a theoretical concept. The **end point** in the titration is when the experimentalist decides that the titration is complete. The experimentalist hopes that the end point will coincide as closely as possible with the equivalence point. Experimentally, the end point in the titration is taken as the point where the graph of "pH *vs.* Volume of NaOH added" rises most steeply: point "E" on the graph.

• At the equivalence point all the HAc initially present has been converted into Ac^-. Accordingly, the pH at the equivalence point in this titration is > 7 (actually, about 9), since the major species present in solution is the weak base Ac^-. *It is therefore a mistake to imagine that the pH at the equivalence point in an acid-base titration is always 7.*

• A titration curve gives us a second experimental method of determining pK_a. At the equivalence point, 1 mol NaOH has been added for every 1 mol acetic acid present initially. Stoichiometrically, the solution contains only Ac^- (and the spectator ion Na^+)[3].

	HAc +	OH^- →	Ac^- +	H_2O
moles before titration	1	0	0	-
moles at equivalence:				
before reaction	1	1	0	-
after reaction	0	0	1	-

Let's compare the situation at the "half equivalence point", where we find $n(OH^-)/n(HAc) = 0.500$. This point ("E/2" on Fig. 1, which is repeated below, see Figure 2) is found experimentally by finding the volume of NaOH at the equivalence point, and finding the point on the graph corresponding to half that value.

	HAc +	OH^- →	Ac^- +	H_2O
moles before titration	1	0	0	-
moles at half equivalence				
before reaction	1	0.5	0	-
after reaction	0.5	0	0.5	-

At the half equivalence point "E/2", $n(Ac^-) = n(HAc)$, and so the pH is equal to the pK_a of the acid being titrated. (If this is not clear, re-read Section 11.7).

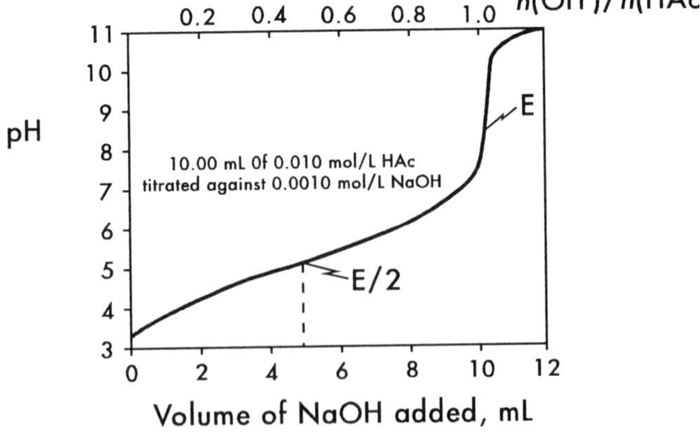

Figure 11.2: The half equivalence point E/2

---

[3] In this discussion, the moles of substances are stoichiometric amounts; the small extents of dissociation of HAc and Ac^- are neglected.

- The region of the graph near the half equivalence point is rather flat. In other words, the pH does not change rapidly when $n(Ac^-)/n(HAc)$ is close to unity (when $n(Ac^-) \approx n(HAc)$). We shall return to this point in Section 11.9.

Titration curves are not restricted to titrations between weak acids and strong bases. Figure 3 shows the pH change during a titration between strong acid and strong base (*e.g*, HCl *vs*. NaOH), and Figure 11.4 shows the pH change during a titration between weak base and strong acid (*e.g*, HCl *vs*. NH_3). The net ionic equations for these reactions follow. Notice that the first of these equations is the net ionic equation for the reaction between any strong acid and any strong base.

$$H^+(aq) + OH^-(aq) \rightarrow H_2O(\ell)$$

$$NH_3(aq) + H^+(aq) \rightarrow NH_4^+(aq)$$

In Figure 3, notice that the pH at the equivalence point is 7.0, since the only ions present in solution are the neutral (spectator) ions Na^+ and Cl^-. In this titration there is no significance to the half equivalence point. Notice also the very steep rise in the curve at the equivalence point. In fact, the stronger the acid, the steeper the slope when it is titrated against a given base. When the acid is very weak, the change in pH as a function of added base at the equivalence point is shallow, and the equivalence point may be difficult to find.

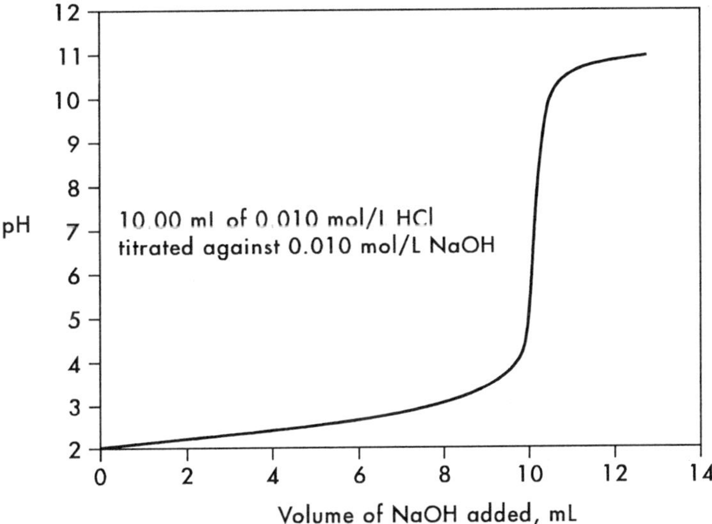

Figure 11.3: Titration curve for the titration between HCl (in the flask) and NaOH (titrant)

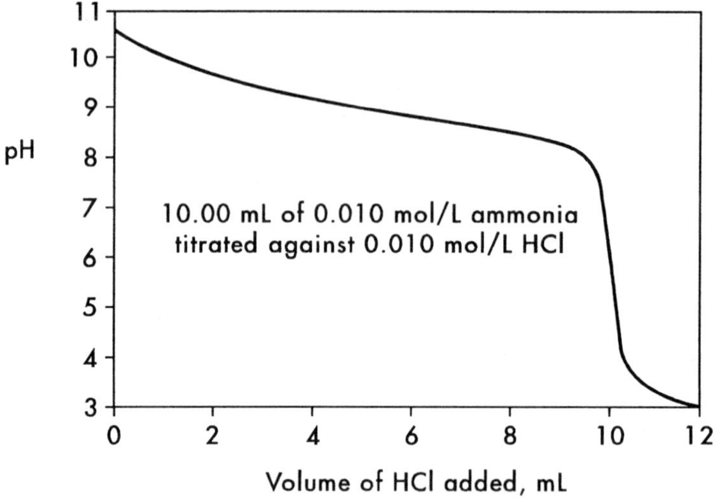

Figure 11.4: Titration curve for the reaction between NH_3 (in the flask) and HCl (titrant).

In Figure 4, notice that the pH decreases as acid is added over the course of the titration. The pH at the equivalence point is < 7 (actually about 5), since the major species present at equivalence is the weak acid NH_4^+. The half equivalence point affords pH = pK_a *of* NH_4^+ (**not** pK_b of NH_3). This is understandable since at the half equivalence point $n(NH_4^+) = n(NH_3)$.

The previous paragraphs are summarized in Table 11.5.

Table 11.5: Properties of pH titration curves

Titration type	pH at equivalence	pH at half equivalence
Strong acid/strong base	7.0	not significant
Weak acid/strong base	> 7	pK_a of conjugate acid
Weak base/strong acid	< 7	pK_a of conjugate acid

11.8.2 Speciation diagrams

In Figures 11.1 - 11.3 we examined what happened to the pH of a solution as an acid or a base was progressively neutralized. For a given degree of neutralization — that is, a given ratio of $[A^-] : [HA]$ — the pH is fixed by the constant value of K_a:

$$K_a = \frac{[H^+][A^-]}{[HA]} \qquad \text{or} \qquad K_a = \frac{[H^+][B]}{[BH^+]}$$

We can turn this argument around: *at a given pH, the ratio of the concentrations of conjugates is fixed.* A graph of this ratio as a function of pH — or, more often, of the fraction of each species as a function of pH — is called a **speciation diagram**. Figure 5 shows the speciation of acetic acetate and acetate ion as a function of pH; the "y" axis is the fraction of a particular species. At low pH, acetic acid is the predominant species; as the pH rises, a significant fraction of acetate ion is present. At pH = pK_a, [HAc] = $[Ac^-]$ = 0.5, which is the "crossover" point on the graph. Above this pH, Ac^- is the predominant species.

Example: *Explain how you would calculate the fraction of acetate ion in an aqueous solution of acetic acid (0.010 mol L^{-1}) at a given pH.*

Answer: *Set up an equilibrium calculation as set out in Section 11.3.*

	HAc	$\rightleftharpoons$ *H^+(aq)* +	*Ac^-(aq)*
initial concentrations	*0.010*	-	-
equilibm. concentrations	*0.010-x*	x	x

$$K_a = [H^+][Ac^-]/[HAc] \text{ (all quantities at equilibrium)}$$
$$= x^2/(0.010-x)$$

Solve for x; fraction of Ac^- = x/c(HAc), where c(HAc) is the initial (stoichiometric) concentration. Likewise, the fraction of HAc is (1-fraction of Ac^-).

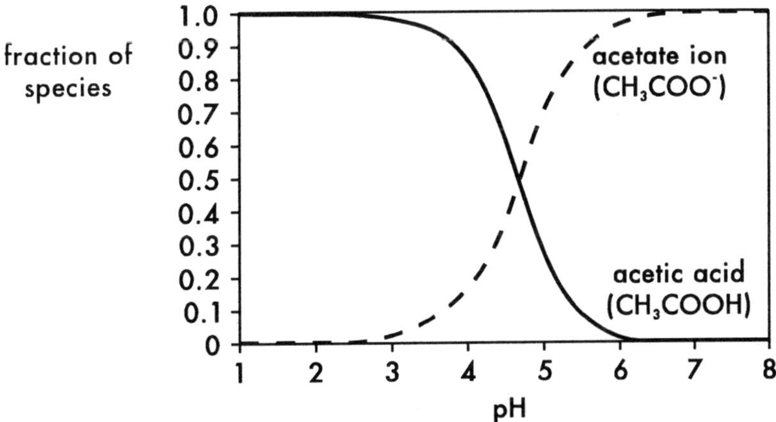

Figure 11.5: Speciation diagram for acetic acid as a function of pH

Figure 6 shows the speciation of all carbonate species (H_2CO_3, HCO_3^-, and CO_3^{2-}) as a function of pH. This time there are two crossover points on the graph; the one at pH 6.4 corresponds to pH = $pK_a(H_2CO_3)$, while that at pH 10.3 corresponds to pH = $pK_a(HCO_3^-)$.

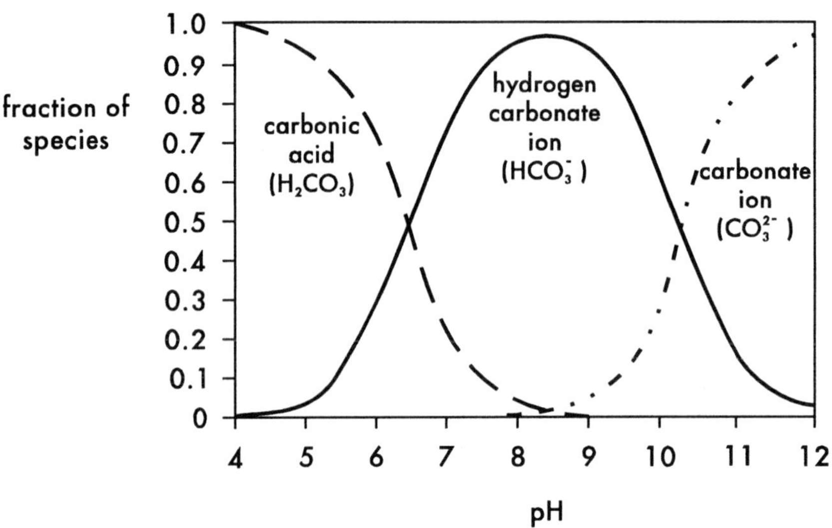

Figure 11.6: Speciation diagram for carbonic acid as a function of pH

In summary, the pK_a of an acid can be found from a speciation diagram; it occurs at a "cross-over" point on the diagram, when the concentrations of the conjugates are equal. At this point, pH = pK_a. For polyprotic acids, more than one crossover point will be present.

11.8.3 Equilibrium constants for acid-base reactions

Let us look again at the net ionic equations for the three acid-base reactions considered in Section 11.8.1; these were HCl + NaOH; HAc + NaOH; and HCl + NH_3.

a) HCl + NaOH

$H^+(aq) + OH^-(aq) \rightarrow H_2O(\ell)$

b) HAc + NaOH

$HAc(aq) + OH^-(aq) \rightarrow Ac^-(aq) + H_2O(\ell)$

c) HCl + NH_3

$NH_3(aq) + H^+(aq) \rightarrow NH_4^+(aq)$

Equation (a) is the reverse of the reaction which defines K_w. It follows that K_c for this reaction is $1/K_w$ and therefore has the value $1.0 \times 10^{+14}$ (mol $L^{-1})^{-2}$ at 298 K. This is a very large equilibrium constant, and we draw two conclusions: (i) the reaction between H^+(aq) and OH^-(aq) proceeds to completion; (ii) the reaction between any strong acid and any strong base has the same net ionic equation as equation (a), and has the same equilibrium constant of $1/K_w$.

Equation (b) is the reverse of the reaction which defines K_b for the acetate ion (you should see at once that it is related somehow to K_b **not** to K_a, since the equation includes the chemical species OH^- but not H^+). It follows that K_c for reaction (b) is $1/K_b$ for the acetate ion. We can calculate this equilibrium constant, because $K_b(Ac^-) = K_w/K_a(HAc)$. Therefore the equilibrium constant for reaction (b) is equal to $K_a(HAc)/K_w$, and has the value $1.8 \times 10^{+9}$ (mol $L^{-1})^{-1}$ at 298 K. Again this is a very large equilibrium constant, and we conclude that the reaction between HAc(aq) and OH^-(aq) proceeds to completion. More generally, the reaction between any weak acid and any strong base has equilibrium constant of K_a(weak acid)/K_w. Such reactions usually proceed to completion, unless the acid in question is especially weak (K_a is especially small).

Equation (c) is the reverse of the reaction which defines K_a for the ammonium ion (again, note that it is related somehow to K_a since the equation includes the chemical species H^+). It follows that K_c for reaction (c) is $1/K_a$ for the ammonium ion. We can calculate this equilibrium constant, because $K_a(NH_4^+) = 5.6 \times 10^{-10}$ mol L^{-1}. Therefore the equilibrium constant for reaction (c) has the value $1.8 \times 10^{+9}$ (mol $L^{-1})^{-1}$ at 298 K. Again this is a very large equilibrium constant, and we conclude that the reaction between H^+(aq) and NH_3(aq) proceeds to completion. More generally, the reaction between any strong acid and any weak base has equilibrium constant of $1/K_a$(conjugate weak acid). Such reactions usually proceed to completion, unless the base in question is especially weak.

Example: *Calculate the equilibrium constant for the reaction between lactic acid ($K_a = 1.4 \times 10^{-4}$ mol L^{-1}) and potassium hydroxide.*

Answer: *Write the net ionic equation. Note that you do not need the to know the formula for lactic acid — call it HLac.*

$$HLac + OH^- \rightleftharpoons Lac^- + H_2O$$

The equation contains OH^- and is therefore related to a K_b equation. In this case it is the reverse of the K_b equation.

$$K_c = 1/K_b \text{ for } Lac^-$$

But $K_b(Lac^-) = K_w/K_a(HLac)$, *so* $K_c = K_a(HLac)/K_w$

$$K_c = (1.4 \times 10^{-4})/(1.0 \times 10^{-14}) = 1.4 \times 10^{10} \text{ L mol}^{-1}$$

11.8.4 Calculation of pH at the equivalence point

At the equivalence point of an acid-base titration, equal moles of acid and base have reacted together. In the case of a monoprotic acid such as formic acid $HCHO_2$ titrated against NaOH:

$$HCHO_2(aq) + OH^-(aq) \rightarrow CHO_2^-(aq) + H_2O(\ell)$$

When the acid is polyprotic, more than one equivalence point may occur. An example is given by the titration of phosphoric acid against NaOH.

$$H_3PO_4(aq) + OH^-(aq) \rightarrow H_2PO_4^-(aq) + H_2O(\ell)$$
$$H_2PO_4^-(aq) + OH^-(aq) \rightarrow HPO_4^{2-}(aq) + H_2O(\ell)$$
$$HPO_4^{2-}(aq) + OH^-(aq) \rightarrow PO_4^{3-}(aq) + H_2O(\ell)$$

An indicator must be carefully selected in order to stop the titration at the correct equivalence point (Section 11.8.5).

In all these situations, the reactants have reacted "to completion" (large K_c) at the equivalence point. In order to calculate the pH at the equivalence point, we must determine what species are present.

Example: *What species are present at the equivalence point in each of the following titrations?*

(a) $HCHO_2$ with KOH

(b) NH_3 with $HClO_4$

(c) NaH_2PO_4 with NaOH (1 equivalent)

Answer: *Write the relevant net ionic equations*

(a) $HCHO_2(aq) + OH^-(aq) \rightarrow CHO_2^-(aq) + H_2O(\ell)$

At equivalence, CHO_2^- is the only acid or base present. CHO_2^- is a weak base, so we expect $pH > 7$.

(b) $NH_3(aq) + H^+(aq) \rightarrow NH_4^+(aq)$

At equivalence, NH_4^+ is the only acid or base present. NH_4^+ is a weak acid, so we expect $pH < 7$.

(c) NaH_2PO_4 is a source of $H_2PO_4^-$. The significance of the statement 1 equivalent is that only one of its two acidic hydrogens is being neutralized.

$$H_2PO_4^-(aq) + OH^-(aq) \rightarrow HPO_4^{2-}(aq) + H_2O(\ell)$$

At equivalence HPO_4^{2-} is the only acid or base present. HPO_4^{2-} is a weak base, so we expect $pH > 7$.

Once the product of neutralization has been identified, you must determine whether it is an acid or a base. The pH can then be calculated as in Section 11.3 and 11.4.

11.8.5 Acid-base indicators

In Chapter 3, we noted that coloured indicators are often used to signal the "end point" in volumetric analysis. Acid-base titration using an indicator is fast and cheap. However, the indicator must be carefully selected according to the properties of the analyte and the titrating reagent.

Acid-base indicators are themselves weak acids, but have the special property that the acidic and basic conjugates differ in colour[4]. We will call the indicator "HIn" and its conjugate base "In^-". One such indicator, phenolphthalein, is colourless in solutions of pH < 8, and pink in solutions of pH > 8. Phenolphthalein has the following properties:

HIn is colourless; In^- is pink; $pK_{In} \approx 8$

The working of the indicator is understood from the definition of K_a for the indicator. We will call this acid dissociation constant K_{In} in order to distinguish it from K_a of the acid being titrated.

$$HIn(aq) + H_2O(\ell) \rightleftharpoons H_3O^+(aq) + In^-(aq)$$

$$K_{In} = [H^+][In^-]/[HIn] \quad \text{or} \quad K_{In}/[H^+] = [In^-]/[HIn]$$

When $[H^+] > K_{In}$ (same as $pH < pK_{In}$), the ratio $[In^-]/[HIn]$ will be less than unity. There will be more HIn than In^- present, and so the solution will take on the appearance of the predominant species HIn (colourless in this case). Conversely, when $[H^+] < K_{In}$ (same as $pH > pK_{In}$), the ratio $[In^-]/[HIn]$ will be greater than unity. There will be more In^- present than HIn, and so the solution will take on the appearance of the predominant species In^- (pink in this case). The indicator will therefore change colour when pH ≈ *its own* pK_a (that is pK_{In}, <u>not</u> the pK_a of the acid being titrated).

This is how to select an indicator.

1. From the type of titration, decide what chemical species will be present in stoichiometric amounts at the equivalence point. If a weak base (*e.g.* Ac^-) is being formed, the pH will be > 7; if a weak acid is being formed (*e.g.* NH_4^+), the pH will be < 7.

---

[4] Although the indicator is an acid, the amount added as an indicator is so small that it does not affect the volume of titrant added to reach the equivalence point. In very precise work, a "blank" titration is performed, in which the indicator is present but the analyte is absent. The volume of titrant needed just to make the indicator change colour can then be determined.

2. Choose an indicator whose pK_{In} is in the approximate range you expect the pH at equivalence.

Remember: *match the pK_a of the indicator to the expected pH of the solution at the equivalence point, NOT to the pK_a of the acid being titrated.*

In the example of the titration of H_3PO_4 with NaOH discussed in the previous section, you can appreciate that a different indicator would have to be used depending on whether you wished to make $H_2PO_4^-$, HPO_4^{2-}, or PO_4^{3-} the product of the reaction.

Table 11.6 gives the pK_{In} values for some useful acid-base indicators, and a suggestion as to useful applications of that indicator.

Table 11.6: pK_{In} values for acid-base indicators

Indicator	pK_{In}	Colour change[a]	Typical titration application
tropeolin	2.0	red to yellow	strong acid + very weak base
methyl orange	3.4	red to orange	strong acid + weak base[b,c]
methyl red	4.9	red to yellow	strong acid + weak base[b]
bromothymol blue	7.1	yellow to blue	strong acid + strong base
phenolphthalein	9.3	colourless to red	weak acid + strong base[b]
alazarin yellow	11.2	yellow to purple	very weak acid + strong base

a colour change as pH increases
b also useful for strong acid + strong base titration
c used in the determination of alkalinity (Section 11.12.3)

11.9 Buffer solutions

11.9.1 What is a buffer solution?

A buffer solution is the name given to a solution which resists a change in pH upon addition of (small) amounts of either strong acid or strong base. Buffer solutions are of great importance in both biochemistry and in environmental chemistry.

Many biological materials, such as proteins, are only stable over limited ranges of pH. Biochemists wishing to study their properties must be careful to maintain them in solutions whose pH is carefully controlled - buffer solutions. Likewise, biological fluids in living cells are maintained at constant pH in order to maintain the biological activity of proteins and other biological molecules. For example, blood is maintained at about pH 7.4; serious illness or death occur if the pH varies by more than about 0.5 pH units from this value.

In the environment there is much concern about the effects of "acid precipitation" upon forests, lakes, and aquatic life. Certain lakes and streams are "well buffered"; they resist acidification, and their animal and plant populations suffer relatively little from the effects of acidic precipitation. Other, poorly buffered lakes do not resist a drop in pH when rain- and snow-fall are acidic. The injurious results of acid precipitation are discussed in Chapter 13. In the next sections we consider how a buffer solution works. Before reading further, make sure that you understand Sections 11.7 and 11.8.3.

In chemical terms, a buffer solution always has the characteristic that it simultaneously contains appreciable quantities of both a weak acid and a weak base. The commonest type of buffer solution contains appreciable quantities of a weak acid and its conjugate weak base (*e.g.*, HAc and Ac^-; NH_4^+ and NH_3; H_2CO_3 and HCO_3^-). "Appreciable" in this context means first, that sufficient amounts of the buffer components are present to neutralize the amounts of acid or base that are likely to be added; and second, that the ratio [conjugate base]/[conjugate acid] is between about 0.1 and 10. A common misconception is that a buffer solution *always* contains equal amounts of the conjugate acid and conjugate base. The latter is a specific example of a buffer solution, which is often called a 1:1 buffer.

There are three ways to prepare a buffer solution.

1. Mix appropriate amounts of the weak acid and the conjugate weak base (*e.g.*, mix acetic acid with sodium acetate; mix ammonium chloride with ammonia).

2. Partly neutralize the weak acid with strong base (*e.g.*, mix sodium hydroxide (limiting reactant) with excess acetic acid; mix sodium hydroxide (limiting reactant) with excess ammonium chloride).

3. Partly neutralize the weak base with strong acid (*e.g.*, mix hydrochloric acid (limiting reactant) with excess sodium acetate; mix hydrochloric acid (limiting reactant) with excess ammonia).

Methods (2) and (3) depend on the fact that the reaction between a weak acid and a strong base (or a strong acid and a weak base) proceeds to completion (K_c is large): Section 11.8.3. As a result of partial neutralization, the final solution contains both of a pair of conjugates in comparable concentration.

11.9.2 How does a buffer solution work?

A buffer solution resists a change in pH upon addition of small quantities of either H^+ or OH^-. The following concepts are needed in order to understand how resistance to pH change takes place.

- The pH of any solution is governed by the quantities of H^+ and OH^- that it contains, but *not* by the quantities of other (weak) acids and bases, except to the extent that these weak acids and bases react with water to give H^+ and OH^-: Section 11.4.3.

- The buffer solution contains appreciable concentrations of a weak acid and its conjugate weak base, the ratio of which determines the pH (Figures 11.5 and 11.6).

- Strong base reacts with the weak acid component of the buffer solution "to completion", and likewise strong acid reacts to completion with the weak base component of the buffer. Thus the weak acid component of the buffer serves to "soak up" small amounts of added strong base, while the weak base component of the buffer neutralizes small amounts of added strong acid.

Example: *Write net ionic equations to show what happens when a small amount of:*
(a) NaOH is added to a buffer solution containing NH_3 and NH_4^+
(b) HCl is added to a buffer solution containing NaH_2PO_4 and Na_2HPO_4
Answer: *In each case, identify the acidic and basic conjugates in the buffer solution.*
(a) NH_4^+ is the acidic conjugate. It will react with strong base OH^-. NH_3 is the basic conjugate which will be unreactive in this case.

$$NH_4^+(aq) + OH^-(aq) \rightarrow NH_3(aq) + H_2O(\ell)$$

(b) $H_2PO_4^-$ is the acidic conjugate and HPO_4^{2-} is the basic conjugate. The basic conjugate reacts with the added strong acid.

$$HPO_4^{2-}(aq) + H^+(aq) \rightarrow H_2PO_4^-(aq)$$

11.9.3 Buffer solutions and titration curves

As stated already, a buffer solution contains appreciable concentrations of a weak acid and its conjugate weak base. One method of preparing such a solution is to partly neutralize a weak acid with strong base; in other words, the solution formed part way through a titration between a weak acid and a strong base is precisely the same as a buffer solution. Consider the example of the partial neutralization of acetic acid (1 mol) by sodium hydroxide (x mol, where $x < 1$).

	HAc	+	OH^-	→	Ac^-	+	H_2O
Moles before reaction	1		x		zero		-
Moles after reaction	1-x		≈ 0		x		-

Turning back to Figures 11.1 and 11.4, we should note that partial neutralization corresponds to the flat part of the curve "pH *vs.* Added Titrant". Thus, *a buffer solution is present when a weak acid (or weak base) solution is partly neutralized, and this corresponds to the flat region of a pH titration curve.*

We can understand what happens in qualitative terms when a small amount of H^+ or OH^- is added to a buffer solution. In terms of the titration curve, addition of strong acid or base moves the reaction a little way along the titration curve. However, since the profile pH *vs.* added H^+ (or OH^-) is rather flat in this region, the pH does not change appreciably (Figure 7).

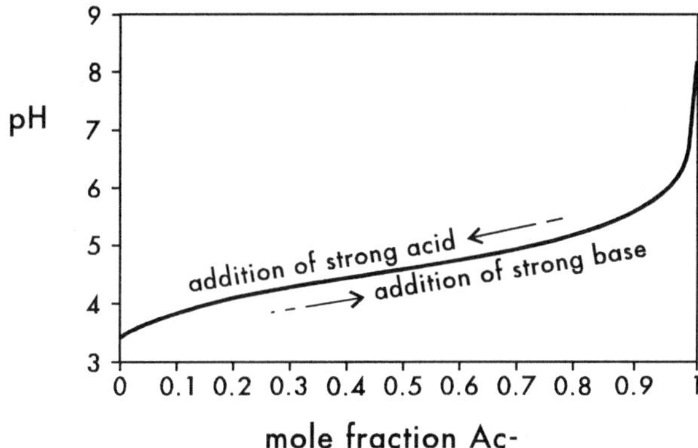

Figure 11.7: Effect of adding a small amount of $H^+(aq)$ or $OH^-(aq)$ to a buffer solution made from HAc and Ac^-.

An alternative way of saying the same thing is that added H^+ is neutralized by the basic component of the buffer, while added OH^- is neutralized by the acidic component of the buffer. Let HA and A^- be the acidic and basic conjugates in the buffer.

On addition of H^+:

$$H^+ + A^- \rightarrow HA$$

H^+, which does affect the pH, is replaced by HA, which does not.

On addition of OH^-:

$$OH^- + HA \rightarrow A^-$$

OH^-, which affects the pH indirectly by regulating the concentration of H^+, is replaced by A^-, which does not affect pH.

11.9.4 Buffer capacity

So far, we have indicated that a buffer solution can resist pH change upon addition of small amounts of strong acid or strong base. The question arises, "how much is small?". The answer depends upon the composition of the buffer. As we shall see in the next section, the pH of a buffer solution depends only on the **ratio** of the buffer components $[A^-]/[HA]$, and not on their absolute amounts. However, the **buffer capacity**, the amount of acid or base which a given buffer solution can absorb for a given pH change, depends critically upon its concentration.

Suppose we start out with a buffer solution containing equal concentrations of NH_3 and NH_4^+, and we have determined that in this particular application the ratio $[NH_3]/[NH_4^+]$ must not fall below 0.8. The net ionic equation follows.

$$H^+(aq) + NH_3(aq) \rightarrow NH_4^+(aq)$$

Let the initial concentrations of NH_3 and NH_4^+ ammonia both equal *Am* and that of the HCl added equal *x*.

	$H^+(aq)$ +	$NH_3(aq)$ →	$NH_4^+(aq)$
conc. before reaction	x	Am	Am
conc. after reaction	≈ 0	$Am-x$	$Am+x$

We set $(Am-x)/(Am+x) = 0.8$; the solution of this equation is $x = 0.1Am$ to one significant figure. Thus the amount of HCl which can be added for a given change of the ratio of the buffer components depends directly on *Am*, the initial concentrations of the buffer components. For the example given *x*, the maximum concentration of HCl that can be added under the stated conditions, is 0.1 times the concentration of the basic conjugate NH_3.

Rule: *The pH of a buffer solution depends only on the **ratio** of the buffer components $[A^-]/[HA]$, but the **buffer capacity** depends directly upon their **absolute** concentrations.*

In the example just discussed, a buffer containing $[NH_3] = [NH_4^+] = 0.1$ mol L^{-1} could absorb 0.01 mol L^{-1} HCl before the ratio $[NH_3]/[NH_4^+]$ fell below 0.8. However, a buffer of the same initial pH, but having $[NH_3] = [NH_4^+] = 0.01$ mol L^{-1} could only absorb 0.001 mol L^{-1} HCl for the same pH change.

11.10 Calculations involving buffer solutions

11.10.1 Calculating the pH of a buffer solution

A buffer solution contains comparable concentrations of a weak acid and its conjugate weak base. Its chemistry is therefore exactly the same as a solution containing a common ion (Section 11.7). Like the solution containing a common ion, the common ion represses the ionization of the weak acid. The consequence of this is that the pH is governed by the stoichiometric concentrations of the conjugates, because you can neglect the small extent of dissociation which occurs (review Section 11.7). For this reason, it is simpler to calculate the pH of a buffer solution than to calculate the pH of a solution containing only a weak acid, or only a weak base.

In the general case, $K_a = \dfrac{[H^+][\text{basic conjugate}]}{[\text{acidic conjugate}]}$

This equation can be rearranged.

$$[H^+] = \frac{K_a\,[\text{acidic conjugate}]}{[\text{basic conjugate}]}$$

In other words, the $[H^+]$ concentration of a buffer solution depends only upon K_a for the acidic conjugate, and the ratio of concentrations of the conjugates.

A useful form of the equation just derived is obtained by taking the negative logarithm of both sides.

$$pH = pK_a + \log_{10}\left\{\frac{[\text{basic conjugate}]}{[\text{acidic conjugate}]}\right\}$$

This equation is often known as the **Henderson-Hasselbalch equation**.

Example: *Calculate the pH of a buffer solution containing 0.10 mol L^{-1} acetic acid, and 0.20 mol L^{-1} sodium acetate.*

Answer: *The acidic conjugate is HAc and the basic conjugate is Ac^-. Substitute the values into the Henderson-Hasselbalch equation. Acetic acid has K_a 1.8 x 10^{-5} mol L^{-1}; pK_a is 4.74.*

$$pH = pK_a + \log_{10}\{[Ac^-]/[HAc]\}$$
$$= 4.74 + \log_{10}(0.20/0.10) = 4.74 + 0.30 = 5.04$$

11.10.2 1:1 and other buffers

We have defined a buffer as containing appreciable (or we could say comparable) concentrations of both of a conjugate acid, conjugate base pair. What do appreciable or comparable mean in this context?

Comparable does not necessarily mean that the concentrations of conjugates are exactly equal. Equal concentrations of conjugates gives a 1:1 buffer. A buffer of this composition has pH = pK_a.

Most buffers are prepared in the laboratory with the ratio of the conjugates between 0.1 and 10. In terms of the titration curve, this places the buffer in the flatter region of the pH *vs.* mole ratio curve, and means that the buffer has reasonable capacity to absorb either H^+ or OH^-. In terms of the Henderson-Hasselbalch equation, the ratio 0.1 < mole ratio < 10 means that useful buffers usually have pH = $pK_a \pm 1$. Of course, if the buffer has more of the conjugate base component, it will resist pH change better when H^+ is added than when OH^- is added; if it contains more of the conjugate acid component, it will resist pH change better when OH^- is added than when H^+ is added.

11.10.3 Calculating the pH after strong acid or strong base is added to a buffer solution

This calculation must be done in two parts.

(i) The reaction of strong acid with weak base proceeds to completion, and likewise the reaction of strong base with weak acid proceeds to completion (Section 11.8.3). Strong base reacts (completely) with the acidic conjugate of the buffer, and strong acid reacts completely with the basic conjugate of the buffer. We must carry out a stoichiometric calculation (in moles) to find the concentrations of all species after the reaction.

(ii) Now we must determine what is present in solution. If the amount of strong base or strong acid was small, compared with the initial amounts of the acidic and basic conjugates, the solution will still be a buffer solution — it will still contain appreciable amounts of both conjugates. If so, use the Henderson-Hasselbalch equation to calculate the pH. If not, and the buffer capacity has been exceeded, the pH of the solution will be regulated by the excess concentration of either H^+ or OH^-, as appropriate.

Example: *Calculate the pH after 0.0050 mol KOH is added to 0.100 L of a buffer solution initially containing NH_3 (0.21 mol L^{-1}) and NH_4Cl (0.27 mol L^{-1}).*

Answer: *There will be a chemical reaction between the strong base OH^- and the weak acid NH_4^+. Work out moles of substance before and after reaction.*

$$n(NH_4^+) = 0.27 \text{ mol L}^{-1} \times 0.100 \text{ L} = 0.027 \text{ mol}$$
$$n(NH_3) = 0.21 \text{ mol L}^{-1} \times 0.100 \text{ L} = 0.021 \text{ mol}$$

	NH_4^+ +	OH^- →	NH_3 +	H_2O
moles before rxn	0.027	0.0050	0.021	-
moles after rxn	0.022	≈ 0	0.026	-

After the reaction, both conjugates are present. The solution is still a buffer solution. Use the Henderson-Hasselbalch equation to calculate pH. NH_4^+ (the acidic conjugate) has $pK_a = 9.24$. Note that we can use either the ratio of moles of the conjugates (0.026/0.022) or the ratio of their concentrations (0.26/0.22).

$$\begin{aligned} pH &= pK_a + \log_{10}\{[NH_3]/[NH_4^+]\} \\ &= 9.24 + \log_{10}\{0.026/0.022\} \\ &= 9.24 + 0.073 = 9.31 \end{aligned}$$

11.11 Practically important buffer solutions

11.11.1 Blood

The chief system responsible for maintaining the pH of blood is H_2CO_3/HCO_3^-. The acidic component, H_2CO_3, has $pK_a = 6.38$. Normal blood has pH = 7.4; since this is on the alkaline side of pH 6.38, it follows that blood contains more HCO_3^- than H_2CO_3. In fact, application of the Henderson-Hasselbalch equation shows us that the ratio $[HCO_3^-]/[H_2CO_3]$ is approximately 12.

The pH of blood is maintained in an ingenious way. Respiration in the tissues converts carbohydrates into CO_2(aq); since CO_2(aq) {= H_2CO_3(aq)} (Footnote 5) is the

---

[5] Strictly, this value is the overall equilibrium constant for the pair of reactions below:

$$CO_2(aq) + H_2O(\ell) \rightleftharpoons H_2CO_3(aq)$$

$$H_2CO_3(aq) + H_2O(\ell) \rightleftharpoons H_3O^+(aq) + HCO_3^-(aq)$$

However, for the purposes of this discussion, CO_2(aq) and H_2CO_3(aq) may be regarded as interchangeable.

acidic component of the buffer solution, increased [CO_2] should lower the pH. This does not happen, because the excess CO_2(aq) is expired as CO_2(g), thereby maintaining the ratio [HCO_3^-]/[H_2CO_3] at the correct value. Under conditions of fatigue, lactic acid (source of H^+) accumulates in the blood, and the pH is regulated by expiring CO_2.

$$HCO_3^-(aq) + H^+(aq,\text{ from lactic acid}) \rightarrow H_2CO_3(aq) \rightarrow H_2O(\ell) + CO_2(g)$$

Clinical laboratories routinely run tests on the composition of blood, analysing for both cations (Na^+, K^+, Ca^{2+}) and anions (principally HCO_3^- and Cl^-). The ratios of these species are frequently diagnostic of the medical condition from which the patient is suffering.

Respiratory acidosis (pH too low) can occur if the blood contains excessive quantities of acid. A common, and frequently fatal, condition in ruminants is "bloat" or "grain overload", where the animal consumes an immoderate quantity of grain, which is broken down to lactic acid in the rumen. The absorption of lactic acid from the rumen can overwhelm the animal's ability to regulate its pH by CO_2 exhalation, even though the animal pants to try to do so.

Under other circumstances, excessive panting can cause the opposite condition, respiratory alkalosis. Too much CO_2 is exhaled, causing the ratio [HCO_3^-]/[H_2CO_3], and hence the pH, to rise. This is seen in humans under conditions of hysteria. Fortunately, hysteria is self correcting; exhaustion leads to cessation of panting, and respiration restores the proper concentration of H_2CO_3.

11.11.2 Antacids

Stomach acid has pH near 1.5 ([H^+ (aq)] $\approx$ 0.03 mol L^{-1}), as noted in Table 11.1. In order to protect the stomach wall from this low pH and the digestive enzymes in the stomach, the stomach wall is lined with a thick mucous. Ulceration of the stomach wall occurs if the mucous layer is compromised. Many people who suffer from acid indigestion make use of antacid preparations, which serve to neutralize excess stomach acid. The need is for a base which will be non-toxic, and have no adverse side effects. Some examples follow.

Carbonates

$NaHCO_3$ (bicarbonate of soda) neutralizes stomach acid effectively:

$$HCO_3^-(aq) + H^+(aq) \rightarrow H_2CO_3(aq) \rightarrow H_2O(\ell) + CO_2(g)$$

It suffers from two disadvantages; the evolution of $CO_2(g)$ causes belching, and the rapid action of the very soluble sodium hydrogen carbonate causes "acid rebound" — the condition where the stomach over-compensates for raising the pH by secretion of additional acid.

Calcium carbonate also neutralizes acid effectively, but is slower acting because the insoluble calcium carbonate must gradually dissolve. One commercial formulation is simply calcium carbonate, which is pressed into a tablet under pressure with the aid of an inert binding agent.

$$CaCO_3(s) + 2\ H^+(aq) \rightarrow Ca^{2+}(aq) + CO_2(g) + H_2O(\ell)$$

Hydroxides

Magnesium hydroxide (milk of magnesia) is insoluble, and does not release $CO_2(g)$. However, high concentrations of Mg^{2+} predispose the user to diarrhea.

$$Mg(OH)_2(s) + 2\ H^+(aq) \rightarrow Mg^{2+}(aq) + 2\ H_2O(\ell)$$

Aluminum hydroxide behaves similarly to magnesium hydroxide. Aluminum salts have generally been held to be non-toxic, although there is currently some controversy as to whether there is relationship between aluminum intake and Alzheimer's disease (senile dementia). To date, however, no information is available to suggest that users of aluminum-containing antacids are predisposed to this condition: Section 10.8. One commercial formulation is an insoluble hydroxo-carbonate of the composition $NaAl(OH)_2CO_3$.

$$NaAl(OH)_2CO_3(s) + 4\ H^+(aq) \rightarrow Al^{3+}(aq) + Na^+(aq) + 3\ H_2O(\ell) + CO_2(g)$$

11.11.3 Lakes

This subject will be covered only briefly, since it is taken up in more detail in Chapter 13. As in blood, the chief buffering component in natural water is H_2CO_3/HCO_3^-. H_2CO_3 originates by equilibration with the $CO_2(g)$ of the atmosphere (Chapter 6), and HCO_3^- can be supplied by partial dissolution of the underlying rock. This can only happen if the underlying rocks contain carbonate ions (usually chalk, marble, or limestone). Igneous rocks such as granite are highly insoluble, and do not furnish significant concentrations of HCO_3^-.

In environmental chemistry, the usual concern is that natural water be able to resist **acidification**, as when the water is assaulted by acid rain. Since the buffer system in question is H_2CO_3/HCO_3^-, it follows that acid will be absorbed by the basic component of the buffer, namely HCO_3^-.

$$HCO_3^-(aq) + H^+(aq) \rightarrow H_2CO_3(aq) \rightarrow H_2O(\ell) + CO_2(g)$$

A high concentration of HCO_3^- gives a high buffer capacity towards acidification (typically seen in limestone areas). Conversely, low buffer capacity and poor resistance to acidification is seen in lakes where the underlying rocks cannot provide a source of HCO_3^- (*e.g.*, the granite rocks of the Canadian Shield and Scandinavia).

11.12 Alkalinity

11.12.1 Estimation of alkalinity

Alkalinity is a convenient measure of the ability of a natural water to resist acidification. As seen in the last paragraph, resistance to acidification is favoured when the water contains a high concentration of the weak base HCO_3^-. The alkalinity of a water sample is its total concentration of all bases (of which the most important is usually HCO_3^-). Alkalinity is routinely measured to determine the quality of both natural waters and industrial waste waters.

Alkalinity is measured experimentally by titrating the bases present in the water sample against strong acid. It cannot be measured simply by determining the pH, because the major bases present such as HCO_3^- are weak bases which contribute little OH^- to the solution. The titration is followed either by the use of a pH meter, or by using methyl orange as an indicator. The end point of the methyl orange titration (near pH 4.3) corresponds closely to the equivalence point for the conversion of hydrogen carbonate ion to H_2CO_3.

$$HCO_3^-(aq) + H^+(aq) \rightarrow H_2CO_3(aq)$$

Alkalinity is often reported in the units "moles H^+ per liter", meaning the number of moles of acid which must be used to neutralize all the weak bases in the water sample to pH 4.3. Note that the alkalinity of the sample gives no clue as to the chemical identities of the bases that were present. For natural water samples, though not for industrial waste water samples, it is usually safe to assume that the predominant base will be HCO_3^-.

11.12.2 Alkalinity and pH

It comes as a surprise to many people that you cannot readily relate the pH of water to its alkalinity. Water in limestone areas has high alkalinity (typically > 1 mmol H^+ L^{-1}), yet is almost neutral (pH 7 to 8). Conversely, a 10^{-4} mol L^{-1} solution of NaOH would have pH 10, but alkalinity of only 0.1 mmol H^+ L^{-1}. The reason why water can have high alkalinity but near-neutral pH is that alkalinity is a measure of the total concentration of bases in the water: weak bases such as HCO_3^- (pK_b = 7.6) contribute only a little OH^- to an aqueous solution. Refer back to the speciation diagram for carbonate (Figure 6) if this point is not clear.

11.13 Enthalpy changes and acid-base equilibria

Like other equilibrium constants, those we have met in this chapter vary with temperature. The values quoted have generally been appropriate to 25°C. As predicted by Le Chatelier's Principle, exothermic reactions have equilibrium constants which decrease at higher temperatures, and endothermic processes become more favourable at higher temperatures.

In general, the neutralization of an acid by a base is exothermic, an observation that will be familiar in the laboratory. Consider these three reactions, all of which are exothermic.

[1a] $H^+(aq) + OH^-(aq) \rightarrow H_2O(\ell)$

[2a] $HF(aq) + OH^-(aq) \rightarrow F^-(aq) + H_2O(\ell)$

[3a] $H_3O^+(aq) + NH_3(aq) \rightarrow NH_4^+(aq) + H_2O(\ell)$

The corresponding reverse reactions must all be endothermic.

[1b] $H_2O(\ell) \rightarrow H^+(aq) + OH^-(aq)$

[2b] $F^-(aq) + H_2O(\ell) \rightarrow HF(aq) + OH^-(aq)$

[3b] $NH_4^+(aq) + H_2O(\ell) \rightarrow H_3O^+(aq) + NH_3(aq)$

Reactions [1b], [2b], and [3b] define, respectively, K_w, K_b for the fluoride ion, and K_a for the ammonium ion. Since most acid-base neutralizations are exothermic, the dissociations of the corresponding weak acids and weak bases must be endothermic. Consequently, the equilibrium constants K_w, K_a, and K_b all increase at higher temperatures. An exception occurs in the case of carboxylic acids such as acetic acid. The heats of dissociation of carboxylic acids are very small, and as a result the K_a values of carboxylic acids vary very little with temperature.

Most compilations of acid-base equilibrium constants refer to 298 K. These values are appropriate for laboratory work in the range 20 - 25°C. For natural waters, where the temperature is often 4 - 10°C, or for blood (37°C), it may be necessary to look up equilibrium constants relevant to these particular applications.

11.14 Summary of how to work pH problems

First determine what is present in the solution, with reference to Section 11.6. There are only a few options.

- The solution contains only a strong acid or a strong base. Calculate the pH directly, by using the K_w expression.

- The solution contains only a weak acid. This is a K_a calculation.

- The solution contains only a weak base. This is a K_b calculation.

- The solution contains both a weak acid and its conjugate base. The solution is a buffer. Calculate the pH by use of the Henderson-Hasselbalch equation.

- The solution contains a weak acid and extra H^+ or a weak base and extra OH^-. Since the weak acid or base contributes little H^+ or OH^- to the solution, the pH is determined almost entirely by the concentration of H^+ or OH^-.

- The solution contains an acid and a base, but is not a buffer. A chemical reaction will occur. Work out the products of the chemical reaction in moles. The resulting solution will be one of the previous types and a second calculation will be required.

Problems

Section 11.1

1. Identify the acidic and basic species in each of the equilibria below (all species (aq) unless noted otherwise).

   a) $HNO_2 + H_2O(\ell) \rightleftharpoons H_3O^+ + NO_2^-$
   b) $NH_3 + HF \rightleftharpoons NH_4^+ + F^-$
   c) $CH_3NH_2 + H_2O(\ell) \rightleftharpoons CH_3NH_3^+ + OH^-$
   d) $H_2S + OH^- \rightleftharpoons HS^- + H_2O(\ell)$

2. Give the conjugate acid of each of the following bases

   a) HPO_4^{2-}
   b) CO_3^{2-}
   c) H_2SO_4
   d) CH_3NH_2

3. Give the conjugate base of each of the following acids

   a) NH_3
   b) $CH_3CH_2NH_3^+$
   c) HPO_4^{2-}
   d) OH^-

4. Which of the following pairs are conjugates?

   a) H_2O and OH^-
   b) NH_4^+ and NH_2^-
   c) NO_2 and HNO_2
   d) H_3PO_4 and PO_4^{3-}

Section 11.2

1. Give the pH of the following solutions:

   a) 0.010 mol L^{-1} HCl
   b) 2.0×10^{-4} mol L^{-1} NaOH
   c) 1.5×10^{-3} mol L^{-1} $HClO_4$
   d) 2.1×10^{-3} mol L^{-1} $Ba(OH)_2$
   e) 3.8×10^{-5} mol L^{-1} HNO_3

2. Give the concentration of $[H^+(aq)]$ in solutions having the following pH

   a) 1.30
   b) 10.624
   c) 5.3
   d) 12.01

3. Write net ionic equations to show that:

   a) CN^- acts as a base in water
   b) $H_2PO_4^-$ acts as an acid in water
   c) CO_3^{2-} acts as a base in water
   d) C_5H_5N acts as a base in water

4. At 50°C K_w has the value 5.5×10^{-14} (mol $L^{-1})^2$. What are the concentrations of [H^+ (aq)] and [OH^- (aq)] in pure water at 50°C?

5. Ammonia gas condenses to a liquid below -33°C. The liquid undergoes self-ionization, with an ion product equilibrium constant of 1.9×10^{-33} (mol $L^{-1})^2$ at -50°C.

   a) Write a net ionic equation showing self-ionization.
   b) Calculate the concentrations of the relevant ions at -50°C.

6. The pH of 10^{-3} mol L^{-1} HCl is 3 and of 10^{-5} mol L^{-1} HCl is 5. Thus, the pH of 10^{-8} mol L^{-1} HCl is (choose one):

   A: 8 B: 7
   C: 6 D: between 6 and 7

Section 11.3

1. Write the expressions defining K_a for the following acids: $HClO_2$; HCO_3^-; $CH_3NH_3^+$; $Cu(H_2O)_4^{2+}$; $H_2PO_4^-$

2. Calculate the pH of each of the following solutions (K_a values are found in Appendix 3).

   a) 1.2×10^{-2} mol L^{-1} acetic acid
   b) 5.0×10^{-3} mol L^{-1} chlorous acid
   c) 3.6×10^{-4} mol L^{-1} NaH_2PO_4
   d) 2.7×10^{-3} mol L^{-1} $CH_3NH_3^+Cl^-$

3. Calculate the pH and percent dissociation of the following acidic solutions

   a) 4.1×10^{-3} mol L^{-1} formic acid
   b) 6.2×10^{-4} mol L^{-1} NH_4Cl
   c) 5.0×10^{-2} mol L^{-1} HIO_3
   d) 0.1 mol L^{-1} $CuCl_2$ (K_a for Cu^{2+}, aq = 3.0×10^{-8} mol L^{-1})

4. Calculate the mass of $NH_4Cl(s)$ that must be added to 1.00 L of water to obtain a solution whose pH is 4.47.

5. Calculate the stoichiometric concentration of $NaHSO_4$ in a solution whose pH is 2.32.

6. Calculate the concentration of CN^- ions in a solution labelled "0.15 mol L^{-1} HCN, aq".

7. Calculate K_a for lactic acid given that a 2.65×10^{-3} mol L^{-1} solution has pH 3.73.

8. Calculate K_a for the pyridinium ion ($C_5H_5NH^+$) given that a 0.020 mol L^{-1} solution of C_5H_5NHCl has pH 3.47.

9. Calculate the concentrations of all ions in a solution labelled 0.0015 mol L^{-1} KH_2PO_4.

10. Calculate the pH after 100.0 mL of 1.5×10^{-3} mol L^{-1} HCl is mixed with 25 mL of 2.5×10^{-2} mol L^{-1} $HClO_4$.

Section 11.4

1. Write expressions defining K_b for the following species: CO_3^{2-}; CN^-, CH_3NH_2, HPO_4^{2-}

2. Calculate the concentration of OH^- and the pH of the following solutions

   a) 0.010 mol L^{-1} NH_3
   b) 1.2×10^{-3} mol L^{-1} NaCN
   c) 5.8×10^{-4} mol L^{-1} Na_3PO_4

3. Calculate the pH and the percent of undissociated acid in each of the following solutions:

   a) 7.1×10^{-3} mol L^{-1} Na_2CO_3
   b) 1.8×10^{-2} mol L^{-1} C_5H_5N

4. Calculate the mass of $NaNO_2$ that must be added to 0.500 L of water to give a solution having pH 8.24.

5. Calculate the volume of 0.100 mol L^{-1} NH_3 solution that must be added to water to give 1.00 L of solution having pH 10.65.

6. What is the stoichiometric concentration of a solution of CH_3NH_2 whose pH is 9.89?

7. Calculate K_b for the azide ion (N_3^-), given that a 0.100 mol L^{-1} solution of NaN_3 has pH 8.86.

8. Calculate K_b for hydrazine, given that a 0.076 mol L^{-1} solution of hydrazine has pH 10.44.

9. a) Calculate K_a for the ion $CH_3CH_2NH_3^+$ given that a 0.14 mol L^{-1} solution of $CH_3CH_2NH_2$ has pH 11.97.

   b) Now calculate the pH of a solution made by dissolving 21.6 g of $CH_3CH_2NH_3Cl$ in enough water to prepare 0.800 L of solution.

Section 11.6

1. Would the following solutions of salts react acidic, neutral, or alkaline in water? Explain, where appropriate, by writing a net ionic equation.

   a) KNO_3
   b) NH_4Cl
   c) Na_3PO_4
   d) $[Fe(OH_2)_6]Cl_3$
   e) $NaC_2H_3O_2$
   f) CsOH
   g) $Ba(NO_3)_2$

2. Can large (*e.g.* 1.0 mol L^{-1}) concentrations of the following ions be present simultaneously in aqueous solution? If not, write a net ionic equation for the reaction that occurs.

   a) H^+ and $H_2BO_3^-$
   b) Na^+ and NO_3^-
   c) NH_4^+ and OH^-
   d) H^+ and Cl^-
   e) NH_4^+ and Br^-
   f) $H_2PO_4^-$ and PO_4^{3-}

3. Identify the following species as acidic, basic, or neutral in aqueous solution.

   a) HSO_4^-
   b) CN^-
   c) $Ni(H_2O)_6^{2+}$
   d) $C_5H_5NH^+$
   e) ClO_4^-
   f) ClO_2^-

Section 11.7

1. Calculate the pH of a solution containing 0.015 mol L^{-1} NaH_2PO_4 and 0.010 mol L^{-1} Na_2HPO_4.

2. Calculate the pH of a solution containing 0.100 mol L^{-1} NH_3 as well as 0.010 mol L^{-1} NH_4Cl.

3. Calculate the mass of sodium acetate that must be added to 0.500L of a 0.150 mol L^{-1} solution of acetic acid to give a solution having pH 5.00.

4. Calculate K_a for the pyridinium ion $C_5H_5NH^+$, given that a solution containing 0.0500 mol L^{-1} C_5H_5N and 0.030 mol L^{-1} C_5H_5NHCl has pH 5.47.

5. Calculate the pH of a solution containing 0.10 mol L^{-1} NH_3 and 0.010 mol L^{-1} NaOH.

Section 11.8

1. Sketch the form of the curve for the titration of 10.0 cm^3 of 0.10 mol dm^{-3} formic acid against 0.10 mol dm^{-3} sodium hydroxide solution.

   Label the axes, and identify

   i) equivalence point ii) methyl red endpoint
   iii) bromothymol blue endpoint iv) phenolphthalein endpoint

   It is not necessary to calculate these points exactly.

2. Write balanced net ionic equations and calculate equilibrium constants for reactions between

   a) sodium hydroxide and hypochlorous acid
   b) potassium formate and hydrochloric acid
   c) sodium hydrogen carbonate and sodium hydroxide
   d) potassium dihydrogen phosphate and perchloric acid
   e) hypochlorous acid and ammonia

3. Use the rule of multiple equilibria to calculate equilibrium constants for these reactions:

   a) $HC_2H_3O_2 + NH_3 \rightleftharpoons NH_4^+ + C_2H_3O_2^-$
   b) $HOCl + CN^- \rightleftharpoons HCN + OCl^-$
   c) $SO_4^{2-} + HSO_3^- \rightleftharpoons HSO_4^- + SO_3^{2-}$
   d) $H_2S + S^{2-} \rightleftharpoons 2HS^-$

4. Does NH_4F react acidic, neutral or alkaline in aqueous solution?

5. a) Sketch the form of the curve for the titration of 20.00 mL of 0.10 mol L^{-1} CH_3NH_2 against 0.10 mol L^{-1} HCl.

   b) Calculate the pH at the equivalence point.

   c) Mark on the curve the point where $pH = pK_a$ of $CH_3NH_3^+$.

6. For the reaction of question 1, calculate the pH of the solution after the addition of the following volumes of NaOH solutions.

   a) 1.0 cm^3 b) 5.0 cm^3
   c) 10.0 cm^3 d) 12.0 cm^3

7. Calcium oxide CaO (2.48 g) is treated with 100.0 cm^3 of 1.260 mol dm^{-3} hydrochloric acid.

$$CaO(s) + 2H^+(aq) \longrightarrow Ca^{2+}(aq) + H_2O$$

   The resulting solution is diluted to 250 cm^3. Calculate the concentrations of all ions in the solution.

8. Write net ionic equations, calculate equilibrium constants and calculate the concentrations of all species (apart from water) when equal volumes of 0.10 mol dm^{-3} solutions of the following are mixed:

   a) hydrochloric acid and barium hydroxide
   b) sulfuric acid and potassium hydroxide
   c) ammonium sulfate and sodium hydroxide
   d) sulfuric acid and sodium formate

9. Suggest suitable indicators for each of the following titrations:

   a) $NaHCO_3$ *vs.* HCl b) Na_2HPO_4 *vs.* NaOH
   c) lactic acid. *vs* NaOH d) NH_3 *vs.* $HClO_4$

10. What colour would be observed if a little bromothymol blue indicator were added to 0.01 mol dm^{-3} solutions of the following?

   a) $HC_2H_3O_2$
   b) CH_3NH_2 ($K_b = 3.7 \times 10^{-4}$ mol L^{-1})
   c) $Al(NO_3)_3$ (K_a for hydrated Al^{3+} is 7.2×10^{-6} mol L^{-1})

Section 11.9

1. Which of the following solutions could be combined in appropriate proportions to prepare buffer solutions:

   a) $HC_3H_5O_2$ and $KC_3H_5O_2$ b) $NaNO_2$ and NaOH
   c) NaH_2PO_4 and Na_2HPO_4 d) CH_3NH_2 and CH_3NH_3Cl
   e) $HC_2H_3O_2$ and HCl f) CH_3NH_3Cl and NaOH

   Where a buffer solution could be prepared indicate roughly the pH range where buffer action could be expected.

2. Calculate the pH of each of the following buffer solutions.

   a) 0.100 mol L^{-1} each of NaH_2PO_4 and Na_2HPO_4
   b) 0.10 mol L^{-1} NH_3 and 0.13 mol L^{-1} NH_4Cl
   c) 0.072 mol L^{-1} Na_2CO_3 and 0.055 mol L^{-1} $NaHCO_3$
   d) 25 mL of 0.100 mol L^{-1} acetic acid and 40 mL of 0.120 mol L^{-1} sodium acetate.

3. Calculate the pH of the buffer solutions made by mixing:

   a) 10 cm^3 of 0.10 mol L^{-1} KOH and 15 cm^3 of 0.10 mol L^{-1} acetic acid
   b) 150 mL of 0.20 mol L^{-1} NH_4Cl and 100 mL of 0.15 mol L^{-1} NaOH
   c) 1.00 L of 0.15 mol L^{-1} NH_3 and 100 mL of 0.80 mol L^{-1} HCl
   d) 1.00 L of 0.20 mol L^{-1} $NaHCO_3$ and 2.00 g of solid NaOH.

4. Which of the following aqueous solutions could in principle be used to prepare a buffer solution? More than one may be chosen.

   a) $CH_3NH_3^+ Cl^-$ and CH_3NH_2 b) K_2CO_3 and NaOH
   c) Na_2HPO_4 and NaH_2PO_4 d) HCl and $NaNO_2$

5. Pyridine is a weak base with K_b 1.4 x 10^{-9} mol L^{-1}. If a buffer solution were prepared from pyridine and its conjugate acid, the pH would be in the range

   A: 4-6 B: 6-8 C: 8-10 D: 10-14

6. Twenty cm^3 of 0.10 mol dm^{-3} Na_2SO_3 and 20 cm^3 of 0.10 mol dm^{-3} $NaHSO_3$ are mixed. The pH of the mixture is 7.25. For what acid can you calculate K_a from these data, and what is the value of the K_a?

7. Write net ionic equations and calculate equilibrium constants for the reactions which occur when:

   a) HCl is added to a formic acid-sodium formate buffer
   b) KOH is added to an ammonia-ammonium chloride buffer
   c) $HClO_4$ is added to a KH_2PO_4-K_2HPO_4 buffer
   d) $Ba(OH)_2$ is added to a Na_2CO_3-$NaHCO_3$ buffer

8. An acetate buffer is prepared having total [acetate] = [acetate ion] + [acetic acid] = 0.220 mol dm^{-3}.
   The pH of the solution is 4.92.

   a) Calculate the concentrations of all species (except water) in the buffer solution (assume Na^+ is the counter ion to acetate).
   b) Solid NaOH 0.70 g is added to 250 cm^3 of the buffer solution. Write a net ionic equation for the reaction which occurs and calculate the pH of the solution following the addition.

9. Sodium azide (0.100 mole), 250 mL of 0.200 mol L^{-1} hydrochloric acid, and water were used to prepare 0.500 L of solution, whose pH was 4.72. Calculate K_a for hydrazoic acid HN_3. (Hint: First write the net ionic equation for any reaction that occurs.)

10. How would you prepare a buffer solution of pH 10.00 given in each case the following:

    a) phenol (solid, C_6H_6O, $K_a = 1.3 \times 10^{-10}$) and 0.10 mol dm^{-3} NaOH solution
    b) solutions (0.20 mol dm^{-3}) of ammonia and ammonium chloride
    c) solutions (0.25 mol dm^{-3}) of sodium carbonate and hydrochloric acid.

Section 11.12

1. Calculate the alkalinity of a water sample if 12.64 mL of 2.05×10^{-2} mol L^{-1} HCl are needed to titrate a 250.0 mL water sample to pH 4.3.

2. Calculate the alkalinity of a water sample if 8.62 mL of 4.60×10^{-3} mol L^{-1} HCl are needed to titrate a 500 mL sample to a methyl orange endpoint.

3. A water sample has pH 8.44 and a total Ca^{2+} concentration of 155 ppm. For this question, assume that the only ions present in the water are Ca^{2+}, HCO_3^-, and CO_3^{2-}.

   a) What are the concentrations of CO_3^{2-} and HCO_3^- in moles per liter.
   b) What volume of 5.02 x 10^{-2} mol L^{-1} HCl is needed to titrate 1.00 L of this water to pH 4.3?
   c) What is the total alkalinity of the water?

4. A water sample obtained from an area of dolomitic limestone $CaCO_3.MgCO_3$ has pH 7.2 and total alkalinity 2.3 x 10^{-3} mol L^{-1} of H^+.

   a) Calculate the concentrations of the major ions in the water. Assume 25°C.
   b) What would be meant if this water was described as being well buffered towards acid?
   c) A 100 mL sample of this water is titrated against 0.0105 mol L^{-1} EDTA (hardness determination). What volume of EDTA solution will be used? State any assumptions you need to make.

5. Two identical 250.0 mL samples of freshly drawn well water have <u>approximate</u> pH 6.6-6.8. One sample is titrated against 0.0510 mol L^{-1} NaOH solution to a phenolphthalein end point (pH 8.3); 11.66 mL of titrant are needed. The other sample is titrated against 0.1000 mol L^{-1} HCl solution to a methyl orange endpoint (pH 4.3); 12.25 mL of titrant are needed. Calculate the concentrations, in mol L^{-1}, of H_2CO_3, HCO_3^-, CO_3^{2-} and Ca^{2+} in the well water. [Hint: Consider carefully which reactions occur in the titrations, and assume that Ca^{2+} is the only cation present besides H^+.]

12 SOLUBILITY EQUILIBRIA

12.1 Introduction

In this chapter we consider the solubility of substances in solvents. As usual, water will be the solvent of most interest; however, the concepts set out here apply equally to non-aqueous solvents. Solubility is a very important concept for understanding the fate of xenobiotics in the environment.

Defintions

***Solute**: the substance which dissolves. It may be a solid, a liquid, or a gas.*

***Solvent**: the liquid in which the solute dissolves*

***Solution**: the combination of solvent and solute*

***Solubility**: the maximum amount of solute which dissolves at equilibrium.*

***Saturated solution**: a solution which is at equilibrium with the undissolved solute. The concentration of the solute is equal to the (equilibrium) solubility.*

12.2 Solubility properties of gases

12.2.1 Henry's Law

The equilibrium solubility of any gas in any solvent is determined by the pressure of the gas in contact with the solvent. Quantitatively, the amount of gas which dissolves is proportional to the partial pressure of the gas. This statement is known as **Henry's Law**, and can be written mathematically as follows:

$$[X, \text{solvent}] = \text{constant} \times p(X\ (g))$$

[X, solvent] represents the concentration of solute gas in the solvent. The proportionality constant is an equilibrium constant, often known as the Henry's Law constant, symbol K_H. The usual units for K_H are mol L^{-1} atm $^{-1}$. However, other units are in use, and it is wise always to check the units of K_H found in any particular compilation. Many environmentalists use K_H in the units mol m $^{-3}$ atm $^{-1}$.

$$K_H = [X,solvent] / p(X,g)$$

An everyday example of Henry's Law in action is seen when a soft drink is opened. The drink is manufactured by dissolving CO_2 in the drink at a pressure of about 2 atm. The unfilled space in the bottle or can contains CO_2 at this pressure. When the drink is opened it comes in contact with the air where $p(CO_2)$ = 3.5 x 10 $^{-4}$ atm, so CO_2(aq) comes out of solution to restore equilibrium. This causes the pleasant effervescence of the drink, which eventually goes flat.

The condition known by deep sea divers as "the bends" is explicable in terms of Henry's law. The pressure under water increases by 1 atm for approximately 10 m depth. Therefore, at 30 m, for example, air has to be breathed at at least 4 atm (1 atm atmospheric pressure plus 1 atm per 10 m depth). If compressed air is breathed, then $p(N_2)$ = 0.8 x p(total) = 3.2 atm, instead of 0.8 atm at the surface. Consequently, the diver's blood will contain four times as much dissolved nitrogen at equilibrium at 30 m depth as at the surface. When the diver returns to the surface, the excess nitrogen escapes from the blood, just like the effervescence of a soft drink when it is opened. This is both painful and dangerous. "The bends" are avoided by having divers breathe a mixture of helium and oxygen, rather than air (because helium is less soluble in water than nitrogen), and by having the diver undergo gradual decompression in a special chamber after prolonged work at great depth.

Table 12.1 gives the Henry's Law constants for some common gases at 25 °C.

Table 12.1: Values of K_H at 298 K

Gas	K_H, mol L^{-1} atm^{-1}	Gas	K_H, mol L^{-1} atm^{-1}
H_2	7.8 x 10^{-4}	CO	9 x 10^{-4}
N_2	6.5 x 10^{-4}	O_2	1.3 x 10^{-3}
CO_2	3.4 x 10^{-2}	O_3	1.3 x 10^{-2}
NO	1.9 x 10^{-3}	NO_2	6.4 x 10^{-3}
SO_2	1.2	NH_3	5.3

Example: *Calculate the concentration of $H_2(aq)$ when hydrogen gas at a pressure of 2.7 atm is equilibrated with water.*

Answer:

$$K_H = [H_2,aq]/p(H_2)$$

$$[H_2,aq] = K_H \times p(H_2)$$

$$= (7.8 \times 10^{-4} \text{ mol L}^{-1} \text{ atm}^{-1})(2.7 \text{ atm})$$

$$= 2.1 \times 10^{-3} \text{ mol L}^{-1}$$

The data in Table 1 suggest that the solubilities of gases in water vary widely. Non-polar substances such as CH_4, N_2, and O_2 have small Henry's Law constants and hence have low solubilities in water. At the other end of the scale, very polar gases such as NH_3 are extremely soluble in water. For gases such as HCl, high solubility is due to the complete chemical reaction of the gas with water.

$$HCl(g) \rightleftharpoons HCl(aq) \rightarrow H^+(aq) + Cl^-(aq)$$

The differing solubilities of gases are of practical importance in connection with scavenging trace gases out of the troposphere by rainfall (see also Chapter 13). Dissolution in rainfall is also called "wet deposition".

Non-polar substances such as hydrocarbons and halogenated hydrocarbons such as chlorofluorocarbons (CFCs, Chapter 15) have very low solubility in water and therefore may persist for long periods in the atmosphere, in the absence of other sinks, such as chemical reaction in the atmosphere. By contrast, gases such as SO_2, gaseous hydrogen peroxide, and gaseous nitric acid have high water solubility, and are readily scavenged out of the atmosphere by raindrops. In the case of SO_2 and HNO_3, scavenging from the gas phase leads to deposition of "acid rain", Chapter 13. Wet deposition represents their chief means of removal from the atmosphere. The solubility properties of atmospheric gases must always be considered as part of the evaluation of the tropospheric lifetime of a gaseous pollutant.

We now discuss the solubility properties of O_2 and CO_2 in a little detail.

12.2.2 Solubility of oxygen in water

The atmosphere contains 0.21 atm O_2, so from the quoted value of K_H at 25 °C it is possible to write:

$$[O_2\text{ (aq)}] = K_H \times p(O_2\text{ (g)})$$

$$= (1.3 \times 10^{-3} \text{ mol L}^{-1} \text{ atm}^{-1})(0.21 \text{ atm})$$

$$= 2.7 \times 10^{-4} \text{ mol L}^{-1}$$

Converting to mg/L, we have

$$[O_2\,(aq)] = 2.7 \times 10^{-4}\ \text{mol L}^{-1} \times 32\ \text{g mol}^{-1} \times (1000\ \text{mg/1 g})$$
$$= 8.7\ \text{mg L}^{-1}$$

This value applies to oxygen which is in equilibrium with the atmosphere. From the definition of ppm for aqueous solutions (Chapter 2), this is the same as 8.7 ppm.

☞ ***Aside:*** *This calculation ignores a small correction, namely that p(total) = 1 atm includes the contribution from $p(H_2O,g)$ which is about 0.03 atm at 25 °C and 100% relative humidity. Consequently, our calculation overestimates $[O_2,aq]$ by about 3%.*

Dissolved oxygen is essential for the survival of aquatic animals. Most fish species, for example, require 5-6 ppm of dissolved oxygen. Without sufficient oxygen, the fish suffocate. Fish kills are common for any reason the oxygen supply is depleted. Oxygen depletion can be due to thermal pollution or the presence of oxidizable substances in the water.

Thermal pollution results when water is used for cooling (*e.g.*, in electricity generating plants), causing it to be returned to a river or lake at a higher temperature. The problem occurs because the solubility of oxygen in water is lower at higher temperature; in fact, the solubilities of all gases in all solvents decrease as the temperature is raised. Consequently, warm water is less oxygenated than cold water. Thermal pollution is more often a problem in summer, when water temperatures are high to begin with, and especially if the oxygen concentration is simultaneously lowered by other forms of pollution.

Oxidizable substances may be present in the water from a variety of sources, including sewage, factory effluents, algal blooms, and agricultural run-off. These oxidizable substances are used as carbon sources by microorganisms, and in the course of their metabolism the microorganisms remove oxygen from the water. Oxygen depletion occurs if reoxygenation from the atmosphere cannot keep pace with the rate of removal. Flowing water, especially where it flows over waterfalls or rapids, is much more highly oxygenated than stagnant water, because reoxygenation is faster when the water is moving. Stagnant, oxygen-depleted water tends to have a foul odour due to the activities of anaerobic microorganisms, which reduce nitrogen and sulfur compounds to odorous NH_3, H_2S, and their organic derivatives.

Algal blooms occur when a water body is supplied with excessive amounts of nutrients; the algae grow quickly, but when they die oxygen is required to oxidize their biomass back to carbon dioxide and water. Untreated or partially treated sewage, factory effluents, especially from food processing (meat packing plants, vegetable and fruit canneries), and animal feedlots (manure seepage) are common sources of oxidizable organic compounds, which are readily utilized by microorganisms. Ironically, some of the more persistent water pollutants that we read about, such as **polychlorinated biphenyls** and certain **pesticides** do not pose a problem of oxygen depletion; their persistence arises precisely because they are so slow to undergo biological oxidation.

Because the concentration of oxygen in the water is of such crucial importance to aquatic life, environmental scientists have devised several different descriptors of the oxygen status of a water body. These are dissolved oxygen concentration, total organic carbon, chemical oxygen demand, and biochemical oxygen demand.

Dissolved Oxygen is the actual concentration of oxygen in the water. Its usual units are ppm (mg L^{-1}) of O_2. Dissolved oxygen is usually measured with an "oxygen meter", which operates by allowing the oxygen to diffuse through a porous membrane into a solution where it causes an electrochemical reaction to occur. The dissolved oxygen concentration is proportional to the electric current induced in the meter. Dissolved oxygen can also be measured by titration (Winkler's method), employing the stoichiometric oxidation of Mn^{2+}(aq) to MnO_2(s). The amount of MnO_2 formed is then determined by "iodometric titration" (Section 3.7). The relevant equations follow.

$$Mn^{2+}(aq) + \tfrac{1}{2}\, O_2(aq) + 2\, OH^-(aq) \rightarrow MnO_2(s) + H_2O(\ell)$$

$$MnO_2(s) + 4\, H^+(aq) + 2\, I^-(aq) \rightarrow Mn^{2+}(aq) + I_2(aq) + 2\, H_2O(\ell)$$

$$I_2(aq) + 2\, S_2O_3^{2-}(aq) \rightarrow S_4O_6^{2-}(aq) + 2\, I^-(aq)$$

Total organic carbon (TOC). This is measured by oxidizing all the organic matter in the water to CO_2, and then analysing the CO_2 formed. The experiment involves passing the sample over a catalyst with oxygen or other oxygen source, and measuring the CO_2 that is formed instrumentally, using an instrument called a gas chromatograph. TOC is usually reported in ppm of carbon. The TOC gives an indication of how much organic matter is present in the water, from which you can estimate how much oxygen might be needed to oxidize all this organic carbon. Of course, TOC is unable to distinguish those substances which are rapidly oxidized, and hence likely to deplete O_2, and those which are less reactive. The advantage of the method is that it can be made rapid and semi-automatic.

Chemical oxygen demand (COD). This is determined by reacting the water sample with a fixed amount of $Na_2Cr_2O_7$ / H_2SO_4 under stated conditions of time and temperature, and then titrating the **unreacted** $Na_2Cr_2O_7$ against a standardized Fe^{2+} solution. This is thus an example of "back titration", Section 3.7. Each mole of $Cr_2O_7^{2-}$ consumed is equivalent, in acidic solution, to 1.5 moles of O_2. In other words, one mole of $Cr_2O_7^{2-}$ can oxidize as much organic material as 1.5 mol of O_2. The following equation shows the equivalence between one mole of $Cr_2O_7^{2-}$ and 3 mol of (O) in acidic solution (all species are (aq)).

$$Cr_2O_7^{2-} + 8\ H^+ \rightarrow 2\ Cr^{3+} + 4\ H_2O + 3\ (O)$$

The next equation shows the titration reaction between the standard Fe^{2+} solution and the $Cr_2O_7^{2-}$ left over.

$$Cr_2O_7^{2-} + 14\ H^+ + 6\ Fe^{2+} \rightarrow 2\ Cr^{3+} + 6\ Fe^{3+} + 7\ H_2O(\ell)$$

Comparing chemical oxygen demand with total organic carbon, the COD method is a little closer to the "real life" situation, because $Cr_2O_7^{2-}$ does not react indiscriminately with all organic compounds. The ones more resistant to reaction with $Cr_2O_7^{2-}$ tend to be the same as those which resist biological oxidation. However, the correlation is not exact.

Biochemical oxygen demand (BOD). BOD is measured by incubating the water sample with aerobic microorganisms under stated conditions of time and temperature (usually 5 days, 25 °C). The dissolved oxygen is measured before and after, and the difference is the BOD. An important consideration in setting up the experimental protocol is that the oxidizable material must be the limiting reactant. If it is not, then all the oxygen will have been consumed before the test ends, and the analyst will not know how much more would have been consumed if only there had been more oxygen available. To avoid this problem, samples having very high BOD need to be diluted before running the test so that oxygen will not be the limiting reactant.

TOC, COD, and BOD are all arbitrary measures of the oxygen status of a water sample, since not all organics oxidize with equal ease. As already mentioned, slowly oxidizable compounds cause less oxygen depletion and hence are less of a threat to aquatic life, because O_2(aq) can be replenished from the air while oxidation is proceeding. Total organic carbon includes all carbon compounds regardless of their oxidation rates. Chemical oxygen demand goes some way to compensating for this, because acidic dichromate to some extent mimics the relative ability of organics to oxidize biologically. Biochemical oxygen demand would seem at first to be the ideal approach, since oxidation is accomplished by microorganisms; however, both the choice of the time and temperature for the test, and the selection of the microorganisms themselves, are arbitrary parameters. All these measures thus have their uses, in terms of convenience of analysis, but none can reflect accurately what happens in a real environmental system.

12.2.3 BOD reduction of sewage and industrial waste water

The immediately preceding section shows that high BOD is associated with oxygen depletion of the water. Sewage contains high concentrations of biologically oxidizable organics, as do many industrial waste waters, especially from the food processing industry. These waters must be treated before being released to the environment. Treatment involves the creation of conditions where microbial oxidation can take place rapidly.

Microbial activity is always present in nutrient-rich water, unless the waste stream also contains toxic substances which kill off the microorganisms. In order to maximize microbial activity, the water should be highly oxygenated. The microorganisms reduce BOD in two separate ways: by using organic compounds as an oxidizable energy source (shown below for ethyl alcohol as an example), and by incorporating them into microbial biomass.

$$C_2H_6O + 3\ O_2 \rightarrow 2\ CO_2 + 3\ H_2O$$

The stoichiometry of this reaction is identical to that of combustion, although the reaction pathways are very different.

Incorporation of nutrients into microbial biomass is promoted by providing nutrients in the correct proportions. As noted in Section 10.3, biomass generally requires the ratio C: N: P to be about 100: 15: 1. If the proportions of N or P in a given industrial waste stream are low, it may be necessary to supplement them by the use of fertilizers in order to stimulate optimal microbial activity.

The following methods are used to treat sewage and/or waste water.

1. Aeration in open lagoons, by bubbling air through the lagoon. This method has been used in the past for industrial waste streams. It is falling out of favour because it is difficult to maintain the whole lagoon in an aerated state, and anaerobic sections of the pond rapidly become foul smelling due to the release of organic sulfur and nitrogen compounds. Furthermore, in Canada the cold temperatures in winter greatly reduce rate of microbial oxidation.

2. Trickling filters, which are used in sewage treatment. In the trickling filter, a long perforated boom rotates slowly over a bed of sand and gravel. The fine spray of water falling from the boom becomes highly aerated, and microorganisms on the sand oxidize the nutrients in the water. Trickling filters take up considerable space, and are not used where land is at a premium.

3. Activated sludge reactors are confined vessels where microbial activity can be optimized by the provision of ideal temperatures for microbial growth. The size of the reactor is determined by the residence time needed for reduction of the BOD to a predetermined target level. Such reactors are used for the treatment of both sewage and industrial waste water.

12.2.4 Solubility of carbon dioxide in water

The solubility behaviour of CO_2 in water is inherently more complex than that of oxygen because CO_2 reacts with water to give the weak acid H_2CO_3. The series of equilibria is shown below (all species except $CO_2(g)$ are (aq).

$$CO_2(g) \rightleftharpoons CO_2 \overset{H_2O}{\rightleftharpoons} H_2CO_3 \overset{-H^+}{\rightleftharpoons} HCO_3^- \overset{-H^+}{\rightleftharpoons} CO_3^{2-}$$

We will regard $CO_2(aq)$ and undissociated $H_2CO_3(aq)$ as interchangeable, and always consider only the sum of their concentrations.

Example: *Calculate the concentration, in ppm, of $CO_2(aq)$ at equilibrium, knowing the partial pressure of CO_2 in the troposphere (350 ppmv or 3.5×10^{-4} atm) and the value of K_H for the dissolution of CO_2 in water at 25°C (3.4×10^{-2} mol L^{-1} atm^{-1}).*

Answer:

$$\begin{aligned} c(CO_2,aq) &= K_H \times p(CO_2) \\ &= (3.4 \times 10^{-2} \text{ mol L}^{-1} \text{ atm}^{-1})(3.5 \times 10^{-4} \text{ atm}) \\ &= 1.2 \times 10^{-5} \text{ mol L}^{-1} \end{aligned}$$

From the molar mass of CO_2 (44 g mol^{-1}) we can obtain the solubility of CO_2 in ppm.

$$\begin{aligned} c(CO_2,aq) &= (1.2 \times 10^{-5} \text{ mol L}^{-1}) \text{ 44 g mol}^{-1} \\ &= 5.3 \times 10^{-4} \text{ g L}^{-1} = 0.53 \text{ mg L}^{-1} \\ &= 0.53 \text{ ppm} \end{aligned}$$

The difference in the solubility behaviour of CO_2 in water compared with O_2 is that H_2CO_3 is a weak acid. Even in pure water, a little of the H_2CO_3 dissociates, making the water slightly acidic. This is conveniently illustrated by the problem posed next.

Example: *What is the pH of a sample of water in equilibrium with atmospheric carbon dioxide?*

Answer: *To solve this problem, we would first identify the relevant chemical equation, namely, the acid dissociation of* H_2CO_3.

$$H_2CO_3(aq) \rightleftharpoons H^+(aq) + HCO_3^-(aq)$$

Our usual solution of equilibrium problems involves identifying "initial" and "equilibrium" concentrations of all chemical species. In this example, we do not need to write out the "initial" concentrations, because the concentration of H_2CO_3 *remains constant, with the value* 1.2×10^{-5} *mol* L^{-1} *(see above), since it is in equilibrium with the atmospheric reservoir of* $CO_2(g)$ *at all times. Therefore we proceed as follows.*

	$H_2CO_3(aq)$	$\rightleftharpoons$	$H^+(aq)$	+	$HCO_3^-(aq)$
equilib conc	1.2×10^{-5}		x		x

Restating the previous point, the "x mol L^{-1}*" which dissociates does not deplete the* H_2CO_3. *We do not write* $(1.2 \times 10^{-5} - x)$ *because at equilibrium, the* H_2CO_3 *is replenished from the atmosphere.*

For H_2CO_3, $K_a = [H^+][HCO_3^-] / [H_2CO_3]$

$K_a = 4.2 \times 10^{-7}$ *mol* L^{-1} *at 25 °C.*

$4.2 \times 10^{-7} = x^2/(1.2 \times 10^{-5})$

Hence $x = [H^+] = 2.2 \times 10^{-6}$, *and pH = 5.65.*

This result is important in that pure water in equilibrium with the air is predicted not to have pH 7; it is slightly acidic due to the presence of dissolved CO_2. This prediction is borne out experimentally. As a result, even unpolluted rainwater is at pH ~ 5.6 rather than 7, and hence is slightly acidic. We predict that as the atmospheric $p(CO_2)$ increases, the acidity of rainwater should increase slightly, and its pH should drop slightly. Chapter 13 will say more about the pH of rainwater.

The concentration of dissolved CO_2 in natural water can be related to alkalinity (Section 11.12). In this context it is useful to define the "total carbonate" content of water in equilibrium with atmospheric CO_2 as $\{[H_2CO_3] + [HCO_3^-] + [CO_3^{2-}]\}$. As we shall show shortly, the total carbonate concentration is pH-dependent. The relationship of total carbonate to alkalinity is that alkalinity[1] is given by $\{[HCO_3^-] + 2\,[CO_3^{2-}]\}$, while total carbonate is given by $\{[H_2CO_3] + [HCO_3^-] + [CO_3^{2-}]\}$. Except at very high pH (pH > 10), the concentration of CO_3^{2-} is negligible compared with that of HCO_3^-. Thus in most natural waters, both the alkalinity and the total carbonate can be approximated by the concentration of hydrogen carbonate ion.

For $CO_2(g)$ in equilibrium with pure water, the total carbonate is easily

---

[1] Alkalinity = $[HCO_3^-] + 2\,[CO_3^{2-}]$. The reason for the "2" is that each mole of CO_3^{2-} requires 2 mol H^+ for neutralization to H_2CO_3.

calculated from information available from the preceding problem.

$$\begin{aligned} \text{Total carbonate} &\sim [H_2CO_3] + [HCO_3^-] \\ &= (1.2 \times 10^{-5} \text{ mol L}^{-1}) + (2.2 \times 10^{-6} \text{ mol L}^{-1}) \\ &= 1.4 \times 10^{-5} \text{ mol L}^{-1} \end{aligned}$$

As the pH rises, the equilibrium between H_2CO_3 and HCO_3^- is drawn to the right (Le Chatelier's Principle) and there is an increase in the total amount of dissolved carbonate in equilibrium with the atmosphere.

$$H_2CO_3(aq) \rightleftharpoons H^+(aq) + HCO_3^-(aq)$$

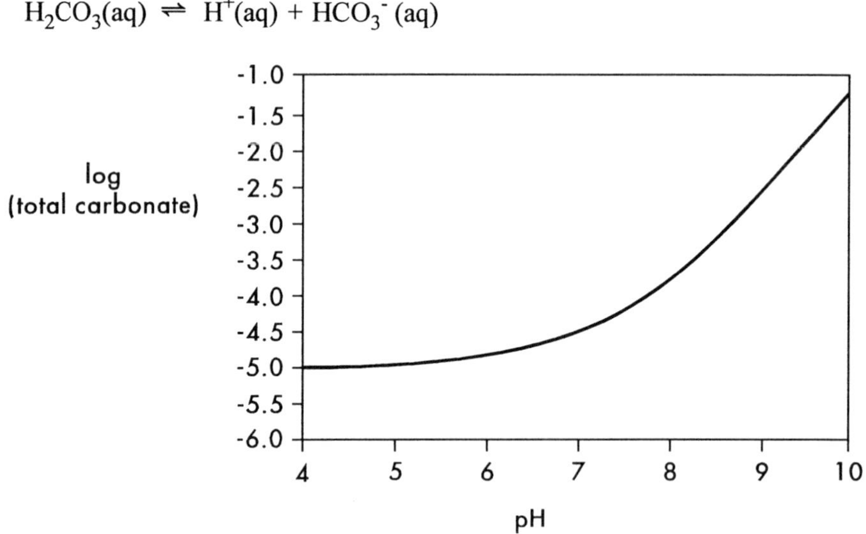

Figure 12.1: Calculated equilibrium concentration of total dissolved carbonate (log scale) as a function of pH.

The atmosphere acts as an inexhaustible reservoir of $CO_2(g)$, maintaining its pressure at 3.5×10^{-4} atm. The concentration of $H_2CO_3(aq)$ also remains constant at 1.2×10^{-5} mol L^{-1}. However, the total carbonate in solution, that is ($[H_2CO_3] + [HCO_3^-] + [CO_3^{2-}]$), increases as shown in Figure 1. In Section 12.3, we will relate the issue of total carbonate to that of well-buffered and poorly-buffered natural waters.

12.2.5 Solubility of non-polar gases in water

Non-polar gases have small Henry's Law constants and hence low solubilities in water. Their low solubilities can be explained by reference to methane as an example.

Methane has a very low solubility in water. We often use "like dissolves like" as a rule of thumb in the context of solubility, so we would say that non-polar methane has no attraction for the polar water molecules of the solvent. This is basically an enthalpic argument, which implies that $\Delta H°$ for the process $CH_4(g) \rightarrow CH_4(aq)$ must be unfavourable (positive). This turns out not to be the case: $\Delta H°$ for dissolution of all gases in all solvents has a negative (favourable) value. Thus methane's low solubility in water is not because water molecules and methane molecules have no mutual attraction. There is another factor at issue, which is the entropy change $\Delta S°$. Entropy changes will be discussed in detail in Chapter 14. For the present, we will introduce the concept by noting that entropy is a measure of the extent of disorder in a physical or chemical system. Highly ordered conditions have low entropy, and are thermodynamically less favoured than disordered conditions, which have more entropy.

When a gas dissolves in a solvent, the dissolved solute has less entropy than the gas, because its molecules are confined to the solution. In addition, water as solvent has the special property of ordering itself around non-polar solute molecules, a phenomenon which has been likened to forming a miniature iceberg around the solute, and so the solvent water also suffers a reduction in entropy. High water solubility is seen only when the enthalpy of solution is negative enough to compensate for the loss of entropy of the system.

The value of K_H, and thus the solubilities of all gases, decrease as the temperature is raised. This is consistent with the negative (exothermic) value of the enthalpy change for the process $X(g) \rightarrow X(aq)$. The reverse process $X(aq) \rightarrow X(g)$ is endothermic, and becomes increasingly favoured at higher temperatures (Le Chatelier's Principle).

12.2.6 Air stripping: Henry's Law in action

Many industrial processes generate waste streams which contain low concentrations of low molar mass contaminants, many of which are subject to government regulation concerning the concentrations which may legally be discharged into waterways. Some of the contaminants are potentially volatile. Air stripping is a technology for removing these contaminants from the aqueous phase. Typically, the aqueous solution is sprayed as a fine mist down a tall cylindrical tower; at the same time air is blown up the tower. Equilibrium is reached, or approached, between the aqueous and gas phases, the efficiency of the process depending upon the Henry's Law constant.

$$X(aq) \rightleftharpoons X(g)$$

Transfer to the gas phase is highly efficient for contaminants such as trichloroethylene which have intrinsically low water solubility, but is even practical for water-soluble gases such as ammonia, provided that good contact between the gas and aqueous phases is achieved by means of a large tower, and fine droplets (to maximize the surface area of the aqueous phase). Transfer out of the aqueous phase is also promoted at higher temperatures; this is often practical at no cost if the waste process stream is already warm.

Air stripping is very cheap, but has the disadvantage that it replaces water pollution by air pollution. For this reason air stripping, by itself, is becoming unacceptable as a technology even though it combats water pollution. Emerging technologies combine air stripping with treatment of the contaminated air stream: two possibilities are *(i)* passing the warm air stream over an oxidation catalyst in order to convert contaminants into innocuous substances such as CO_2, H_2O, and N_2; *(ii)* "biofiltration", in which the exhaust gases are passed through a bed of biologically active material (often a slime of bacteria on an inert support). The microorganisms then effect the final oxidation of the contaminant.

12.3 Solubilities of ionic solids in water

12.3.1 Solubility rules

Common experience tells us that inorganic substances vary greatly in their solubility in water. Common salt (NaCl) is quite soluble, but limestone and marble (both $CaCO_3$) have no noticeable solubility, and we call them "insoluble". As a rule of thumb, "soluble" is often taken to mean that a concentration of 0.1 mol L^{-1} or more can be achieved. However, even these so-called insoluble substances are able to dissolve in water to a slight extent.

Rules

Solubility rules for ionic compounds

1. *Almost all salts of the Group IA metals and of NH_4^+ are soluble in water.*
2. *All nitrates are soluble.*
3. *Most chlorides, bromides, and iodides are soluble, the exceptions being those of Ag^+ and Hg_2^{2+}.*
4. *Most sulfates are soluble, the major exceptions being those of barium, lead, mercury, bismuth and tin. Calcium sulfate is sparingly soluble.*
5. *Most carbonates, oxides and hydroxides, phosphates, and sulfides are insoluble. Exceptions are the metals of Rule 1; calcium and barium hydroxides are also fairly soluble.*

Learning solubility rules seems like a dull activity, but is very useful in understanding where different elements are found in the environment. This is important in terms of exploration for minerals; soluble elements tend to be found in the oceans; other elements are found as one or more of their least soluble salts. For example, much of the sodium, potassium and magnesium in the Earth's crust resides in the oceans. Solid deposits of these metals are usually found in the form of chlorides, which are believed to have originated as long dried-up seas. Likewise, most of the chlorine, bromine, and iodine in the crust is found in the oceans.

The major ores of many commercially important metals are sulfides. These metals include copper, lead, cadmium, mercury, nickel, and zinc, whose sulfides are insoluble in water. The problem of acidic precipitation (Chapter 13) stems directly from the processing of these ores in order to obtain the metal from the sulfide. The major ores of iron and aluminum are oxides, while much of the calcium in the crust exists as the insoluble carbonate (chalk, marble and limestone are all forms of calcium carbonate). Calcium carbonate rocks are sedimentary, and have a biological origin: the exoskeletons of long dead marine animals. As another example, insoluble calcium phosphate is the phosphorus source of greatest commercial importance. Calcium phosphate (also called rock phosphate) is the starting material for phosphate fertilizers.

12.3.2 The solubility product equilibrium constant

Let's start our discussion with some hypothetical ionic substance AB, which is composed of cations A^+ and anions B^-. When AB dissolves in water, the following equilibrium is established.

$$AB(s) \rightleftharpoons A^+(aq) + B^-(aq)$$

The equilibrium constant governing this process is called a "solubility product constant", K_{sp}. For the hypothetical substance AB:

$$K_{sp} = [A^+(aq)][B^-(aq)]$$

The concentration of the pure solid AB(s) is omitted, as usual, from the equilibrium constant expression. As a generalization, the larger the value of K_{sp}, the higher the solubility. Selected solubility product constants are provided in Appendix 4. The units of K_{sp} are $(mol\ L^{-1})^n$, where "n" is the total number of moles of ions formed when 1 mol of AB dissolves.

Example: *What are the units of K_{sp} for calcium phosphate $Ca_3(PO_4)_2$?*
Answer: *First write out the equilibrium expression, and then define K_{sp}.*

$$Ca_3(PO_4)_2(s) \rightleftharpoons 3\,Ca^{2+}(aq) + 2\,PO_4^{3-}(aq)$$

$$K_{sp} = [Ca^{2+}]^3[PO_4^{3-}]^2$$

There are five concentration terms, so the units of K_{sp} are $(mol\ L^{-1})^5$.

12.3.3 Calculating solubility from K_{sp} and vice versa

At the outset of this discussion, we must stress that this will not be a complete treatment. Solubility equilibria are complicated by other factors which we shall not consider. Consequently, the calculation of solubility from K_{sp} may be seriously in error, especially if the substance is fairly soluble and dissociates to form ions of high charge ($> \pm1$).

The simplest calculation involves a 1:1 electrolyte such as AgCl. The calculation takes a little more care when the ratio of ions is other than 1:1. The next two examples illustrate the method of calculation.

Example: *Calculate the solubility of AgCl, given that $K_{sp} = 1.6 \times 10^{-10}\ (mol\ L^{-1})^2$*
Answer: *Write out the equation, and define the solubility as 's'.*

	$AgCl(s)$	$\rightleftharpoons$	$Ag^+(aq)$	+	$Cl^-(aq)$
equilib'm conc	-		s		s

Remember that the "concentration" of the pure solid AgCl is irrelevant.
Define K_{sp} and substitute:

$$K_{sp} = [Ag^+,aq][Cl^-,aq]$$

$$1.6 \times 10^{-10}\ (mol\ L^{-1})^2 = s^2$$

$$s = 1.3 \times 10^{-5}\ mol\ L^{-1}$$

Since AgCl is a 1:1 electrolyte, the solubility of AgCl, the concentration of $Ag^+(aq)$, and the concentration of $Cl^-(aq)$ all have the value $1.3 \times 10^{-5}\ mol\ L^{-1}$.

Example: *Calculate the solubility of CaF_2, given that*

$$K_{sp} = 5.3 \times 10^{-9}\ (mol\ L^{-1})^3$$

Answer: *Again write out the equation. This time, when you define the solubility of CaF_2 as 's', remember that when 's' mol L^{-1} of CaF_2 dissolve they form 's' mol L^{-1} of Ca^{2+} but '2s' mol L^{-1} of F^-.*

	$CaF_2(s)$	$\rightleftharpoons$	$Ca^{2+}(aq)$	+	$2\ F^-(aq)$
equilib'm conc	-		s		2s

Define K_{sp} and substitute:

$$K_{sp} = [Ca^{2+}(aq)][F^-(aq)]^2$$

$$5.3 \times 10^{-9} = (s)(2s)^2 = 4s^3 \text{ and } s = 1.1 \times 10^{-3}\ mol\ L^{-1}.$$

Thus the equilibrium solubility of CaF_2 and the concentration of $Ca^{2+}(aq)$ are both 1.1×10^{-3} mol L^{-1}, but the concentration of $F^-(aq)$ is 2.2×10^{-3} mol L^{-1}.

The calculation of K_{sp} from solubility data is straightforward, as seen in the example below. In this case the solubility was reported in grams of solute per 100 mL of water; we shall take this to be approximately the same as 100 mL of solution. The error is small, except when the salt has appreciable solubility. The present example would be a case where a significant error would result.

Example: *Scandium sulfate ($Sc_2(SO_4)_3$) has solubility 10.3 g per 100 mL water at 25°C. Calculate K_{sp}.*

Answer: *First work out the solubility of scandium sulfate in mol L^{-1}.*

$$M(Sc_2(SO_4)_3) = 378\ g\ mol^{-1}$$

$$c(Sc_2(SO_4)_3) = \frac{10.3\ g}{0.100 L} \times \frac{1}{378\ g\ mol^{-1}} = 0.272\ mol\ L^{-1}$$

Now write out the K_{sp} equilibrium expression and indicate the concentrations of the ions.

	$(Sc_2(SO_4)_3)(s)$	$\rightleftharpoons$	$2\ Sc^{3+}(aq)$	+	$3\ SO_4^{2-}(aq)$
concn.	-		2×0.272		3×0.272

Define K_{sp} and evaluate numerically

$$K_{sp} = [Sc^{3+}]^2\ [SO_4^{2-}]^3$$

$$= (2 \times 0.272\ mol\ L^{-1})^2.(3 \times 0.272\ mol\ L^{-1})^3$$

$$= 0.16\ (mol\ L^{-1})^5$$

How do we determine whether the concentration of a solution represents the equilibrium solubility? This question can be approached from an experimental or a theoretical perspective. Theoretically, we determine whether the concentration of solute is the same as that calculated from the equilibrium constant. Experimentally, a saturated solution is one in which the solution is in equilibrium with the undissolved solute. This definition applies to both ionic and non-ionic solutes. Thus if we are handed a reagent bottle labelled "saturated KCl solution", we should be able to see crystals of KCl(s) present, in addition to the solution phase. Only if the solid is present can the reaction possibly be at equilibrium.

12.3.4 Common ion effect

In Section 11.7 we saw that the dissociation of a weak acid was suppressed in the presence of a source of strong acid (H^+) or in the presence of a salt of the same acid. In just the same way, the dissolution of a slightly soluble salt is inhibited if it dissolves into a solution which already contains one of its component ions. For example, $CaSO_4$ is less soluble in either dilute $CaCl_2$ solution or in dilute Na_2SO_4 solution than it is in pure water.

$$CaSO_4(s) \rightleftharpoons Ca^{2+}(aq) + SO_4^{2-}(aq)$$

Either $Ca^{2+}(aq)$ (for example, in $CaCl_2$ solution) or $SO_4^{2-}(aq)$ (in Na_2SO_4 solution) drive the solubility equilibrium towards the left, and hence lower the solubility of $CaSO_4$.

Example: *Recalculate the solubility of calcium fluoride in a solution that contains 0.010 mol L^{-1} NaF instead of in pure water.*

Answer: *We proceed as in a regular K_{sp} calculation, only remembering that the aqueous solution contains 0.010 mol L^{-1} of fluoride ion before any of the CaF_2 dissolves.*

Step 1: Write out the solubility equation, and set up table of initial and equilibrium concentrations.

	$CaF_2(s)$	$\rightleftharpoons$	Ca^{2+} (aq)	+	$2F^-$ (aq)
initial conc.	-		0		0.010
equilib. conc.	-		s		0.010 + 2s

Note that at equilibrium, the total F^- concentration is the 0.010 mol L^{-1} which was present initially, plus the 2s mol L^{-1} of F^- that were formed for each "s" mol L^{-1} of CaF_2 that dissolved.

Step 2: Define K_{sp} and evaluate. The value of K_{sp}, 5.3×10^{-9} (mol $L^{-1})^3$ was given before.

$K_{sp} = [Ca^{2+}][F^-]^2 = (s)(0.010 + 2s)^2$

Assume $2s << 0.010$, then $5.3 \times 10^{-9} \approx s\,(0.010)^2$

$s \approx 5.3 \times 10^{-5}$

This is indeed $<< 0.010$, justifying the approximation.

Since "s" was defined as the solubility of CaF_2, we see that the presence of 0.010 mol L^{-1} of the common ion F^- reduced the solubility of CaF_2 from 1.1×10^{-3} to 5.3×10^{-5} mol L^{-1}.

12.3.5 Effect of acidity on aqueous solubility

Most hydroxides, carbonates, phosphates, and sulfides are only slightly soluble in water; earlier, we saw that these insoluble salts are responsible for immobilizing many cations in the Earth's crust. These salts contain "basic anions", that is, anions which are the conjugate bases of weak acids (Section 11.6). It follows that these anions have an intrinsic affinity for H^+. The general case of the reaction of a basic anion B^- reacting with H^+ is given below.

$$B^- + H^+ \rightleftharpoons HB$$

K_c for this reaction is $1/K_a$ for the weak acid HB, and hence K_c is larger the weaker the acid HB.

It follows that insoluble salts which contain basic anions tend to dissolve in acid. This can be stated alternatively by saying that salts of weak acids experience higher solubility in acidic solutions. Let's examine the case of the reaction of calcium carbonate with hydrochloric acid. The overall reaction is:

$$CaCO_3(s) + 2\ HCl(aq) \rightleftharpoons CaCl_2(aq) + CO_2(g) + H_2O(\ell)$$

The overall equation obscures the chemistry involved. We first write a net ionic equation, by removing the spectator chloride ions (Equation A).

$$\text{A}\quad CaCO_3(s) + 2\ H^+(aq) \rightleftharpoons Ca^{2+}(aq) + CO_2(g) + H_2O(\ell)$$

Equation A can be broken down into a solubility equilibrium, two acid-base equilibria, and a Henry's Law equilibrium.

$$(i)\quad CaCO_3(s) \rightleftharpoons Ca^{2+}(aq) + CO_3^{2-}(aq)$$

$$(ii)\quad CO_3^{2-}(aq) + H^+(aq) \rightleftharpoons HCO_3^-(aq)$$

$$(iii)\quad HCO_3^-(aq) + H^+(aq) \rightleftharpoons H_2CO_3(aq)$$

$$(iv)\quad H_2CO_3(aq) \rightleftharpoons CO_2(g) + H_2O(\ell)$$

Addition of these four equations reconstitutes Equation A, and from the Rule of Multiple Equilibria, (Section 8.10) the equilibrium constant for the overall reaction is obtained as the product of the equilibrium constants for equations *(i)* through *(iv)*.

$$K = K_c(i)K_c(ii)K_c(iii)K_c(iv)$$
$$= K_{sp}(1/K_{a}, HCO_3^-)(1/K_{a}, H_2CO_3)(1/K_H)$$
$$= \frac{(6.0 \times 10^{-9} \text{ mol}^2 \text{ L}^{-2}).(4.8 \times 10^{-11} \text{ mol L}^{-1})^{-1}}{(4.2 \times 10^{-7} \text{ mol L}^{-1})(3.4 \times 10^{-2} \text{ mol L}^{-1} \text{ atm}^{-1})}$$
$$= 8.7 \times 10^{9} \text{ atm L mol}^{-1}.$$

The large value of K_c shows how favourable is the dissolution process.

Example: *Calculate the solubility of $CaCO_3$ in equilibrium with 350 ppmv of CO_2 at pH 9.00 and at pH 7.00.*
Answer: *The overall equation is Equation A. Write out this reaction, and indicate the equilibrium concentrations/pressures. Because the pH and $p(CO_2)$ are each held constant, the only unknown is $[Ca^{2+}]$. At pH 9.00:*

A	$CaCO_3(s)$	$+ 2H^+(aq)$	$\rightleftharpoons$	$Ca^{2+}(aq)$	$+ CO_2(g)$	$+ H_2O(\ell)$
conc/press	-	1.0×10^{-9}		x	3.5×10^{-4}	

Define K_c and substitute:

$$K_c = [Ca^{2+}]p(CO_2)/[H^+]^2$$
$$8.7 \times 10^9 = x(3.5\times10^{-4})/(1.0\times10^{-9})^2$$
$$x = (8.7 \times 10^9)(1.0 \times 10^{-18})/(3.5 \times 10^{-4})$$
$$= 2.5 \times 10^{-5} \text{ mol L}^{-1}$$

Correspondingly at pH 7.00, $[H^+] = 1.0 \times 10^{-7}$ mol L^{-1}

$$8.7 \times 10^9 = x(3.5\times10^{-4})/(1.0\times10^{-7})^2$$
$$x = (8.7 \times 10^9)(1.0 \times 10^{-14})/(3.5 \times 10^{-4})$$
$$= 2.5 \times 10^{-1} \text{ mol L}^{-1}$$

This calculation illustrates how sharply the solubility of a slightly soluble salt can increase as the pH falls.

We shall explore the effects of environmental acid on insoluble salts more thoroughly in Chapters 13 and 17. For now, we preview those discussions by noting that acidification of natural waters by deposition of acidic gases (Chapter 13) or as a result of mining operations (Chapter 17) has the undesirable side effect of mobilizing metals into the water from the underlying rock. Many of these important minerals contain toxic elements such as Cu, Cd, Pb, Hg, etc. These can be a source of toxic metals in drinking water sources; however, we must also note that there are areas in the world where the water naturally contains relatively high concentrations of these metal ions.

12.3.6 Practical aspects of scale formation from hard water

Equation A of the previous section illustrated the dissolution of $CaCO_3$ in strong acid. From the environmental perspective, a much more important reaction is the solubilization of limestone ($CaCO_3$) by the weak acid H_2CO_3. The product is calcium hydrogen carbonate, which is much more soluble than $CaCO_3$. This reaction is responsible for the hardness of water in limestone areas (refer back to Section 10.4). The relevant equation (B) follows, and is broken down into its component equilibria.

$$CaCO_3(s) + H_2CO_3(aq) \rightleftharpoons Ca(HCO_3)_2(aq)$$

B $$CaCO_3(s) + H_2CO_3(aq) \rightleftharpoons Ca^{2+}(aq) + 2\ HCO_3^-(aq)$$

(i) $$CaCO_3(s) \rightleftharpoons Ca^{2+}(aq) + CO_3^{2-}(aq)$$

(ii) $$CO_3^{2-}(aq) + H^+(aq) \rightleftharpoons HCO_3^-(aq)$$

(iii) $$H_2CO_3(aq) \rightleftharpoons HCO_3^-(aq) + H^+(aq)$$

Addition of these three equations reconstitutes Equation B.

$$K_c = K_c(i)K_c(ii)K_c(iii)$$
$$= K_{sp}(1/K_a\ ,\ HCO_3^-)(K_a\ ,\ H_2CO_3)$$
$$K_c = (6.0 \times 10^{-9}\ mol^2\ L^{-2})(4.8 \times 10^{-11}\ mol\ L^{-1})^{-1}(4.2 \times 10^{-7}\ mol\ L^{-1})$$
$$= 5.3 \times 10^{-5}\ (mol\ L^{-1})^2$$

Over time, dissolution of limestone by CO_2-laden rainwater and stream water has been responsible for the formation of caves and gorges. Higher concentrations of calcium in water are favoured by raising the concentration of H_2CO_3 (or equivalently $p(CO_2)$ of the air in contact with water). This process occurs underground, where biological action can raise $p(CO_2)$ to values as high as 0.05 atm, compared with 0.00035 atm in the open atmosphere. Consequently, hard water from wells may contain substantially higher concentrations of Ca^{2+} than would be expected on the basis of $c(H_2CO_3)$ having the value 1.0×10^{-5} mol L^{-1} that we calculated in Section 12.2. In addition to its high hardness (Ca^{2+}), such water also has high alkalinity (HCO_3^-). Hardness and alkalinity generally go together, an exception being when hardness is due to contact with $CaSO_4$ (gypsum). In that case, alkalinity remains low, because the sulfate anion is virtually non-basic in aqueous solution ($K_b = 8.3 \times 10^{-13}$ mol L^{-1}).

Scale is deposited from hard water containing calcium hydrogen carbonate because reaction B is reversible. When water is heated, the solubility of $CO_2(aq)$ — same as $H_2CO_3(aq)$ — is reduced, and $CO_2(g)$ is formed, shifting the equilibrium below to the right (reverse of equation B).

$$Ca^{2+}(aq) + 2\ HCO_3^-(aq) \rightleftharpoons CaCO_3(s) + CO_2(g) + H_2O(\ell)$$

Calcium carbonate therefore deposits as scale in the hot water system.

We can now understand why lime softening is a successful treatment method for hard water for use on an industrial scale (Section 10.5). Lime, $Ca(OH)_2$, provides a source of OH^-, which undergoes an acid-base reaction with HCO_3^-; the calcium ions precipitate the carbonate ions that form.

overall reaction: $Ca(OH)_2 + Ca(HCO_3)_2 \rightarrow 2\ CaCO_3(s) + 2\ H_2O$

component reactions: $2\ OH^-(aq) + 2\ HCO_3^-(aq) \rightarrow 2\ CO_3^{2-}(aq) + 2\ H_2O(\ell)$

$2\ Ca^{2+}(aq) + 2\ CO_3^{2-}(aq) \rightarrow 2\ CaCO_3(s)$

12.3.7 Complexation reactions

In the previous two sections we found that acid had the effect of increasing the solubility of certain salts because $H^+(aq)$ reacted with one component of the equilibrium — in this case, a basic anion. The phenomenon in which one partner in the dissolution equilibrium is also involved in another equilibrium process is called complexation. The reaction of a slightly soluble salt with acid is therefore seen to be a specific example of a more general phenomenon. Here are some additional examples.

1. Acid increases the solubility of silver acetate by complexing with the acetate ions (to form acetic acid)

overall equilibrium: $AgAc(s) + H^+(aq) \rightleftharpoons Ag^+(aq) + HAc(aq)$

component equilibria: $AgAc(s) \rightleftharpoons Ag^+(aq) + Ac^-(aq)$

$Ac^-(aq) + H^+(aq) \rightleftharpoons HAc(aq)$

2. Ammonia increases the solubility of silver chloride by complexing with the silver ions to form the ion $Ag(NH_3)_2^+$.

overall equilibrium: $AgCl(s) + 2\ NH_3(aq) \rightleftharpoons Ag(NH_3)_2^+(aq) + Cl^-(aq)$

component equilibria: $AgCl(s) \rightleftharpoons Ag^+(aq) + Cl^-(aq)$

$Ag^+(aq) + 2\ NH_3(aq) \rightleftharpoons Ag(NH_3)_2^+(aq)$

3. Fluoride ion increases the solubility of aluminum in natural waters by complexing with Al^{3+} to form AlF^{2+}

overall equilibrium: $Al(OH)_3(s) + F^-(aq) \rightleftharpoons AlF^{2+}(aq) + 3\ OH^-(aq)$

component equilibria: $Al(OH)_3(s) \rightleftharpoons Al^{3+}(aq) + 3\ OH^-(aq)$

$Al^{3+}(aq) + F^-(aq) \rightleftharpoons AlF^{2+}(aq)$

12.3.8 Precipitation reactions

Precipitation is the reverse of dissolution, and so it involves the same equilibrium process. Frequently we need to know whether a precipitate will form when two solutions are mixed. For example, you might want to make up a solution containing a mixture of ionic nutrients in order to grow microorganisms. If some of the ions are incompatible in the sense that they will precipitate, then the nutrients will not be present at the concentrations you planned. Solubility rules allow you to determine qualitatively whether to expect precipitation to occur. What you have to do is to consider the various combinations of cations with anions, and whether any of the combinations represent an insoluble salt.

Example: *In which of the following cases do you expect precipitation to occur when equal volumes of 0.1 mol L^{-1} solutions are mixed?*

A calcium chloride and silver nitrate
B sodium chloride and calcium nitrate
C potassium sulfide and cadmium chloride
D barium hydroxide and copper (II) sulfate

Answer: *The possibilities to consider are the "crossover" products in each case.*

A: The crossover products are calcium nitrate and silver chloride. All nitrates are soluble (solubility rule 2), but silver chloride is insoluble (solubility rule 3). Therefore a precipitate of silver chloride forms.

B: The crossover products are calcium chloride and sodium nitrate. All nitrates are soluble (solubility rule 2) as are all salts of sodium, an alkali metal (solubility rule 1). No precipitate forms.

C: The crossover products are cadmium sulfide and potassium chloride. All potassium salts are soluble (alkali metal, solubility rule 1), but most sulfides are insoluble (solubility rule 5). Therefore a precipitate of cadmium sulfide forms.

D: The crossover products are barium sulfate and copper (II) hydroxide. Both barium sulfate (solubility rule 4) and most hydroxides (solubility rule 5) are insoluble. Therefore a precipitate forms containing both barium sulfate and copper (II) hydroxide.

Calculations can be carried out to determine quantitatively whether a precipitate will form when two solutions are mixed. Our approach is to determine whether the concentrations present in solution are compatible with the K_{sp} expression. This is done by defining a reaction quotient Q_c (Q_{sp} in this case), and comparing the values of Q_{sp} and K_{sp}, in the same way as was done for gas phase equilibria in Section 8.8.

Example: *Magnesium carbonate has K_{sp} 3.5 x 10^{-8} (mol $L^{-1})^2$ at 25°C. Show by calculation whether a precipitate will form when 1.0 L of 2.0 x 10^{-4} mol L^{-1} $MgCl_2$ is mixed with 1.0 L of 3.0 x 10^{-4} mol L^{-1} Na_2CO_3.*

Answer: *First write out the precipitation reaction, and note that it is the reverse of the equation defining K_{sp} for $MgCO_3$.*

$$Mg^{2+}(aq) + CO_3^{2-}(aq) \rightarrow MgCO_3(s)$$

Remember that when the solutions are mixed the volume changes, so the concentrations of $Mg^{2+}(aq)$ and $CO_3^{2-}(aq)$ also change. In this case the total volume doubles, and the concentrations are halved. In more complex cases, you may have to work out the final concentrations step by step.

Now use the concentrations of Mg^{2+} (1.0 x 10^{-4} mol L^{-1}) and CO_3^{2-} (1.5 x 10^{-4} mol L^{-1}) to calculate Q_{sp}.

$$Q_{sp} = [Mg^{2+}][CO_3^{2-}] = (1.0 \times 10^{-4}\ mol\ L^{-1})(1.5 \times 10^{-4}\ mol\ L^{-1})$$
$$= 1.5 \times 10^{-8}\ (mol\ L^{-1})^2$$

Therefore, $Q_{sp} < K_{sp}$. The solution is too dilute for a precipitate to form.

12.3.9 Phosphate removal from sewage

The problem of eutrophication was discussed in Section 10.3, where it was attributed to abnormally high concentrations of phosphate, derived in large part from sewage. This problem has been attacked in two ways: legislation on the maximum amounts of phosphates that may be added to detergent formulations, and "advanced" or "tertiary" treatment of sewage.

Conventional sewage treatment involves two stages. Primary treatment is a simple settling process to remove solids. Secondary treatment is the BOD reduction described briefly in Section 12.2. This incidentally reduces phosphate levels somewhat, as the microorganisms incorporate some phosphorus into their biomass.

In tertiary treatment for phosphate, most of the remaining phosphate is removed by precipitation. The favoured treatment agents are lime, filter alum, and ferric chloride. Lime treatment is used most commonly.

In lime treatment, $Ca(OH)_2$ is added. The hydroxide ions bring the pH to *ca.* 9, while the elevation of the concentration of Ca^{2+} causes precipitation of calcium phosphate, which is very insoluble. Since the phosphate anion is basic, the solubility of calcium phosphate is lower at high pH; hence lime affords an economical method of simultaneously raising the pH and providing calcium for precipitation. In the equation below, note that HPO_4^{2-} (not PO_4^{3-}) is the predominant form of phosphate ion at pH 9.

$$5\ Ca^{2+}(aq) + 3HPO_4^{2-}(aq) + 4\ OH^-(aq) \rightarrow Ca_5(PO_4)_3OH(s) + 3\ H_2O(\ell)$$

Numerous calcium-phosphate salts are known. The most insoluble, which is also the most stable, is hydroxylapatite $Ca_5(PO_4)_3OH$, which has K_{sp} approximately 10^{-56} $(mol\ L^{-1})^9$. However, other materials such as $Ca_3(PO_4)_2$ are the kinetic products of precipitation. Since these other products are somewhat more water-soluble than hydroxylapatite, their formation is to be avoided if the removal of phosphate is to be optimized. In order to assure the formation of hydroxylapatite, the solution is "seeded" with a small amount of finely powdered hydroxylapatite, which provide nuclei for the further growth of hydroxylapatite crystals.

When filter alum, $Al_2(SO_4)_3$, is used as the precipitant, the product is $AlPO_4$ ($K_{sp} = 1 \times 10^{-21}$ $(mol\ L^{-1})^2$). At pH 5, $H_2PO_4^-$ is the major phosphate species, and so the alkalinity of the sewage is reduced when $AlPO_4$ is formed.

$$Al^{3+}(aq) + H_2PO_4^-(aq) + 2\ OH^-(aq) \rightarrow AlPO_4(s) + 2\ H_2O(\ell)$$

Likewise, $FeCl_3$ affords $FePO_4$, although this is converted during sewage sludge digestion into $Fe_3(PO_4)_2$, $K_{sp} = 8 \times 10^{-34}$ $(mol\ L^{-1})^5$. The latter reaction occurs when the sludge is heated under anaerobic conditions, and involves a reduction of Fe(III) to Fe(II); the reactant for this process is shown as 3 [H], implying reduction by an unknown substance.

$$Fe^{3+}(aq) + H_2PO_4^-(aq) + 2\ OH^-(aq) \rightarrow FePO_4(s) + 2\ H_2O(\ell)$$
$$3\ FePO_4(s) + 3\ [H] \rightarrow Fe_3(PO_4)_2(s) + H_2PO_4^-(aq) + H^+(aq)$$

Both $Al_2(SO_4)_3$ and $FeCl_3$ are used at lower pH (about 5) than lime, because at higher pH the very insoluble hydroxide $Al(OH)_3$ or $Fe(OH)_3$ is formed instead of the phosphate. The solubility behaviour of aluminum salts is discussed further in Section 13.5.

12.4 Partition equilibria

12.4.1 Definition of a partition constant

An equilibrium constant may be written for the distribution of a solute between two **immiscible** (mutually insoluble) liquids. An example of a pair of immiscible liquids is oil and water, and for practical purposes, water is usually one of the liquids in question. The **partition coefficient** - also called a distribution coefficient - is defined below.

$$K_{part} = \frac{\text{concentration of solute in solvent 1}}{\text{concentration of solute in solvent 2}}$$

It is usual (but not essential) for the same units of concentration to be chosen for each solvent, in which case the partition coefficient is dimensionless.

Applications of partition equilibria constitute the remainder of this section.

12.4.2 Extraction of non-polar compounds from water

Non-polar compounds have low solubility in water, which is a very polar solvent. For example, when antibiotics such as penicillin are made industrially by fermentation, the penicillin is produced at very low (ppm) concentrations in an aqueous "broth". In order to isolate the antibiotic, it is extracted out of the aqueous phase into an organic solvent in which it is much more soluble *i.e.*, $K_{part} >> 1$. Because K_{part} is large, a small volume of organic solvent can be used to remove a high proportion of the penicillin from a much larger volume of water. Subsequently, the organic solvent can be removed, leaving behind the pure penicillin. This concept is used routinely to extract non-polar substances from aqueous solutions, where frequently they occur at low concentrations.

Extraction is also used during the analysis of water for environmental contaminants. Substances such as PCBs and dioxins occur in the aqueous environment at exceedingly low concentrations, yet because of concerns about their possible toxicity these trace amounts must be measured. Present analytical capability does not allow samples of lake or river water to be analysed directly for PCBs or dioxins, because there is too little of the pollutant to detect. Instead, the contaminant is first extracted into a non-polar organic solvent such as hexane, for which $K_{part} >> 1$. The volume of the extraction solvent is less than that of the water, thereby concentrating the analyte. Additionally, hexane has a low boiling point, and so the extract can be evaporated down to an even smaller volume (a few microliters if necessary). The efficient extraction of the analyte from the aqueous phase into hexane, combined with the reduction in volume of the hexane extract, concentrates the analyte sufficiently to allow analysis to be carried out.

Example: *o-Phenantholine is four times more soluble in toluene than in water. Calculate the fraction of phenantholine remaining in the aqueous phase after a solution containing 85 mg solute in 100 mL water is extracted with 70 mL of toluene.*

Answer: *The interpretation of the solubility information is that $K_{part} = 4$. Define x as the mass of solute (mg) in toluene at equilibrium, and define concentration in the units mg/mL. Remember that any units of concentration can be used, but the concentration in water and the concentration in toluene must have the same units if K_{part} is to be dimensionless.*

solubility in toluene = x mg/70 mL

solubility in water = (85-x) mg/100 mL

$$K_{part} = \frac{\text{conc. in toluene}}{\text{conc in water}} = \frac{(x)/70 \text{ mg/mL}}{(85-x)/100 \text{ mg/mL}}$$

Hence x = 63 mg

Fraction remaining in the aqueous phase = 22 mg/85 mg = 0.26

Example: *A pesticide is present at a concentration of 1.8 ppb in water. A 1.00 L sample of the water is extracted quantitatively using three successive 100 mL portions of hexane. The hexane extracts are combined, and evaporated to a final volume of 500 µL. What is the concentration of pesticide in the final solution?*

Answer: *Original mass of pesticide = $1.8\ \mu g\ L^{-1} \times 1.00\ L$*

= 1.8 µg

Since extraction was quantitative, the final solution also contains 1.8 µg of solute.

conc = 1.8 µg/500 µL = $3.6 \times 10^{-3}\ g\ L^{-1}$

This is the same as $3.6\ mg\ L^{-1} \equiv 3.6$ ppm $\equiv$ 3600 ppb. Hence the concentration has been increased by a factor of 2000.

One pitfall to be aware of when carrying out extractions is that some solutes may exist in ionic forms in certain ranges of pH. This phenomenon occurs when the substance contains one or more acidic or basic groups, and is common when drugs and other biologically active molecules must be extracted from water or aqueous media such as blood. In these circumstances it is essential to know the pK_a of the acid, in order to predict the pH range in which extraction will be possible. Aspirin ($HC_9H_7O_4$) represents an example of this sort; it has $pK_a \approx 3$, and at pH > pK_a exists predominantly as the conjugate base $C_9H_7O_4^-$. The conjugate base, which is ionic, cannot be extracted from water into a non-polar solvent. Extraction into a non-polar solvent is only possible at pH < pK_a, when Aspirin exists in the molecular (non-ionic) form $HC_9H_7O_4$. Therefore, in order to extract Aspirin from biological fluids, it would first be necessary to acidify the sample to pH < 3.

***Example**: Pentachlorophenol (PCP, HC_6Cl_5O) is a weak acid having pK_a 5.5. It is a common contaminant in soil and water, because of its use as a wood preservative. Suggest how you could extract PCP into the non-polar solvent toluene from a sample of water from a pond having high alkalinity.*

***Answer**: Pond water of high alkalinity usually has pH 7-8 (Section 11.12). At this pH, which is $> pK_a$, PCP will exist almost entirely as the conjugate base $C_6Cl_5O^-$. Since this is ionic, it will not extract out of water. Before extraction, the water from the pond must be acidified to pH < 5.5.*

12.4.3 Bioconcentration

You may have read that fish in the Great Lakes basin and in industrialized rivers such as the Hudson and Mississippi (U.S.), the Danube and the Rhine (Europe) have been found to contain relatively high concentrations of organic chemicals of industrial origin. Many of these substances are non-polar. Some of their names may be familiar from news reports: DDT, polychlorinated biphenyls (PCBs), dioxins, Mirex. Relatively high concentrations in this context means parts per billion, or at worst, parts per million. Nevertheless fears about the toxicity of such substances has caused legislators to place restrictions upon the amounts of such fish that may safely be eaten.

Fish living in waters polluted with non-polar pollutants tend to accumulate the xenobiotics by a process called **bioconcentration**. This occurs when a non-polar substance in the water partitions into the fat of the fish. The fat acts as a non-polar extraction solvent or, put in another way, the value K_{part} for the ratio c(fat)/c(water) is high. For this reason aquatic organisms generally bioconcentrate non-polar solutes from the water in which they live.

In practice, it is more common to define a **bioconcentration factor** as the ratio of the average concentration of the solute in the whole organism to the concentration of the solute in water.

$$\text{Bioconcentration Factor (BCF)} = \frac{\text{concentration of solute in organism}}{\text{concentration of solute in water}}$$

The BCF is related to K_{part} according to the percent by weight of fat in the organism (assuming that we are still considering the case of a non-polar solute which is predominantly found in the fatty tissue).

$$\text{BCF} \approx K_{part} \times \text{\% by weight of fat}$$

The concentrations of these solutes, which include pesticides and industrial chemicals, may be thousands, or even millions of times greater in aquatic organisms than in the water in which the organisms live.

It is difficult to carry out experiments using the actual fat of organisms, because fat is hard to work with experimentally, and also varies in composition from species to species, and even slightly from individual to individual, depending upon diet. Instead, a reference solvent is often used to mimic the chemical behaviour of fat. The most common such solvent is octanol (an eight carbon alcohol). Octanol: water partition coefficients K_{ow} are easily obtained in the laboratory and can be used to *predict* the bioconcentration factor where this has not been measured experimentally.

$$K_{ow} = \frac{\text{concentration of solute in octanol}}{\text{concentration of solute in water}}$$

A common strategy is to assume that the aquatic organism contains about 5% fat by weight. Under these assumptions we can write the following approximation.

$$BCF \approx 0.05 \times K_{ow}$$

This approximation is useful for predicting bioconcentration factors when they have not been measured experimentally. For example, suppose you wished to estimate the likely BCF of a new industrial chemical in trout. The value of K_{ow} could be measured easily in the laboratory, allowing a quick estimate of the bioconcentration factor to be obtained.

Bioconcentration can cause the level of a potentially toxic substance to be much higher in the organism than in the water in which it lives. This is why it is possible for the water in the Great Lakes to be considered safe for human consumption, even though the consumption of the fish living in that water has been restricted.

Example: *The value of K_{ow} for polychlorinated biphenyls (PCB) is about 3×10^6. Estimate the average concentration of PCB in a mature lake trout, assumed to be at equilibrium with the water, if the concentration of PCB in the water is 13 ppt. Assume the PCB is found only in the fat and that the fat comprises 7% of the body weight of the trout.*

Answer: *First calculate the concentration of PCB in the fat, using K_{ow} as a model system;* i.e., *assume c(fat) would be the same as c(octanol).*

$$K_{ow} = \frac{\text{concentration of solute in octanol}}{\text{concentration of solute in water}}$$

$$c(\text{octanol}) = c(\text{fat}) = K_{ow} \times c(\text{water}) = (3 \times 10^6) \times (13\ \text{ppt})$$

$$= 3.9 \times 10^7\ \text{ppt}\ (39\ \text{ppm})$$

However, the fat comprises only 7% of the trout's body weight.

$$c(\text{trout}) = 0.07 \times 39\ \text{ppm} = 3\ \text{ppm}.$$

In this example, we assumed that octanol could be used as a model solvent for fat, and hence that "concentration in fat" would be equal to "concentration in octanol". In practice, this somewhat oversimplifies the situation, and more complex relationships between c(fat) and K_{ow} have been developed.

Biomagnification is a phenomenon related to bioconcentration. It is the effect whereby organisms high in the food chain tend to accumulate more of a persistent pollutant than those lower down in the same ecosystem. Consider the following food chain:

plankton → plankton-eating fish → fish-eating birds

Each organism higher in the food chain consumes the pollutant burden of the organism on which it feeds. If the pollutant is persistent -- not readily metabolized -- it will biomagnify (or bioaccumulate). It is possible for toxic levels of the pollutant to be reached at the top of the food chain. Biomagnification of persistent organochlorine insecticides is suspected to have been the cause of reduced reproductive success among fish-eating birds in the 1960s and 1970s. Replacement of these insecticides by less persistent substances, which can be used at lower concentrations, has allowed the levels of substances such as DDT to decline in North American waters. Fish caught in the Great Lakes basin have shown a steady decline in the concentrations of organochlorine compounds over the past 15 years. (Concern exists however that the decline may not be permanent, as Third World use of older pesticides is increasing; these materials can reach North America and Europe by transport through the atmosphere.)

12.4.4 Pharmacokinetics and toxicokinetics

In the previous section, we concentrated upon the equilibrium aspects of bioconcentration, viewing it solely as a partition equilibrium. However, there is also a kinetic aspect to this issue. Fish taken from, for example, the Great Lakes characteristically show increasing pollutant burdens with age, suggesting that even the older fish have not reached equilibrium with the water in which they swim. Consumption of larger, older fish therefore represents a greater risk to human health than eating younger, less polluted ones.

Let us apply kinetic ideas to the phenomenon of bioconcentration. It is possible to describe the uptake of non-polar pollutants into organisms such as fish by means of the kinetic scheme below: "c" is the concentration of the pollutant.

c(aq)	→	c(fish)	uptake, rate constant k_1
c(fish)	→	c(aq)	clearance, rate constant k_2
c(fish)	→	reaction	metabolism, rate constant k_3

Analysis of the dependence of the concentration of the toxic substance with time is called toxicokinetics.

An exactly equivalent scheme can be written to describe the behaviour of drugs in the body. The drug is ingested (k_1), excreted (k_2), and metabolized (k_3). The behaviour is then described as pharmacokinetics.

Equation [1] describes the dependence of c(fish) with time.

$$[1]\quad dc(fish)/dt = k_1 c(aq) - (k_2 + k_3)c(fish)$$

For aquatic life living in a large lake, we can assume that c(aq) is constant *i.e.*, uptake of the pollutant by aquatic life does not appreciably change the concentration in the water. This will be true for the fish in a large lake, but might not be true for a fish in a laboratory aquarium. Eq. [1] simplifies when the steady state concentration c(fish, eq) has been reached, when dc(fish)/dt = zero, eq. [2].

$$[2]\quad k_1 c(aq) = (k_2 + k_3)c(fish, eq)$$

From this equation, we can see that the bioconcentration factor can be related to the rate constants for uptake and clearance of the toxicant (compare Section 8.11).

$$BCF = c(fish, eq)/c(aq) = k_1/(k_2 + k_3)$$

If the steady state has not been reached, eq. [1] must be integrated. The integrated form is given by eq. [3].

$$[1]\quad dc(fish)/dt = k_1 c(aq) - (k_2 + k_3)c(fish)$$

$$\frac{dc(fish)}{\{k_1 c(aq) - (k_2 + k_3)c(fish)\}} = dt$$

Integrating between limits:

$$[3]\qquad \ln\left[\frac{\{k_1 c(aq) - (k_2 + k_3)c(fish)\}_{\text{time } t(2)}}{\{k_1 c(aq) - (k_2 + k_3)c(fish)\}_{\text{time } t(1)}}\right] = -(k_2 + k_3)\{t(2) - t(1)\}$$

For the case where the fish is initially uncontaminated, c(fish) = zero at t = zero, and eq. [3] simplifies to eq. [4].

$$[4]\qquad \ln\left[\frac{\{k_1 c(aq) - (k_2 + k_3)c(fish)\}}{\{k_1 c(aq)\}}\right] = -(k_2 + k_3).t$$

This can also be written as:

$$\ln\left[1 - \frac{(k_2 + k_3)c(\text{fish})}{k_1 c(\text{aq})}\right] = -(k_2 + k_3)t$$

Figure 12.2 shows an example of the uptake of a chlorinated dioxin by young trout from water in a laboratory experiment. Uptake continued for 28 days with the dioxin concentration in the water kept constant during that time. After 28 days, the trout were transferred to clean water, and the loss of the toxic substance from the trout was followed for a further 28 days. For this toxicant, k_3 was essentially zero (very slow metabolism), and from these data, it was possible to determine the rate constants k_1 and k_2 and the bioconcentration factor. The values, to one significant figure, were $k_1 = 200$ day $^{-1}$ and $k_2 = 0.1$ day $^{-1}$, leading to a bioconcentration factor ($\equiv k_1/k_2$) of approximately 2000.

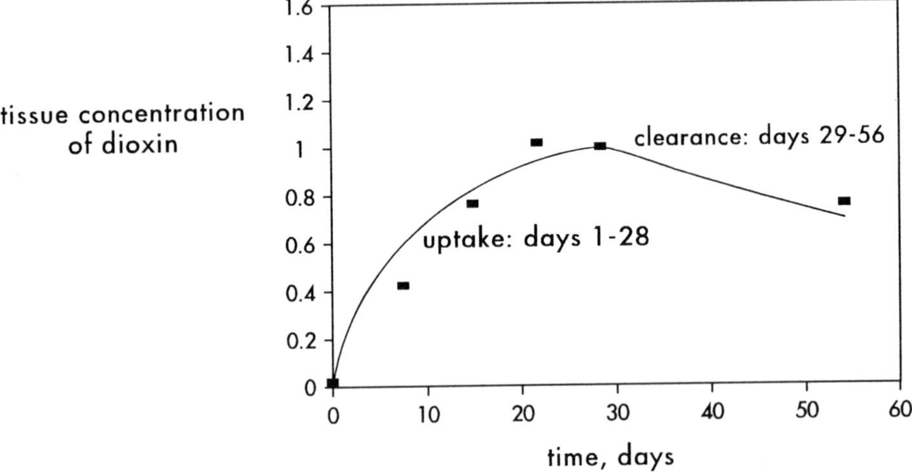

Figure 12.2: Uptake and clearance of 1,3,6,8-tetrachlorodibenzo-*p*-dioxin by juvenile trout in water

Problems

Section 12.2

1. Use the Henry's Law constants in Table 12.1 to calculate the ratio $[N_2(aq)]/[O_2(aq)]$ when air is equilibrated with water at 25°C.

2. Calculate the concentration of CO(aq) in equilibrium with 5.2 atm of CO(g).

3. At 5°C the weight percent of CH_4 in an aqueous solution, which is equilibrated with CH_4 (g) at 1.00 atm, is 0.00341%. Calculate K_H for methane at 5°C.

4. At 60°C the solubility of N_2 in water is 8.2 cm^3 of CH_4 (g) per liter (corrected to STP), when $p(N_2)$ = 101.3 kPa. Calculate K_H for N_2 at 60°C.

5. Dry air at 1.0 atm is passed over an electric discharge, converting 0.85% of the O_2 to O_3. The ozonized air is equilibrated with water at 25°C in a vessel containing 12,000 L of air and 1,000 L of water. Calculate the concentration of O_3 (aq) at equilibrium.

6. A sample of water is equilibrated with the atmosphere at 0°C and then analysed by the Winkler method (equations below).

$$Mn^{2+} + 2OH^- + \tfrac{1}{2}O_2 \rightarrow MnO_2\,(s) + H_2O$$
$$MnO_2(s) + 4H^+ + 2I^- \rightarrow Mn^{2+} + I_2 + 2H_2O$$
$$I_2 + 2Na_2S_2O_3 \rightarrow Na_2S_4O_6 + 2NaI$$

A 50.00 cm^3 sample of oxygenated water is treated by the above reactions and the I_2 liberated is titrated against 0.01136 mol L^{-1} $Na_2S_2O_3$, of which 8.11 cm^3 are required to reduce all the I_2. Calculate the solubility of O_2 in water at 0°C in mol L^{-1}, and hence the Henry's Law constant for oxygen at 0°C.

7. The COD of a water sample is determined as follows: A test water sample (100.0 mL) and a control using pure water (100.0 mL) are separately heated with 25.00 mL of $Na_2Cr_2O_7$ in 50% H_2SO_4 for 2 hours. A 25.00 mL aliquot is withdrawn from each solution and titrated against a solution of ferrous ammonium sulfate: $[Fe^{2+}] = 4.024 \times 10^{-3}$ mol L^{-1}.

$$6Fe^{2+} + Cr_2O_7^{2-} + 14H^+ \rightarrow 6Fe^{3+} + 2Cr^{3+} + 7H_2O$$

The titers of the test water are 9.77 mL and 26.40 mL of Fe^{2+} solution, respectively. Calculate the COD of the test water sample.

8. a) A raw sewage sample has organic matter content of 720 mg L^{-1}. Assume for this problem that the organic matter can be treated as if it were glucose $C_6H_{12}O_6$. What is the O_2 requirement for the complete oxidation of 1.2 x 10^5 L of this sewage? Give your answer in mg of O_2.

   b) The 1.2 x 10^5 L of sewage in part (a) is accidentally discharged into a lake of capacity 3.5 x 10^6 m^3. Assuming uniform mixing, what is the additional BOD (in mg L^{-1}) that is placed on the waters of the lake as a result?

   c) Give two reasons why a large discharge of sewage would be more damaging to the aquatic life of a very warm lake (*e.g.* 25°C) than a very cold one (*e.g.* 4°C).

9. a) A waste sample has BOD 80 mg L^{-1} and is to be discharged into a lake whose dissolved oxygen content is 8.1 ppm. How many liters of waste can be added to each liter of lake water if the dissolved oxygen of the lake must be guaranteed not to fall below 6.3 ppm?

   b) In practice much more waste can be added to the lake without serious risk of the dissolved oxygen falling below 6.3 ppm. Explain.

10. The TOC of material entering an activated sludge reactor is 350 ppm and the phosphate concentration is 7.8 ppm. Calculate the concentration of nitrogen, in the form of ammonia, that should be present for optimum BOD reduction.

11. Calculate the concentration of CO_2 (aq) and the pH of a sample of club soda which is made from pure water bottled under a pressure of 2.2 atm of CO_2 (g).

12. Calculate the total carbonate in equilibrium with atmospheric CO_2 at 350 ppm (a) at pH 7.0 (b) at pH 9.0.

13. A waste water stream contains 15 ppm of ammonia. Calculate the volume of air for each liter of waste water that would be needed to reduce c(NH_3 (aq)) to 0.50 ppm at 25°C if complete equilibrium between the phases were achieved.

14. Calculate the pH of a solution of ammonia in equilibrium with p(NH_3) of 0.050 atm at 25°C.

15. a) A sewage treatment plant is designed to process 3.0×10^6 L of sewage daily. What capacity is required for the primary settling lagoon if the residence time is to be 6 hours?

    b) Suppose that the raw sewage at the plant has BOD of 850 ppm. If a 90% reduction in BOD is achieved during secondary treatment, what volume of oxygen will be required (assume 15°C)?

    c) At what average rate must oxygen be transferred from the atmosphere to the sewage?

Section 12.3

1. Predict whether the following salts are soluble or insoluble in water: $MgSO_4$, $(NH_4)_2S$, $Ca_3(PO_4)_2$, $ZnCl_2$, AgBr, $Pb(OH)_2$

2. Write net ionic equations to describe the reactions, if any, which occur when the solutions below are mixed.
   a) NH_4Cl and KNO_3
   b) $CuSO_4$ and $(NH_4)_2S$
   c) $CdCl_2$ and $Ba(OH)_2$
   d) $MgSO_4$ and Na_2CO_3

3. Three flasks contain separately solutions of $MgCl_2$, $AgNO_3$, and $MnSO_4$. All are colourless. Describe how you would distinguish them if you had available test solutions of NaCl, Na_2S, and Na_2CO_3.

4. Write net ionic equations to describe the reactions, if any, that occur when the following solutions are mixed.
   a) $BaCl_2$ and H_2SO_4
   b) KNO_3 and $CaCl_2$
   c) Na_2S and $CuSO_4$
   d) $Al_2(SO_4)_3$ and NaOH
   e) $Ca(OH)_2$ and $Ca(HCO_3)_2$

5. Using data of Appendix 4, calculate the solubility, in mol L^{-1} of:
   a) $PbCl_2$
   b) $CaCO_3$
   c) ZnS
   d) $Pb(OH)_2$

6. The concentration of CrO_4^{2-} (aq) in a saturated solution of $SrCrO_4$ is 6 x 10^{-3} mol L^{-1}. Calculate K_{sp} for $SrCrO_4$.

7. Calculate the solubilities in water at 25 °C:
   a) of $CaCO_3$ in ppm of Ca^{2+}
   b) of BaF_2 in ppm of BaF_2

8. A solution of $Pb(NO_3)_2$ has concentration 0.022 mol dm^{-3}. How many moles of Na_2SO_4 (s) must be added to 2.00 dm^3 of this solution to precipitate 99.9% of all the lead ions?

9. What mass of $Mg(OH)_2$ will dissolve
   a) in 0.500 dm^3 of pure water
   b) in 0.500 dm^3 of a solution of NaOH whose concentration is 0.010 mol dm^{-3}.

10. Show by calculation whether precipitation will occur
    a) when 0.100 dm^3 of 1.0 x 10^{-4} mol dm^{-3} NaCl is added to 0.015 dm^3 of 5.6 x 10^{-5} mol dm^{-3} $AgNO_3$.
    b) when saturated solutions (100 cm^3 each) of $CaSO_4$ and Li_2CO_3 are mixed. K_{sp} for Li_2CO_3 = 1.7 x 10^{-3} (mol $L^{-1})^3$.

11. Calculate ΔH° for dissolution of the following salts, in order to determine whether their solubility will increase or decrease as the temperature is raised
    a) $CaCO_3$
    b) CaF_2
    c) $Cu(OH)_2$

12. What volume of 0.100 mol dm^{-3} NaF must be added to 20.0 cm^3 of 0.300 mol dm^{-3} $BaCl_2$ before precipitation is observed?

13. a) Iodic acid HIO_3 has K_a = 1.7 x 10^{-1} mol L^{-1}. Calculate the pH of a 0.30 mol dm^{-3} solution of potassium iodate.
    b) Given that $Cu(IO_3)_2$ has K_{sp} = 1.4 x 10^{-7} (mol $L^{-1})^3$, show by calculation whether a clear solution will be formed if 0.50 g of $CuSO_4.5H_2O$ is added to 100 cm^3 of the KIO_3 solution described in part (a).

14. Calculate the solubility of $Fe(OH)_3$ at pH 1.0, pH 2.0, and 3.0.

15. Calculate the equilibrium concentrations of soluble phosphate produced when $Ca_3(PO_4)_2$ dissolves in pure water.

16. A lake has pH 7.25 and contains 0.04 ppm phosphorus and 75 ppm of Ca^{2+} (aq). Is it saturated with respect to hydroxylapatite?

17. A lake water sample has the following partial analysis: total carbonate, 86 ppm; nitrate, 0.12 ppm, ammonia, 0.04 ppm, phosphate (as PO_4^{3-}), 0.08 ppm. Which is the limiting nutrient?

18. The residence time of the water in Lake Erie is 2.7 years. If the input of phosphorus to the lake is halved, how long will it take for the concentration of phosphorus in the lake water to fall by 10%?

19. A sewage sample contains, after secondary treatment, 8.8 ppm phosphorus as total phosphate. It is brought to pH 9.0, and $[Ca^{2+}]$ = 4.7 mmol L^{-1} by the addition of lime. What fraction of the phosphate is precipitated if the precipitate is (i) hydroxylapatite; (ii) $Ca_3(PO_4)_2$?

20. Filter alum $Al_2(SO_4)_3$ is often used to remove phosphate ion from waste water. A wastewater of pH 5.62 containing 25 ppm total phosphate is treated with alum until the equilibrium concentration of Al^{3+} is 4.0 x 10^{-9} mol L^{-1}. What fraction of the phosphate is precipitated as $AlPO_4$(s)? Consider only the equilibria below:

$$AlPO_4(s) \rightleftharpoons Al^{3+}(aq) + PO_4^{3-}(aq)$$

$$H_2PO_4^{-}(aq) \rightleftharpoons HPO_4^{-}(aq) + H^{+}(aq)$$

$$HPO_4^{2-}(aq) \rightleftharpoons PO_4^{3-}(aq) + H^{+}(aq)$$

13 ACID RAIN

13.1 What is "acid rain"?

Acid rain, and its companions acid snow and acid fog, represent an environmental issue which has been in the public consciousness for many years. Acid rain causes damage to crops, to forests, to environmentally sensitive lakes, and also to buildings and engineering structures made of stone (limestone) and metal (iron and steel). Acid rain is not a new phenomenon: air pollution caused by burning coal had been recognized as a public nuisance in cities at least two centuries ago, as coal replaced wood as the principal energy source of the new industries of the Industrial Revolution. Even the term "acid rain" is not new; it was apparently first used in a book published in England in 1872.

Acid rain is usually associated with heavy industry, whether in developed regions such as North America and Western Europe, or in more recently industrialized or industrializing areas such as Eastern Europe, China and India. In this chapter, we will examine the sources of acid rain, the chemistry involved, and the effects of this pollution on the environment. We will conclude by studying some of the measures that have been proposed and taken to alleviate this problem.

As explained in Chapter 12, unpolluted rainwater has a pH close to 5.6, in consequence of the raindrops being in equilibrium with the atmospheric concentration of carbon dioxide.

$$CO_2(g) + H_2O(\ell) \rightleftharpoons H_2CO_3(aq) \rightleftharpoons H^+(aq) + HCO_3^-(aq)$$

Restating the point, even completely clean rain does *not* have a pH of 7.0, because it is not absolutely pure water; it contains equilibrium amounts of the atmospheric gases.

Acidic precipitation is generally defined as having pH lower than about 5.0; pH 4 to 4.5 is not uncommon, and isolated examples of rain and fog having pH lower than 2 have been recorded. To put this in context, vinegar and lemon juice have pH 3.0 and 2.2 respectively. Since even unpolluted rainwater is slightly acidic, it may be helpful to think of "acid rain" as rain that is more acidic than normal.

13.2 Sources of acid rain

The main causes of acid rain are **sulfur oxides** and **nitrogen oxides** in the atmosphere. These gases are present in trace amounts (ppbv) even in natural, unpolluted air; they generally become a problem only when they occur in higher than normal amounts as a result of human activities. However, the absolute amounts of these gases are very small, even in polluted air, amounting to no more than a very few ppmv. Acid rain results when these gases are oxidized in the atmosphere and return to the ground dissolved in raindrops.

SO_2 falls as H_2SO_3 and H_2SO_4:

$$SO_2 + H_2O \rightarrow H_2SO_3$$

$$SO_2 \xrightarrow{\text{oxidize}} SO_3 \rightarrow H_2SO_4$$

NO_x falls as HNO_3 (for more detail, see Chapter 9):

$$NO \xrightarrow{\text{oxidize}} NO_2$$

$$NO_2 + OH \rightarrow HNO_3$$

In most areas, sulfur oxides are the major contributor to acid precipitation, but the nitrogen oxides predominate on the U.S. west coast where acid rain and photochemical smog (Chapter 9) are closely linked. In a few regions of the world, notably parts of Alaska and New Zealand, highly acidic rain falls naturally as a result of the emission of HCl and SO_2 from volcanoes.

13.2.1 Sulfur oxides

Two important activities which lead to the release of sulfur oxides into the atmosphere are **coal burning** and the **roasting of metal sulfide ores**.

Coal typically contains 2-3% of sulfur by mass, and this sulfur is oxidized along with the carbon when the coal burns.

$$\text{S (as organosulfur compounds or metal sulfide)} + O_2(g) \rightarrow SO_2(g)$$

Not all coals are alike however. In North America, many of the coals from southern West Virginia, Kentucky, and the Canadian and American Rockies are low in sulfur while other coals from the American Midwest, northern Appalachia, and Nova Scotia have sulfur contents at the upper end of the range. The air pollution associated with the Industrial Revolution of the nineteenth century in Europe and North America was mainly due to ready access to high-sulfur coals in these areas. High sulfur coal is likewise the source of severe air pollution today in the Peoples' Republic of China and Eastern European countries such as the former East Germany, Poland, the Czech Republic and Slovakia.

The iron and steel industry and coal-burning thermal power plants use a great deal of coal. While 2-3% of sulfur may not sound like much, the sheer volume of coal consumed in these industries leads to prodigious emissions of sulfur dioxide.

Example: *Calculate the volume of SO_2 emitted at STP upon burning 10,000t of coal having 2.3% sulfur by weight.*

Answer: *First calculate the mass of sulfur*

$$(2.3\% \text{ of } 10{,}000\,t = 2.3 \times 10^2\,t)$$

Then calculate amount of SO_2 in mol:

$$n(SO_2) = n(S) = 2.3 \times 10^2\,Mg\,/\,(32\,g\,mol^{-1}) = 7.2\,Mmol$$

Finally calculate the volume of SO_2 at STP, using the molar volume of an ideal gas ($22.4\,L\,mol^{-1}$):

$$V(SO_2) = 7.2\,Mmol \times 22.4\,L\,mol^{-1} = 1.6 \times 10^2\,ML$$

Since a megaliter is not a customary SI unit, change to L (1.6×10^8 L) or to m^3 ($1.6 \times 10^5\,m^3$).

Many commercially important metals occur in nature as sulfide ores (see also Chapter 17). These metals include nickel (NiS), copper (Cu_2S and $CuFeS_2$), zinc (ZnS), lead (PbS), and mercury (HgS). The first step in recovering these metals from their ores consists of roasting the ore in air to give the metal oxide, which is subsequently reduced to the element, usually with coke.

$$2\,MS\,(M = Ni, Zn, Pb) + 3\,O_2(g) \rightarrow 2\,MO + 2\,SO_2(g)$$

$$MO + C(s) \rightarrow M + CO(g)$$

With Cu_2S and HgS, the metal is formed directly.

e.g. $$Cu_2S(s) + O_2(g) \rightarrow 2\,Cu(s) + SO_2(g)$$

Since metal extraction is a very large-scale industry, the quantities of sulfur dioxide that are released through roasting are correspondingly very large. Even though some of the SO_2 is captured and converted into sulfuric acid at some of the largest smelting locations, thousands of tonnes of SO_2 are still released daily. The nickel smelter at Sudbury, Canada, remains the world's largest single point source emission of SO_2, despite pollution control measures which are discussed later in this chapter.

Example: *The nickel ore at Sudbury typically contains 1.3% nickel and 10% sulfur by weight. Calculate the mass of SO_2 that would be formed for each tonne of nickel produced if no anti-pollution measures were practised.*

Answer: *The mass ratios of Ni: S are 1.3: 10. Hence:*

$$\text{mass of S} = (1\text{ t Ni}) \times (10\text{ t S}/1.3\text{ t Ni}) = 7.7\text{ t}$$

$$n(S) = n(SO_2) = 7.7\text{ Mg}/32\text{ g mol}^{-1} = 0.24\text{ Mmol}$$

$$\text{mass}(SO_2) = 0.24\text{ Mg} \times \mathbf{M}(SO_2)$$

$$= 0.24\text{ Mmol} \times 64\text{ g mol}^{-1} = 15\text{ Mg} = 15\text{ t}$$

Note this answer: if all the ore were roasted and no pollution control measures were taken, the mass of SO_2 emitted would be 15 times the output of the primary product, nickel.

13.2.2 Nitrogen oxides

As discussed in Chapter 9, combustion sources are the principal cause of acidic precipitation due to nitrogen oxides. This phenomenon is linked to that of photochemical smog, since the formation of HNO_3 is one of the sinks by which the free radical species NO_2 is removed from the atmosphere.

$$N_2 + O_2 \rightleftharpoons NO \xrightarrow{O_3} NO_2 \xrightarrow{OH} HNO_3$$

To date, public perception of acidic emissions from coal burning and metal smelting has been directed mostly towards SO_2. However, these processes also produce NO_x, because the *nitrogen oxides are always formed in small amounts whenever air is heated* (Sections 8.7 and 9.1). Analyses of air polluted by acidic emissions show the amount of nitric acid to be typically one third of the total acid. In North America, the fraction of acidic deposition in the form of nitrate is growing steadily, as efforts to curb SO_2 emissions achieve greater success. Control of NO_x emissions has so far been less successful. NO_x is more difficult to control for two reasons: lack of convenient technology to control point source emissions such as thermal electricity plants, and the large number of non-point sources such as cars and trucks in urban areas.

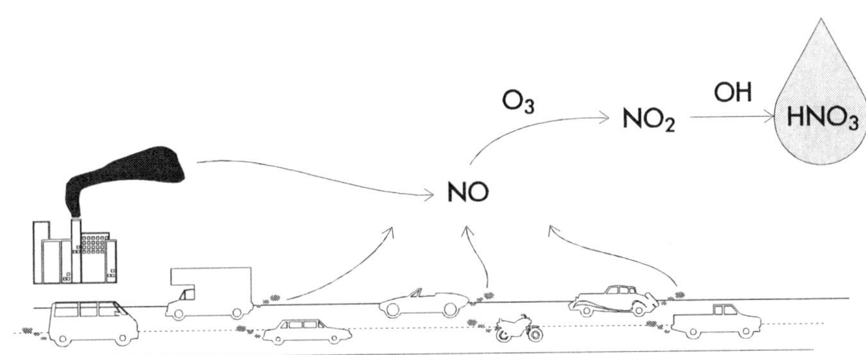

Combustion sources, especially non-point sources, are the principal cause of acid precipitation from nitrogen oxides.

13.3 Chemistry of acid rain

Unpolluted rainwater, as we have noted already, has pH close to 5.6 as a result of equilibration of raindrops with the *ca.* 350 ppmv of CO_2 in the troposphere. This yields the weak acid H_2CO_3 for which $K_a = 4.2 \times 10^{-7}$ mol L^{-1} at 25 °C.

In absolute terms, the concentrations of NO_2 and/or SO_2 are small - no more than 1 ppmv or so - even in highly polluted air. As shown in the previous section, these gases are ultimately precipitated in rain as HNO_3, H_2SO_3 and H_2SO_4. These gases have a much greater impact on the pH of rain than CO_2: they are more soluble in water and are also stronger acids. H_2SO_4 and HNO_3 dissociate completely in water, while H_2SO_3 has $K_a = 1.7 \times 10^{-2}$ mol L^{-1} at 25 °C. This is shown quantitatively below. Throughout this section $SO_2(aq)$ and $H_2SO_3(aq)$ will be taken as interchangeable, *cf.* $CO_2(aq)$ and $H_2CO_3(aq)$.

For CO_2:

1. $CO_2(g) + H_2O(\ell) \rightleftharpoons H_2CO_3(aq)$

2. $H_2CO_3(aq) \rightleftharpoons H^+(aq) + HCO_3^-(aq)$

3. $CO_2(g) + H_2O(\ell) \rightleftharpoons H^+(aq) + HCO_3^-(aq)$

Equation (3) is the sum of Equations (1) and (2), so by the rule of multiple equilibria, the equilibrium constant for reaction (3) is the product of the equilibrium constants of reactions (1) and (2).

Reaction 1: $K = K_H = 3.4 \times 10^{-2}$ mol L^{-1} atm^{-1}

Reaction 2: $K = K_a = 4.2 \times 10^{-7}$ mol L^{-1}

Reaction 3: $K_c = K_H \times K_a = 1.4 \times 10^{-8}$ mol^2 L^{-2} atm^{-1}

An entirely analogous set of equilibria may be written for SO_2:

For SO_2:

$$SO_2(g) + H_2O(\ell) \rightleftharpoons H_2SO_3(aq)$$

$$H_2SO_3(aq) \rightleftharpoons H^+(aq) + HSO_3^-(aq)$$

$$SO_2(g) + H_2O(\ell) \rightleftharpoons H^+(aq) + HSO_3^-(aq)$$

$$K_c = K_H \times K_a = (1.2 \text{ mol L}^{-1} \text{ atm}^{-1}) \times (1.7 \times 10^{-2} \text{ mol L}^{-1})$$

$$= 2.1 \times 10^{-2} \text{ mol}^2 \text{ L}^{-2} \text{ atm}^{-1}$$

Summarizing, the equilibrium constant for the overall reaction is larger in the case of SO_2 than of CO_2 because:

1. SO_2 is more soluble in water than CO_2;
2. H_2SO_3 is a stronger acid than H_2CO_3.

Consequently, a small concentration of $SO_2(g)$ has a greater influence on the pH of rain than a much larger concentration of $CO_2(g)$.

Example: *Calculate the pH of rainwater in equilibrium with 0.12 ppmv of SO_2 in the troposphere.*

Answer: *First calculate the pressure of SO_2 in atm, remembering that the total pressure in the troposphere is 1 atm.*

$$p(SO_2) = \frac{0.12 \text{ ppm} \times 1 \text{ atm}}{10^6 \text{ ppm}} = 1.2 \times 10^{-7} \text{ atm}$$

Now set up the chemical equation with equilibrium concentrations, assuming that the solution of some of the SO_2 into the raindrops does not affect the pressure of SO_2 in the gas phase.

$$SO_2(g) + H_2O(\ell) \rightleftharpoons H^+(aq) + HSO_3^-(aq)$$

$$1.2 \times 10^{-7} \qquad - \qquad x \qquad x$$

Define K_c and substitute the values:

$$K_c = \frac{[H^+][HSO_3^-]}{p(SO_2)} = x^2/1.2 \times 10^{-7} = 2.1 \times 10^{-2} \text{ (mol L}^{-1})^2 \text{ atm}^{-1}$$

$$x^2 = (1.2 \times 10^{-7})(2.1 \times 10^{-2}) = 2.5 \times 10^{-9}$$

$$x = [H^+] = 5.0 \times 10^{-5} \text{ mol L}^{-1}$$

$$pH = 4.30$$

Thus 0.12 ppmv of $SO_2(g)$ in equilibrium with rainwater will produce a pH of 4.30 in the water, compared with the pH 5.6 produced by 350 ppmv of $CO_2(g)$.

Assumptions:

1. SO_2 is the only gas contributing to the acidity.

2. Dissolution of $SO_2(g)$ in the aqueous phase does not deplete it from the gas phase; this assumption is reasonable if the volume of the aqueous phase is small.

3. Dissolution is fast enough that equilibrium is attained.

4. Neglect any further dissociation of HSO_3^- to SO_3^{2-}. This is reasonable because K_a for HSO_3^- is 10^6 times smaller than K_a for H_2SO_3.

The chemistry of rain acidified by the sulfur oxides is complicated in that the sulfur may be deposited either as $H_2SO_3(aq)$ as shown above, or it may first be oxidized to $SO_3(g)$ and then precipitate as $H_2SO_4(aq)$. Deposition may occur either in the aqueous form (wet deposition) or in association with particulate matter (dry deposition) in which case much of the sulfur will deposit in the form of sulfite or sulfate ions rather than the free acids.

13.4 Mechanism of oxidation of SO_2

The situation is very complex, because oxidation can occur by three quite separate routes: homogeneously in the gas phase, homogeneously in the aqueous phase of raindrops, clouds or fog and heterogeneously on the surface of particles. The prevailing atmospheric conditions, especially the humidity and the concentration and composition of particulate matter, determine the relative importance of these processes.

13.4.1 Homogeneous gas phase oxidation

The hydroxyl radical is the active agent for gas phase oxidation of SO_2. Recall from Section 9.1 that the hydroxyl radical (OH) is formed in the troposphere by solar photolysis of ozone. The hydroxyl radical adds to SO_2 in a manner analogous to its addition to NO_2.

$$SO_2 + OH \rightarrow HSO_3 \qquad k_2 = 9 \times 10^{-13}\ \text{cm}^3\ \text{molec}^{-1}\ \text{s}^{-1}$$

Although the rate constant for this reaction is moderately large, the reaction is slow, because OH and SO_2 are both trace tropospheric constituents.

Unlike HNO_3, which is a stable molecule with an even electron count, the HSO_3 formed in the preceding reaction is a free radical (odd electron) species. HSO_3 is not to be confused with the anion HSO_3^-, which is produced in the acid dissociation of H_2SO_3: HSO_3 and HSO_3^- differ in charge, and are therefore chemically distinct species, just like OH and OH^-. HSO_3 is oxidized rapidly by molecular O_2.

$$HSO_3 + O_2 \rightarrow SO_3 + HO_2$$

Example: *Calculate the half-life of SO_2 if gas phase oxidation is the only sink for SO_2, and the steady state concentration of OH is 2×10^6 radicals cm^{-3}.*

Answer: *If the concentration of OH is constant, the reaction is pseudo-first order.*

$$rate = k[SO_2][OH] = k'[SO_2], \text{ where } k' = k[OH]$$

Calculate k':

$$k' = (9 \times 10^{-13}\ cm^3\ molec^{-1}\ s^{-1})(2 \times 10^6\ molec\ cm^{-3})$$
$$= 1.8 \times 10^{-6}\ s^{-1}$$

(Note: correct units for a (pseudo)-first order rate constant)

Calculate half-life: $t_{1/2} = 0.693/k' = 4 \times 10^5\ s = 5$ *days*

Because this reaction is rather slow ($t_{1/2} \approx 5$ days), homogeneous gas phase oxidation of SO_2 is the major oxidation pathway only when no faster reaction is available. Aqueous phase oxidation can be faster (next section), but only under humid conditions. Homogeneous gas phase oxidation is therefore the predominant reaction under dry, even arid, conditions.

13.4.2 Homogeneous aqueous phase oxidation

SO_2 can be oxidized in the aqueous phase of clouds or raindrops by H_2O_2, which is formed by the disproportionation of HO_2 radicals. The reaction pathway is shown below. Recall that one of the reactions by which HO_2 radicals are formed is the tropospheric oxidation of carbon monoxide (Section 9.2).

$$2\ HO_2(g) \rightarrow H_2O_2(g) + O_2(g)$$
$$SO_2(aq) + H_2O_2(aq) \rightarrow H_2SO_4(aq)$$

or

$$HSO_3^-(aq) + H_2O_2(aq) \rightarrow HSO_4^-(aq) + H_2O(\ell)$$

Many of the details of the oxidation of SO_2 in rain or clouds are not yet understood. Disproportionation of HO_2 may occur in the gas phase, followed by dissolution of the H_2O_2 in the water droplet ($K_H = 10^5$ mol L^{-1} atm^{-1}); alternatively, the HO_2 radicals may first enter the aqueous phase ($K_H = 2 \times 10^3$ mol L^{-1} atm^{-1}), and then react together. The kinetics of oxidation of H_2SO_3 (*i.e.*,

SO_2(aq)) are complex, because H_2SO_3 and its ionized form HSO_3^- react at different rates in aqueous solution.

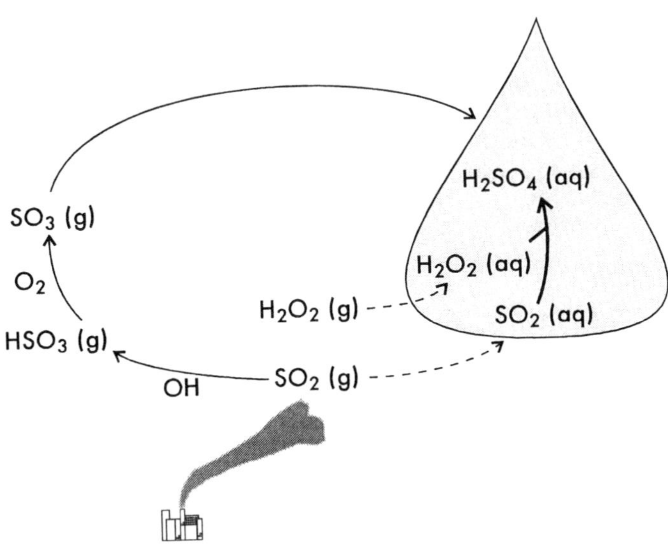

Oxidation of SO_2 can occur by several routes.

13.4.3 Heterogeneous oxidation on particles

This reaction is probably similar to that involved in the industrial oxidation of SO_2, discussed below, but the mechanism is not known in detail.

$$SO_2 + \tfrac{1}{2}\,O_2 \xrightarrow{\text{catalyst}} SO_3 \rightarrow H_2SO_4 \ (\text{or } SO_4^{2-})$$

Salts of vanadium, manganese, and iron are all effective catalysts for this oxidation, and all are widely distributed environmentally. For example, significant amounts of vanadium are present in both coal and Venezuelan oil. The particulate matter formed when these fuels burn contains vanadium. Thus the emissions formed by burning coal frequently contain both the reactant (SO_2) and the catalyst for its oxidation (vanadium-bearing particles).

Particles containing both vanadium and sulfate have been found in locations as remote as the Canadian High Arctic. Initially these were suspected to originate on the eastern seaboard of North America. Later information showed that the true source is coal burned in Eastern Europe, the particles having reached northern Canada over the North Pole.

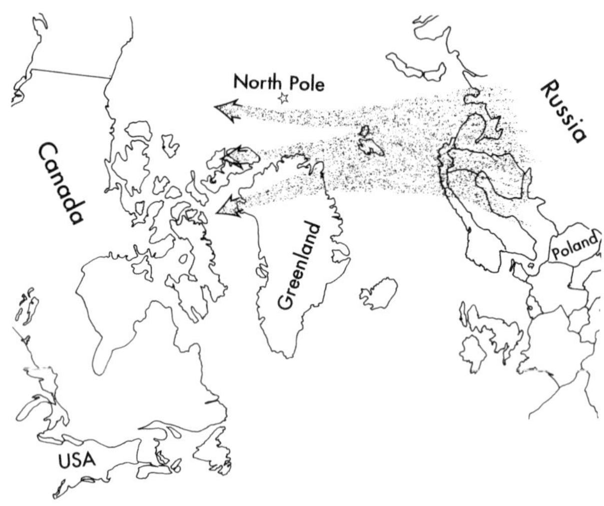

Trans-Polar pollution, in the form of particles, originates in Eastern Europe.

13.4.4 Rates of oxidation and deposition of sulfur oxides

The processes involved here are oxidation of SO_2, and deposition of both SO_2 and SO_3. These can be represented schematically.

$$SO_2 \xrightarrow{\text{oxidation, } k_1} SO_3$$

$$SO_2 \xrightarrow{k_2} \text{deposition}$$

$$SO_3 \xrightarrow{k_3} \text{deposition}$$

The rates of oxidation and deposition of SO_2 vary considerably with the conditions. In dry air, such as occurs over western Canada and the U.S. Southwest, rates of oxidation as low as 0.2% per hour have been recorded. Under these conditions, homogeneous gas phase oxidation by the OH radical is the predominant pathway, and the half-life for oxidation is several days, as shown in a preceding calculation. At the other extreme, rates up to 30% per hour havebeen reported under very humid conditions, when the oxidation occurs mainly in the aqueous phase. More typical rates of oxidation and also of deposition are both in the range 1 to 10% per hour.

Example: *Calculate the half-life of SO_2 in the atmosphere if the combined rates of oxidation and removal are 3% per hour.*

Answer: *The interpretation of the question in the context of the scheme above is that $(k_1 + k_2) = 0.03\ h^{-1}$. Since this is a composite first order rate constant, we can calculate the half-life directly.*

$$t_{½} = 0.693/k = 0.693/0.03 = 23\ h = 1\ day.$$

Since the half-lives of SO_2 and SO_3 in the atmosphere are only a few days, this means that acid precipitation can be expected over whatever distance is travelled by an air mass from a pollution source during a week or so. Assuming a wind speed of as little as 20 km h^{-1}, such an air mass will travel nearly 3500 km in a week. Acid precipitation can thus be considered a regional pollution problem. It is more than a local pollution issue, as the "plume" of pollution can travel many hundreds of km, but it is not a global problem like that of the chlorofluorocarbons (Chapter 15), whose atmospheric lifetimes are so long that they become globally distributed. Acidic gases are deposited too rapidly for such universal mixing. Likewise, acidic emissions do not stay in the atmosphere long enough to migrate to the stratosphere; acid rain chemistry is exclusively a tropospheric phenomenon.

13.5 Effects of acidic emissions

13.5.1 Effects on vegetation

Plants are much more susceptible than animals to the effects of acid rain. The effects on plants which must be considered are those of the gaseous pollutants themselves, and that of lowered pH. Whereas the damaging effects of photochemical smog (Chapter 9) have been focused mainly on agricultural crops, forests have been the major concern regarding acidic precipitation.

Sulfur dioxide inhibits plant growth at concentrations less than 0.1 ppmv. Concentrations between 0.1 and 1 ppmv can cause observable injury to plants and trees even if the length of exposure is only a few hours. Such levels are reached quite frequently in areas subject to acid rain. Combinations of gaseous pollutants often act synergistically; this is observed for the combinations SO_2/NO_2 and SO_2/O_3.

Excessive acidity is also harmful to plants. Soil chemistry is altered as the pH changes, and this problem is most serious for poorly buffered soils, many of which tend to be naturally acidic. Few plants can tolerate acidic soils and, among other effects, the germination of seeds and the growth of seedlings may be inhibited. Leaves show observable damage below pH 3.5.

Much controversy has surrounded the issue of whether acid rain is responsible for damage to forests. Forests in Scandinavia, eastern North America, and the Black Forest in Germany have all experienced reduced productivity in recent years. The Black Forest has been especially hard hit, with some commentators predicting its "probable destruction" over the next few decades. The observed symptoms are consistent with long term acidification, namely a yellowing of the leaf, called chlorosis. Continued exposure causes the needles of conifers to turn reddish-brown; eventually they drop and the trees die. Acid rain has been assumed to be the likely cause of these problems, although recent information suggests that the presence of volatile organic compounds of industrial origin (Section 9.9) may also be responsible.

When clouds encounter acidic gases in the lower troposphere, it is the base of a cloud which becomes the most acidic. Mists and fog therefore pose a special threat to forests, which are often swathed in mist in upland areas.

13.5.2 Effects on health

Acidic gases are irritants to the respiratory tract. At atmospheric levels no more than 1 to 2 ppmv, sulfur dioxide is absorbed high in the respiratory tract, and does not penetrate deep into the lungs to reach the far more sensitive alveoli. Even so, respiratory irritation may occur at 1-2 ppmv, especially in elderly people and those already suffering from lung diseases such as emphysema. Respiratory distress is more pronounced if particulate matter is also present in the air: this phenomenon, known as "London smog", was very prevalent in the U.K. in the days when coal was in general use for domestic heating. In one particularly severe week of smog in 1952, the death rate in London, England was 4000 higher than the average number.

Penetration of acidic gases to the alveoli occurs when the concentration reaches 25 ppmv or so. Although these levels may be encountered by workers in industries such as smelting, tanning, paper-making, and sulfuric acid manufacture, the irritant effects of the gas at these concentrations (wheezing, coughing, tearing) give a severe enough warning that actual injury is rare.

13.5.3 Effects on buildings etc.

Limestone ($CaCO_3$) has been a preferred building material for many centuries. Even under conditions of very clean air it is subject to slow attack, by the same chemical processes which carve out caves and gorges (Chapter 12).

$$CaCO_3(s) \rightleftharpoons Ca^{2+}(aq) + CO_3^{2-}(aq)$$
$$CO_3^{2-}(aq) + H^+(aq) \rightleftharpoons HCO_3^-(aq)$$
$$H_2CO_3(aq) \rightleftharpoons H^+(aq) + HCO_3^-(aq)$$

overall reaction $$CaCO_3(s) + H_2CO_3(aq) \rightleftharpoons Ca^{2+}(aq) + 2\ HCO_3^-(aq)$$

The equilibrium constant for this reaction has the rather small value of 5.3×10^{-5} $mol^2\ L^{-2}$, as was shown in Chapter 12. Acidic precipitation greatly increases the tendency towards dissolution. Dissolution can then be represented by the overall reaction below, which has a much larger equilibrium constant (3.0×10^8, no units).

$$CaCO_3(s) \rightleftharpoons Ca^{2+}(aq) + CO_3^{2-}(aq)$$
$$CO_3^{2-}(aq) + H^+(aq) \rightleftharpoons HCO_3^-(aq)$$
$$HCO_3^-(aq) + H^+(aq) \rightleftharpoons H_2CO_3(aq)$$

overall reaction $$CaCO_3(s) + 2\ H^+(aq) \rightleftharpoons Ca^{2+}(aq) + H_2CO_3(aq)$$

The damage done to historical momunments is seen clearly in Figure 1, where the loss of detail in the faces is very evident. Fine stone carving is most at risk as the outer layers of the stone flake off. This is called "sulfation"; the calcium carbonate is replaced by calcium sulfate, which is both more water-soluble than calcium carbonate (K_{sp} 5×10^{-4} $(mol\ L^{-1})^2$ *vs.* 6×10^{-9} $(mol\ L^{-1})^2$) and has less structural strength.

$$CaCO_3(g) + SO_2(g) + \tfrac{1}{2}O_2(g) \rightarrow CaSO_4(g) + CO_2(g)$$

Iron and steel are well known for their susceptibility to corrosion (details in Section 17.4), and the protection of such structures with paint costs billions of dollars annually. The chemistry of corrosion under atmospheric conditions is extremely complex, leading eventually to hydrated Fe_2O_3. Hydrogen ions catalyse the reaction, explaining why acid precipitation causes increased rates of corrosion. Zinc is often used to protect steel from corrosion (galvanized steel: see Section 17.4), but even galvanized steel is subject to accelerated corrosion by acid. Under unpolluted conditions, zinc offers good atmospheric protection; by contrast, under polluted industrial conditions galvanized steel may last as little as five years.

13.5.4 Effects on natural waters

The extent of acid precipitation and hence the magnitude of the problem of acidification is seen in data from the Great Lakes basin, which suggest annual wet deposition of sulfur and nitrogen of almost 400,000 and 200,000 tonnes respectively.

Gettysburg.
Left, General John Reynolds (exposed to acid precipitation), right, General Abner Doubleday (protected).
Gettysburg National Military Park, USA.

Parthenon.
Left, Karyatide (exposed to acid precipitation), right, Karyatide (protected).
Parthenon, Athens, Greece.

Figure 13.1: Etching of stonework caused by acidic precipitation

Acidification is mainly a problem in areas where the underlying rocks provide poor buffering capacity: *i.e.*, the water in contact with these rocks has low alkalinity. Rocks such as granite offer little protection by way of buffering, whereas carbonate rocks such as chalk and limestone are able to react with the acid and hence to neutralize it. The overall reaction with limestone is the same reaction as was met in the previous section. There it was an example of an environmental problem caused by acid precipitation; in the context of natural waters, the same reaction offers protection against the adverse effects of acidic precipitation!

$$2\ H^+(aq) + CaCO_3(s) \rightleftharpoons Ca^{2+}(aq) + CO_2(g) + H_2O(\ell)$$

The $H^+(aq)$ actually reacts with $HCO_3^-(aq)$ that is contributing to the alkalinity of the water, and $CaCO_3(s)$ goes into solution to restore equilibrium. As a result of the solid $CaCO_3$ going into solution, the pH of, for example, the lake is not changed significantly by the addition of the acidic rainwater. Lakes and streams in limestone areas are therefore fairly insensitive to acidic precipitation.

Acidic precipitation causes problems in areas of granitic rocks, namely Northern and Eastern Canada, the Northeastern United States, and Scandinavia. In these areas, the lakes are poorly buffered, because of the insolubility of the underlying rocks. Lakes in these areas have "normal" pH about 6.5 to 7, and total alkalinity in the range 10^{-4} mol H^+ L^{-1}. Today, many such lakes record pH levels of 5.0 and lower. By contrast, lakes in areas of carbonate rocks commonly have pH 7 to 7.5, and total alkalinity 10^{-3} mol H^+ L^{-1} or higher. Their higher total alkalinity explains their greater resistance to acidification, in addition to which, the alkalinity lost to H^+ can be replaced by dissolution of the rock. This is not possible where the rock is granite.

Acidified lakes do not support the variety of life that can be found in their non-acidified counterparts. Progressive acidification results in a parallel loss of aquatic organisms.

- pH 6.0 Death of snails and crustaceans
- pH 5.5 Death of salmon, rainbow trout, and whitefish
- pH 5.0 Death of perch and pike
- pH 4.5 Death of eel and brook trout

Below pH *ca.* 4 the lake becomes a suitable environment for white moss, which prefers acidic conditions. This plant forms a "felt mat" on the lake bottom. This may grow to 0.5 m or more thick, and prevent the exchange of nutrients between the water and the bottom sediments. It also prevents the sediments from exerting any buffering action. The result is a lake whose waters are crystal clear, but support very few forms of aquatic life.

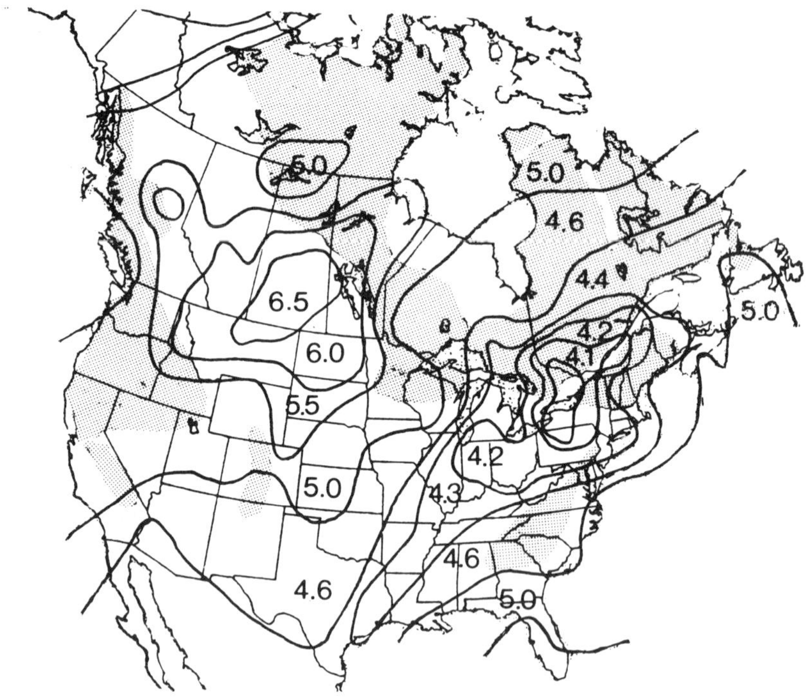

Precipitation in North America is acidic over wide areas. The shading shows regions with underlying granite — these regions are particularly susceptible to acid rain.

The information above shows that loss of game fish can be expected in lakes whose pH has already dropped to pH 5 or lower. By 1976, about half the lakes in the Adirondack Mountains of New York State had no fish in them, whereas forty years earlier almost all these lakes supported a population of sport fish. This observation correlates with comparisons of the alkalinity of Adirondack lakes today *vs.* sixty years ago: of 274 lakes for which data were available, 80% had suffered loss of alkalinity, the median loss being 50 μmol H^+ L^{-1}. The loss of the fish has serious economic consequences for tourism in regions like upstate New York and Northern Ontario.

A special problem for aquatic animals such as fish is that spawning, which generally takes place in the early spring, coincides with what is often the worst "pulse" of acidity of the year: the influx into the lake of the winter's accumulation of acid snow during the annual spring run-off: Figure 2. The result is decreased rates of hatching and reduced viability of the newly hatched fry.

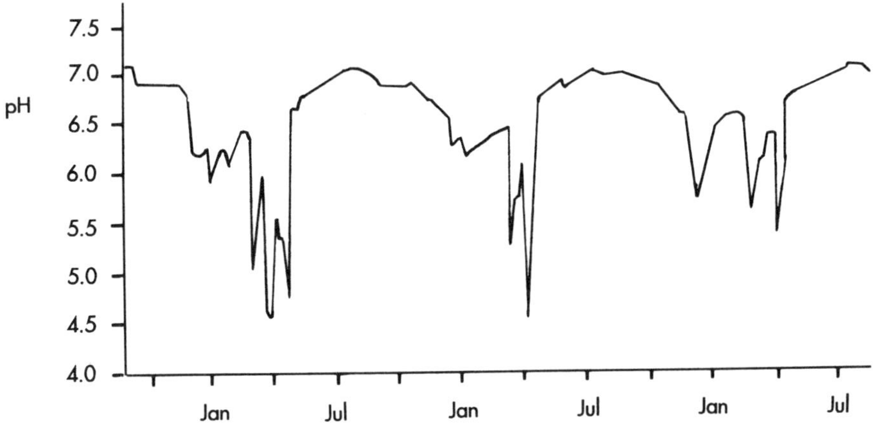

Figure 13.2: Seasonal variation in pH in an Adirondack lake

Experiments on rejuvenating acidified lakes have been conducted by using limestone to neutralize the excess acid. The limestone is sprayed as a powder using aircraft. The method is costly - about $20,000 to $50,000 to maintain a single 40 acre (20 ha) lake for 10 years. The effectiveness of the method has also been questioned; if the source of acidity is not controlled, the lake will quickly deteriorate again.

13.5.5 Acidity and metal ions[1]

We saw in Section 12.3 that salts containing basic anions become more soluble as the pH falls. Consequently, increasing acidity of a body of water tends to increase the aqueous concentrations of metal ions. The composition of the underlying bedrock determines the identities of these metals, which may include toxic elements such as Cd, Pb, and Hg. In each case, the metals are solubilized because of the reaction of H^+ with the basic anion with which the metal is associated. Such a reaction is similar in every respect to the dissolution of $CaCO_3$ in lakes which overlie limestone. Consider, for example, the case of PbS.

$$PbS(s) \rightleftharpoons Pb^{2+}(aq) + S^{2-}(aq) \qquad K_{sp} = 1 \times 10^{-28}\ mol^2\ L^{-2}.$$

In pure water, we calculate $[Pb^{2+}] = 1 \times 10^{-14}$ mol L^{-1}. A precise calculation in a natural water is difficult, because the concentrations of aqueous sulfide and other anions will vary from case to case. However, calculations for pure water suggest that the concentration of $[Pb^{2+}]$ will be millions of times larger at pH 4 than in pure water. Under some conditions, the acidification of water can compromise its suitability for drinking, since strict guidelines apply to the acceptable concentrations of metal ions in water (Section 10.8).

Example: *Calculate the change in the concentration of iron when a water body is acidified from pH 6.0 to pH 4.0. Assume that the underlying rock contains hydrated Fe_2O_3, and that this has the same K_{sp} as $Fe(OH)_3(s)$, $1 \times 10^{-38}\ (mol\ L^{-1})^4$.*

Answer: *Write the solubility equilibrium:*

$$Fe(OH)_3(s) \rightleftharpoons Fe^{3+}(aq) + 3\ OH^-(aq)$$

From the definition of K_w, at pH 6.0, $[OH^-] = 1.0 \times 10^{-8}\ mol\ L^{-1}$, and at pH 4.0, $[OH^-] = 1.0 \times 10^{-10}\ mol\ L^{-1}$.

From the definition of K_{sp}: $K_{sp} = [Fe^{3+}][OH^-]^3$

Therefore at pH 6.0,
$$[Fe^{3+}] = K_{sp}/[OH^-]^3 = 1 \times 10^{-38}\ (mol\ L^{-1})^4/(1.0 \times 10^{-8}\ mol\ L^{-1})^3 = 1 \times 10^{-14}\ mol\ L^{-1}$$

Likewise at pH 4.0,
$$[Fe^{3+}] = K_{sp}/[OH^-]^3 = 1 \times 10^{-38}\ (mol\ L^{-1})^4/(1.0 \times 10^{-10}\ mol\ L^{-1})^3 = 1 \times 10^{-8}\ mol\ L^{-1}$$

---

[1] Other aspects of the importance of toxic metal ions in water are discussed in Sections 10.9 and 17.5.

Although in this example the concentration of Fe^{3+}(aq) remains small even at pH 4, its concentration is a million times greater at pH 4 than at pH 6 — and at pH 2 would be calculated to be another million times greater still: 1×10^{-2} mol L^{-1}.

One metallic element which has been the subject of attention lately is aluminum. Aluminum is highly toxic towards fish, and there is also the possibility that drinking water high in aluminum may be one of the factors predisposing towards Alzheimer's disease (see Section 10.7).

The aqueous chemistry of aluminum is quite complex. Aluminum is an "acidic cation"; the hydrated Al^{3+} and $AlOH^{2+}$ ions are weak acids having pK_a about 5.5 and 5.6 respectively. The solubility of aluminum (as $Al(OH)_3$ or hydrated Al_2O_3) is at a minimum near pH 6.5. At lower pH, the cationic aluminum species Al^{3+}(aq), $AlOH^{2+}$(aq), $Al(OH)_2^+$(aq) are formed, while above pH 6.5, the solubility increases due to the formation of $Al(OH)_4^-$ (aq).

$$Al(OH)_3(s) + H^+(aq) \rightleftharpoons Al(OH)_2^+(aq) + H_2O(\ell)$$

$$Al(OH)_2^+(aq) + H^+(aq) \rightleftharpoons AlOH^{2+}(aq) + H_2O(\ell)$$

$$AlOH^{2+}(aq) + H^+(aq) \rightleftharpoons Al^{3+}(aq) + H_2O(\ell)$$

$$Al(OH)_3(s) + OH^-(aq) \rightleftharpoons Al(OH)_4^-(aq)$$

The preceding equilibria indicate that Al_2O_3 and $Al(OH)_3$ are **amphoteric**, meaning that they dissolve in both acid and base. In the context of acid precipitation, the important aspect of the solubility is the increase in concentration of cationic aluminum species at lower pH. Figure 3 is a "speciation diagram", which shows the predominant aluminum species in aqueous solution as a function of pH.

Most natural waters have pH 5-8, so they can contain significant amounts of Al^{3+}, $AlOH^{2+}$, $Al(OH)_2^+$, and $Al(OH)_4^-$ if they are in equilibrium with the bedrock. Furthermore, the total solubility of aluminum rises sharply as the pH falls, similar to the marked increase of $Fe(OH)_3$ at low pH in the example calculation earlier in this chapter. This allows us to explain the toxicity of aluminum towards fish. Consider a fish swimming in a lake of pH 5. The fish's blood is close to pH 7.4; its gill membrane is very thin in order to allow the efficient diffusion of oxygen from the water, and so a steep proton gradient is set up across the membrane. Because the concentration of dissolved aluminum decreases steeply as the pH rises, a gelatinous precipitate of $Al(OH)_3$ forms on the fish's gills, leading to death by suffocation.

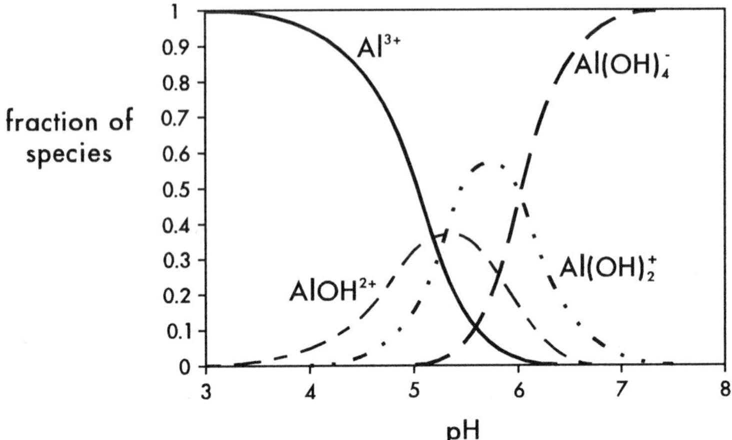

Figure 13.3: Speciation of aluminum over the pH range 3 - 8.

13.5.6 An extreme example of natural acidity

In 1983, scientists at the University of Toronto reported a case of extreme natural acidification in a group of small lakes in the Canadian Arctic. The "Smoking Hills" are situated on the Arctic Ocean, far from any habitation. They consist of cliffs, strata of which comprise combustible shales which have been burning continuously, probably for hundreds of years, and which were likely ignited originally by lightning. Sulfurous smoke from the fires gives the region its name, and over the years it has acidified the small ponds in the immediate area to the extent that those ponds closest to the burning shales have pH 2 and lower. Even though the rock underlying this particular region is largely limestone, the assault of acidity over so many years has long since overcome the capacity of these ponds to resist a change in pH. The data below show very clearly how the metal concentrations in these ponds increase as the pH drops.

The Smoking Hills

Table 13.1: Chemical characteristics of ponds in the Smoking Hills area

	Pond A	Pond B	Pond C	Pond D
pH	1.8	2.8	3.6	8.1
Distance[a]	40	190	670	4400
Acidity[b]	0.12	6.4×10^{-3}	6×10^{-4}	~0
Alkalinity[b]	~0	~0	~0	1×10^{-3}
SO_4^{2-}, ppm	16,000	380	160	110
Be, ppm	230	40	-	-
Mg, ppm	500	22	13	42
Al, ppm	590	18	3.8	-
Ca, ppm	370	45	49	61
Mn, ppm	64	2.8	3.5	-
Fe, ppm	2,600	32	1.1	0.04
Ni, ppm	22	0.2	0.06	-

a: distance in meters from the nearest burning shale
b: in mol L^{-1}

Pond A, which is very close to the burning shale, has pH 1.8, very high metal content, and very high $[SO_4^{2-}]$, reflecting the effect of sulfate deposition. The sulfate concentration falls off with distance from the shales, and the pH rises. Ponds A, B, and C have no residual alkalinity at all; in fact they have residual acidity. The data suggest that much of this acidity is in the form of free H^+(aq): see example at the end of this paragraph. Pond D, which is 4.4 km distant from the shales has almost normal water chemistry: pH 8 and total alkalinity in the mmol L^{-1} range. The effect of pH on total metal ion concentrations is reflected very clearly in the data.

Example: *Calculate the pH in each of ponds A, B, and C if all the acidity were due to H^+ (aq).*

Answer: *If all the acidity were due to H^+ (aq), then pH $= -\log_{10}[H^+]$.*

Pond A: pH(calc) $= -\log_{10}(0.12) = 0.9$

Pond B: pH(calc) $= -\log_{10}(6.4 \times 10^{-3}) = 2.2$

Pond C: pH(calc) $= -\log_{10}(6 \times 10^{-4}) = 3.2$

These values are fairly close to the experimental ones: 1.8, 2.8, and 3.6 respectively, suggesting that much of the acidity is due to free H^+ (aq). The solutions are buffered to some extent by Fe^{3+} (aq) and Al^{3+} (aq).

13.6 Acid rain abatement

This discussion will be restricted to the abatement of acidic emissions from stationary sources. The control of NO_x from mobile sources was discussed in Section 9.6.

There are several possible strategies to prevent environmental damage by a pollutant. They include:

- minimize the production of the pollutant;
- dilute the pollutant so that it is no longer harmful;
- convert the pollutant to a harmless or, better, useable substance.

All these strategies have been explored in the case of sulfur dioxide. However, the two main SO_2-producing industries require different approaches. These, it will be remembered, are coal burning and the smelting of metal ores.

13.6.1 Metal smelters

As noted earlier in the chapter, SO_2 emissions result when the ores are roasted in air to convert the metal sulfide to an oxide. The size of a smelter determines its environmental impact. For example, the Inco Ltd smelter at Sudbury, Ontario processes about 35,000 tonnes of ore per day having a typical analysis as shown below.

Ni	1.3%	Cu	1.3%	Co	0.04%
Fe	20%	S	10%	CaO	4%
MgO	4%	SiO_2	40%	Al_2O_3	10%

The principal sulfides in the ore are FeS_2, $CuFeS_2$, NiS and CuS. The absolute amount of nickel in the ore is rather small, and the amount of sulfur by mass is 7 to 8 times greater than the amount of nickel. We saw earlier that producing 1 tonne of nickel from this ore without any environmental control would cause 15 tonnes of SO_2 to be emitted. Now let's consider the total potential emissions of SO_2 from processing this amount of ore.

Example: *Calculate the total annual SO_2 emissions resulting from roasting 35,000 t of ore each working day, if the ore has the composition specified above. Assume 300 working days per year.*

$$\text{Total mass of S} = 10\% \times (35{,}000\ \text{Mg day}^{-1}) \times (300\ \text{days year}^{-1})$$
$$= 1.05 \times 10^6\ \text{Mg}\ (1.05\ \text{Tg})$$
$$n(S) = n(SO_2) = (1.05\ \text{Tg})/(32\ \text{g mol}^{-1}) = 0.033\ \text{Tmol}$$
$$\text{mass of } SO_2 = n(SO_2) \times \mathbf{M}(SO_2) = 0.033\ \text{Tmol} \times 64\ \text{g mol}^{-1}$$
$$= 2.1\ \text{Tg}$$

This is the same as 2.1 million tonnes of SO_2 per year. To put this number in context, the 1988 emissions from Sudbury were 685,000 tonnes and the Ontario government has mandated that Inco further reduce annual SO_2 emissions to 265,000 tonnes by 1994. Putting this another way, the operation already releases only one third of the total potential SO_2, and will be down to a little over 10% of total possible emissions by 1994. We will examine the technology which makes these reductions possible.

(i) Concentration of the ore

Typical ores contain many different minerals. Some are valuable, but others (SiO_2, Al_2O_3 etc) have no value. The different minerals have different **densities**, and this property allows them to be separated by a process called **oil flotation**. The ore is crushed finely so that as far as possible, individual particles will be composed of single minerals. The crushed ore is vigorously agitated with air and water, to which has been added oil and a small amount of a surfactant (a surfactant is a substance which lowers the surface tension of water; soaps and the active cleaners in detergents are examples of surfactants). The composition of the water/oil/surfactant phase is adjusted so that the less dense particles are carried by surface tension at the air/liquid interface, and the denser particles are allowed to sink. The frothy air/oil/water mixture is then allowed to separate out in a second chamber, whereupon the lighter of the solid particles now sink to the bottom. Refinements to the technology permit more than one such separation to be carried out, so that a number of fractions of different densities can be obtained. Each fraction, of different density, represents concentration of a different mineral phase.

The traditional process at Sudbury has involved roasting the separated sulfide ores, giving a nickel stream and an iron ore recovery stream. Sulfur dioxide results from roasting the concentrated ore.

$$2\ NiS(s) + 3\ O_2(g) \rightarrow 2\ NiO(s) + 2\ SO_2(g)$$
$$4\ FeS_2(s) + 11\ O_2(g) \rightarrow 2\ Fe_2O_3(s) + 8\ SO_2(g)$$

The nickel oxide is reduced to the metal with coke, and the iron oxide shipped off site for iron production. The process is presently being redesigned because of both the poor economics of producing iron this way, and the need to avoid producing so much SO_2. These efforts to reduce SO_2 emissions focus on reducing the amount of pyrite (FeS_2) going to the smelter, by increasing the efficiency of obtaining the nickel concentrate by flotation. The objective is to eliminate roasting this sulfur, and to return it to the ground. A second change, planned for 1994, is to roast the remaining CuS-NiS concentrate with pure oxygen, rather than with air. As seen in the next paragraphs, this change facilitates the subsequent recovery of SO_2, in order to meet the lower SO_2 emission objectives. The use of O_2 rather than air also has the side benefit of reducing NO_x emissions. The capital cost of these improvements is expected to be almost $500 million.

Sudbury "superstack"

(ii) Capture of SO_2

In the Sudbury operation, some sulfur dioxide is recovered as liquid SO_2 for resale and some is oxidized to sulfuric acid, which is shipped off site for use in the production of fertilizers (Chapters 4 and 8). Over 1 million tonnes of sulfuric acid and 100,000 tonnes of liquid SO_2 are shipped from Sudbury every year. A slightly different approach is taken by Cominco, operator of a large lead and zinc smelter in Western Canada, which also produces sulfuric acid, but converts most of it directly into ammonium sulfate for use as fertilizer. This is more economical in Western Canada because of the availability of natural gas at low cost: recall from Chapter 4 that natural gas is the chief chemical starting material in the production of ammonia.

The chemical reactions involved in the conversion of SO_2 to H_2SO_4 are shown below.

$$SO_2(g) + \frac{1}{2}\,O_2 \xrightarrow{V_2O_5\text{, catalyst, }450°C} SO_3(g)$$

$$SO_3(g) \xrightarrow{\text{dissolve in conc. }H_2SO_4} \underset{\text{oleum}}{H_2SO_4.SO_3(\ell)} \xrightarrow{H_2O} H_2SO_4$$

The exothermic conversion of SO_2 to SO_3, $\Delta H° = -99$ kJ mol^{-1}, is carried out at moderate temperature: a compromise between maximizing the equilibrium concentration of SO_3 (low temperature, Le Chatelier's Principle) and attaining equilibrium rapidly (favoured by high temperature): recall Section 8.7. The substitution of oxygen for air in the roasting process will enhance the conversion of SO_2 to SO_3, by providing a "higher strength" SO_2 gas stream. This is seen from the following arguments.

1. The equilibrium constant for the oxidation of SO_2 is given below:

$$K_p = \frac{p(SO_3)}{\{p(SO_2)p(O_2)^{\frac{1}{2}}\}}$$

If air is used instead of oxygen, the reactants and product will all be diluted fivefold, because of the 4 mol N_2 that accompany every mol O_2. The efficiency of SO_3 production is given by $p(SO_3)/p(SO_2)$.

$$p(SO_3)/p(SO_2) = K_p p(O_2)^{\frac{1}{2}}$$

Since $p(SO_3)/p(SO_2)$ is proportional to $p(O_2)^{\frac{1}{2}}$, the fivefold higher $p(O_2)$ increases the equilibrium value of $p(SO_3)/p(SO_2)$ by a factor of 2.3.

2. In addition, there is a kinetic advantage to providing a higher strength gas stream because the rate of reaction is proportional to $p(SO_2)$.

Other environmental considerations require the conversion of SO_3 to H_2SO_4 to be carried out in two steps: see also Section 4.2. The SO_3 stream is dissolved in concentrated sulfuric acid to give "oleum"; this reaction occurs rapidly and efficiently. Water is then added to produce the conventional "concentrated" sulfuric acid (93% H_2SO_4 in this case). Recall from Chapter 4 that the dissolution of SO_3 in water to give sulfuric acid directly is rather slow, and it is difficult to ensure that all the SO_3 is trapped by the water. The two-step approach is taken for environmental reasons: to prevent the formation of a corrosive fog of sulfuric acid droplets around the plant when any "escaped" SO_3 reacted with atmospheric moisture.

The attraction of sulfuric acid manufacture is that it converts a polluting waste product (SO_2) into a useful commodity (H_2SO_4). In practice, the economics of the process are very unfavourable, because the capital costs of the sulfuric acid manufacturing part of the operation are very expensive (more than \$100 million for a large plant), and because the selling price of smelter grade sulfuric acid is very low, currently about \$20-50 per tonne if bought in bulk. Smelter grade sulfuric acid must compete with the purer "virgin" sulfuric acid, which is manufactured from elemental sulfur and which sells in bulk for only \$60-100 per tonne. Elemental sulfur is very cheaply available, both from underground deposits in Texas and as a byproduct of the "sweetening" of sour natural gas in the Western Canadian gasfields (Chapter 4).

(iii) Dilution

The environmental impact of acidic emissions can also be reduced by dilution (dispersion of the stack gases after they leave the plant). A 400 m high "superstack" is used at Inco's Sudbury operation in order to release waste gases at high elevation; this has greatly protected the vegetation in the immediate locality of Sudbury. However, tall stacks are not a pollution reduction measure; they merely disperse the pollution over a wider area, without reducing the total mass of pollutant (SO_2 in this case) released.

13.6.2 Coal-burning power plants

Coal, as already mentioned, contains up to 2 to 3% sulfur by mass. Consider the combustion of 1 tonne of a coal containing 3% of sulfur, using the exactly stoichiometric quantity of air. The amount of carbon in 1 t of coal is about 80,000 mol (10^6 g/12 g mol $^{-1}$), which will react with 80,000 mol O_2 to produce 80,000 moles of CO_2. Also present in the stack gases will be 4 x 80,000 = 320,000 moles of N_2, since every 1 mol O_2 in air is accompanied by 4 mol N_2. Correspondingly, up to 1000 mol S (0.03 x 10^6 g/32 g mol $^{-1}$) will be burned, affording 1000 mol of SO_2, which would be emitted to the environment.

The 1000 mol of SO_2 therefore represents less than 1% of the total volume of gases released to the environment; this is too little for conversion of the SO_2 to sulfuric acid to be economical. Consequently, abatement measures have been directed towards dilution (tall stacks), and to preventing emission. The options are *(i)* switch to low sulfur coal, *(ii)* remove sulfur from the coal, *(iii)* chemical removal of SO_2 from the effluent gases.

As noted already, low sulfur coal tends to be found far from the sites of traditional industries. Transportation costs mitigate against switching, even though some coals have sulfur content as low as 0.3%. Removing sulfur from coal is often called **coal cleaning**. Coal cleaning is usually cheaper than removing SO_2 after combustion. The technology for coal cleaning involves oil flotation, the same technology as discussed in the previous section. As in the case of concentration of metal ores, coal cleaning depends on the fact that most of the sulfur in coal occurs as pyrite, FeS_2, which has a higher density (4.5 g cm $^{-3}$) from that of coal (2.3 g cm $^{-3}$). Although oil flotation removes pyrite efficiently, it does not remove any sulfur which is "organically bound", that is, covalently bonded as part of the structure of coal.

SO_2 emissions can be minimized by capturing the SO_2 before it is released from the stack. Since SO_2 is an acidic gas, the method of choice is to trap it with base. Industrially, base does not mean the NaOH or KOH which we might choose in the laboratory: they are too expensive. Lime, $Ca(OH)_2$, and limestone, $CaCO_3$, are the most commonly used. Most often, the stack gases are "scrubbed" with base by passing a fine spray of the lime or limestone slurry in water down the stack as the hot combustion gases are passing upwards. This process is called "Flue Gas Desulfurization".

$$SO_2(g) + Ca(OH)_2(aq) \rightarrow CaSO_3(s) + H_2O(\ell)$$
$$SO_2(g) + CaCO_3(s) \rightarrow CaSO_3(s) + CO_2(g)$$

Scrubbing equipment is expensive to install and maintain, and the cost of the large amount of lime or limestone is considerable. Not the least of the problems is the disposal of the thousands of tonnes per year of the $CaSO_3$ byproduct, which is obtained as an aqueous slurry, at least 98 percent water. This must be "dewatered" in huge holding tanks and eventually disposed of on the land: it represents a serious environmental problem itself. For all these reasons, electric power utilities tend not to install scrubbers unless they are required to do so through legislation.

Fluidized bed combustion with admixture of limestone is an alternative to scrubbing. Fluidized bed combustion involves burning finely ground coal on a screen through which air is passed from underneath, keeping the coal particles suspended until they burn. The small size of the particles increases the rate and efficiency of combustion. If pulverized limestone is added to the finely ground coal, the SO_2 formed through combustion reacts with the limestone (or the CaO produced from it at the high temperature) and is converted to a mixture of $CaSO_3$ and $CaSO_4$. Once again, the disposal of the byproduct is a difficulty, since it has no commercial value.

13.7 Political and economic considerations

We saw in Section 13.4 that an acidic plume may deposit SO_2 and SO_3 over thousands of km^2, and hence that acidic emissions can, and do, cross national boundaries. The United States complains about the pollution from copper smelters in Mexico, Canada complains about pollution from power generating plants in the Ohio Valley and the U.S. Midwest, and Sweden complains about emissions in Britain and West Germany. The costs of acidic precipitation, in terms of structural, ecological and environmental damage, run to many billions of dollars each year. In the U.S. Northeast alone the annual cost of acidic emissions has been estimated at 5 **billion** dollars, made up as follows.

• Forests	\$1.75 x 10^9
• Agriculture	\$1.00
• Corrosion of buildings, bridges etc.	\$2.00
• Tourism and fishing	\$0.25
Total	\$5.0 x 10^9

Unfortunately, SO_2 reduction strategies are themselves very expensive. In the U.S. it has been estimated that reducing SO_2 emissions by power utilities by 70% could cost \$10 billion or more, which could increase electricity prices by over 20%. Because of the transboundary nature of much of the pollution, it is politically difficult to persuade SO_2-producing jurisdictions to pay for emission reductions which largely benefit others.

The 1990 U.S. Clean Air Act is a very significant piece of legislation in the North American context, following many years during which the governments of Canada and the Northeastern U.S. states had pressed the U.S. federal government for legislation to limit acidic emissions. The Act calls for an overall reduction of both NO_x and SO_2. SO_2 emissions are to be cut from 17 million tons to 7 million tons, with the following targets for coal fired power stations: 1995: 1 kg per 10^9 J; 2000: 0.5 kg per 10^9 J. Many other governments have also established targets for reducing SO_2 emissions, usually from mid-1980s baseline levels; these include Ontario (50% reduction by 1994 by the province's four major polluting industries), France and Germany (50% each), and Sweden (60%). The U.S. legislation proposes the introduction of transferable vouchers as a means of encouraging the reduction of acidic emissions. Power plants, for example, which reduce their acidic emissions below target levels may be allowed to sell their "unused pollution" to companies which have failed to meet their targets. This proposal is very controversial, since it may be argued that the voucher system offers a licence to pollute. Nevertheless the idea is being seriously considered by other governments as a means of encouraging emission reduction.

Problems

Sections 13.1 - 13.3

1. a) Calculate the mass of SO_2 that is formed concurrently with the production of 1.0 t of Zn from ZnS.

   b) If all this SO_2 escaped, what volume of air (at STP) would be contaminated to the extent of 1.0 ppmv?

   c) If alternatively, all the SO_2 were trapped with lime, and oxidized to $CaSO_4$, what mass of $CaSO_4$ would be produced?

2. a) NO_2 is slightly soluble in water, $K_H = 6.4 \times 10^{-3}$ mol L^{-1} atm $^{-1}$ at 25°C. Calculate the concentration of NO_2 at equilibrium with raindrops, if the air contains 2.0 ppmv of NO_2(g).

   b) Is NO_2 likely to be scavenged efficiently from the air by rainfall?

   c) If the equilibrium concentration of NO_2(aq) were converted completely to $HNO_2 + HNO_3$, what would be the pH of the rain in contact with 2.0 ppmv of NO_2(g)?

$$2NO_2(aq) \rightarrow HNO_2(aq) + HNO_3(aq)$$

3. It has been suggested that p(CO_2) may reach 600 ppmv within a century. What would be the pH of rainwater in equilibrium with 600 ppmv of CO_2(g)?

4. Calculate the pH when air containing 1.0 ppmv of SO_2 equilibrates with water at 25°C under the following conditions:

   a) 1.00 L of air and 50.0 cm^3 of water come to equilibrium in a closed vessel

   b) mist equilibrates with an unlimited supply of air.

5. Calculate the pH of rainwater in equilibrium with 0.25 ppmv acidic emissions if the acidic emission is in the form of SO_2 (g).

6. A power plant burns 10,000 t of coal per day. Calculate the mass of acidic gas emission if

   a) the coal contains 2.63% S by mass and the stack gases also contain 15 ppmv of NO_x (take the molar mass of NO_x as 38 g mol $^{-1}$ on average).

   b) Calculate the reduction in emissions if the plant changes to cleaned coal with a sulfur content of 0.36%

Section 13.4

1. The reaction SO_2 (g) + OH (g) $\rightarrow$ HSO_3 (g) has $k = 9 \times 10^{-13}$ cm^3 $molec^{-1}$ s^{-1} at 295 K.

   a) Calculate the half life of SO_2 (g) if this is the only sink and [OH] = 1×10^{7} molec cm^{-3}

   b) Calculate the rate of reaction in molec cm^{-3} s^{-1} under conditions where p(SO_2) = 0.26 ppmv.

2. Consider two competing mechanisms for oxidizing SO_2.

   SO_2 (g) + OH (g) $\rightarrow$ HSO_3 (g) $\quad k_1 = 9 \times 10^{-13}$ cm^3 molec^{-1} s^{-1}

   SO_2 (aq) + H_2O_2 (aq) $\rightarrow$ H_2SO_4 (aq) $\quad k_2 = 1 \times 10^3$ L mol^{-1} s^{-1}

   Given $K_H(SO_2) = 1.2$ mol L^{-1} atm^{-1} and $K_H(H_2O_2) = 1 \times 10^5$ mol L^{-1} atm^{-1}, calculate the relative reaction rates under the following conditions: [OH, g] = 5×10^6 molec cm^{-3}; p(H_2O_2 (g)) = 1.0 ppbv; p(SO_2 (g)) = 1.0 ppmv; amount of liquid water in the atmosphere = 0.010 g L^{-1}.

3. Under very humid conditions, SO_2 oxidation has been observed to occur at rates up to 30% per hour. Under these conditions the following is the major reaction pathway.

   SO_2(aq) + H_2O_2 (aq) $\rightarrow$ H_2SO_4 (aq) $\quad k = 1 \times 10^3$ L mol^{-1}s^{-1}

   Assuming that SO_2 (g) and SO_2 (aq) are at all times equilibrated, what does this suggest about the concentration of H_2O_2 (aq) under these conditions?

4. a) Calculate the half-life of NO_2 (g) in the atmosphere if [OH] = 2.4×10^6 molec cm^{-3}.

   NO_2 (g) + OH (g) $\rightarrow$ HNO_3 (g) $\quad k = 2.0 \times 10^{-11}$ cm^3 molec^{-1}s^{-1}

   b) If, in a laboratory experiment, [OH] is maintained constant at 2.4×10^6 molec cm^{-3} and 10.0 L of air initially by containing 3.5 ppmv of NO_2 is maintained in contact with 0.010 L of water, calculate the pH of the water after 4.5 h.

Section 13.5

1. The reaction below represents the overall reaction for "sulfation" of limestone.

   $CaCO_3$ (s) + H_2SO_4 (aq) $\rightleftharpoons$ $CaSO_4$ (s) + H_2O (ℓ) + CO_2 (g)

   a) Calculate the equilibrium constant for this reaction, assuming that H_2SO_4 is fully dissociated.

   b) If the pH of the aqueous phase is 5.0 and p(CO_2 (g)) = 350 ppmv, is the system at equilibrium when [SO_4^{2-}] = 1.0 mmol L^{-1}?

a

b

c

d

Damage caused by acid rain and its companions, acid snow and acid fog. See Section 13.5, page 382-4.

a Eroded statue, Salisbury Cathedral, England. Fine stone carvings are at considerable risk of damage as the outer layers of the stone flake off.

b A biologically 'dead' lake, near Goteborg, Sweden. Acidic precipitation is a particular problem in areas of granitic rock. Lakes in these areas are poorly buffered, because of the insolubility of the underlying rock.

c,d Tree crown damage, and dead trees, in the Erzgebirge Mountains (Germany/Czech Republic). The heavy use of high-sulfur lignite coals in this region has hit German forests especially hard.

Top, Swedish experiments on reviving acidified lakes use limestone, $CaCO_3$, to neutralize the excess acid. The limestone is sprayed as a powder. The method is costly - Sweden has been spending 100 million Kroner per year (about 25 million US dollars) on lake liming. Despite this considerable expense, fewer than a third of the threatened lakes have been treated. See Section 13.5, page 387.

Bottom, the Smoking Hills, Northwest Territories, Canada. These shales have been burning for hundreds, maybe thousands, of years. Acidic emissions have caused the pH levels of nearby ponds to drop as low as 1.8 See Section 13.5.6, page 390.

2. The aqueous chemistry of aluminum may be oversimplified as follows:

$$Al(OH)_3\ (s) \rightleftharpoons Al^{3+}\ (aq) + 3OH^-\ (aq) \quad K_{sp} = 1.3 \times 10^{-33}\ (mol\ L^{-1})^4$$

$$Al^{3+}\ (aq) + H_2O\ (\ell) \rightleftharpoons AlOH^{2+}\ (aq) + H^+\ (aq) \qquad K_a = 1.0 \times 10^{-5}\ mol\ L^{-1}$$

a) Calculate the total concentration of soluble aluminum (Al^{3+} + $AlOH^{2+}$) in equilibrium with $Al(OH)_3$ at pH 5.28

b) Show, by calculation, why precipitation of $Al(OH)_3(s)$ can occur on the gills of a fish (gill pH = 7.4) which swims in a lake whose pH is 5.28.

3. a) $AlPO_4$ (s) has $K_{sp} = 1.0 \times 10^{-21}\ (mol\ L^{-1})^2$. Using the data of problem 3, and K_a data for H_3PO_4, calculate whether 100 ppm of Al^{3+} would be sufficient to precipitate 90% of the phosphate from a solution initially containing 1.0×10^{-4} mol L^{-1} total phosphate at pH 6.00.

b) Explain the significance of this result by reference to Section 12.3.9.

4. a) A 1.0 L sample of lake water is titrated against 1.05×10^{-3} mol L^{-1} HCl to a methyl orange endpoint. The volume of HCl required is 8.48 cm^3. What is the total alkalinity of the lake in mol H^+ per liter?

b) Given [CO_2, aq] = 1.0×10^{-5} mol L^{-1} and pH = 6.33, what are the concentrations of HCO_3^- and CO_3^{2-} in the above lake?

c) Is the above lake well buffered or poorly buffered? Could this lake be located close to your home? Explain.

d) The lake contains 2.0×10^6 m^3 of water. During the spring runoff 4.2×10^4 m^3 of water having pH 4.15 (acidity assumed to be entirely present as H^+ (aq)) are added to the lake in one day. What is the pH of the lake at the end of the day? (Remember: initial pH was 6.33 and assume [CO_2 (aq)] is constant at 1.0×10^{-5} mol L^{-1}).

5. The equilibrium constant for the association of F^- (aq) with aluminum is given below

$$Al^{3+}\ (aq) + F^-\ (aq) \rightleftharpoons AlF^{2+}(aq) \qquad K_c = 2.5 \times 10^6\ L\ mol^{-1}$$

Recalculate the concentration of total dissolved aluminum at pH 5.28 (Problem 2) if the water contains 1.0 ppm of fluoride ion.

6. Mercury oxide has solubility 5.3 ppm. Its dissolution in water can be represented approximately as the reaction below.

   $HgO\ (s) + H_2O\ (\ell) \rightleftharpoons Hg^{2+}\ (aq) + 2OH^-\ (aq)$

   a) Calculate K_c for this reaction.

   b) Calculate the solubility of mercury (Hg^{2+} (aq)) in water at pH 6.0 and pH 8.0

   c) Suggest an environmental significance to this calculation.

7. Iron occurs as a carbonate mineral $FeCO_3$, which has $K_{sp} = 3.1 \times 10^{-11}$ $(mol\ L^{-1})^2$.

   a) Calculate the concentration of iron (in ppm) in well water drawn from a limestone area in which the $CaCO_3$ also contains $FeCO_3$. Take c(Ca^{2+}, aq) = 120 ppm. Hint: use the solubility properties of $CaCO_3$ to estimate $[CO_3^{2-}\ (aq)]$.

   b) Make a suggestion as to what other species besides $FeCO_3$ (s), Fe^{2+} (aq) and CO_3^{2-} (aq) might be present.

8. Magnesium hydroxide has $K_{sp} = 5.7 \times 10^{-12}$ $(mol\ L^{-1})^3$.

   a) Estimate the concentration of Mg^{2+} (aq) in sea water (pH 8.1) assuming that sea water is saturated with respect to magnesium.

   b) Compare your answer from part a) with the experimental value in Table 10.2. Can you suggest any reasons for the discrepancy?

9. $Al(OH)_3$ dissolves in dilute sodium hydroxide solution

   $Al(OH)_3\ (s) + OH^-\ (aq) \quad Al(OH)_4^-\ (aq) \qquad K_c = 0.89$

   Calculate the solubility of aluminum in ppm at pH 8.00 and pH 10.00.

Section 13.6

1. A lake containing $5.2 \times 10^6\ m^3$ of water is at pH 4.80. What mass of limestone needs to be sprayed onto the surface of this lake to bring the pH to 5.50, if all the acidity is in the form of free H^+ (aq)?

2. At 450°C the reaction

   $SO_2\ (g) + ½O_2\ (g) \rightarrow SO_3\ (g)$

   has $K_p = 24\ atm^{-½}$. SO_2 (initial pressure 2.0 atm) and air (initial pressure 20 atm) are passed over a catalyst at 450°C. Under these conditions 97% of the SO_2 is converted to SO_3. Did the reaction reach equilibrium?

3. Calculate the equilibrium value of p(SO_3 (g)) under the conditions of Problem 2, using initial p(SO_2 (g)) = 2.0 atm, and either air (5.0 atm) or pure oxygen (5.0 atm).

4. a) Calculate the total number of moles of gases produced per day by a coal-burning power station which burns 8,500 t of coal per day.
   b) If the sulfur content of the coal is 2.25% by weight calculate (i) the mass of SO_2 formed per day (ii) the concentration of SO_2 (g) in ppmv in the gases leaving the stack.

5. a) Calculate the mass of lime needed per day to react with 95% of all the SO_2 produced under the conditions of Problem 4.
   b) What mass of $CaSO_4$ is produced per year if the stoichiometry follows the equation below?
   $Ca(OH)_2 + SO_2 + ½O_2 \rightarrow CaSO_4$

6. In 1985, emissions of SO_2 and NO_x in southern Ontario were 4.95×10^5 t and 4.42×10^5 t respectively. How many moles of H^+ correspond to the total of these emissions over southern Ontario?

7. Jeffries (1991) measured the ratio SO_4^{2-}: HCO_3^- in lakes in central Ontario. Lakes having $[SO_4^{2-}]/[HCO_3^-] > 1$ were considered to be seriously acidified. Explain how this conclusion was reached.

8. The 1995 U.S. target for SO_2 emissions from power stations is 1 kg SO_2 per 10^9J. Express this in terms of tonnes of SO_2 per kWh of electricity.

14 ENERGETICS IN CHEMICAL REACTIONS II: FREE ENERGY CHANGES

14.1 The Gibbs-Helmholtz equation

In Chapter 5, we considered the enthalpy changes associated with chemical reactions. Recall that the enthalpy change is the heat change at constant temperature and pressure.

We might be tempted to think that the sign of the enthalpy change was the only criterion for deciding whether a physical process or chemical reaction would be thermodynamically favoured. This is definitely not the case, as may be seen by the following examples. Ice melts above 0 °C and water boils above 100 °C, yet both these processes are endothermic ($\Delta H° = +6.0$ and $+44$ kJ mol $^{-1}$ respectively). Salt dissolves in water, and this process is also endothermic ($\Delta H° = +4.0$ kJ mol $^{-1}$).

Chemists and physicists of the nineteenth century came to realize that a second factor, in addition to enthalpy change, must be considered in order to make a judgement on whether a given physical or chemical process would proceed. This second factor is the **entropy change** (ΔS) associated with the process. We shall discuss entropy further in the next section. For now, the important concept to grasp is that thermodynamics include two factors: an enthalpic and an entropic contribution. Together, these combine to give the **free energy** change for the process (ΔG), and this is stated mathematically as the **Gibbs-Helmholtz** relationship, eq. [1].

[1] $\Delta G = \Delta H - T\Delta S$

A thermodynamically favoured process has $\Delta G < 0$, and is called *spontaneous*. A spontaneous process may occur, although kinetic considerations determine whether it will proceed at a measurable rate. However, a non-spontaneous process ($\Delta G > 0$) will definitely not occur; in fact, the process will be spontaneous in the reverse direction. Finally, a system for which $\Delta G = 0$ is at equilibrium, and no further change is anticipated[1].

Definition

Spontaneous processes *have ΔG < 0, non-spontaneous processes have ΔG > 0, and systems at equilibrium have ΔG = 0.*

As we shall see in Section 14.5, the concept of equilibrium is crucial to the theory of free energy. This has two consequences for us. First, we must modify our concept of a reaction that proceeds to completion, and think of it as a reaction with an extremely high (though not infinite) equilibrium constant. Second, free energy change is not directly applicable to photochemical reactions, where the ground state of the absorbing molecules, the excited state, and the photons are not at equilibrium. As noted in Chapter 5, photon energy is considered to be a source of internal energy ΔE, rather than of free energy.

Let's look again at equation [1], realizing that a spontaneous process requires ΔG to be negative.

[1] $\Delta G = \Delta H - T\Delta S$

There are two terms to consider: the enthalpic term and the entropic term. An exothermic process (ΔH negative) contributes to a negative ΔG. However, *the entropic contribution favours spontaneity when the entropy change is positive.* Moreover, since the entropic contribution is the term ($-T.\Delta S$), its contribution to ΔG increases as the temperature rises.

The interplay between ΔG, ΔH, and ΔS results in four possibilities (Table 14.1), given specified concentrations of reactants and products.

Table 14.1: Influence of the signs of ΔH and ΔS on reaction spontaneity

1. ΔH negative; ΔS positive: reaction is spontaneous at all temperatures.
2. ΔH positive; ΔS negative: reaction is non-spontaneous at all temperatures (reverse of (1)).
3. ΔH positive; ΔS positive: reaction is non-spontaneous at low temperature and becomes increasingly spontaneous as the temperature is raised.
4. ΔH negative; ΔS negative: reaction is spontaneous at low temperature and becomes less spontaneous as the temperature is raised (reverse of (3)).

[1] Recall from Section 8.1 however that reactions do not "stop" at equilibrium. Instead, the reaction at equilibrium proceeds at equal rates in the forward and the reverse directions. Therefore, when ΔG = 0, no change in the composition of the reaction mixture occurs, although individual molecules exchange identities between reactants and products.

14.2 Spontaneity

It is important to define exactly what we mean by the term "spontaneous change". A useful definition is that

Definition

*a **spontaneous change** has a natural tendency to occur without being subjected to any external influences.*

Think about the following examples.

• A balloon (or a cylinder of compressed gas) is opened to the atmosphere, and all the gas inside escapes. The reverse change can be brought about, but only if an external influence (a pump) is used to pump gas back into the balloon or cylinder.

• The Niagara River flows downwards over Niagara Falls. In principle, the water could be made to flow upwards, (if a sufficiently large pump were used), but energy would have to be supplied from outside.

• A hot cup of coffee spontaneously cools down until it reaches the temperature of the room. A cold cup of coffee can be warmed up, but only by expending energy from an outside source.

There is an additional point to notice about the three examples just given. In terms of the Law of Conservation of Energy (also known as the First Law of Thermodynamics), there is no reason why these processes could not proceed in the non-spontaneous direction. It is the distinction between spontaneous and non-spontaneous that determines in which direction these processes will actually proceed. Consider water falling over Niagara Falls. This process changes potential energy into kinetic energy. Energy would be conserved if kinetic energy were changed back into potential energy. However, the first process is spontaneous and is observed experimentally, while the second is non-spontaneous and is not observed.

Just as in these physical examples, chemical reactions have a spontaneous tendency to proceed in a definite direction (unless they are already at equilibrium). The reaction may proceed quickly or slowly, depending on the rate constant for the reaction at the given temperature, but the spontaneous tendency is in one particular direction. Outside energy must be supplied in order to reverse the spontaneous tendency of the reaction.

14.3 Entropy

The concept of entropy is intuitively less obvious than that of enthalpy, mainly because everyone has day-to-day experience with heat flow. There are two different ways of approaching entropy, as discussed in the next two sections.

14.3.1 Entropy and disorder

Entropy can be taken as a measure of the disorder of a system. For example, consider the room in which you are sitting. The molecules which make up the air in the room can each occupy any location inside the room. In principle, all of them could be arranged neatly in one corner, with the rest of the room a perfect vacuum. This arrangement, although *possible*, is overwhelmingly improbable, and is never seen in real life. If all the molecules were arranged neatly, they would be highly ordered; another way of saying this is that their entropy would be low. The situation where all the molecules in the air of the room are randomly located in the room is a less orderly (more disordered) state, and correspondingly has higher entropy.

Rule: *States having high entropy are less ordered than states of lower entropy*

A slightly different example is that of two identical flasks connected by a stopcock which is initially closed. One flask contains gaseous nitrogen, the other gaseous oxygen. When the stopcock is opened the gases mix, and eventually their concentrations become equal. Mixing is not driven by an enthalpy change, but because of the increased entropy of the final, mixed state, compared with the initial, unmixed state.

In the nineteenth century, Boltzmann developed the relationship between entropy and order, eq. [2].

[2] $S = k \ln W$

In eq. [2], k is the Boltzmann constant (which is the same as the ideal gas constant, R, only divided by the Avogadro constant; therefore $k = 8.314$ J mol^{-1} K^{-1}/6.022 x 10^{23} mol^{-1}, or 1.38 x 10^{-23} J K^{-1}). W is the number of ways in which the particles in the system can be arranged, yet still have the same total energy. The greater the number of possible equi-energetic arrangements of the particles, the larger their entropy. In the example of the flasks containing gaseous nitrogen and oxygen, there were more equi-energetic arrangements when the gases could move into both flasks, as opposed to when they were each confined to one of them.

Rule: *Entropy increases as the disorder of the system increases.*

A corollary of Boltzmann's equation is that a perfectly ordered crystal at 0 K must have zero entropy[2], since there is only one possible arrangement of all the atoms, ions, or molecules in the crystal (W = 1; lnW = 0). This property of a perfect crystal gives a reference point to the entropy scale. Consequently the entropies of substances can be determined absolutely, unlike enthalpies which, as we saw in Chapter 5, can only be measured as enthalpy *changes*. Tabulated values of entropies, such as the standard molar entropies in Appendix 1, are absolute values; they are not "entropies of formation", unlike tabulations of standard molar enthalpies and free energies of formation.

For a given substance, entropy increases with temperature, as the atoms or molecules move, vibrate and rotate more vigorously. For different substances at the same temperature, entropy increases with molecular complexity, and according to the state of matter.

Example: *One mole of $H_2(g)$ has greater entropy than 1 mol H(g), due to the greater complexity of a molecule, which can vibrate and rotate and thus take up many different internal positional arrangements. (However 1 mol $H_2(g)$ has lower entropy than 2 mol H(g): as discussed later in this section).*

For a given substance, the gaseous state always has the greatest entropy, followed by the liquid, and then the solid. Gaseous molecules (or atoms) are free to explore every part of their container. In a liquid the molecules are confined to only part of the container, but are free to move randomly within the volume of the liquid. In a solid the molecules are most ordered, because they are constrained to occupy fixed positions within a crystal lattice[3].

As a rough generalization, the entropies of substances of similar molecular complexity increase in the following order:

---

2 This is known as the Third Law of Thermodynamics: the entropy of a perfectly ordered crystal is zero at 0 K. In order to derive this result, we must remember that placing the same number of identical atoms or molecules into the same perfect crystal lattice, but in a different order, does not constitute a different arrangement, since the individual atoms or molecules are indistinguishable from each other.

3 This analysis of solids is an oversimplification to the extent that plastics and glasses are generally non-crystalline. Their entropies are higher than those of crystalline solids of similar molecular complexity, but lower than those of true liquids.

- perfect crystal (0 K) ... zero entropy
- disordered solids, and crystals at T > 0 K
- liquids ... much larger entropy than solids
- gases ... much larger entropy than liquids

When the concept of disorder is applied to chemical reactions, it is possible to predict whether an increase or a decrease in the entropy of the system is likely. We have to consider the following points.

- Are there more, or fewer, product species than reactant species?

- Are there more, or fewer, product substances than reactant substances in a more disordered state of matter: (g) *vs.* (ℓ) or (s); (ℓ) *vs.* (s)?

Example: *Predict whether ΔS is positive or negative for the following chemical reactions.*

a) $C_2H_6(g) + 3\frac{1}{2}\ O_2(g) \rightarrow 2\ CO_2(g) + 3\ H_2O(\ell)$

b) $2\ KClO_3(s) \rightarrow 2\ KCl(s) + 3\ O_2(g)$

c) $C(s) + 2\ S(s) \rightarrow CS_2(\ell)$

Answer:

a) There are 2 mol product gases and 4½ mol reactant gases. ΔS is predicted to be negative.

b) There are 3 mol product gases and no reactant gases. ΔS is predicted to be positive.

c) The product is a liquid, and the reactants are solids. ΔS is predicted to be positive.

Another aspect of the extent of disorder is seen when comparing molecules of different complexity. Consider again the difference between 1 mol of H(g) and 0.5 mol of $H_2(g)$, each of which contains 6×10^{23} hydrogen atoms. In the first case the atoms are free to move at random, independently of each other. In the second case, each hydrogen atom is tethered to one other hydrogen atom through a covalent bond, and the two must travel together. There is thus more disorder (higher entropy) in 1 mol of H(g) than in 0.5 mol of $H_2(g)$.

We can now understand why molecules such as H_2 tend to dissociate into atoms at very high temperatures. The enthalpy change for this process is a positive contribution to ΔG (equal to the H-H bond energy), while the entropic contribution (-TΔS) is negative (because the entropy of 2 mol H(g) is greater than that of 1 mol $H_2(g)$). This is case (3) of Table 14.1. However, a very high temperature is needed before the TΔS term approaches the ΔH term in magnitude, because ΔH° for dissociation is so high. Therefore significant quantities of hydrogen atoms are at equilibrium with $H_2(g)$ only at very high temperature.

14.3.2 Entropy and heat

From Boltzmann's equation [2], it is clear that the units of entropy are J K^{-1}, and correspondingly, the **molar entropy** of any substance has the units J mol^{-1} K^{-1}. This is also consistent with eq. [1].

[1] ΔG (J or kJ mol^{-1}) = ΔH (J or kJ mol^{-1}) - $T\Delta S$ (K x J mol^{-1} K^{-1})

Entropy is therefore closely related to work and energy. For example, if air escapes from a cylinder under pressure, it can do useful work, and simultaneously its entropy increases. It can also be pumped back into the cylinder, which requires work to be done on the gas, which in turn requires the expenditure of energy, and simultaneously its entropy decreases.

Both dimensionally and in physical terms, entropy is closely related to heat. This can be seen most easily in the example of a solid melting to give a liquid, for example ice → water. At the melting point T_m of ice, both solid and liquid coexist at equilibrium and at this temperature $\Delta G = 0$. Therefore the enthalpy change and the entropy change can be related at this particular temperature using eq. [1]: $\Delta H = T_m\Delta S$, or equivalently, $\Delta S = \Delta H/T_m$. We will explore this relationship more fully in Section 14.4, after the concept of a standard molar free energy change has been introduced.

In the previous section it was noted that the entropies of substances increase with temperature. This is because the addition of heat raises the entropy (disorder) of the substance by increasing the thermal motions of the molecules or ions. Thus S(NaCl (s)) at 298 K > S(NaCl (s)) at 100 K; S(H_2O (ℓ)) at 85 °C > S(H_2O (ℓ)) at 25 °C. The entropy of a perfect crystal is zero at 0 K, because at 0 K all thermal motion stops, and the crystal attains its maximum degree of order[4].

Entropy change can be related quantitatively to heat. For a system at equilibrium at a given temperature, the addition of a small amount of heat (q) increases the entropy of the system by a definite amount.

$q = T\Delta S$, since the system is at equilibrium and $\Delta G = 0$

Note that the equation $q = T\Delta S$ does not calculate the temperature change ΔT, but the entropy change ΔS; the heat input is assumed to be small enough that the temperature does not change significantly. Since $\Delta S = q/T$, the entropy change for a given heat input becomes less as the temperature is raised. Thus although heat input has the effect of increasing the entropy of the system,

---

[4] Not all crystals are perfectly ordered, even at 0 K. For example, a crystal of CO retains disorder down to the lowest temperatures achievable, because some of the molecules align themselves C-O in the crystal, while others are aligned O-C. Many other more complicated examples of this phenomenon are known.

the influence of a given amount of heat input (q) on a particular system depends on the temperature. If the temperature is low, the components of the system have relatively low entropies, and a given input of heat may produce a sizable increase in ΔS. At higher temperatures where more disorder already exists, the same input of heat causes a smaller increase in ΔS.

Example: *Calculate the entropy change at 298 K and at 400 °C when 1 mol CO(g) is heated by a heat input of 1.0 J mol^{-1}.*
Answer: *ΔS = q/T*
At 298 K, ΔS = 1.0 J mol^{-1}/298 K = 3.4 x 10^{-3} J mol^{-1} K^{-1}
At 400 °C (763) K, ΔS = 1.0 J mol^{-1}/673 K = 1.5 x 10^{-3} J mol^{-1} K^{-1}

14.3.3 The Second Law of Thermodynamics

The Second Law of Thermodynamics is perhaps one of the most insightful descriptions of the properties of the physical universe. It has been stated in many ways, of which the following is a very useful one.

Law: *The* ***total*** *entropy change for a spontaneous process is always positive*

At first sight, this definition seems inconsistent with what we have said about spontaneity in Section 14.1. In case (4) of Table 14.1 it was stated that at certain temperatures a reaction could be spontaneous even though ΔS was negative. Indeed, the freezing of a liquid when $T < T_m$ is just such a situation.

In order to reconcile the previous paragraph with the statement of the Second Law, we must note the word ***total*** as it applies to the entropy change. From time to time in previous paragraphs, we have used the expression "the entropy of the system" — that is, the entropies of the reactants and products of the process under investigation. In laboratory terms, the "system" could be taken to be the reaction vessel inside which the reaction takes place. However, the ***total*** entropy change is not just the difference in entropy between reactants and products; it includes the entropy change in the surroundings (and the surroundings theoretically include the whole universe!). What does this mean?

Let's pursue our example of Table 14.1, case (4), where ΔH and ΔS are both negative. Remember that ΔH and ΔS both refer to the *system*. As the reaction progresses, heat is liberated (ΔH is negative) and flows into the surroundings. In fact, since the reaction takes place at constant temperature and pressure (definition of enthalpy, Chapter 5), **all** the heat released is eventually transferred to the surroundings. This transfer of heat increases the entropy of the surroundings by stimulating additional motion, and hence disorder, in the molecules which make up the surroundings. Thus we can write the following relationships. The subscript "reaction" means the same as "system".

(a) $\Delta S_{total} = \Delta S_{system} + \Delta S_{surroundings}$
(b) $-\Delta H_{reaction} = q_{surroundings} = T\Delta S_{surroundings}$

Equation (b) is valid because the system and the surroundings reach thermal equilibrium. Substitution gives eq. [3].

$$[3]\ \Delta S_{total} = \Delta S_{system} + \Delta S_{surroundings}$$
$$= \Delta S_{reaction} - \Delta H_{reaction} / T$$

Notice that equation [3] can be directly related to the Gibbs-Helmholtz equation [1], since the right hand side of eq.[1] is equal to $- T.\Delta S_{total}$.

$$[1]\ \Delta G_{reaction} = \Delta H_{reaction} - T\Delta S_{reaction}$$
$$\Delta G_{reaction} = -T\Delta S_{total} = -T(\Delta S_{reaction} + \Delta S_{surroundings})$$

Several consequences follow from the Second Law.

1. The Second Law predicts that there is **no** Law of Conservation of Entropy. Quite the reverse; for spontaneous processes the total entropy always increases. Ordered systems (for example, living organisms) are possible only at the expense of increasing the disorder of the universe as a whole. Organisms do this by metabolizing nutrients; the heat released as a byproduct of metabolism increases the entropy of the surroundings sufficiently that ΔS_{total} increases. Compare energy with entropy.

Rule: ***Energy:*** *Energy is neither created nor destroyed; it is simply converted from one form into another (First Law, when work is considered as a form of energy)*

Entropy: *The entropy of the universe is continuously increasing (another statement of the Second Law)*

2. Not all forms of energy are equivalent. In particular, heat is a low grade form of energy, since its main effect is to increase entropy. It is predicted that eventually the universe will wind down, as all its fuel is consumed and degraded into heat. Once the temperature of the universe becomes equal everywhere, no further useful work will be possible.

3. These last two paragraphs have tremendous environmental significance, not in terms of the ultimate fate of the universe, but of energy considerations today. Most useful machines are examples of what thermodynamicists call "heat engines": these devices include automobiles and other transportation devices, and electricity generating plants (other than hydroelectric plants). Heat engines all produce useful work (which may subsequently be converted into electrical energy) by burning a fuel to produce heat, and then taking advantage of a **difference** in temperature -- for example, the temperature differential between the interior of the car engine and the outdoors. In a power station, the heat released is used to convert liquid water into steam at high temperature and pressure; the steam performs useful work by driving the turbines of an electric generator. In a car, the heat released causes the combustion products to be formed at high temperature and pressure; the gaseous combustion products perform useful work by moving pistons inside the engine, the motion of which is converted to motion at the vehicle's wheels.

Quite apart from mechanical and frictional losses of energy inherent in machinery, there is a fundamental restriction on the efficiency of any heat engine.

$$\text{\% efficiency} = \frac{\text{(temperature difference between engine and surroundings)}}{\text{temperature of the engine}}$$

Engine efficiency is therefore promoted by operating at as high a temperature as possible since the temperature of the surroundings is generally fixed. We saw one consequence of this in Chapter 9, where the desire to limit NO_x formation (an endothermic reaction favoured by high temperature) is at odds with the objective of maximum engine efficiency.

It is thus inevitable that only a limited fraction of all the chemical energy present in a fuel can be converted to useful work using a heat engine. Put another way, only a limited fraction of the energy is "free" energy, usable energy; the rest is wasted as heat, which simply increases the entropy of the surroundings. The reason for the continued interest in fuel cells (Chapter 16) is that they convert chemical energy directly into electricity, without the wasteful intermediate step of burning the fuel to produce heat. Much higher efficiencies are theoretically possible using fuel cells than heat engines.

14.4 Standard molar free energy functions

14.4.1 Standard molar free energy of formation

The standard molar free energy of formation of any substance is defined analogously to its standard molar enthalpy of formation (Section 5.3). It has the symbol $\Delta G°_f$.

Definition

*The **standard molar free energy of formation** of any substance at 25 °C is the free energy change when 1 mol of the substance in its standard state is formed from its elements, and the elements are in their usual physical states at 25 °C and 1 atm.*

The term "standard state" refers to a pure solid or a pure liquid. If the substance is a gas, the standard state refers to a pressure of 1 atm. For solutions, the standard state refers to a concentration of 1 mol L^{-1}. The issue of standard state is discussed further in Section 14.5.

In principle, $\Delta G°_f$ can be defined at any temperature. However almost all tabulations list values of $\Delta G°_f$ appropriate to 25 °C, and that practice is followed in Appendix 1 of this book.

By analogy with the standard molar enthalpy of formation, we can make the following points about $\Delta G°_f$.

- The value of $\Delta G°_f$ is zero for all elements which are in their usual physical states at 25 °C and 1 atm.

- The value of $\Delta G°_f$ for the hydrogen ion (H^+ (aq)) is set to zero by convention, and this allows the values of $\Delta G°_f$ for all other aqueous ions to be assigned.

- As with standard molar enthalpy of formation, it is most important to be careful to look up the correct species (*e.g.*, $Br_2(\ell)$ *vs.* Br_2(g); OH(g) *vs.* OH^- (aq), etc).

14.4.2 Standard molar free energy change

Arguments identical to those in Section 5.4 can be used to deduce eq. [4].

[4] $\Delta G°(\text{reaction}) = \sum \Delta G°_f(\text{products}) - \sum \Delta G°_f(\text{reactants})$

The use of this equation is *restricted to reactions taking place at or near 298 K.* This restriction is because tabulated free energies of formation are invariably quoted for the temperature 298 K.

This is an important difference between $\Delta G°_f$ and $\Delta H°_f$. In the case of enthalpy change, the variation of $\Delta H°$ with temperature is usually small, and is generally neglected except in the most precise work. It is therefore possible to calculate enthalpy changes for reactions occurring at temperatures other than 298 K, using tabulated values of $\Delta H°_f$, without introducing serious error into the calculation. By contrast, tabulated values of $\Delta G°_f$ can **only** be used for processes occurring at 298 K (or very close to 298 K). This is because $\Delta G°$ contains the contribution ($-T\Delta S°$), which depends directly upon T.

Example: *Calculate the standard molar free energy change for the reaction below, taking place at 298 K.*

$$CH_4(g) + 2\,O_2(g) \rightarrow CO_2(g) + 2\,H_2O(\ell)$$

Answer:

From eq. [4]:

$$\Delta G°(\text{reaction}) = \sum \Delta G°_f(\text{products}) - \sum \Delta G°_f(\text{reactants})$$

$$\Delta G°(\text{rxn}) = \Delta G°_f(CO_2\,(g)) + 2\,\Delta G°_f(H_2O\,(\ell)) - \Delta G°_f(CH_4\,(g)) + 2\,\Delta G°_f(O_2\,(g))$$

Substitute values from Appendix 1:

$$\Delta G°(\text{rxn}) = (-394.359) + 2(-237.178) - (-50.75) - 2(\text{zero})$$

$$\Delta G°(\text{rxn}) = -817.97\ \text{kJ mol}^{-1}$$

Note particularly the units of the answer: kJ mol^{-1}. In this context "per mole" means per mole of the equation as written. In this example it means per 1 mole of methane reacted, per 2 moles of liquid water formed, per 1 mole of carbon dioxide formed, etc.

The foregoing concepts are summarized in Table 14.2.

Table 14.2: Dependence of thermodynamic functions on temperature

Enthalpy change, $\Delta H°$:	Approximately independent of temperature; temperature dependence considered only in more advanced work
Entropy change, $\Delta S°$:	Not strongly dependent upon temperature; temperature dependence considered only in more advanced work
Free energy change, $\Delta G°$:	Varies strongly with temperature. Care must always be taken to use values appropriate to the temperature involved

Again by analogy with the standard molar enthalpy of reaction (Section 5.2) we can make generalizations about $\Delta G°$(reaction).

- The value of $\Delta G°$ for the reverse of a reaction is numerically equal to $\Delta G°$ for the forward reaction but with the sign reversed.

- When a reaction can be carried out by a pathway involving several steps, $\Delta G°$ for the *overall* change is equal to the sum of the $\Delta G°$'s for the individual steps. This statement is analogous to Hess' Law of heat summations, applied to $\Delta H°$.

A different approach must be taken if $\Delta G°$(reaction) is needed for a reaction taking place a temperature other than 298 K (an example is a reaction occurring in the stratosphere, where the temperature is typically 200 - 250 K: see Chapter 15). In this case we use the Gibbs-Helmholtz equation, eq. [1]. Since both $\Delta H°$(reaction) and $\Delta S°$(reaction) are approximately independent of temperature, each can be calculated separately, and then combined to give $\Delta G°$.

Step 1: $\Delta H°(\text{reaction}) = \sum \Delta H°_f(\text{products}) - \sum \Delta H°_f(\text{reactants})$
Step 2: $\Delta S°(\text{reaction}) = \sum S°(\text{products}) - \sum S°(\text{reactants})$
Step 3: $\Delta G°(\text{reaction}) = \Delta H°(\text{reaction}) - T\Delta S°(\text{reaction})$

Example: *Calculate the standard molar free energy change for the reaction below, taking place at 800 K.*

$$CH_4(g) + 2\ O_2(g) \rightarrow CO_2(g) + 2\ H_2O(g)$$

Answer:

Step 1:

$\Delta H°(\text{reaction}) = \sum \Delta H°_f(\text{products}) - \sum \Delta H°_f(\text{reactants})$

$\Delta H°(\text{rxn}) = \Delta H°_f(CO_2\ (g)) + 2\ \Delta H°_f(H_2O\ (g)) - \Delta H°_f(CH_4\ (g)) + 2\ \Delta H°_f(O_2\ (g))$

Substitute values from Appendix 1:

$\Delta H°(\text{rxn}) = (-393.509) + 2(-241.818) - (-74.81) - 2(\text{zero})$
$= -802.34\ \text{kJ mol}^{-1}$

Step 2:

$\Delta S°(\text{reaction}) = \sum \Delta S°(\text{products}) - \sum \Delta S°(\text{reactants})$

$\Delta S°(\text{rxn}) = \Delta S°(CO_2\ (g)) + 2\ \Delta S°(H_2O\ (g)) - \Delta S°(CH_4\ (g)) + 2\ \Delta S°(O_2\ (g))$

Substitute values from Appendix 1:

$\Delta S°(\text{rxn}) = (213.63) + 2(188.716) - (186.155) - 2(205.029)$
$= -51.51\ \text{J mol}^{-1}$

Step 3:

$\Delta G°(\text{reaction}) = \Delta H°(\text{reaction}) - T\Delta S°(\text{reaction})$
$= -802.34\ \text{kJ mol}^{-1} - (800\ \text{K})(-51.51 \times 10^{-3}\ \text{kJ mol}^{-1})$
$= -798.22\ \text{kJ mol}^{-1}$

14.4.3 Phase transitions

At a given pressure, pure substances have fixed melting points and fixed boiling points. This is understandable from eq. [1]:

[1] $\Delta G = \Delta H - T\Delta S$

Consider the melting of a solid, such as ice. In order for melting to be spontaneous, ΔG must be negative. ΔH_{fusion} is always positive (endothermic). Upon melting ΔS is positive, since S(liquid) > S(solid). Above the melting point T_m, $(T\Delta S) > \Delta H$, and the solid melts. Below T_m, $(T\Delta S) < \Delta H$, the spontaneous process is liquid → solid, while above T_m, the spontaneous process is solid → liquid. When $T = T_m$, $(T\Delta S) = \Delta H$, the system is at equilibrium, and both solid and liquid can coexist. Exactly similar arguments apply to the boiling point.

For the case of a pure solid melting to a pure liquid, the "reactants" and "products" are both in their standard states at temperature T_m. Under these conditions ΔH is a standard molar enthalpy change ($\Delta H°$), ΔS is a standard molar enthalpy change ($\Delta S°$), and ΔG is a standard molar free energy change $\Delta G°$.

Example: *The standard molar enthalpy of fusion of ice is +6.0 kJ mol*$^{-1}$*. Calculate the standard molar entropy of fusion of 1 mol of ice at 1 atm pressure.*

Answer: *The melting point of ice at 1 atm pressure is 273 K, and at this temperature* $\Delta G = \Delta G° = 0$ *(solid ice and liquid water can coexist in equilibrium).*

$\Delta G° = 0;\ \Delta H° = +\ T\Delta S°$

$\Delta S° = \Delta H°/T = +\ 6.0 \times 10^3\ J\ mol^{-1}/273\ K = 22\ J\ mol^{-1}\ K^{-1}$

Similar arguments can be applied to boiling of a liquid or sublimation of a solid, but with an important restriction: $\Delta G°$ is a standard molar free energy change only when the gas is formed in its standard state (1 atm pressure). In the case of boiling, for example, this would apply to the "normal" (1 atm) boiling point, which in the case of water is 100 °C[5].

---

[5] For any change of physical state or chemical reaction at equilibrium, it is always true that since $\Delta G = 0$, the ΔH term of eq. [1] can be equated with the $T\Delta S$ term. However, ΔH and ΔS will only be standard molar quantities under special conditions. Therefore such a calculation is only useful in the special case that all substances are present in their standard states.

14.5 ΔG, ΔG°, and the equilibrium constant

14.5.1 Relationship between ΔG° and the equilibrium constant

The intimate relationship between free energy and reversibility makes it not surprising that free energy change should be related to an equilibrium constant. Equation [5] is the form of this relationship.

[5] $\Delta G° = - RT\ln K$

This is a most important association; among other things, it relates the equilibrium constant at a given temperature to a thermodynamic property of the reaction system, namely ΔG°.

Numerous other thermodynamic relationships follow from eq. [5]. Although they cannot be explored in detail in this text, we can at least mention them.

- By combining equations [1] and [5] it is possible to obtain ΔH° and ΔS° for a reaction from the variation of the equilibrium constant with temperature (eq. [6]). A plot is made of ln*K* *vs.* T^{-1}; ΔH° is obtained from the slope of the graph, and ΔS° from the intercept.

$$\Delta H° - T\Delta S° = -RT\ln K$$

[6] $\therefore \ln K = -(\Delta H°/RT) + (\Delta S°/R)$

- Equation [6] provides a method of determining *K* at a new temperature, once *K* is known at an initial temperature. The method assumes ΔH° and ΔS° to be constant. Using subscript (1) for the initial temperature and subscript (2) for the new temperature we obtain eq. [7].

[7] $\ln(K_2/K_1) = -(\Delta H°/R)\{1/T_2 - 1/T_1\}$

Notice the important point that in order to calculate *K* at a new temperature given *K* at a different temperature we only need to know ΔH°; *we do not need to know ΔS°*. This point will be reexamined in Section 14.5.3.

A digression must be made at this point to discuss the nature of the equilibrium constant in eq. [5]. The problem becomes apparent when you notice that the dependent variable is ln*K* which, like all logarithms, has to be dimensionless: in other words, a pure number. Yet the equilibrium constants we have met (K_p, K_c, K_a, K_b, K_w, K_{sp}, K_H) all have units: how can you take their logarithm? We must make the following distinction.

- K_p, K_c, K_a, K_b, K_w, K_{sp}, and K_H are all examples of *experimental* equilibrium constants. They are measured experimentally, and their units (of pressure, concentration etc) reflect the units that were used to measure the pressures or concentrations of the reactants and products of the reaction when it reached equilibrium.

- *K* (notice the italics) is the *thermodynamic* equilibrium constant. It is obtained from fundamental thermodynamic properties of the reactants and products through eq. [5] or [6], and is dimensionless.

The thermodynamic quantity corresponding to the pressure or concentration of each reactant or product is called its **activity** which, like *K*, is a dimensionless quantity. At the level of approximation that we shall use in this book, activity is related to experimental quantities of substances as follows.

- activity of a gas $= \dfrac{\text{actual pressure of the gas in atm}}{\text{standard state of 1 atm}}$

- activity of a solute $= \dfrac{\text{actual concentration of the solute in mol L}^{-1}}{\text{standard state of 1 mol L}^{-1}}$

- activity of a pure solid or pure liquid = 1.000

Note especially points (1) to (3) below (which apply at the level of the approximation used in this book).

1. In precise work, the standard state of a solute is taken as 1 mol kg^{-1} of solvent, because masses of solvent can be measured more precisely than volumes. For aqueous solutions, in which we are most interested in this book 1 mol kg^{-1} and 1 mol L^{-1} are approximately equal at low concentration.

2. The activity of a reaction component is numerically equal to its pressure or concentration, but without the units[6]. The activity is thus the ratio between the experimental quantity and the "standard state" (1 atm for gases, 1 mol L^{-1} for solutes). For example, when we write $a(Ca^{2+}) = 0.32$, this is the same

---

[6] In more advanced work, it is necessary to take into account that activity is only numerically equal to pressure or concentration under conditions of very low pressure (gases) or very high dilution (solutions). At higher concentrations or pressures an additional factor, the **activity coefficient** must be included.

activity = pressure (concentration) x activity coefficient

The activity coefficient has the value unity at very low pressure and concentration, and decreases in magnitude as pressure or concentration increases. In this course, we are implicitly setting the activity coefficient equal to 1.000, an approximation which is true only at very low concentration or pressure.

as writing $a(Ca^{2+}) = (0.32 \text{ mol L}^{-1})/(1 \text{ mol L}^{-1})$. We then approximate by regarding $a(Ca^{2+}) = 0.32$ as numerically equal to $c(Ca^{2+}) = 0.32 \text{ mol L}^{-1}$.

3. The equilibrium constant K is numerically equal to K_p or K_c, as appropriate, but without the units. Thus, taking the Haber process as an example:

$$N_2(g) + 3\ H_2(g) \rightleftharpoons 2\ NH_3(g)$$

$$K_p = \frac{\{p(NH_3)\}^2}{p(N_2)\{p(H_2)\}^3} \qquad \text{units of } K_p\text{, atm}^{-2}$$

$$K = \frac{\{a(NH_3)\}^2}{a(N_2)\{a(H_2)\}^3} \qquad \text{units of } K\text{, none}$$

Therefore $\Delta G° = -RT\ln K \approx -RT\text{"}\ln K_p\text{"}$, where the quotation marks are intended to emphasize that it is the numerical value of K_p that is being used. We can use eq. [5] to calculate K and hence estimate the experimental parameters K_p, K_c, etc, starting with tabulated thermodynamic quantities. Alternatively, given experimental values for an equilibrium constant we can determine ΔG° for the reaction[7]

4. The assignment of unit activity to pure solids and pure liquids provides the justification for leaving these substances out of equilibrium constant expressions. For example, consider the evaporation of liquid water.

$$H_2O(\ell) \rightleftharpoons H_2O(g)$$

$$K = \frac{a(H_2O\ (g))}{a(H_2O\ (\ell))} = \frac{a(H_2O\ (g))}{1} \approx \frac{p(H_2O\ (g)),\ \text{atm}}{1 \times 1\ \text{atm}}$$

K = "K_p" (again using the quotes to denote the numerical value of K_p)

$$\therefore\ K_p = p(H_2O\ (g))$$

---

[7] The issue of standard state is rather important in thermodynamics. As we saw in Chapter 8, K_p and K_c are usually different numerically. The tabulated values of ΔG° are all based on the standard states defined above: 1 atm for gases, and 1 mol L^{-1} for solutes in solution. If you were to use K_c for a gas phase reaction to calculate ΔG° for that reaction, the numerical value of ΔG° would be wrong because you had chosen the wrong standard state (1 mol L^{-1} instead of 1 atm).

Example: *Explain (no calculation) how you would estimate K_p for the following reaction at 1000 K.*

$CaCO_3(s) \rightarrow CaO(s) + CO_2(g)$

Answer: *First calculate $\Delta G°$ for the reaction. Since $T \neq 298$ K, this must be done by looking up $\Delta H°_f$ and $S°$ for each reactant and product to obtain $\Delta H°(rxn)$ and $\Delta S°(rxn)$, and then using the Gibbs-Helmholtz equation.*

Second, calculate K by use of eq. [5]. Finally, assume K ≈ "K_p", and add the units of K_p (atm, in this example).

14.5.2 ΔG vs. ΔG°

Care must be taken to distinguish ΔG from ΔG°. ΔG° is a *standard* molar free energy change; it is the free energy change which occurs when each reactant and product substance is present at unit activity (approximately, under the standard conditions of 1 atm for gases, and 1 mol L^{-1} for solutions). In practice, these conditions rarely apply (we encountered an exception in the cases of solids melting, and liquids boiling at 1 atm pressure). Consequently ΔG° is best thought of as the value that is obtained by calculation from tabulated data.

To take a previous example, suppose we are interested in the evaporation of liquid water at 298 K.

$$H_2O(\ell) \rightleftharpoons H_2O(g)$$

ΔG° can be calculated from the free energies of formation of $H_2O(g)$ and $H_2O(\ell)$ — the value is + 8.6 kJ mol $^{-1}$. This can be interpreted as follows. "The free energy change for the vaporization of water at 298 K is + 8.6 kJ mol $^{-1}$ when the liquid phase consists of pure water, and the vapour phase contains water vapour at a pressure of 1.00 atm."

Does this statement make physical sense? Yes: at 298 K, the equilibrium vapour pressure of water is 0.031 atm (Section 6.2.2), and so a pressure of 1.00 atm is much greater than the equilibrium vapour pressure. Evaporation under these conditions is non-spontaneous to the extent of + 8.6 kJ mol $^{-1}$. Put another way, the condensation of water vapour to form the liquid is the spontaneous process under these conditions, to the extent of - 8.6 kJ mol $^{-1}$. Yet another interpretation is that it would require the input of at least 8.6 kJ mol $^{-1}$ from an outside source to cause liquid water to vaporize to $H_2O(g)$ at 1 atm pressure under these conditions. (At the risk of seeming repetitious, be sure that you realize that the preceding three sentences are all different ways of looking at the *same* phenomenon.)

In contrast with ΔG°, ΔG is the free energy change for the reaction under the particular conditions under investigation. Equation [8] gives a method of calculating ΔG.

[8] $\Delta G = \Delta G° + RT\ln Q$

Q has been met before; it is the reaction quotient (Section 8.8). Q has the same form as the equilibrium constant K, but the activities (or, approximately, pressures or concentrations) are the actual values under the conditions being investigated. These are not necessarily the same as the values at equilibrium. Remember that the condition for equilibrium is $\Delta G = 0$; under these conditions $Q = K$, and eq. [8] simplifies to give eq. [5].

For the example of water evaporation at 25 °C, let us calculate ΔG, when $p(H_2O, g) = 0.024$ atm. Notice that this value is less than the equilibrium value of 0.031 atm, so we expect an answer showing that evaporation will be spontaneous under these conditions. We already know that $\Delta G° = +8.6$ kJ mol^{-1} at 25 °C. Substitution into eq. [8] gives the following result.

$$[8]\ \Delta G = \Delta G° + RT\ln Q$$
$$= +8.6 \text{ kJ mol}^{-1} + (8.314 \text{ J mol}^{-1} \text{ K}^{-1})(298 \text{ K})\ln(0.024 \text{ atm}/1 \text{ atm})$$
$$= +8.6 \text{ kJ mol}^{-1} + (-9.24 \times 10^3 \text{ J mol}^{-1}) = -0.6 \text{ kJ mol}^{-1}$$

The result is that ΔG is negative (spontaneous evaporation), as expected.

At this point, we can review Section 14.4.3. In summary, ΔH° = T.ΔS° when a solid melts or a liquid boils at 1 atm, but not usually for a chemical reaction at equilibrium. When a solid melts, the reactant is a pure solid and the product is a pure liquid. Both have unit activity, and so $K = 1$, and $\Delta G = \Delta G° = 0$. Similar arguments apply when a liquid boils at its normal boiling point to give a vapour *at 1 atm pressure*. However, other equilibrium constants in general can take any values, large and small. So while ΔG is always zero at equilibrium, ΔG° is only zero for the particular situation where $K = 1$.

Only if $\Delta G° = 0$, $K = 1$ is it true that $\Delta H° = T\Delta S°$.

Equations [5] and [8] can be combined to give a more formal explanation of why the reaction quotient can be used to predict the direction in which a reaction will proceed towards equilibrium (*cf.* Le Chatelier's Principle, Section 8.8).

$$[9]\ \Delta G = \Delta G° + RT\ln Q$$
$$= -RT\ln K + RT\ln Q = RT\ln(Q/K)$$

When $Q/K < 1$, $\ln(Q/K)$ is negative, ΔG is negative, and the reaction is spontaneous in the "forward" direction. Conversely, when $Q/K > 1$, $\ln(Q/K)$ and ΔG are both positive, and the reaction is spontaneous in the "reverse" direction.

Notice two other points concerning eq. [9].

- When $Q = K$, the reaction is at equilibrium. Under these conditions, $\Delta G = 0$, and equation [9] reduces back to eq. [5].

- When all reactants and products are at unit activity, $Q = 1$, the logarithmic term becomes zero, and $\Delta G = \Delta G°$.

We can summarize the relationship between ΔG and the progress of a reaction through Figure 1.

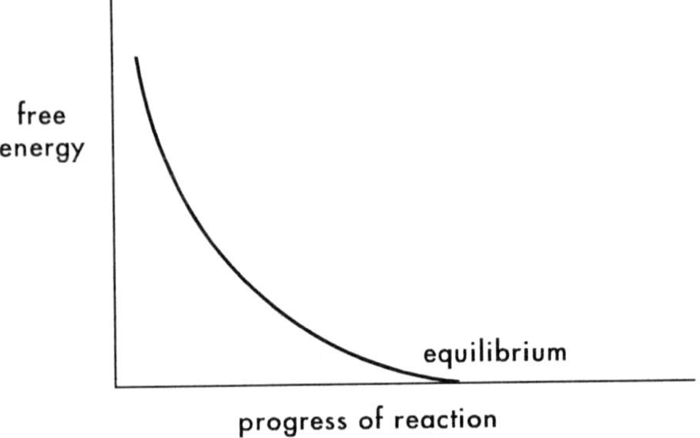

Figure 14.1: Trend in ΔG with the progress of a reaction

The minimum point in the curve, where $\Delta G = 0$, represents equilibrium. Starting at the left hand ordinate (100% reactants, no products yet formed), ΔG is infinitely negative: there is an infinite tendency for the reaction to proceed in the forward direction ($Q/K = 0$). As more and more products form, Q/K increases but is still less than unity, and the value of ΔG becomes less negative, rising to zero at equilibrium ($Q/K = 1$). Beyond equilibrium, $Q/K > 1$, and ΔG becomes increasingly positive; in other words, the reverse reaction becomes increasingly spontaneous.

14.5.3 Thermodynamics, temperature, and Le Chatelier's Principle

In Section 8.7, we found that the equilibrium constant for an endothermic process would increase at higher temperatures, while exothermic processes would be relatively favoured at lower temperatures. It was argued, since heat could be considered a "constraint" upon a system at equilibrium, that the reaction would proceed in such a direction as to remove the constraint. We can now offer a more formal explanation.

Recall eq. [5] and [6].

[5] $\Delta G° = - RT\ln K$

[6] $\therefore \ln K = -(\Delta H°/RT) + (\Delta S°/R)$

These equations can be written in the exponential form.

[10] $K = \exp(-\Delta G°/RT)$

or

[10a] $K = \exp(-\Delta H°/RT)\exp(+\Delta S°/R)$

$= \text{constant} \times \exp(-\Delta H°/RT)$

Thus K is independent of temperature in the entropic term, but temperature dependent in the enthalpic term. This explains why in order to calculate K at a new temperature, given the value of K at a different temperature, it was only necessary to know ΔH° and not ΔS° (eq. [7]).

Equation [10] shows why the sign and magnitude of ΔH° determines the effect on equilibrium constants. In the simplest case, suppose that ΔH° is zero. The first exponential term in eq. [10a] becomes unity ($e^0 = 1$) and K is independent of temperature, consistent with Le Chatelier's Principle. The dissociation of acetic acid in water is an example of a reaction whose equilibrium constant is almost independent of temperature (ΔH° = -0.25 kJ mol^{-1}: Appendix 1).

Now suppose that the reaction is endothermic, and ΔH° is positive. The first term of eq. [10a] is a negative exponent, which becomes less negative as T increases. Consequently, K increases with increasing temperature, consistent with the prediction of Le Chatelier's Principle. The larger the positive value of ΔH°, the greater the magnitude of the change in K for a given temperature change.

Finally consider the case where ΔH° is negative. Now the first term of eq. [10a] is a positive exponent, which decreases as T increases. K therefore decreases with increasing temperature, again consistent with the prediction of Le Chatelier's Principle. The more exothermic the reaction, the greater the magnitude of the change in K for a given temperature change[8].

---

[8] An equivalent explanation is provided using the logarithmic equation [6]. Since ln$K$ is proportional to -ΔH°/T, we draw the following conclusions.

- when ΔH° is positive (endothermic reaction), -ln$K$ becomes smaller (ln$K$ becomes larger) with increasing T. The equilibrium constant increases with temperature.

- when ΔH° is negative (exothermic reaction), +ln$K$ becomes smaller with increasing T, and the equilibrium constant decreases with increasing temperature.

14.6 Some thermodynamic properties of aqueous solutions

14.6.1 Solubility of gases in water

It was noted in Section 12.2 that many common gases are only slightly soluble in water, and that the solubilities of all gases in all liquids decrease with increasing temperature. These phenomena can be explained by reference to thermodynamic principles. Consider the general process $X(g) \rightleftharpoons X(aq)$. [X,aq] defines the solubility of X under given conditions, and is related to the equilibrium constant K_H.

$$K_H = [X\ (aq)]/p(X\ (g))$$

Making the approximation that the numerical value "K_H" = K, we can relate the solubility to thermodynamic parameters, through eq. [6].

[6] $\ln K = -\Delta G°/RT = -(\Delta H°/RT) + (\Delta S°/R)$

Low solubilities of common gases in water. Solubility is often "explained" by the maxim that "like dissolves like". Water is a very polar liquid, and so it is argued that polar, and especially ionic, solutes dissolve in water because they somehow "fit in" to the polar environment. According to this argument, non-polar substances (and not just gases) do not dissolve appreciably in water because the water molecules have more attraction for each other than for the molecules of the solute. This argument, while commonly put forward, is incorrect; it implies that the major factor which determines solubility is the enthalpy change for the process:

$$\text{solute}(s,\ \ell,\ \text{or } g) \rightleftharpoons \text{solute}(aq)$$

In the case of gases, the standard molar enthalpy change for the dissolution of all gases in water is negative: a favourable contribution to $\Delta G°$. If enthalpy change were the only factor to consider, all gases would be highly soluble in water. However, it is not; the favourable standard molar enthalpy change is opposed by a very unfavourable standard molar entropy change. $\Delta S°$ for the change $X(g) \rightarrow X(aq)$ is negative, implying that the solution is more ordered than the separated components solvent and solute. This ordering of the solution has two origins. First, the motion of the solute molecules is inevitably more restricted when they are confined to the aqueous solution than when they are free to travel unhindered in the gas phase. Second, when a non-polar solute dissolves in water, the solvent water molecules order themselves around the solute molecule, a process that has been likened to the formation of a miniature iceberg in the solution[9]. Thus not only is the solute more ordered in solution, so also is the solvent, when the solvent is water and the solute is a non-polar gas such as N_2, O_2 or CH_4.

Consider the example $CH_4(g) \rightarrow CH_4(aq)$. In this case, $\Delta H° = -14$ kJ mol^{-1} and $\Delta S° = -103$ J mol^{-1} K^{-1}. Hydration is therefore exothermic, but the entropy change is highly unfavourable. At 298 K, the term ($-T\Delta S°$) contributes +30.7 kJ mol^{-1} to $\Delta G°$, more than offsetting the exothermic $\Delta H°$. This results in a low solubility at 25 °C: 1.3 x 10^{-3} mol L^{-1}, when $p(CH_4) = 1$ atm.

For polar gases, the entropy factor is still unfavourable, because ordering of solvent and solute still occurs. However, the enthalpy of solution is more exothermic in such cases, making a larger negative contribution to ΔG. An extreme example is seen in the case of gases like HCl, where dissolution involves complete dissociation into ions. The reaction $HCl(g) \rightarrow H^+(aq) + Cl^-(aq)$ has $\Delta H° = -75$ kJ mol^{-1} and $\Delta S° = -130$ J mol^{-1} K^{-1}. Although the entropy change is even more unfavourable than in the case of methane ($-T\Delta S° = +39$ kJ mol^{-1} at 298 K), the highly exothermic nature of the reaction causes $\Delta G°$ to be negative, and HCl so has a very high solubility in water (12 mol L^{-1} at 25 °C).

Solubility decreases with increasing temperature. This is readily explicable from eq. [10]. $\Delta H°$ for the dissolution of gases in solvents is universally exothermic, and so the equilibrium constant for dissolution decreases with increasing temperature, as discussed in the last section[10]. This is true whether $\Delta H°$ is slightly exothermic (*e.g.*, CH_4) or highly exothermic (*e.g.*, HCl).

14.6.2 Solubility of solids and liquids in water

Similar considerations apply to solids and liquids as to gases when they dissolve in water — the only difference is that solids and liquids, unlike gases, have intermolecular or interionic forces holding the solid or liquid together. These enthalpic forces must be overcome when the substance dissolves (positive contribution to $\Delta H°_{solv}$); against this however, the molecules or ions gain

---

[9] This part of the argument is counter-intuitive, and requires some thought. The intuitive concept would be that placing a "foreign" molecule into water would be a source of disorder. In fact, the reverse is true; the foreign molecule orders the water molecules around itself, and the result is a *decrease* in solvent entropy.

[10] An interesting point to note is that $\Delta G°$ for dissolution becomes more positive with increasing temperature because of the increased contribution of the positive term ($-T.\Delta S°$) when $\Delta S°$ is negative. However, *K* is not determined directly by changes in $\Delta G°$, but in $\Delta G°/T$, as seen in eq. [10].

enthalpic stabilization by water (negative contribution to ΔH°_{solv}). The combined effect may leave ΔH°_{solv} either positive or negative.

From the entropic point of view, the entropy of the *solute* increases when the crystal dissolves, leaving the individual molecules or ions free to travel independently through the solution. Again, this increase in entropy is opposed by the ordering of the solvent which accompanies solvation, and the combined effect may leave ΔS°_{solv} either positive or negative.

In summary, the stabilization of the molecules or ions in aqueous solution must more than compensate for the loss of inter-particle attraction if high solubility is to result. Both enthalpic and entropic factors must be considered. Again, it may be useful to compare non-polar solutes with polar (especially ionic) ones.

The "like dissolves like" argument is applied to the solubility properties of solids and liquids as well as to gases. Non-polar substances such as hydrocarbons have low solubility in water ("oil and water don't mix"), but higher solubility in other non-polar solvents, especially organic solvents. Some authors refer to this phenomenon as the "hydrophobic[11] effect". The hydrophobic effect is invoked to explain why many non-polar solutes can be extracted out of water into a non-polar solvent, and why toxic non-polar substances become associated mainly with the lipid (fatty) tissues of organisms (Section 12.4). As in the case of gas solubility, the low solubility of non-polar solutes in water is mostly due to entropic rather than enthalpic considerations: the water molecules become ordered around a non-polar solute molecule.

As noted in Section 12.3, ionic solids vary greatly in their solubility in water. Furthermore, some ionic solids increase in solubility as the temperature rises, and some decrease. Let's look at some representative examples (Table 14.3).

Table 14.3: Enthalpies and entropies of solution of ionic solids at 298 K

Compound	ΔH°(soln), kJ mol^{-1}	ΔS°(soln), J mol^{-1} K^{-1}
$NaCl$	+4	+43
NH_4Cl	+15	+75
$CaCl_2$	-81	-45
$AlCl_3$	-329	-263
$PbCl_2$	+23	-13
$CaCO_3$	-13	-203
$BaSO_4$	+26	-103

11 "Hydrophobic" literally means water-hating; contrast "hydrophilic": water-loving.

NaCl and NH_4Cl are examples of salts that have high water solubility despite having endothermic heats of solution. This property is exploited in the laboratory where cooling baths made of ice/NaCl or ice/NH_4Cl are used to produce temperatures < 0 °C; the dissolution of the salt removes heat from the water and produces a cooling effect. The endothermicity of solution is offset by an increase in disorder upon dissolving, and both these salts have quite high solubility (> 1 mol L^{-1}) in water.

Calcium chloride and aluminum chloride are examples of extremely soluble salts. The explanation is to be found in the high enthalpy of solvation, associated with the enthalpic stabilization of the highly charged metal cation (Al^{3+} > Ca^{2+}) by water. The highly charged cations strongly order the water molecules around them, however, and this is the source of the very unfavourable $\Delta S°_{solv}$. Even so, the value of $\Delta H°_{solv}$ is so overwhelmingly negative that high solubility results.

Lead chloride combines an endothermic value of $\Delta H°_{solv}$ with a very small negative $\Delta S°_{solv}$. This compound has low solubility at 25 °C, but the solubility rises with increasing temperature. Calcium carbonate and barium sulfate are both examples of highly insoluble salts. At first glance, barium sulfate appears to be similar to lead chloride ($\Delta H°_{solv}$ is positive and $\Delta S°_{solv}$ is negative). However, the solubility of $BaSO_4$ is much lower than that of $PbCl_2$, because $\Delta H°_{solv}$ is more positive and $\Delta S°_{solv}$ is more negative. Calcium carbonate is very insoluble because of the extremely negative value of $\Delta S°_{solv}$, even though $\Delta H°_{solv}$ is exothermic.

Problems

Sections 14.1 - 14.3

1. Consider a chemical reaction having $\Delta H°$ negative and $\Delta S°$ negative. Which of the statements below is/are true?

   A: The total entropy of the products is greater than the total entropy of the reactants.

   B: The entropy charge in the reaction makes a positive contribution to $\Delta G°$.

   C: The reaction becomes more spontaneous as the temperature is raised.

   D: For suitable numerical values of $\Delta H°$ and $\Delta S°$, the reaction may be spontaneous at a given low temperature, yet non-spontaneous at some higher temperature.

2. Use the values of $\Delta H°_f$ and S° in Appendix 1 for CH_3OH (ℓ) and CH_3OH (g) to estimate the normal (1 atm) boiling point of methanol, CH_3OH.

3. Metallic tin exists in two forms, which have the following thermodynamic properties.
   Sn, white $\Delta H°_f$, 0.0 kJ mol $^{-1}$; S° 51.55 J mol $^{-1}$ k $^{-1}$
   Sn, grey $\Delta H°_f$, -2.09 kJ mol $^{-1}$; S° 44.14 J mol $^{-1}$ k $^{-1}$
   a) Which form of tin is stable at 298 K?
   b) Over what range of temperatures is grey tin stable?

4. Pure sulfuric acid has the following thermodynamic properties and melts at 10.36°C
   H_2SO_4 (s) $\Delta H°_f$ -816.00 kJ mol $^{-1}$ S° ?
   H_2SO_4 (ℓ) $\Delta H°_f$ -813.99 kJ mol $^{-1}$ S° 156.9 J mol $^{-1}$ K $^{-1}$
   Calculate the standard molar entropy of solid sulfuric acid.

5. Elemental iodine has the following thermodynamic properties
   I_2 (s) $\Delta H°_f$ 0 S° 116.14 J mol $^{-1}$ K $^{-1}$
   I_2 (g) $\Delta H°_f$ 65.52 kJ mol $^{-1}$ S° 260.6 J mol $^{-1}$ K $^{-1}$
   Calculate the temperature at which I_2(s) is at equilibrium with I_2 (g) at p = 1.00 atm.

6. State whether you expect the reactants or the products of the following reactions to have greater entropy.
   a) $3O_2$ (g) $\rightarrow$ $2O_3$ (g)
   b) Zn (s) + 2 HCl (aq) $\rightarrow$ $ZnCl_2$ (aq) + H_2 (g)
   c) CH_4 (g) + $2O_2$ (g) $\rightarrow$ CO_2 (g) + $2H_2O$ (ℓ)
   d) CH_4 (g) + $2O_2$ (g) $\rightarrow$ CO_2 (g) + $2H_2O$ (g)
   e) Br_2 (g) + $3Cl_2$ (g) $\rightarrow$ $2BrCl_3$ (ℓ)

7. The heat of fusion of ice is +6.0 kJ mol $^{-1}$. Calculate the entropic contribution to ΔG when an iceberg of mass 10,000 t melts in an ocean whose temperature is 3.8°C.

8. The temperature inside the engine of a car is 800 K and the temperature outside is 30°C. What is the theoretical efficiency of this machine in the absence of any mechanical or frictional losses?

9. What would be the advantage in terms of theoretical efficiency in changing from 600°C to 700°C as the temperature of the steam driving the turbines of an electrical generator, if the temperature of the cooling water were 35°C?

Section 14.4

1. For which of the following substances is $\Delta G°_f$ not equal to zero?

   a) O_3 (g) b) Cl_2 (g) c) O_2 (ℓ) d) Br_2 (g)

   e) I_2 (s) f) H_2O (ℓ)

2. Calculate ΔG° for the following reactions, all assumed to be occurring at 298 K.

   a) $CH_4 (g) + 2O_2 (g) \rightarrow CO_2(g) + 2H_2O (\ell)$

   b) $O_3 (g) + NO (g) \rightarrow NO_2 (g) + O_2 (g)$

   c) $HNO_3(aq) + NaOH (aq) \rightarrow NaNO_3 (aq) + H_2O (\ell)$

   d) $CaSO_4 (s) \rightarrow Ca^{2+} (aq) + SO_4^{2-} (aq)$

   e) $HCl (g) \rightarrow HCl (aq)$

3. Calculate ΔG° for the following reactions, which are assumed to occur at the temperatures indicated

   a) $3O_2 (g) \rightarrow 2O_3 (g)$ T = 220 K

   b) $NO (g) + O_3 (g) \rightarrow NO_2 (g) + O_2 (g)$ T = -10°C

   c) $SO_2 (g) + ½O_2 (g) \rightarrow SO_3 (g)$ T = 450°C

   d) $N_2 (g) + 3H_2 (g) \rightarrow 2NH_3 (g)$ T = 500°C

   e) $CaCO_3 (s) \rightarrow CaO (s) + CO_2 (g)$ T = 1000°C

Section 14.5

1. Use thermodynamic tables to calculate ΔG°, and hence estimate the equilibrium constant for each of the reactions below.

   a) $HF (aq) \rightleftharpoons H^+ (aq) + F^- (aq)$ T = 25°C

   b) $CuSO_4.5H_2O (s) \rightleftharpoons CuSO_4 (s) + 5H_2O (g)$ T = 120°C

   c) $Ni (s) + 4CO (g) \rightleftharpoons Ni(CO)_4 (g)$ T = 150°C

   d) $CH_3OH (\ell) \rightleftharpoons CH_3OH (g)$ T = 25°C

   e) $NO (g) + O_3 (g) \rightleftharpoons NO_2 (g) + O_2 (g)$ T = 300 K

   f) $Cl (g) + O_3 (g) \rightleftharpoons ClO (g) + O_2 (g)$ T = 220 K

2. At 220 K the reaction below has rate constant 8.7×10^{-12} cm^3 $molec^{-1}$ s^{-1}

   $Cl\ (g) + O_3\ (g) \rightarrow ClO\ (g) + O_2\ (g)$

   Use your answer from Problem 1 (f) to estimate the rate constant for the reverse reaction at this temperature.

3. Calculate $\Delta G°$, and hence ΔG for the reactions below occurring under the stated conditions.

   a) $CS_2\ (\ell) \rightleftharpoons CS_2\ (g)$ T = 298 K,

      $p(CS_2\ (g)) = 0.038$ atm

   b) $COCl_2\ (g) \rightleftharpoons CO\ (g) + Cl_2\ (g)$ T = 400 K,

      $p(COCl_2\ (g)) = 0.62$ atm, $p(CO, g) = 1.1$ atm; $p(Cl_2, g) = 0.14$ atm

   c) $N_2\ (g) + 3H_2\ (g) \rightleftharpoons 2NH_3\ (g)$ T = 700 K,

      $p(NH_3\ (g)) = 6.5$ atm, $p(N_2, g) = 1.6$ atm, $p(H_2, g) = 25$ atm

   d) $CaCO_3\ (s) \rightarrow CaO\ (s) + CO_2\ (g)$ T = 1000°C,

      $p(CO_2\ (g)) = 850$ ppmv

   e) $CaCO_3\ (s) + SO_2\ (g) + ½O_2\ (g) \rightarrow CaSO_4\ (s) + CO_2$ T = 650°C,

      $p(SO_2\ (g)) = 150$ ppmv, $p(O_2) = 0.050$ atm, $p(CO_2) = 350$ ppmv

4. On Mars, p(total) is 0.006 atm, and 95% of this is CO_2. The summertime temperature is 220 K and the wintertime temperature 125 K. In the wintertime polar icecaps of solid CO_2 form. Calculate the temperature at which CO_2 first condenses in the Martian fall, using the thermodynamic data below.

   Fusion of $CO_2(s)$: $\Delta H°$: 8.37 kJ mol^{-1} $\Delta S°$: 38.64 J mol^{-1} K^{-1}

   Vaporization of $CO_2(\ell)$: $\Delta H°$: 16.24 kJ mol^{-1} $\Delta S°$: 75.03 J mol^{-1} K^{-1}

Section 14.6

1. The solubility in weight percent of methane in water is given below, (*p*(methane) = 760 torr).

t, °C	Solubility	t, °C	Solubility
0	0.00396	30	0.00191
5	0.00341	40	0.00159
10	0.00296	50	0.00136
15	0.00260	60	0.00115
20	0.00232	70	0.00093
25	0.00209	80	0.00070

   a) Calculate ΔH° and ΔS° for the dissolution of CH_4 in water and verify the statement in the text that the process CH_4 (g) → CH_4 (aq) is enthalpically favoured and entropically disfavoured.

   b) Look up the thermodynamic constants of CH_4 (g) to obtain $\Delta H_f°$ and S° for CH_4 (aq).

2. Use the data below to calculate ΔH° and ΔS° for the reaction N_2 (g) → N_2 (aq), and hence explain why the solubility of N_2 (g) in water decreases with temperature. $\Delta H_f°(N_2\ (aq)) = -14.7$ kJ mol^{-1}; $S°(N_2\ (g)) = 0.19$ kJ mol^{-1} K^{-1}; $S°(N_2\ (aq)) = 0.08$ kJ mol^{-1} K^{-1}.

3. ClO_2 (used in drinking water treatment) has the following thermodynamic properties.

   ClO_2 (g) $\Delta H°_f$ 102.5 kJ mol^{-1} S° 256.7 J mol^{-1} K^{-1}

   ClO_2 (aq) $\Delta H°_f$ 74.9 kJ mol^{-1} S° 164.9 J mol^{-1} K^{-1}

   a) Calculate the Henry's Law constant for ClO_2 in water at 10°C.

   b) Calculate the concentration of ClO_2 (aq) in ppm, in equilibrium with 10.0 ppmv of ClO_2 (g) at 10°C.

4. The table below gives the solubility of N_2 (g) in water at various temperatures. The data all relate to $p(N_2) = 1.00$ atm, and are expressed as cm^3 of N_2 (corrected to STP) per liter of water.

t, °C	$V(N_2)$	t, °C	$V(N_2)$
0	23.3	30	12.8
5	20.6	35	11.8
10	18.3	40	11.0
15	16.5	50	9.6
20	15.1	60	8.2
25	13.9		

Calculate ΔH° and ΔS° for the dissolution of N_2 in water.

5. Calculate ΔG° for the following solubility equilibria, and use the result to estimate K_{sp} for the salts in question.

   a) $CaCO_3$ (s) $\rightleftharpoons$ Ca^{2+} (aq) + CO_3^{2-} (aq)

   b) $AlPO_4$ (s) $\rightleftharpoons$ $AlPO_4$ (aq)

   c) Ag_2S (s) $\rightleftharpoons$ Ag_2S (aq)

6. Estimate the solubilities of the following salts by the use of thermodynamic data

   a) AgCl

   b) BaF_2

7. CO has the following thermodynamic properties in a solution of pure acetic acid (HAc (ℓ))

   ΔH°_f -110.5 kJ mol $^{-1}$ S° 132.6 J mol $^{-1}$ K $^{-1}$

   Calculate the solubility of CO in acetic acid, when the solution is in equilibrium with 0.10 atm of CO (g) at 298 K.

15 STRATOSPHERIC OZONE DEPLETION

15.1 The stratosphere and the "ozone layer"

The loss of ozone from the upper atmosphere (technically, the stratosphere) is of current societal concern. In this chapter, we study the reactions by which ozone is formed and destroyed in the stratosphere. We shall examine the relationship between stratospheric ozone and **chlorofluorocarbons** (CFCs), and the efforts that are currently under way to provide alternatives to these harmful, but very useful chemicals.

In discussing atmospheric problems concerning ozone, it is vital to distinguish between ground level ozone and stratospheric ozone. At ground level, abnormally high concentrations of ozone are undesirable, because of the toxicity of ozone towards living organisms (Section 9.3). The pollution due to photochemical smog is associated with abnormally high ozone levels. In the upper atmosphere (stratosphere) the reverse situation applies. Ozone performs the important function of shielding the Earth's surface from damaging ultraviolet radiation, and pollution is associated with abnormally low ozone levels. The toxicity of ozone is, of course, irrelevant in the stratospheric context, because life does not exist in the stratosphere.

The stratosphere typically extends from about 15-50 km altitude. The temperature increases fairly steadily from about 215 K at 15 km to 270 K at 50 km, while p(total) decreases over the same interval of altitude from 10^{-1} to 10^{-4} atm (Figure 1).

Ozone is a minor, but chemically very important, component of the stratosphere; however its concentration is always low, never exceeding much more than 10 ppmv, even at its maximum near 35 km altitude. *It is important to realize that the major components of the stratosphere are N_2 and O_2, just as in the troposphere.*

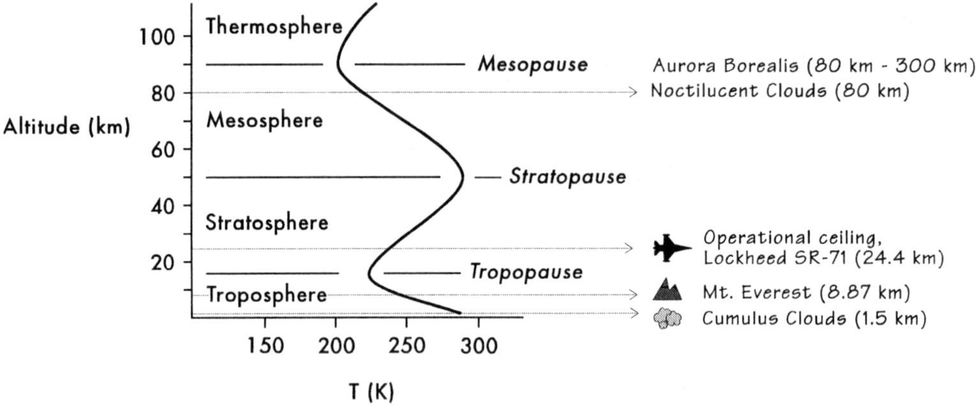

Figure 15.1: Variation of temperature (K) with altitude in the atmosphere.

Example: *Calculate the concentration of ozone molecules per cubic centimeter at 35 km altitude, assuming* $p(O_3) = 10$ *ppmv,* $p(\text{total}) = 0.006$ *atm and* $T = 240$ *K.*

Answer: *First calculate the partial pressure of* O_3*:*

$$p(O_3) = 10 \text{ ppmv} \times (1 \text{ atm}/10^6 \text{ ppmv}) \times p(\text{total}) = 6 \times 10^{-8} \text{ atm}$$

Use the ideal gas equation to convert pressure to concentration

$$n/V = P/RT = \frac{6 \times 10^{-8} \text{ atm}}{0.0821 \text{ L atm mol}^{-1} \text{ K}^{-1} \times 240 \text{ K}}$$

$$= 3 \times 10^{-9} \text{ mol L}^{-1}$$

Convert to molecules per cubic centimeter:

$$n/V = (3 \times 10^{-9} \text{ mol L}^{-1}) \times (6.022 \times 10^{23} \text{ molec mol}^{-1}) \times 1 \text{ L}/1000 \text{ cm}^{-3}$$

$$= 2 \times 10^{12} \text{ molec cm}^{-3}$$

Although news media stories refer to depletion of the ozone layer, the term "ozone layer" is misleading, because it implies a distinct region of the atmosphere in which ozone is a major atmospheric constituent. In reality, ozone is a trace component throughout the atmosphere. Even in the unpolluted troposphere its concentration is *ca.* 10^{12} molec cm $^{-3}$, rising to about 5×10^{12} molec cm $^{-3}$ at 25 km altitude[1]. To restate an important point, ozone is not the major chemical constituent of the stratosphere, even though it is very important to the chemistry of the stratosphere. To indicate how little ozone there really is in the atmosphere, if all of it were compressed into a single layer at STP, that layer would be just 3 mm thick.

Example: *Taking the radius of the Earth as 6000 km, calculate the mass of ozone in the hypothetical ozone layer just referred to.*

Answer: *Since the depth of the layer (3 mm) is so small compared with the radius of the earth (6,000 km), the volume of the ozone layer is equal to the area of the surface ($4\pi R^2$) multiplied by the depth (ΔR).*

$$V = 4\pi R^2 \Delta R = (4)(\pi)(6000\ km)^2.(3\ mm) \times (1\ km/10^6\ mm)$$
$$= 1.4 \times 10^3\ km^3$$

Use the molar volume at STP to convert km^3 of atmospheric gases to moles of ozone.

$$n(O_3) = (1.4 \times 10^3\ km^3)(10^4\ dm/1\ km)/(22.4\ dm^3\ mol^{-1})$$
$$= 6.1 \times 10^{13}\ mol$$
$$mass\ (O_3) = (6.1 \times 10^{13}\ mol)(48\ g\ mol^{-1})$$
$$= 3 \times 10^{15}\ g = 3 \times 10^9\ tonnes$$

15.2 Formation and destruction of ozone

15.2.1 Mechanism for formation and destruction of ozone without catalysis

Under the influence of sunlight, oxygen (O_2) is continually being changed into ozone (O_3) and ozone is likewise converted back to ordinary oxygen. Each day 350,000 tonnes of ozone are made - and destroyed - in the atmosphere. It is a misconception to think that the natural process is $O_2 \rightarrow O_3$, and that ozone is destroyed because of air pollution. If that were so, natural processes would have long since changed all the O_2 in the atmosphere to O_3! What certain air pollutants can do is to speed up the rate of loss of ozone so that its **steady state concentration** declines. This discussion will be deferred to Section 15.3.

The principal reactions involved in the formation and destruction of ozone are given in reactions [1]-[4]. The sequence [1], [2] is the mechanism for forming

---

[1] The maximum concentration of ozone occurs at about 25 km altitude when concentration is expressed in the units molec cm^{-3}; this does not coincide with the maximum (35 km) when the units are ppmv, because p(total) changes with altitude.

ozone, while the sequence [3], [4] is the mechanism for destroying ozone.

	Reaction	$\Delta H°$, kJ mol^{-1}
[1]	$O_2 \xrightarrow{h\nu,\ \lambda < 240\ nm} 2\ O$	498 - E(photon)
[2]	$O + O_2 \rightarrow O_3$	-105
[3]	$O_3 \xrightarrow{h\nu,\ \lambda < 325\ nm} O_2 + O$	105 - E′(photon)
[4]	$O + O_3 \rightarrow 2\ O_2$	-392

Notice that oxygen atoms are instrumental in both forming ozone (eq. 2) and destroying it (eq. 4). Oxygen atoms are very reactive, and exist for much less than a second in the stratosphere before undergoing reaction. This has an important consequence; reactions [1]-[4] all come to a halt at sunset. Reactions [1] and [3] stop because they require sunlight, and reactions [2] and [4] stop because no more oxygen atoms are being formed photochemically. As a result the concentration of ozone during the night is essentially the same as at the end of the day. Later, we shall see the significance of this observation in the context of polar "ozone holes".

Reactions [1] through [4] are all energy-releasing once the energies of the photons are included in reactions [1] and [3]. The enthalpy changes in reactions [2] and [4] are calculated using standard molar enthalpies of formation; those of reactions [1] and [3] also require the energy of the photon to be included.

Example: *Calculate the enthalpy change for equation [1] when the wavelength of the photons is 220 nm, and $T = 230$ K. Carry out this calculation (a) assuming $\Delta E = \Delta H°$; (b) making the correction $\Delta H° = \Delta E + \Delta nRT$.*

Answer: *From Chapter 5:*

$$\Delta E(photons) = (1.2 \times 10^5/\lambda)\ kJ\ nm\ mol^{-1} = (1.2 \times 10^5/220) = 5.5 \times 10^2\ kJ\ mol^{-1}$$

(a) If $\Delta E = \Delta H°$, and $\Delta H°$ for dissociation of $O_2(g)$ is 495 kJ mol^{-1}, then

$$\Delta H° = 498 - 5.5 \times 10^2 = -52\ kJ\ mol^{-1}$$

(b) If $\Delta H° = \Delta E + \Delta nRT$, we must calculate ΔnRT

$$\Delta n = +1\ (1\ mol\ O_2 \rightarrow 2\ mol\ O)$$

$$\Delta nRT = (1)(8.314\ J\ mol^{-1}\ K^{-1})(230K) = 1.9 \times 10^3\ J\ mol^{-1} = 2\ kJ\ mol^{-1}$$

$$\therefore \Delta H° = -52 + 2 = -50\ kJ\ mol^{-1}$$

We note that inclusion of the term ΔnRT makes only a small correction to the final answer.

Light absorption by O_2 and O_3 is responsible for shielding the Earth's surface from radiation having $\lambda < 300$ nm. Ordinary oxygen is the main absorber near 200 nm, while ozone is most important in the range 230 - 320 nm. The shielding property of ozone is due to its ability to absorb radiation of these wavelengths, which simultaneously converts it back to O_2, eq. [3]. Life as we know it could not exist at the Earth's surface without the protection of the atmosphere to shield us from the highly energetic "deep UV" solar radiation having $\lambda < 300$ nm (Figure 2).

In Figure 2, the upper curve is the solar spectrum as it would be observed from outer space; the lower curve is what we experience at the Earth's surface. The Sun's radiative emission occurs right across the ultraviolet, visible, and infrared regions of the spectrum. However, radiation having $\lambda <$ *ca.* 290 nm does not penetrate to the surface. The difference between the two curves (hatched region in Fig. 2) represents the radiation absorbed in the atmosphere (mostly in the stratosphere) by O_2 and O_3. The net result of Reactions [1] through [4] is the conversion of the energy of solar photons into heat, and as a result the stratosphere is warmer than the upper parts of the troposphere: refer back to Figure 1.

Example: *The standard enthalpy of formation of $O_3(g)$ is $+142$ kJ mol^{-1}. Calculate the daily heat input to the stratosphere if 350,000 tonnes of ozone are converted onto O_2 daily.*

Answer: *The destruction of ozone is the reverse of the reaction which defines $\Delta H°_f$, namely:*

$O_3(g) \rightarrow 3/2\ O_2(g) \qquad \Delta H° = -142$ kJ mol^{-1}

$M(O_3) = 48$ g mol^{-1}

$n(O_3$ destroyed$) = 350{,}000$ Mg/48 g mol^{-1} $= 7.3 \times 10^3$ Mmol

heat released $= (7.3 \times 10^3$ Mmol$) \times (142$ kJ mol$^{-1})$

$= 1.0 \times 10^6$ GJ $(1.0 \times 10^{15}$ J$)$

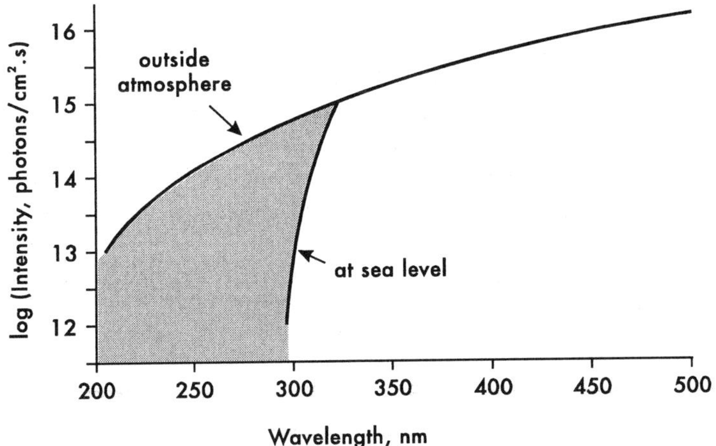

Figure 15.2: Solar spectrum at the Earth's surface and outside the atmosphere

To summarize the preceding paragraphs: *the stratosphere is a region of the atmosphere where the reaction 3 $O_2 \rightleftarrows 2\ O_3$ is an important chemical process. As a result, the stratosphere is warmer than the upper troposphere, and radiation having λ < 290 nm is absorbed, preventing it from penetrating to the Earth's surface.*

15.2.2 *Equilibrium* vs *steady state*

As indicated by reactions [1]-[4] and the summary equation 3 $O_2 \rightleftarrows 2\ O_3$, ozone formation is a reversible process. At any time, ozone exists at a steady state concentration, the balance between the rate of formation and the rate of destruction. However, the steady state concentration of ozone does not correspond to equilibrium, because the formation of ozone requires solar energy. We can appreciate this distinction by referring back to Section 14.5.

Recall that the condition for equilibrium is $\Delta G = 0$. The stratospheric system 3 $O_2 \rightleftarrows 2\ O_3$ is an example of a steady state which does not correspond to equilibrium; the input of solar energy maintains it far from equilibrium, with O_2 "pumped" uphill energetically until the rate of ozone production is balanced at the steady state by its rate of reversion to O_2. The observation of non-equilibrium steady states is common environmentally, where sunlight can constitute a source of energy which is external to the "system".

Let's take care to distinguish the terms "equilibrium" and "steady state".

Definition

Equilibrium *is the condition where ΔG for the reaction is zero; the reaction has no spontaneous tendency to occur in either direction.*

A ***steady state*** *is achieved whenever the concentration of a substance does not change with time ($d[c]/dt = 0$); the steady state need not correspond to equilibrium: the system may remain far from equilibrium for an indefinite period if energy is continuously supplied from outside.*

Equilibrium is thus the special case of a steady state in which $\Delta G = 0$ in addition to the constancy of composition of the mixture.

☞ *Caution: Always be careful to distinguish an environmental steady state from thermodynamic equilibrium.*

Example: *(a) Use thermodynamic data to calculate the equilibrium constant for the reaction: $3/2\ O_2(g) \rightleftharpoons O_3(g)$ at a stratospheric temperature of 218 K. (b) Use this value to estimate the equilibrium partial pressure of O_3 in the stratosphere if p(total) is 0.010 atm.*

Answer: *We shall calculate K (numerically equal to K_p) making use of the relationship $\Delta G° = -RT.\ ln(K)$. Therefore we need to calculate $\Delta G°$. This must be done by adding the $\Delta H°$ and $\Delta S°$ contributions; we cannot use tabulated $\Delta G°_f$ values because the temperature is not 298 K.*

Standard enthalpies of formation: O_2, zero; O_3, +142 kJ mol^{-1}

Standard molar entropies: O_2, 205.0; O_3, 238.8 J mol^{-1} K^{-1}

$$\Delta H° = \Delta H°_f(O_3\ (g)) - 3/2\ \Delta H°_f(O_2\ (g)) = 142\ kJ\ mol^{-1}$$
$$= 1.42 \times 10^5\ J\ mol^{-1}$$
$$\Delta S° = S°(O_3\ (g)) - 3/2\ S°(O_2\ (g)) = 238.8 - (1.5 \times 205.0)$$
$$= -68.7\ J\ mol^{-1}\ K^{-1}$$
$$\Delta G° = \Delta H° - T\Delta S°$$
$$= 1.42 \times 10^5\ J\ mol^{-1} - (218\ K)(-68.7\ J\ mol^{-1}\ K^{-1})$$
$$= 1.57 \times 10^5\ J\ mol^{-1}$$

$$ln\ K = \frac{-1.57 \times 10^5\ J\ mol^{-1}}{8.314\ J\ mol^{-1}\ K^{-1} \times 218\ K} = -86.6$$

$$K = e^{-86.6} = 2.4 \times 10^{-38}$$

Since this is a gas phase reaction the standard state was 1 atm, so the units of the experimental equilibrium constant are atm$^{-1/2}$:

$$K_p = 2.4 \times 10^{-38}\ atm^{-1/2}$$

(b) $K_p = p(O_3)/p(O_2)^{3/2}$

Knowing K_p and the pressure of O_2 (= 0.21 x p(total)), the equilibrium pressure of O_3 is the only unknown.

$$p(O_3) = K_p p(O_2)^{3/2}$$
$$= 2.4 \times 10^{-38}\ atm^{-1/2}(0.21 \times 0.010\ atm)^{3/2}$$
$$= 2.1 \times 10^{-42}\ atm$$

We can now compare the (equilibrium) pressure of 2.1×10^{-42} atm with the *ca.* 10 ppmv of O_3 actually present in the stratosphere. If p(total) = 0.010 atm, then 10 ppmv corresponds to $p(O_3) = 1 \times 10^{-7}$ atm. We see that the experimental partial pressure of O_3 is about 35 orders of magnitude larger than the equilibrium value. Hence the reaction is maintained at a steady state which is very far from equilibrium. The source of energy which maintains this non-equilibrium steady state is sunlight (reactions [1] and [3]).

15.2.3 Chlorine atoms as catalysts for the destruction of ozone

Recent research has shown that reactions [3] and [4] account for only half the rate of destruction of O_3 in the stratosphere. The remainder of the ozone is destroyed by catalytic processes, one of which has the chlorine atom (Cl) as its catalyst.

[5] $Cl + O_3 \rightarrow ClO + O_2$

[6] $ClO + O \rightarrow Cl + O_2$

Adding [5] and [6] together gives eq. [4]:

[4] $O + O_3 \rightarrow 2\ O_2$

Therefore the sequence [5], [6] is another way of carrying out reaction [4], and increases the rate of destruction of ozone. Notice that the chlorine atom consumed in reaction [5] is regenerated in reaction [6], allowing the cycle to be repeated over and over. The chlorine atom is therefore a catalyst for reaction [4]. Cyclic reactions of this sort are called **chain reactions**. In this particular chain reaction, up to 20,000 ozone molecules can be catalytically destroyed before the chlorine atom is finally removed by some other reaction.

We can compare the energetics of the direct reaction [4] with the catalysed reaction [5]+[6] in terms of reaction enthalpies and activation energies. The data are summarized in Table 15.1.

Table 15.1: Enthalpy changes and activation energies for uncatalysed and catalysed decomposition of ozone

Reaction	[4]	[5]	[6]
$\Delta H°$, kJ mol^{-1}	-391.8	-162.5	-229.3
E_{act}, kJ mol^{-1}	18	2	1

As expected from Section 5.2, $\Delta H°[4] = \Delta H°[5] + \Delta H°[6]$. The activation energy for reaction [4] (18 kJ mol^{-1}) is considerably higher than those of either reaction [5] or [6]. Because each of the catalysed reactions is intrinsically faster than the uncatalysed route, a low concentration of chlorine atoms can substantially influence the overall sink strength for ozone. These ideas are summarized in the energy diagram (Figure 3): compare Section 7.9.

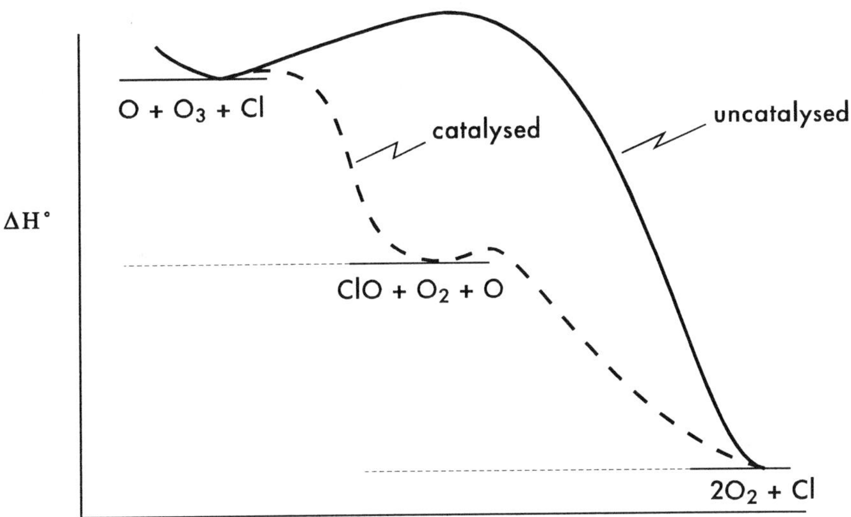

Figure 15.3: Energetics of the catalysed and uncatalysed mechanisms of ozone destruction.

The kinetics of chain reactions are more complicated than those of ordinary, non-chain reactions. Although the kinetics of chain reactions are beyond the scope of this text, we can note here that the overall rate of the reactions ([5] + [6]) is determined chiefly by the rates of **initiation** and **termination** (the introduction and removal, respectively, of the active catalyst, which is the chlorine atom). These processes determine the concentration of the catalytic agent, Cl. In the case of reactions [5] and [6], initiation of the chain is caused by photolysis of chlorine-containing molecules such as HOCl and CFCs (see below) to give free chlorine atoms. The chains are interrupted by conversion of the chlorine atoms to stable chlorine compounds such as HCl and $ClONO_2$ (chlorine nitrate).

15.3 Chlorofluorocarbons in the stratosphere

Chlorofluorocarbons (CFCs)[2] contain chlorine, fluorine, and carbon. They

[2] CFCs are often known by the trade name Freons. Freon is DuPont registered trademark. However, CFCs are now made world-wide by many different companies, as the original production patents have long since expired.

Key properties of refrigerator fluids

	bp(°C)	crit. temp/press
CFC-12	-30	111/40 atm.
NH_3	-33	132/112 atm.
SO_2	-10	157/77 atm.

have been manufactured since the 1930's. The most important CFCs commercially are CFC-11 ($CFCl_3$) and CFC-12 (CF_2Cl_2)[3]. The non-toxic, non-flammable CFC-12 was introduced originally as the operating fluid in refrigerators, displacing the highly toxic and odorous SO_2 and NH_3 which had hitherto been used for this purpose. Refrigerant fluids must be gaseous at room temperature, but easily compressible to a liquid. CFC-12 for example has normal (1 atm) boiling point -30 °C, and since its critical temperature is 111 °C, it can be liquified at ordinary temperatures. Its molar enthalpy of vaporization is 35 kJ mol^{-1}, large enough to give it a good capacity to carry heat from the inside of the refrigerator to outside. CFCs made the refrigerator safe enough to be operated as a domestic appliance in every home, where previously it had been restricted to industrial use.

Other uses for CFCs include blowing agents for expanded plastic foams and propellants for aerosol sprays. CFC-11 ($CFCl_3$, b.p. 24 °C) is preferred in these applications. Again the lack of toxicity, lack of flammability, lack of odour, and ease of liquefaction made CFCs ideal for these uses.

The combined annual production of CFC-11 and CFC-12 in the mid-1980s was over 700,000 tonnes, most of which (> 600,000 t) was estimated to be released to the atmosphere. Tropospheric concentrations of CFC-11 at ground level were *ca.* 50 pptv when they were first discovered in the atmosphere in 1971, and had risen to *ca.* 150 ppt by 1979. As noted in Chapter 6, CFCs are important tropospheric greenhouse gases, in addition to their connection to the problem of stratospheric ozone depletion, as discussed in this chapter.

CFC-11 and CFC-12 are completely unreactive in the troposphere. They are unaffected by tropospheric solar radiation ($\lambda > 290$ nm) and they contain no C-H bonds, which would render them reactive towards the hydroxyl radical (Section 9.1). However, upon migration upwards to the stratosphere, they are susceptible to gradual photolysis, eq. [7].

---

[3] The "Rule of 90" enables one to determine the composition of a CFC from the "code number" such as CFC-12. Add 90 to the code number; in the case of CFC-12, this gives 102. The three digits 1, zero, and 2 are the numbers of carbon, hydrogen, and fluorine atoms in the molecule. The remaining atoms must be chlorine. Another example: CFC-141: 141 + 90 = 231, so this molecule contains 2 carbons, 3 hydrogens, 1 fluorine, and by difference 2 chlorines. You cannot tell from the formula $C_2H_3FCl_2$ which isomer is involved, and letters a, b, etc are used to differentiate them.

$$[7] \quad CF_2Cl_2 \xrightarrow{h\nu,\ \lambda < 250\ nm} CF_2Cl + Cl$$

The chlorine atom thus released can participate in the catalytic mechanism for destroying ozone (eq. [5] and [6]).

Example: *Show by calculation that radiation of wavelength 250 nm is sufficiently energetic to bring about reaction [7].*

Answer: *We need to determine whether photons of wavelength 250 nm are sufficiently energetic to cleave a C-Cl bond*

ΔE *(photons)* $= 1.2 \times 10^5/\lambda$ *kJ mol*$^{-1}$

$= 1.2 \times 10^5/250 = 480$ *kJ mol*$^{-1}$

C-Cl Bond Energy from Appendix 2 $= 340$ *kJ mol*$^{-1}$

Conclusion: *the photon energy is greater than that necessary to cleave the bond, hence the reaction is energetically possible.*

The first warning that CFCs might be responsible for depleting stratospheric ozone was made in 1974. At that time, it was not known that reactions [5] and [6] occurred naturally in the stratosphere; chlorine atom assisted destruction of ozone was suggested to occur only as a consequence of the release of CFCs to the atmosphere. It is now known that the Cl/ClO chain reaction (eq. [5] and [6]) proceeds even in the unpolluted stratosphere. The release of CFCs does not introduce a brand-new sink for ozone; rather, it increases the strength of a pre-existing sink.

Environmental concern about CFCs intensified because they are extremely long-lived pollutants. They are completely unreactive in the troposphere, and even in the stratosphere, their decomposition is very slow: the lifetimes of CFC-11 and CFC-12 are estimated to be of the order of 70 and 110 years, respectively. Therefore, environmental damage caused by CFCs is a problem that could persist for many generations.

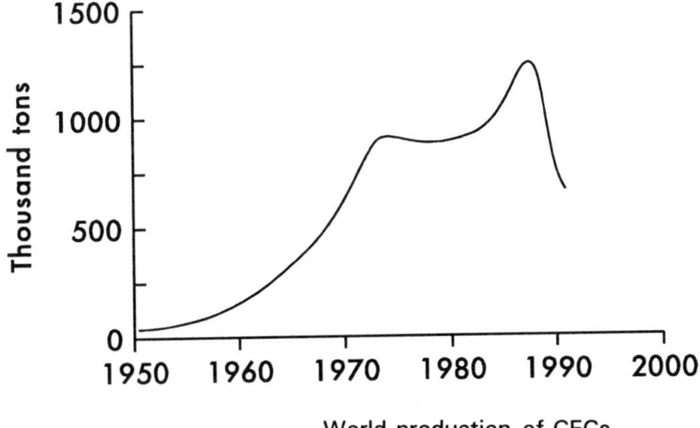

World production of CFCs.

***Example**: Assuming an average half-life of 100 years for CFCs 11 and 12, and a present combined concentration of 230 pptv, calculate how long would be needed for their concentration to fall to 50 pptv, if release to the atmosphere ceased today.*

***Answer**: From the half life we can calculate the first order, or pseudo-first order, rate constant for loss from the atmosphere.*

$$t_{1/2} = 0.693/k$$

$$\text{Hence } k = 0.693/100\,yr = 6.93 \times 10^{-3}\,yr^{-1}$$

Substitute into the first order rate equation and solve for t

$$\ln(c/c_o) = -kt$$

$$t = \{\ln(230\text{ pptv} / 50\text{ pptv})\}/6.93 \times 10^{-3}\,yr^{-1}$$

$$= 220\,yr = 2 \times 10^2\,yr \text{ assuming one significant figure}$$

15.4 Polar ozone holes

In the mid-1980's, evidence began to accumulate about an "Antarctic ozone hole" which developed in the late winter, with local depletions of stratospheric ozone up to 50%. The Antarctic in winter is a unique location in that the air circulation is entirely circumpolar, with almost no admixture of air from lower latitudes. No ozone can be generated or destroyed in the stratosphere during the long dark Antarctic winter while the Sun is below the horizon. Thus the ozone in the atmosphere in the polar springtime is the ozone that was present the previous fall.

The mechanism for the sudden loss of Antarctic ozone in early spring is believed to involve heterogeneous reactions which occur during the late winter on the surface of crystals of $HNO_3 .3H_2O$ and of water ice (H_2O). The crystals form when the temperature of the polar stratosphere is below about -70 °C. Chlorine-containing molecules such as $ClONO_2$ and HCl break down on the surface of these crystals, producing Cl_2 and HOCl.

[9] $HCl + ClONO_2 \rightarrow Cl_2 + HNO_3$

[10] $H_2O + ClONO_2 \rightarrow HOCl + HNO_3$

Unlike CFCs, which absorb only in the deep UV, Cl_2 and HOCl absorb radiation in the visible and near UV. This allows photolysis of Cl_2 and HOCl to begin immediately after the polar sunrise, even when light levels are low. Photolysis releases a sudden "pulse" of chlorine atoms, which vigorously initiate the radical chain reactions [5] and [6] at a time when the polar day is too short to allow ozone to be replaced. As much as 50% of the total ozone column has been observed to be destroyed in one week at some Antarctic locations in early spring.

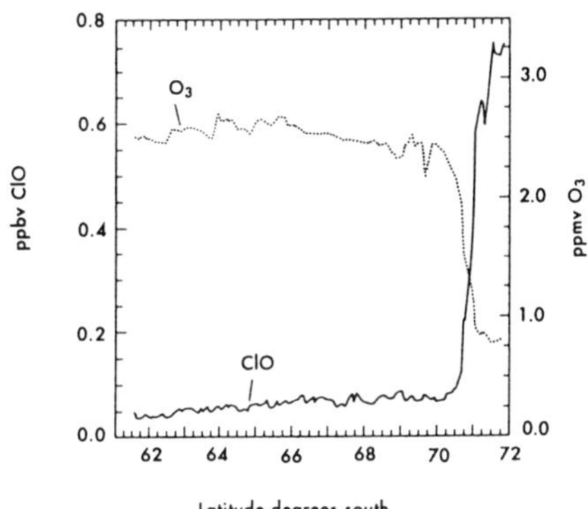

ClO and O_3 concentration over Antarctica, 18 km altitude, as a function of latitude, September 21, 1987.

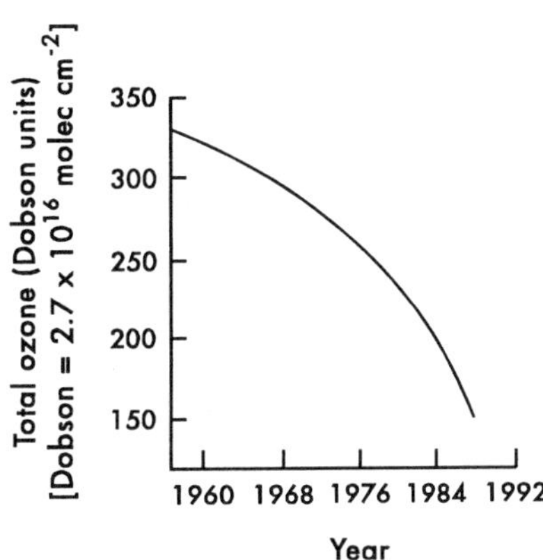

O_3 concentration over Antarctica, 1960 - 1988.

As spring progresses, the ice crystals sublime, new ozone is formed, and the ozone levels return to normal. Recent observations have suggested that a similar phenomenon may occur in the northern hemisphere, perhaps affecting latitudes well south of the Arctic Circle. The unanswered question is to what extent ozone depletion may extend into the temperate zones of the Earth, where humans, other animals, crops and forests would be exposed to higher UV intensities.

15.5 Consequences of ozone depletion

The immediate consequence of a reduction in the total amount of stratospheric ozone would be a greater penetration of short wavelength ultraviolet radiation to the Earth's surface (Figure 4).

Biological effects could also be anticipated if the total "ozone column" above the Earth were reduced, because radiation having λ < 320 nm is damaging biologically. Most biologists today accept that a critical event in the development of life on Earth was the evolution of photosynthetic organisms. This led to the production of significant levels of O_2 in the atmosphere and hence to the protective ozone shield. Only after this shield developed could organisms emerge from the safety of several meters' depth of ocean and colonize the land. It is therefore likely that many species, animals, plants, and microorganisms would be adversely affected by increased doses of high energy radiation. The penetration of additional UV radiation into the polar regions is of concern in terms of the oceanic phytoplankton, whose productivity may be compromised by higher intensities of ultraviolet radiation. This could upset the whole ecology of the oceans.

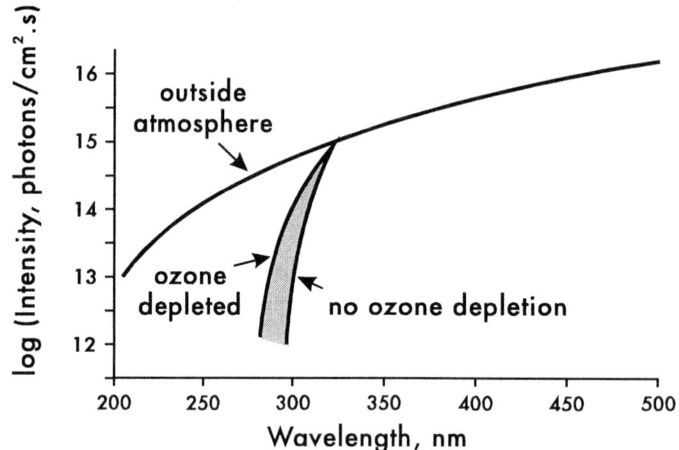

Figure 15.4: Increased penetration of high energy UV radiation to the troposphere as a result of decreased shielding by ozone

For humans, an expected consequence of an increased intensity of short wavelength solar radiation would be higher incidence of skin cancer. Solar radiation having $\lambda < 320$ nm is responsible for causing sunburn and, upon excessive sunlight exposure, skin cancer. Almost all cases of skin cancer can be attributed to over-exposure to sunlight, the exceptions occurring mostly in people suffering from inherited conditions where DNA-repair mechanisms are faulty. Skin cancer is almost exclusively an affliction of light skinned people; dark skin contains a larger quantity of the brown pigment melanin, which is an effective absorber of ultraviolet radiation, preventing it from reaching the living cells underneath the outer, dead, layer of the skin. Susceptible people tend to develop skin cancer in areas that are most exposed to the Sun: forehead, nose, around the neck and V of the throat, and on the arms. Modern lifestyles involve greater exposure of the skin to sunlight than the fashions of yesteryear, and have been accompanied by large increases in the skin cancer rate. In Australia, it has been estimated that a quarter of all Australians now living will be treated for skin cancer at some time in their lives, almost entirely as a result of excessive sun exposure.

Epidemiological data indicate that the incidence of skin cancer among whites is higher at lower latitudes (*i.e.*, closer to the Equator). The explanation of this effect is that at low latitudes the Sun's rays are more direct, because the Sun rises higher in the sky. When the Sun is overhead, solar radiation travels a shorter distance through the atmosphere before reaching the surface, and filtering by ozone is least effective. Conversely, when the Sun is low in the sky, radiation reaches the surface obliquely, and travels through a greater depth of atmosphere. This allows more efficient filtering of UV radiation by ozone. Consequently, the greatest flux of UV radiation is experienced in the tropics and/or at high altitudes near midday. This is why fair skinned people are advised to keep out of the Sun in the middle of the day, especially on holiday in the tropics or in the mountains.

In biochemical terms, the damaging effects of UV radiation are due to its absorption by molecules of DNA, which makes up the genetic code. Absorption of radiation by DNA can induce photochemical reactions in the DNA, causing the genetic code to be misread at the time of cell division[4].

Example: *What is the energy of UV-B radiation (λ 290-320 nm)?*
Recall the definition of UV-B radiation from Section 5.7.

Answer: *From Section 5.7,* $\Delta E(\text{photons}) \approx 1.2 \times 10^5/\lambda$ kJ nm mol^{-1}

At 290 nm, $\Delta E \approx 1.2 \times 10^5/290 = 4.1 \times 10^2$ kJ mol^{-1}

At 320 nm, $\Delta E \approx 1.2 \times 10^5/320 = 3.8 \times 10^2$ kJ mol^{-1}

Comparing these values for the energy of UV-B radiation (380-410 kJ mol^{-1}*) with bond dissociation energies in Appendix 2, we see that UV-B is sufficiently energetic to break many covalent bonds.*

Fair-skinned people can protect themselves from the harmful effects of UV radiation by the use of **sunscreens**. A sunscreen is a chemical substance which is applied to the skin to carry out the same function as melanin, namely to prevent high energy radiation from reaching the living cells in the lower layers of the skin. Suitable sunscreens must be able to absorb UV radiation without undergoing photochemical reactions, since photodegradation would render the sunscreen ineffective. The sunscreen molecules must therefore be efficient at degrading the energy of the solar photons into heat.

15.6 The Montréal Protocol

Political acceptance of scientific predictions of stratospheric ozone depletion occurred in North America before any adverse consequences had actually been observed; this stands in contrast to most other environmental problems with which society has had to deal. The prediction of a threat to stratospheric ozone by CFCs was made in 1974, and the use of CFCs as aerosol propellants was banned in North America as early as 1978.

A significant event in terms of political awareness of ozone depletion was the signing in 1987 of the "Montréal Protocol on substances that deplete the ozone

---

[4] The damaging effects of UV radiation have useful applications also; microorganisms are killed upon UV exposure. This sterilizing action of UV radiation is used in many European communities for the disinfection of drinking water (Section 10.8.2), and also in hospitals and laboratories for sterilizing equipment by means of "germicidal lamps".

layer". This international treaty set targets for CFC production to be cut back to 1986 "baseline" levels by mid-1989, cut to 80% of baseline by 1993, and to 50% of baseline by 1998. Subsequent information has suggested that these proposed cuts in CFC use would be insufficient to prevent substantial loss of stratospheric ozone over the next half century (Figure 5). Notice the long timescale; even with an immediate and total ban on the use of CFCs, the atmosphere would not return to "normal" for over a century.

Many governments in both Europe and North America, as well as major CFC producers, have called for a complete phase-out of all use of CFC-11 and CFC-12, in a 1992 extension to the Montréal Protocol. The United States EPA also suggests phasing out CH_3CCl_3, which is used as an industrial degreasing solvent, but which is not covered by the Montréal Protocol. Phase out of the most serious ozone-depleting CFCs in Canada is required by 1995 under the extended Montréal Protocol.

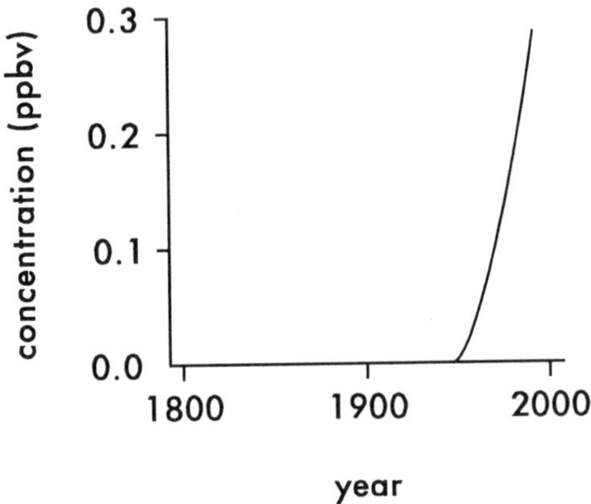

Tropospheric concentration of CFCs since *ca.* 1800

The question arises as to what is a "normal" concentration of stratospheric ozone. Ozone levels fluctuate because of variability of the Sun's output; events here on Earth can also modify ozone levels. A major meteorite impact in 1908 caused significant ozone depletion for 3-4 years; more recently, atmospheric nuclear testing in the 1950s and 1960s depleted the ozone shield at that time.

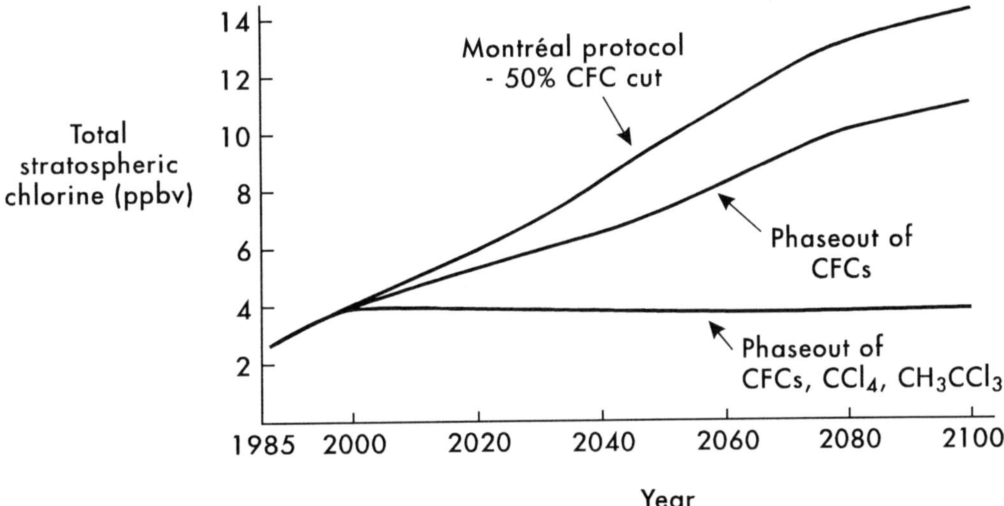

Figure 15.5: Predicted changes in the concentration of stratospheric chlorine for different levels of CFC use over the next century.

Severe volcanic episodes can also lower the levels of stratospheric ozone from "normal". In each of the foregoing examples, heated gases, which inevitably include NO_x, were injected directly into the stratosphere. Under these conditions another chain reaction, analogous to eq. [5]-[6], provides an additional mechanism for ozone loss.

$$NO + O_3 \rightarrow NO_2 + O_2$$
$$NO_2 + O \rightarrow NO + O_2$$

As with eq. [5]-[6], the sum of these two reactions is the same as eq. [4], and NO_x acts as a catalyst for the destruction of ozone.

15.7 CFC replacement compounds

We have seen that the properties of CFCs — non-flammable, non-toxic, and appropriate boiling point — give them important uses in modern society as the working fluids in refrigerators and air-conditioners, and as the blowing agents for certain plastic foams, such as those used for insulation, furniture and bedding.

Merely to ban the use of CFCs from these applications is impractical; therefore, CFC replacements are being sought. The goal is to develop compounds that will have the desirable properties for the critical applications described, while being more "environmentally friendly".

CFCs suffer from two drawbacks.

1. They are so stable chemically that they survive unchanged in the troposphere and eventually migrate to the stratosphere[5].

2. They contain chlorine atoms, which can catalyse the destruction of stratospheric ozone.

Therefore the ideal CFC replacement molecule should be sufficiently unreactive to be used in its intended application, but somewhat reactive in the lower atmosphere so that it will be oxidized before it has time to reach the stratosphere. It should also contain few, or ideally no, chlorine atoms.

Potential CFC replacement compounds are likely to be small halogenated organic molecules. CFCs owe their extreme chemical inertness to being fully halogenated. At the other end of the structural spectrum, hydrocarbons are highly flammable and potentially explosive.

Table 15.2: Properties of hydrocarbon derivatives

CH_3CH_3	CF_3CH_2Cl	CF_2Cl_2
hydrocarbon	partly halogenated hydrocarbon	fully halogenated hydrocarbon
highly flammable	virtually non-flammable	completely non flammable
tropospheric $t_{1/2}$, days	tropospheric $t_{1/2}$, years	tropospheric $t_{1/2}$, decades
ODP[a] = zero	very small ODP	strongly ozone-depleting

a ODP = ozone deleting potential. It is calculated by combining the proportion of the substance expected to reach the stratosphere, rate of chlorine release, and available chlorine.

CFC replacements will probably be compounds with one or two carbon atoms (to keep the boiling point in the right range) and some appropriate combination of hydrogen, fluorine, and chlorine substituents. Totally hydrogenated compounds (*i.e.*, hydrocarbons) are unsuitable because they are flammable and potentially explosive. Totally fluorinated compounds (*i.e.*, fluorocarbons) are unsuitable

[5] Besides their chemical stability, the values of K_H for CFCs are so small that they have little tendency to be "washed out" of the atmosphere by dissolving in raindrops.

because they are even more inert than CFCs; this would make them highly persistent greenhouse gases (Section 6.3), even though their stratospheric ozone depleting potential (ODP) would be zero. Too much chlorine in the molecule tends to make the substance toxic, and also leads to a high ODP.

The chief candidates as CFC replacements are partly fluorinated hydrocarbons with minimal or zero chlorine content. The presence of hydrogen in the molecule confers reactivity in the troposphere, because it allows the attack of OH upon the molecule (Section 9.2). For example, with CFC-134a:

[11] $CF_3CH_2F + OH \rightarrow H_2O + CF_3CHF$

The radical CF_3CHF can then easily undergo reaction with O_2.

$$CF_3CHF + O_2 \rightarrow CF_3\text{-}CHF\text{-}OO$$

As explained in Section 9.2, such a species will rapidly be converted to CO_2, water, and HF through a complex series of reactions.

The key to the tropospheric reactivity of potential CFC replacements is the rate constant (at 298 K) for their reaction with OH (Table 15.2).

Table 15.3: Rate constants for the reaction of CFCs and related compounds with OH

Compound	**k_{298}, cm^3 molec^{-1} s^{-1}**	**Compound**	**k_{298}, cm^3 molec^{-1} s^{-1}**
CH_4	8.4×10^{-15}	$CHFCl_2$	3.0×10^{-14}
C_2H_6	2.7×10^{-13}	CF_2CCl_2, $CFCl_3$	$< 5 \times 10^{-16}$
CH_3CCl_3	1.2×10^{-14}	CF_3CHFCl	1.0×10^{-14}
CF_3CH_2F	8.6×10^{-15}	CH_3CF_2Cl	3.6×10^{-15}

Example: *Calculate the half-life of CFC-134a in the troposphere, taking an assumed global average steady state concentration of OH of 8×10^5 molec cm^{-3}.*

Answer: *If [OH] is maintained at a steady state, reaction [11] follows pseudo-first order kinetics.*

$$\text{rate} = k'[CF_3CH_2F], \text{ where } k' = k[OH]$$

$$k' = (8.6 \times 10^{-15} \text{ cm}^3 \text{ molec}^{-1} \text{ s}^{-1})(8 \times 10^5 \text{ molec cm}^{-3})$$

$$= 7 \times 10^{-9} \text{ s}^{-1}$$

$$t_{1/2} = 0.693/k' = 1 \times 10^8 \text{ s} = 3 \text{ years}$$

The following compounds are presently under consideration as CFC replacements.

1. **As replacement for CFC-11 in foams**:

   CFC-22 (CHF_2Cl, b.p. -41°C);
   CFC-123 (CF_3CHCl_2, b.p. 28 °C);
   CFC-141b (CH_3CFCl_2, b.p. 32 °C)

Because these compounds contain hydrogen as well as chlorine, they each have only 2-5% the capacity to deplete ozone as CFC-11. CFC-11 replacements are particularly important in the furniture industry; hydrocarbon foam expanders are unacceptable in the manufacture of urethane foams for upholstered furniture and bedding, where flame resistance is an important property.

2. **As replacement for CFC-12 in refrigerators and air conditioners**:
   CFC-134a ($CF_3\ CH_2F$, b.p. -26 °C);
   CFC-152 (CH_3CHF_2, b.p. -25 °C)

Both compounds have zero ozone depleting potential (no chlorine in the molecule), although CFC-152 is somewhat flammable. CFC-152 is already commercially available, but CFC-134a is so far available only in "pilot plant" quantities.

CFCs are, or have been, used in other applications such as aerosol propellants and degreasing solvents, and were formerly used in the very controversial foamed fast food containers. North American industry has had over a decade to adjust to the ban on CFCs in aerosols. Butane, dimethyl ether, and pressurized CO_2 have successfully been substituted for CFC-11 as aerosol propellants. The first two of these are both flammable, and require additionally a flame suppressing agent. Methylene chloride (CH_2Cl_2) is used in the case of butane, and water in the case of dimethyl ether.

With phase-out of "hard" CFCs (those with high ozone depleting potential) required in many countries by the end of 1995, production is expected to start declining quickly after late 1993. After 1995, only recycled CFCs will be available. Users of CFCs who have not yet made the switch to alternative products may find themselves in a squeeze during 1994-6 when the old CFCs will be in short supply.

15.8 Halons

"Halons" are partly brominated CFCs which are used in fire extinguishers, especially those used on electrical fires. Their firefighting properties depend on the presence of weak C-Br bonds, which cleave at the high temperature of a fire, and interrupt the chain reactions which characterize combustion. Two common halons are H-1301 and H-1211 (the digits are the numbers of carbon, fluorine, chlorine, and bromine respectively in the molecule).

H-1301: CF_3Br H-1211: CF_2ClBr

Like CFCs, halons are unreactive in the troposphere and capable of photolysis in the stratosphere, where they can destroy ozone by serving as the source of catalytic bromine atoms. The catalytic cycle is analogous to eq. [5] and [6].

$$Br + O_3 \rightarrow BrO + O_2$$
$$BrO + O \rightarrow Br + O_2$$

Halons are even more effective ozone-depleters than CFC-11 and CFC-12, and like them, have long stratospheric lifetimes (H-1301, 70 years; H-1211, 15 years). Under the extended Montréal Protocol, their use is to be phased out by the year 2000. Part of the problem with the use of halons has been the insistence by insurance companies that institutional firefighting equipment such as that used to flood computer installations with halons be tested with a full charge of halon. Far more halon has been released to the atmosphere over the years in testing such equipment than in fighting fires.

15.9 Political considerations

The Montréal Protocol represented a considerable achievement in obtaining agreement for a partial phase-out of the "hard" CFCs. It was argued by the North Americans that they would have a harder time cutting CFC use by 50% than the Europeans, who were still using CFCs for aerosol propellants, and could simply discontinue that use. Events have since overtaken this argument, because both North American and Western European governments now agree on the need for a total phase-out of hard CFCs.

A more troubling issue is how to meet the aspirations of developing countries, whose citizens would naturally like access to modern conveniences such as refrigeration. Two problems exist. First, the older CFCs such as CFC-12 (CF_2Cl_2) are considerably cheaper to make than the proposed replacements, and also use "lower tech" processes. An extra few dollars for a refrigerator may make North Americans grumble, but does not exclude them from the refrigerator market. That may not be true in the Third World.

A second, more important obstacle concerns chemical patents. Patents are the means by which inventors prevent other people from copying their inventions. A chemical patent can cover a novel use for a certain composition of matter, or a novel methodology for manufacturing a composition of matter. A patent is granted only if the concept is original and represents more than an obvious extension of what is called "prior art". Thus you cannot patent the use of CFC-134a in refrigerators; it is obvious to any technically trained person that a substance with this boiling point might have such a use, given the prior art of using CFC-12 as a refrigerant.

A chemical patent, if granted, runs for 17 years from the date of filing the patent, and gives the inventor an exclusive right to exploit the invention. The inventor may licence the patent to other users, and may charge a royalty fee for this privilege. Anyone who benefits from another person's patented invention without permission may be held liable in a court of law for patent infringement, and can be assessed substantial damages. After the expiration of 17 years, the technology passes into the public domain and can legally be exploited by anyone.

The older CFCs such as CFC-11 and CFC-12 were introduced in the 1930s, and all process or use patents on these compounds have long since expired. Anyone can therefore go into business making "generic" CFCs. In contrast, major chemical manufacturers such as Du Pont and Allied Signal (both U.S.) and ICI (U.K.) have invested literally billions of dollars in developing technologies for synthesizing replacements such as CFC-134a efficiently. They have a right, and a legitimate expectation, to recover their investment through the price charged to customers. This causes the following dilemma.

The developed world would like to see the developing countries jump from no technology to advanced, low-pollution technology such as the use of CFC-134a in a single leap, without the intervening step of low technology such as the use of CFC-12 with its attendant high pollution. Unfortunately, low technology is cheaper and more accessible. Moreover, the owners of the advanced technology expect to recapture their development costs. It will do the stratosphere no good if the developed world embraces CFC replacement compounds, while simultaneously the developing world industrializes with "hard" CFCs. A way must be found to enable developing countries to jump into advanced technology while still compensating the owners of that technology for their inventions.

A related issue was raised in Section 13.7 over acidic precipitation. However, the problem is more serious in the case of stratospheric ozone depletion because CFCs have much longer atmospheric residence times (decades) than acidic gases (days). Thus acidic precipitation is a *regional* pollution problem, because acidic gases travel only 10^3 km or so during their atmospheric lifetime. Stratospheric ozone depletion by CFCs is, by contrast, a *global* pollution problem. CFCs are so long lived that they become globally dispersed, no matter where they happen to have been emitted. One country's emissions become everyone's concern.

Problems

Section 15.1 - 15.2

1. Calculate the pressure of ozone in atm and in ppmv at the tropopause (15 km altitude, 217 K), given $[O_3]$ = 1.0 x 10^{12} molec cm $^{-3}$, and p(total) = 0.12 atm.

2. a) Use thermodynamic data to calculate K for the following reaction at the stratopause, where p(total) = 1.0 x 10 $^{-4}$ atm and T = 270 K.

   $$3O_2\ (g) \rightleftharpoons 2O_3\ (g)$$

   b) Calculate the equilibrium pressure of O_3 under these conditions.
   c) Compare the value from part (b) with the experimental $p(O_3)$ of 4 ppmv.

3. Calculate the residence time of ozone in the atmosphere, if the total mass of ozone is 3 x 10^9 t, and 350,000 t are made and destroyed each day.

4. At an altitude of 170 km, T = 1100 K and p(total) = 3 x 10 $^{-10}$ atm. Under these conditions 80% of all oxygen molecules are dissociated into atoms (but most of the N_2 is still molecular).
   a) Calculate *K* for the reaction O_2 (g) $\rightleftharpoons$ 2O (g) under these conditions, using thermodynamic data.
   b) Estimate the percent dissociation of O_2 at equilibrium.
   c) Explain why the value calculated in part (b) is different from the experimental value.

5. Calculate ΔH° for reactions [1] and [3] of the text, for photons of wavelength 220 nm.

6. The reaction O (g) + O_2 (g) $\rightarrow$ O_3 (g) has k = 3.7 x 10^{-16} cm^3 molec $^{-1}$ s^{-1} at 220 K, when p(total) = 0.010 atm (this rate constant depends on the total pressure).
   a) Calculate the rate of reaction when p(O) = 2.1 x 10 $^{-4}$ ppmv.
   b) Calculate the half-life of oxygen atoms under these conditions.
   c) What does the result in (b) imply about the concentration of oxygen atoms in the night-time atmosphere?

7. This question concerns whether radiation of 300 nm is capable of bringing about the reaction

$$O_3\ (g) \rightarrow O_2\ (g) + O\ (g)$$

a) Calculate the energy of 300 nm radiation in kJ mol $^{-1}$.
b) Calculate ΔH° and ΔE for the reaction at -55°C (stratosphere).
c) Estimate whether 300 nm radiation is capable of bringing about this reaction.
d) Estimate the longest wavelength capable of bringing about this reaction, and comment upon your answer.

8. Table 15.3 contains data on flux of photons (per cm^2 of the Earth's surface) for different wavelength ranges. Column A is the flux outside the atmosphere, and Column B is the flux at the Earth's surface when the Sun is directly overhead.

Table 15.3 Solar photon intensity vs wavelength

Wavelength Range	I_o (Photons cm^{-2} s^{-1}) A	B
202-210	1.2×10^{13}	0
210-220	3.4×10^{13}	0
220-230	5.3×10^{13}	0
230-240	5.6×10^{13}	0
240-250	5.7×10^{13}	0
250-260	8.7×10^{13}	0
260-270	2.7×10^{14}	0
270-280	2.5×10^{14}	0
280-290	4.0×10^{14}	0
290-300	6.9×10^{14}	4.2×10^{11}
300-310	1.0×10^{15}	1.8×10^{14}
310-310	1.1×10^{15}	7.5×10^{14}
320-330	1.4×10^{15}	1.3×10^{15}
330-340	1.7×10^{15}	1.6×10^{15}
340-350	1.7×10^{15}	1.7×10^{15}

a) Calculate the rate of absorption of energy (J cm^{-2} s^{-1}) in the atmosphere over the range 200-350 nm.

b) Where in the atmosphere does this absorption occur, and what causes it?

c) Is the value calculated in part (a) applicable to all times of the day at all locations on the Earth's surface?

9. Consider this oversimplified scheme for the destruction of ozone in the atmosphere

(I) $O + O_3 \rightarrow 2O_2$

$k_1 = 1.5 \times 10^{-11} e^{-2218/T}$ cm^3 $molecule^{-1}$ s^{-1}

(IIA) $O_3 + Cl \rightarrow ClO + O_2$

$k_{IIA} = 8.7 \times 10^{-12}$ cm^3 $molecule^{-1}$ s^{-1} at 220 K

(IIB) $O + ClO \rightarrow Cl + O_2$

$k_{IIB} = 4.3 \times 10^{-11}$ cm^3 $molecule^{-1}$ s^{-1} at 220 K

Use the following steady state concentrations, in molecules per cm^3, to answer questions (a)-(e). O = 5.0×10^7; Cl = 1.0×10^5; ClO = 6.4×10^7; $O_3 = 3.2 \times 10^{12}$.

a) Calculate the activation energy of Reaction I.

b) Calculate the rates of Reactions I, IIA, and IIB at 220 K, in the units molecules cm^{-3} s^{-1}.

c) What is the overall rate of the Cycle IIA + 11B? Explain your reasoning.

d) What fraction of the ozone is destroyed under these conditions by the direct Reaction I, rather than by the Cycle II?

e) Why is there currently concern that the importance of Cycle II is increasing, and what would be the significance if this concern proved to be well founded?

10. This question follows on from Problem 9. Suppose, which is unrealistic, that Reactions I, IIA, and IIB are the only sinks for O_3, Cl, and ClO, and assume that the concentrations of O_3 and O are maintained at a steady state, which is reasonable. What will happen to the rates of Reactions IIA and IIB, and what will ultimately be the fraction of ozone destroyed by the direct Reaction I?

11. a) Estimate the heat of formation of the ClO(g) radical from bond energy data: Cl_2 = 243; O_2 = 498; ClO = 205 kJ mol $^{-1}$.
    b) Do you think that this is a very accurate estimate of ΔH°_f? Explain.
    c) What is the importance of the ClO radical in the chemistry of the atmosphere?
    d) Use your estimate of ΔH°_f for ClO to estimate the enthalpy changes for the catalytic cycle for the decomposition of ozone, reactions [5] and [6].

Section 15.3

1. Give chemical formulas for the following
   a) CFC - 12
   b) CFC - 113
   c) CFC - 152

2. Give CFC numbers for the following substances
   a) CHF_2Cl
   b) $ClCH_2\,CF_2Cl$
   c) $CF_3\,CHFCl$

3. a) $CFCl_3$ has $\Delta H^\circ_{vap.}$ 26.88 kJ mol $^{-1}$. Calculate the heat which must be supplied by an aerosol can to vaporize 0.25 g of $CFCl_3$ aerosol propellant.
   b) Does the aerosol can become
      (i) hotter
      (ii) cooler
      (iii) remain at the same temperature when the $CFCl_3$ vaporizes?

4. If p(CFC) ~ 250 pptv, estimate the total mass of CFC in the atmosphere (take M(CFC) as an average value of 130 g mol $^{-1}$).

5. The C-Cl bond strength in CFC-12 is 318 kJ mol $^{-1}$. Estimate the wavelength range over which you would expect this reaction to be possible, and comment on your calculated result.

6. A natural decomposition route for N_2O is a photochemical reaction

$$N_2O \xrightarrow{h\upsilon} N_2 + O^*$$

   a) Calculate the maximum wavelength needed to bring about this reaction from the information: ΔH°_f (N_2O, g) = 82 kJ mol $^{-1}$ ΔH°_f (O, g) = 249 kJ mol $^{-1}$; excitation energy of O* = 188 kJ mol $^{-1}$.

   b) Suggest an explanation for the fact that the reaction

$$N_2O \xrightarrow{h\upsilon} N_2 + O^*$$

   does not occur in the troposphere but does occur in the stratosphere.

   c) Calculate ΔG for the thermal reaction

$$N_2\ (g) + O\ (g) \rightarrow N_2O\ (g)$$

   in the stratosphere at 220 K, assuming p(O (g)) = 2.0 x 10 $^{-4}$ ppmv, total pressure = 0.010 atm, and p(N_2O (g)) = 1.6 x 10 $^{-3}$ ppmv. Explain whether the reaction is likely to be an important atmospheric source of N_2O.

   d) The lifetime of N_2O in the atmosphere is reported to be 100 years. What is the apparent first order rate constant for the disappearance of N_2O?

7. a) Calculate ΔG° for each step in the catalytic cycle [5], [6] at 220 K

   [5] Cl (g) + O_3 (g) → ClO (g) + O_2 (g)

   [6] ClO (g) + O (g) → Cl (g) + O_2 (g)

   b) Would the following system be at equilibrium with respect to reaction [5]?

   [Cl] 1.0 x 10^5 molec cm $^{-3}$, [ClO] 1.2 x 10^6 molec cm $^{-3}$, p(O_3) = 11 ppmv, p(total) = 0.26 atm.

Section 15.7

1. The concentration of CH_3CCl_3 in the atmosphere is currently 0.15 ppbv.

   a) Calculate the mass of CH_3CCl_3 in the atmosphere.

   b) Use the kinetic data in Table 15.2 to estimate the half-life of CH_3CCl_3 in the tropospherical [OH] has the "average" value of 8 x 10^5 molec cm $^{-3}$.

   c) The concentration of CH_3CCl_3 is growing 4% per year. Estimate the rate of release of CH_3CCl_3 into the atmosphere and the rate of reaction of CH_3CCl_3 with atmospheric constituents in tonnes per year.

2. Use the rate constant in Table 15.2 for the reaction of CF_3CH_2F with OH in order to calculate the fraction of this substance which reaches the stratosphere, if the half-life for migration to the stratosphere is 5 years. Assume a constant value of 8×10^5 molec cm^{-3} for [OH].

3. CFC-22 has critical parameters T_c 96°C, P_c 48.5 atm. Calculate the concentration of CFC-22 in mol L^{-1} at its critical point.

4. a) Estimate the longest wavelength of radiation capable of clearing CF_3Br to CF_3 + Br.
   b) Is this radiation present
      (i) in the troposphere only
      (ii) in the stratosphere only
      (iii) in both the troposphere and the stratosphere?
   c) Given that CF_3Br absorbs radiation only at $\lambda < 255$ nm, in which region of the atmosphere do you expect CF_3Br to react?

5. CF_3Br has ΔH_f° = -642.7 kJ mol^{-1}. Use this information, plus data in Appendices 1 and 2 to estimate the standard molar enthalpy of formation of the trifluoromethyl radical, $CF_3(g)$

6. CF_2Cl_2 is prepared by the following reaction (under catalytic conditions)

$$CCl_4\,(g) + 2HF\,(g) \rightarrow CF_2Cl_2\,(g) + 2HCl\,(g)$$

Thermodynamic parameters for CCl_4 (g) and CF_2Cl_2 (g) follow

	CCl_4 (g)	CF_2Cl_2 (g)
ΔH_f° (kJ mol^{-1})	- 102.9	- 477.0
S° (J mol^{-1} K^{-1})	309.7	300.7

   a) Calculate ΔG° and hence K_p for the reaction above at T = 250°C
   b) Is the following reaction mixture at equilibrium: $p(CCl_4)$ = 0.12 atm, p(HF) = 0.70 atm, p(HCl) = 1.4 atm, $p(CF_2Cl_2)$ = 2.1 atm
   c) CCl_4 (g, 0.55 atm) and HF (g, 1.1 atm) are heated to 250°C in the presence of catalyst. What fraction of the CCl_4 is left unreacted at equilibrium?

16

ELECTROCHEMISTRY

16.1 Electrochemistry, oxidation and reduction

Electrochemistry deals with the interaction between chemical reactions and electricity: chemical reactions which produce electricity, and chemical reactions which are driven by the input of electrical energy. The equipment for carrying out an electrochemical reaction is called an **electrochemical cell**. In the simplest case, this consists of a reaction vessel containing an electrically-conducting liquid, which may be either the solution of an electrolyte, or sometimes a molten salt. Two electrically-conducting **electrodes** dip into the solution. The electrodes are connected to each other by means of an external circuit, and the circuit is completed because electricity can be also carried through the reaction mixture by the ions that are present. Figure 1 presents a schematic.

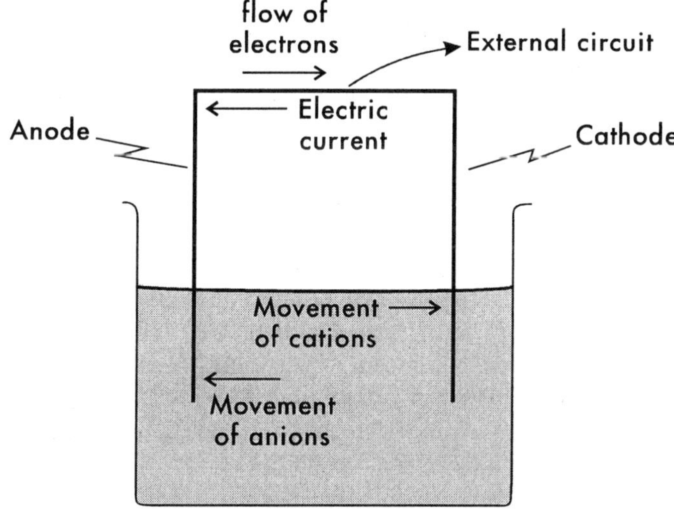

Figure 16.1: Schematic of an electrochemical cell

Electrochemical cells are named differently according to whether electrical energy is an input or an output of the chemical reaction. When the chemical reaction produces electricity, the electrochemical cell is said to be a **galvanic cell**; when electricity is required to drive the chemical reaction, the electrochemical cell is termed an **electrolytic cell**.

Let's consider some examples of galvanic and electrolytic cells. Common batteries (flashlight batteries, watch batteries, heart pacemaker batteries) are all examples of galvanic cells -- they produce electricity by means of a chemical reaction, and they become "dead" when their reactants are used up. Electrolytic cells are more common in industry. The chlor-alkali process (Chapter 4) involves the production of Cl_2 and NaOH by passing an electric current through an aqueous solution of NaCl. The production of the metals sodium, magnesium, and aluminum involves, in each case, passing an electric current through a molten salt of the corresponding metal. Car batteries and rechargeable nickel-cadmium batteries are both galvanic and electrolytic (though not at the same time!). They are galvanic when they are being used to produce electricity (as when you start your car); they are electrolytic when they are connected to the electricity supply for recharging.

The chemical processes which occur at the electrodes are described as **oxidation and reduction**: this area of chemistry is also known as **redox chemistry**, the word redox being a contraction of <u>red</u>uction/<u>ox</u>idation. We will now define oxidation and reduction.

The terms oxidation and reduction have different definitions according to context.

Definitions

Oxidation:

(i) addition of oxygen to, or removal of hydrogen from, the substance

(ii) removal of one or more electrons from the substance

Reduction:

(i) addition of hydrogen to, or removal of oxygen from, the substance

(ii) addition of one or more electrons to the substance

Definitions (i) and (ii) are completely consistent, although the definitions in terms of addition or removal of electrons are broader than those in terms of oxygen or hydrogen addition or removal. The pair of definitions (i) is especially useful in the context of organic chemistry or biochemistry. However, in the context of electrochemistry, only the definitions (ii) are useful.

Example: *In each of the reactions below state whether oxidation or reduction occurred.*

a) $Fe(s) \rightarrow Fe^{2+}(aq) + 2e^-$

b) $CH_4 \rightarrow CO_2$ + other products

c) $Zn(s) + Cu^{2+}(aq) \rightarrow Zn^{2+}(aq) + Cu(s)$

Answer:

a) *The iron gave up two electrons and was therefore* ***oxidized*** *to Fe(II)*

b) *The* CH_4 *lost hydrogen and gained oxygen, and was* ***oxidized*** *to* CO_2

c) *This one is more tricky. The Zn lost two electrons and was* ***oxidized*** *to* Zn^{2+}*; simultaneously, the* Cu^{2+} *gained two electrons and was* ***reduced*** *to Cu.*

The first two examples did not show complete chemical reactions. The third example, which did show a balanced chemical reaction illustrates a very important point, namely that *in any balanced redox reaction, oxidation of one reactant must be accompanied by reduction of another reactant.* Oxidation cannot occur in the absence of reduction, and *vice versa.* The situation is reminiscent of the Brønsted-Lowry definition of acids and bases (Section 11.1), where acidic behaviour can only be manifested in the presence of a base, and *vice versa.*

The two electrodes of a simple electrochemical cell are given different names according to the nature of the chemical reaction that occurs there. The **anode** is the electrode at which oxidation occurs; at this electrode, electrons leave the electrode and move into the external electrical circuit. The **cathode** is the electrode at which reduction occurs; at the cathode, electrons enter the solution from the external electrical circuit. These points are noted on Figure 1: remember that the flow of current in an electric circuit is always taken to be the opposite from the direction of flow of the negatively charged electrons.

The terms we have met so far are summarized in Table 16.1. Before reading further, make sure that you have learned the meaning of each of these words.

Table 16.1: Summary of definitions of electrochemical terms

Galvanic cell:	an electrochemical cell which produces electricity
Electrolytic cell:	an electrochemical cell which is driven by the input of electricity
Oxidation:	a chemical process in which the reactant loses electrons
Reduction:	a chemical process in which the reactant gains electrons
Anode:	the electrode at which oxidation occurs
Cathode:	the electrode at which reduction occurs

Example: *At what electrode do the following processes occur?*

a) $2\ Cl^-(aq) \rightarrow Cl_2(g) + 2\ e^-$

b) $Fe^{3+}\ (aq) + e^- \rightarrow Fe^{2+}\ (aq)$

c) $H_2O_2(aq) \rightarrow H_2O(\ell) + \frac{1}{2}\ O_2(g)$

Answer:

a) Cl^- *loses electrons, and is oxidized to* Cl_2. *Oxidation occurs at an* ***anode.***

b) Fe^{3+} *gains electrons, and is reduced to* Fe^{2+}. *Reduction occurs at a* ***cathode.***

c) H_2O_2 *loses oxygen and is reduced to water. Reduction occurs at a* ***cathode.***

16.2 Half-cells and cell diagrams

16.2.1 Half-cells

Before discussing any half-cells, let's ask what would happen if we were to add metallic zinc to a solution of copper sulfate. This reaction was the subject of example (c) of the last-but-one set of examples.

$$Zn(s) + CuSO_4(aq) \rightarrow ZnSO_4(aq) + Cu(s)$$

This can also be written as a net ionic equation.

$$Zn(s) + Cu^{2+}(aq) \rightarrow Zn^{2+}(aq) + Cu(s)$$

At the level of observation, the silvery zinc is seen to become covered with a red-brown layer of copper. In terms of redox chemistry, the zinc is oxidized to Zn^{2+}, and Cu^{2+} ions are simultaneously reduced to metallic copper. The reaction is spontaneous, and releases energy in the form of heat.

Now let's carry out the same experiment in a slightly different way. Instead of simply mixing the reactants, we keep the {zinc + zinc sulfate} separate from the {copper + copper sulfate} by means of a porous partition. This is the very simple galvanic cell shown in Figure 2.

Compartment A consists of a rod of zinc metal which dips into a solution of $ZnSO_4$; compartment B is a rod of copper metal dipping into a solution of $CuSO_4$. Compartments A and B are connected electrically in two ways; through the external circuit, shown with the voltmeter, and by means of the component LJ, which stands for "liquid junction". A liquid junction is used in an electrochemical cell whenever the contents of the anode and cathode compartments must be kept physically separated, to prevent them from reacting chemically with each other, as we saw happened when the reactants zinc metal and copper sulfate were simply mixed together. The liquid junction allows electrical communication between the anode and cathode compartments without mixing the reactants. Types of liquid junction include a fine porous glass frit, or a tiny hole between electrode compartments, or a tube filled with an inert electrolyte such as KCl(aq) but with a frit or gelatin plug at each end.

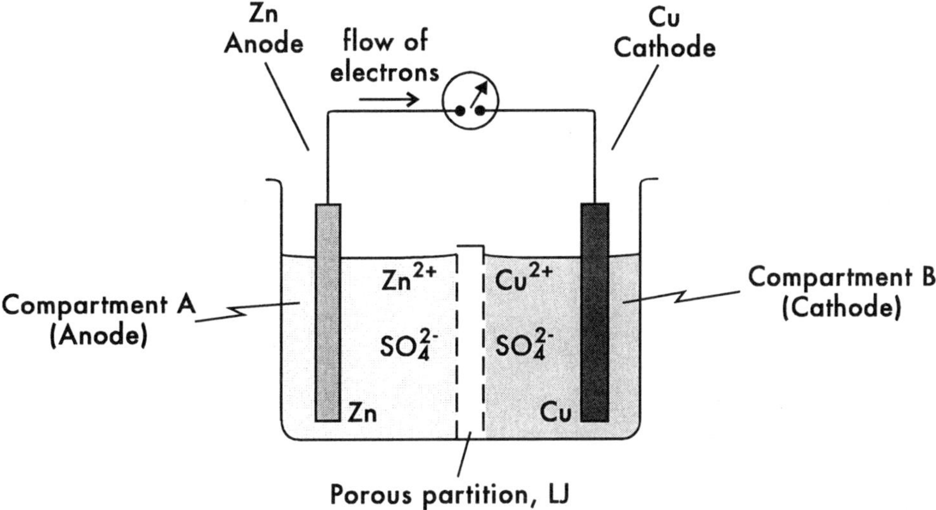

Figure 16.2: A simple galvanic cell

Let us consider in turn the reactions occurring at the anode and the cathode. In this cell, compartment A is the anode, and compartment B is the cathode.

Anode (oxidation): $Zn(s) \rightarrow Zn^{2+}(aq) + 2\,e^-$

Zinc loses two electrons to form Zn^{2+} (aq) and the two electrons leave the solution, moving into the external circuit. As the reaction proceeds, the anode gradually dissolves.

Cathode (reduction): $Cu^{2+}(aq) + 2\,e^- \rightarrow Cu(s)$

Cu^{2+} ions are reduced at the cathode by two electrons which come from the external circuit; the Cu(s) which is formed deposits on the surface of the cathode.

Overall (net) reaction: $Zn(s) + Cu^{2+}(aq) \rightarrow Zn^{2+}(aq) + Cu(s)$

Overall, there has to be a complete electrical circuit. The number of electrons leaving the anode must be balanced by the number re-entering the cell at the cathode; in addition, charge must be carried through the cell compartments and the liquid junction, which is why these cell components must all include electrolytes. The Zn^{2+} ions carry charge away from the anode, while in the cathode compartment Cu^{2+} ions carry charge to the metallic cathode, where they form elemental Cu by reaction with the electrons which have travelled through the external circuit. As we shall see in Section 16.4, the potential of the cell is determined by the concentrations of the electrochemically active species in the two cell compartments.

Suppose we were to remove the cathode compartment of Figure 2 completely and replace it with a silver cathode and an electrolyte of $AgNO_3$(aq). Now look again at the anode and cathode reactions.

Anode (oxidation):	Zn(s)	$\rightarrow$ Zn^{2+} (aq) + 2 e^-
Cathode (reduction):	Ag^+ (aq) + e^-	$\rightarrow$ Ag(s)
Overall reaction:	Zn(s) + 2 Ag^+ (aq)	$\rightarrow$ Zn^{2+} (aq) + 2 Ag(s)

The anode reaction is completely unchanged from the preceding example. However, at the cathode, Ag^+ ions are reduced by electrons which came from the external circuit and Ag(s) is deposited on the surface of the cathode. In order to preserve electrical neutrality, 2 mol Ag^+ must be reduced for each 1 mol of Zn(s) that is oxidized.

The foregoing discussion makes it clear that the anode and the cathode have independent functions in an electrochemical cell. In principle, we can "mix and match" different anodes with different cathodes to set up a whole variety of different electrochemical cells. Therefore, we can conveniently divide any electrochemical cell into two "half cells" and consider the chemistry in each of them independently.

16.2.2 Cell diagrams

Diagrams such as Figure 2 are rather cumbersome to draw for every electrochemical cell in which we might be interested, and so chemists have devised a shorthand method to write down the components of such cells. In order to see how this works, let's examine the **cell diagram** for the cell in Figure 2.

$$\text{Zn(s)} \mid \text{ZnSO}_4\ (x \text{ mol L}^{-1}) \parallel \text{CuSO}_4\ (y \text{ mol L}^{-1}) \mid \text{Cu(s)}$$

This cell could also have been written in terms of the net ionic equation.

$$\text{Zn(s)} \mid \text{Zn}^{2+}\ (x \text{ mol L}^{-1}) \parallel \text{Cu}^{2+}\ (y \text{ mol L}^{-1}) \mid \text{Cu(s)}$$

Rules: *There are strict rules for writing cell diagrams by this shorthand method.*

- *The anode is on the extreme left of the line, and the cathode is on the extreme right. Thus electrons can be imagined to travel through the cell from right to left. Additionally, the electrons return from the anode to the cathode through the external circuit, as shown in Figure 2.*

- *The contents of the cathode compartment are written beside the cathode; the contents of the anode compartment likewise appear next to the anode.*

- *A single vertical line | is used to denote a change of physical state (e.g., the change from the solid metal zinc to the aqueous zinc sulfate solution). If there is a mixture of substances in a given compartment, they are written with a comma between.*

- *The liquid junction is denoted by means of a double vertical line ‖.*

- *The concentrations of all reactants and products must be included, because the potential of an electrochemical cell changes with reactant and product concentration (or pressure, if the substance is a gas).*

Example: *Write the conventional representation for an electrochemical cell in which the cathode reaction is the reduction of Fe^{3+} to Fe^{2+} at an electrode made of platinum, and the anode reaction is the dissolution of a zinc rod to form Zn^{2+}.*

Answer: *The anode reaction is written on the left.*

$$Zn(s) \mid Zn^{2+}(aq) \parallel Fe^{3+}(aq), Fe^{2+}(aq) \mid Pt(s)$$

As we shall see in Section 16.4, the complete description of the cell requires us to provide in addition the concentrations of all these components.

16.3 Reduction potentials

16.3.1 Standard reduction potential

Every electrochemical cell operates at a definite, and calculable, potential, which depends on the specific chemical reactions involved and the concentrations of the reactant and product substances. Reduction potentials provide a means of systematizing this information. Reduction potentials ($\mathscr{E}$) are thermodynamic parameters, comparable with standard molar enthalpies and free energies of formation. They are always listed for half-cells, and the potential of the complete cell is calculated from the individual half cell potentials. By analogy with $\Delta G°$ and ΔG, $\mathscr{E}°$ and $\mathscr{E}$ are respectively the standard reduction potential and the reduction potential of the experimental cell under investigation. Only standard reduction potentials are listed in tables: a listing of standard reduction potentials at 298 K is given in Appendix 4.

Standard reduction potentials are defined in terms of the potential when the oxidized form (reactant) and the reduced from (product) are each at unit activity (or, approximately, 1 atm for gases and 1 mol L^{-1} for solutes in solution). Consider the standard reduction potentials for the two half cells in Figure 2.

Cathode reaction: This is already a reduction reaction.

$$Cu^{2+}(aq, a = 1) + 2\ e^- \rightarrow Cu(s) \qquad \mathscr{E}° = +\ 0.337\ V$$

In words: the reduction potential of the Cu^{2+} /Cu half cell is + 0.337 volts when both Cu^{2+} (aq) and Cu(s) are present at unit activity ($\approx$ 1 mol L^{-1} Cu^{2+}, and pure Cu metal): review the definition of standard states, Section 14.5.1.

Anode reaction: Here the chemical reaction is $Zn(s) \rightarrow Zn^{2+}(aq, a = 1) + 2\ e^-$

This half reaction is the reverse of the reaction recorded in the tables of standard *reduction* potentials.

$$Zn^{2+}(aq, a = 1) + 2\ e^- \rightarrow Zn(s) \qquad \mathscr{E}° = -\ 0.762\ V$$

The reduction potential of the Zn^{2+} /Zn half cell is - 0.762 volts when both Zn^{2+} (aq) and Zn(s) are present at unit activity. When the chemical reaction is reversed, as when oxidation occurs instead of reduction, the sign of $\mathscr{E}°$ changes. Thus the standard half-cell potential for the oxidation of Zn(s) to Zn^{2+}(a = 1) is +0.762V.

16.3.2 Cell potential

The standard reduction potential $\mathscr{E}°$ of an electrochemical cell is obtained by combining the standard reduction potentials of the two half cells of which it is composed. As a first example, consider the cell

$$Zn(s)\ |\ Zn^{2+}(aq, a = 1)\ \|\ Cu^{2+}(aq, a = 1)\ |\ Cu(s).$$

Cathode reaction: This is already a reduction; the cathode reaction is written as it occurs:

$$Cu^{2+}(aq,\ a = 1) + 2\ e^- \rightarrow Cu(s) \qquad \mathscr{E}° = +\ 0.337\ V$$

Anode reaction: This is an oxidation, so the reduction reaction defining $\mathscr{E}°$ in the anode compartment must be reversed:

$$Zn(s) \rightarrow Zn^{2+}(aq,\ a = 1) + 2\ e^- \qquad \mathscr{E}° = +\ 0.762\ V$$

Adding these two reactions gives $\mathscr{E}°$ for the overall process as +1.099 V (add the two half cell potentials above).

Since we had to reverse the reaction, and hence the sign of $\mathscr{E}°$, in the anode compartment but not in the cathode compartment, we can make an important generalization, eq. [1].

[1] $\mathscr{E}°(\text{cell}) = \mathscr{E}°(\text{cathode}) + \mathscr{E}°(\text{anode})$

Be careful to note in eq. [1] that $\mathscr{E}°$(anode) = -("standard reduction potential of the anode reaction)". $\mathscr{E}°$(anode) has the sign of the standard reduction potential reversed.

Example: *Calculate the standard potential for the cell below, using standard reduction potentials from Appendix 4.*

$$Ag(s)\ |\ Ag^+\ (a = 1)\ \|\ Cu^{2+}\ (a = 1)\ |\ Cu(s)$$

Answer: *The standard reduction potentials are + 0.799 V for Ag^+ and + 0.337 V for Cu^{2+}. Ag is the anode of the cell, and Cu is the cathode.*

$$\mathscr{E}°(\text{cell}) = \mathscr{E}°(\text{cathode}) + \mathscr{E}°(\text{anode}) = 0.337\ V + (-\ 0.799\ V) = -\ 0.462\ V$$

You may wonder why it is that we add together $\mathscr{E}°$(cathode) and $\mathscr{E}°$(anode) without considering the number of electrons involved in the half reactions. In the example above $Cu^{2+}(aq) \rightarrow Cu(s)$ is a two electron change, while $Ag(s) \rightarrow Ag^+(aq)$ is a one electron change, yet we just added the values of $\mathscr{E}°$ together. Unlike when we add ΔH°'s (Chapter 5) or multiply K's (Chapter 8), we seem not to take stoichiometry into account when we add potentials. The reason is that the potential of the half cell does not depend on how we write the half reaction (unless we reverse it).

$$Ag^+(aq) + e \rightarrow Ag(s) \qquad \mathscr{E}° = +\ 0.799\ V$$

$$2\ Ag^+(aq) + 2\ e \rightarrow 2\ Ag(s) \qquad \mathscr{E}° = +\ 0.799\ V$$

We shall be able to explain this point after reading Section 16.4.

16.3.3 Standard hydrogen electrode

Experimentally, you cannot measure a half cell potential directly. You can only measure the overall potential of a complete cell. In order to divide up the potential of a complete cell between the two half cells, it is necessary to assign a value arbitrarily to some particular half cell.

Consider the cell below:

$$Zn(s) \mid Zn^{2+}(aq, a = 1) \parallel H^{+}(aq, a = 1), H_2(g, a = 1) \mid Pt(s)$$

$$\mathcal{E} = \mathcal{E}° = +0.762V$$

At the anode, zinc metal dissolves to give Zn^{2+} (aq) under standard conditions (unit activity), while at the cathode H^{+}(aq) is reduced to hydrogen gas, again under standard conditions.

By arbitrarily setting to zero the standard reduction potential of the half cell $H^{+}(aq) \mid H_2(g)$, we determine the standard reduction potential of the $Zn^{2+}(aq) \mid Zn(s)$ reaction as -0.762V.

$$\mathcal{E}°(\text{anode}) = \text{-standard reduction potential of } Zn^{2+} \mid Zn$$

$$\begin{aligned} \mathcal{E}°(\text{anode}) &= \mathcal{E}°(\text{cell}) - \mathcal{E}°(\text{cathode}) \\ &= +0.762V - 0.000V \\ &= +0.762V \end{aligned}$$

The half cell involved in the Standard Hydrogen Electrode is written below.

$$\parallel H^{+} (aq, a = 1) \mid H_2 (g, a = 1), Pt \qquad \mathcal{E}° = 0.000 \text{ V}$$

or, approximately:

$$\parallel H^{+}(aq, 1.000 \text{ mol L}^{-1}) \mid H_2 (g, 1.000 \text{ atm}), Pt \qquad \mathcal{E}° = 0.000 \text{ V}$$

In words: the Standard Hydrogen Electrode (Figure 3) is a half cell in which H^{+} (aq) at unit activity is in contact with an inert platinum electrode and hydrogen gas at unit activity (or alternatively, H^{+}(aq, ~ 1 mol L^{-1}) in contact with H_2(g) at 1 atm). This half cell is assigned a potential of 0.000 V. Consequently, $\mathcal{E}°$ for any half reaction is the same as $\mathcal{E}$(cell) for the electrochemical cell in which the half reaction in question is made the cathode, and the Standard Hydrogen Electrode acts as the anode. The concept of dividing up a cell potential by assigning $\mathcal{E}°(H^{+} \mid H_2)$ as zero is reminiscent of the way in which $\Delta H°_f(H^{+}, aq)$ and $\Delta G°_f(H^{+}, aq)$ were set to zero (Sections 5.3 and 14.4). As we shall see in the next section, these two conventions are completely self-consistent.

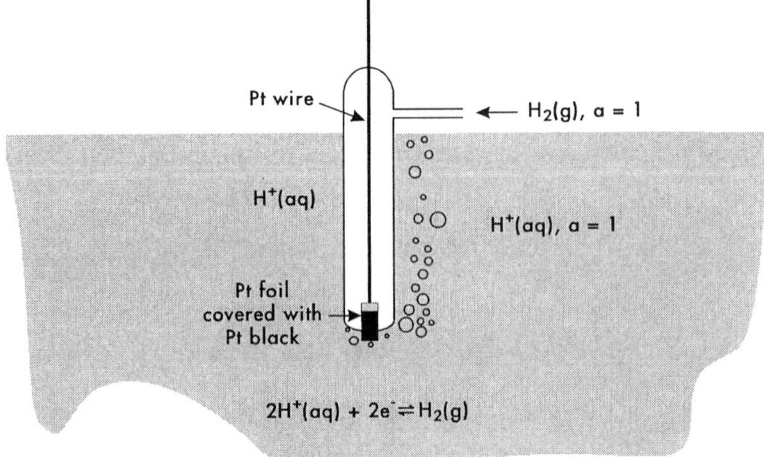

Figure 16.3: Experimental arrangement of a standard hydrogen electrode

16.4 Nernst equation

The chemical reaction taking place in a galvanic cell is always a spontaneous reaction, and some or all of the energy produced in the reaction is released as electricity. An electrolytic reaction uses electricity to drive an otherwise non-spontaneous reaction. There is a linear relationship between free energy and cell (or half cell) potential, eq. [2].

In eq. [2], n is the number of electrons involved in the redox process, and $\mathscr{F}$ is a constant known as the Faraday constant, after the nineteenth century scientist, Michael Faraday. The Faraday constant has the value 96,487 C mol $^{-1}$ (coulombs per mole) and represents the electric charge on 1 mol of electrons. Since charge (C) x potential (V) has the units of energy (1 C x 1 V = 1 J), the product $\mathscr{F}\mathscr{E}$ has the units J. Hence eq. [2] affords ΔG (ΔG°) in the units J mol $^{-1}$.

[2a] $\Delta G = -n\mathscr{F}\mathscr{E}$

[2b] $\Delta G^\circ = -n\mathscr{F}\mathscr{E}^\circ$

Note the following points about eq. [2].

1. Spontaneous reactions have ΔG negative, but $\mathscr{E}$ positive.
2. The assignment $\mathscr{E}^\circ = 0.000$ V for the standard hydrogen electrode is equivalent to setting ΔG°_f for H^+ (aq) to zero (Section 14.4).

The importance of eq. [2] is that electrochemistry provides the most direct method of measuring the free energy changes associated with a particular reaction. Free energy change can therefore be measured experimentally, just as, for example, enthalpy change. In fact, electrochemical measurements are much simpler to make than calorimetric ones (Section 5.8), although their precision is not as high.

It is important to draw attention to the negative sign in eq. [2]. Recalling from Chapter 14 that a spontaneous process is associated with a negative ΔG, a spontaneous reaction will have *positive* $\mathscr{E}$. Let us summarize the properties of ΔG and $\mathscr{E}$ (Table 14.2).

Table 14.2: Summary of the properties of ΔG and $\mathscr{E}$

Spontaneous reaction:	ΔG negative; $\mathscr{E}$ positive; galvanic cell
Non-spontaneous reaction:	ΔG positive; $\mathscr{E}$ negative; electrolytic cell
ΔG°, $\mathscr{E}^\circ$:	The value of ΔG or $\mathscr{E}$ when all reactants and products are at unit activity; in practice, the value found in tables or calculated from tabulated data.
ΔG, $\mathscr{E}$:	The value of ΔG or $\mathscr{E}$ pertaining to the particular reaction conditions under investigation

Equation [2] allows us to relate $\mathscr{E}$ to $\mathscr{E}°$, just as ΔG can be related to ΔG°.

$\Delta G = \Delta G° + RT\ln Q$

[3] $$\mathscr{E} = \mathscr{E}° - \left\{\frac{RT}{n\mathscr{F}}\right\}\ln Q$$

Equation [3] is known as the Nernst equation after its discoverer, Walter Nernst. The Nernst equation can be applied to both electrochemical half cells and to complete cells. For a reaction which has reached equilibrium, $\mathscr{E} = \Delta G = 0$, and hence:

$$\mathscr{E}° = \left\{\frac{RT}{n\mathscr{F}}\right\}\ln K$$

The factor $(RT/n\mathscr{F})$ has the numerical value $0.0257/n$ at 298 K when natural logarithms are used. When common (base 10) logarithms are used, the corresponding value is $0.0591/n$. It is noted here to save the trouble of calculating it repeatedly in the following examples.

As you look at eq. [3], notice that if $Q < 1$, $\mathscr{E} > \mathscr{E}°$. Conversely, if $Q > 1$, $\mathscr{E} < \mathscr{E}°$. Does this make intuitive sense?

- Consider the cathode ‖ Cu^{2+}, x mol L^{-1} | Cu(s).
- When $a(Cu^{2+}) = 1$, $\mathscr{E} = \mathscr{E}° = +0.337$ V.
- In this case, $Q = 1/a(Cu^{2+})$, since Cu^{2+} is a reactant.
- When $a(Cu^{2+}, aq) < 1$ (less Cu^{2+} in solution), it will be more difficult for Cu^{2+} (aq) to be reduced at the cathode: $Q > 1$, $\mathscr{E} < \mathscr{E}°$.
- Alternatively, if $a(Cu^{2+}, aq) > 1$ (more Cu^{2+} in solution), it will be easier for Cu^{2+} (aq) to be reduced; $Q < 1$, $\mathscr{E} > \mathscr{E}°$.

Example: *Calculate $\mathscr{E}$ for the following anode, given that $\mathscr{E}°$ for the reduction of Cu^{2+} (aq) is + 0.337 V.*

Cu(s) | Cu^{2+} (1.4×10^{-4} mol L^{-1}) ‖

Answer: *Since this is an oxidation process, $\mathscr{E}° = -0.337$ V. The chemical reaction is:*

$$Cu(s) \rightarrow Cu^{2+}(aq) + 2e^-$$

From the Nernst equation [3]:

$$\mathscr{E} = \mathscr{E}° - (0.0257/n)\ln([Cu^{2+}])$$
$$= (-0.337) - (0.0257/2)\ln(1.4\times10^{-4})$$
$$= -0.337 - (0.0129)(-8.87) = -0.223 \text{ V}$$

Check whether the answer looks correct. In this case $a(Cu^{2+}) < 1$. Since Cu^{2+} is a product, the reaction will proceed more spontaneously because its concentration is low. In other words $Q < 1$, $\mathscr{E} > \mathscr{E}°$.

The Nernst equation allows us to understand why $\mathscr{E}^\circ$ is independent of the way we choose to write a half-cell reaction, or a complete cell reaction. Consider again the example at the end of Section 16.3.2.

(1) $Ag^+(aq) + e \rightarrow Ag(s) \quad \mathscr{E}^\circ = +\ 0.799\ V$

(2) $2\ Ag^+(aq) + 2\ e \rightarrow 2\ Ag(s) \quad \mathscr{E}^\circ = +\ 0.799\ V$

From equation [3]: $\mathscr{E}^\circ = (0.0591/n)\log_{10}K$ where $K = [Ag^+]^{-1}$

In case (1) $\mathscr{E}^\circ = (0.0591/1)\log_{10}[Ag^+]^{-1} = -0.0591\log_{10}[Ag^+]$

In case (2) $\mathscr{E}^\circ = (0.0591/2)\log_{10}[Ag^+]^{-2} = -0.0296\log_{10}[Ag^+]^2$

$= (-0.0296)(2\log_{10}[Ag^+]) = -0.0591\log_{10}[Ag^+]$

The value of $\mathscr{E}^\circ$ does not change because the difference in the number of electrons involved in the reaction (n) is exactly balanced by the change in the logarithmic term.

Equation [2], which relates potential to free energy change, gives a different perspective on this problem. As we have seen, $\mathscr{E}^\circ$ (and also $\mathscr{E}$) are the same for cases (1) and (2). However, ΔG° (and ΔG) are twice as large for case (2) as for case (1), since in case (2), two moles of Ag^+ are reduced, compared with only one mole in case (1).

[2] $\Delta G(^\circ) = -n\mathscr{F}\mathscr{E}(^\circ)$

Rule: *Potential (in volts) is independent of the amount of reaction at a given set of concentrations, but the free energy change (in joules <u>per mole</u>) is directly proportional to the amount of chemical reaction.*

16.5 Concentration cells and ion selective electrodes

16.5.1 Concentration cells

Consider the following cell.

$Cu(s)\ |\ Cu^{2+}\ (0.010\ mol\ L^{-1})\ \|\ Cu^{2+}\ (0.10\ mol\ L^{-1})\ |\ Cu(s)$

This cell spontaneously develops a potential of + 0.030 V. In physical terms we can explain this by realizing that the entropy of the system will be maximized if the $Cu^{2+}(aq)$ concentrations in the anode and cathode compartments are equal (this is analogous to the hot cup of coffee spontaneously cooling until its temperature matches that of its surroundings: Section 14.2). The spontaneous reaction is for copper to dissolve into the anode compartment and to deposit from the cathode compartment until the concentrations equalize. At that point, the system reaches equilibrium, and $\mathscr{E} = \Delta G = 0$. The potential of the cell is calculated from the Nernst equation.

For each half cell:

[3] $$\mathscr{E} = \mathscr{E}°(½\text{cell}) - \left\{\frac{RT}{n\mathscr{F}}\right\} \ln Q$$

For the cathode:

$$\begin{aligned}\mathscr{E} &= \mathscr{E}°(\text{cathode}) - (0.0257/2)\ln\{1/[Cu^{2+}]\}\\ &= (+0.337) - (0.0129)\ln(1/0.10)\\ &= +0.307 \text{ V}\end{aligned}$$

For the anode:

$$\begin{aligned}\mathscr{E} &= \mathscr{E}°(\text{anode}) - (0.0257/2)\ln\{[Cu^{2+}]\}\\ &= (-0.337) - (0.0129)\ln(0.010)\\ &= -0.278 \text{ V}\\ \mathscr{E}(\text{cell}) &= \mathscr{E}(\text{cathode}) + \mathscr{E}(\text{anode}) = 0.029 \text{ V}\end{aligned}$$

Notice for this or any other concentration cell that $\mathscr{E}°$(anode) = -$\mathscr{E}°$(cathode), and so $\mathscr{E}°$(cell) = 0. The value of $\mathscr{E}$(cell) varies only because there are different concentrations of electrochemically reactive species in the two cell compartments.

16.5.2 Determination of concentration from measurement of cell potential

Let's begin our discussion with two very simple (but impractical!) cells for measuring the concentration of an analyte. First we consider the cell below.

$$\text{Pt, } H_2(g), 1 \text{ atm} \mid H^+(aq), x \text{ mol L}^{-1} \parallel H^+(aq), a = 1 \mid H_2(g), 1 \text{ atm, Pt}$$

This cell could be used to measure pH.

$$\begin{aligned}\mathscr{E}(\text{cathode}) &= \mathscr{E}°(\text{cathode}) = 0.000 \text{ V}\\ \mathscr{E}(\text{anode}) &= \mathscr{E}°(\text{anode}) - 0.0591 \log_{10}([H^+])\\ &= 0.000\text{V} + 0.0591 \text{ pH } \underline{\text{or}}\ \ 0.000\text{V} - 0.0591 \log_{10}(x)\end{aligned}$$

Hence $\mathcal{E}$(cell) = 0.0591 pH

In the same way, the concentration of a metal ion such as Cu^{2+} (aq) could be measured making it the anode of a comparable cell to that just shown.

$$Cu(s) \mid Cu^{2+} (aq),\ x \text{ mol L}^{-1} \parallel H^{+} (aq),\ a = 1 \mid H_2 (g),\ 1 \text{ atm, Pt}$$

This cell could be used to measure the concentration of Cu^{2+}.

$$\mathcal{E}(\text{cathode}) = \mathcal{E}°(\text{cathode}) = 0.000 \text{ V}$$
$$\mathcal{E}(\text{anode}) = \mathcal{E}°(\text{anode}) - (0.0257/2)\ln([Cu^{2+}]) = -0.337 - 0.0129\ln(x)$$

Hence $\mathcal{E}$(cell) = $\mathcal{E}$(cathode) + $\mathcal{E}$(anode) = (-0.337) - (0.0129ln(x)).

The concentration of the analyte (H^+ (aq) or Cu^{2+} (aq) in the examples cited) can be determined by measuring the potential of the corresponding cell.

The electrode (anode in the cases discussed) which contains the ion whose concentration is to be measured is called the **indicator electrode**. The Standard Hydrogen Electrode which provides the other half-cell is called a **reference electrode**. Since the potential of the reference electrode is fixed, changes in the cell potential result only from changes in the concentration of the unknown analyte.

Reference Electrodes

The Standard Hydrogen Electrode seems like an obvious choice for a reference electrode, but is inconvenient for routine use. Gases are inherently more difficult to work with than solutions, and in addition, the surface of the platinum must be carefully prepared if reproducible results are to be obtained. For these reasons, various secondary standards are used as reference electrodes. A reference electrode is a reproducible electrode whose potential is well known. One of the commonest of these is the Saturated Calomel Electrode, usually abbreviated SCE. This electrode consists of liquid mercury (anode), in contact with a solution which is saturated with both calomel (mercury(I) chloride) and potassium chloride.

$$\parallel KCl(aq, sat), Hg_2Cl_2 (aq, sat) \mid Hg(\ell) \qquad \mathcal{E} = +\ 0.244 \text{ V at 298K}$$

16.5.3 Ion selective electrodes

An ion selective electrode is a variant of the simple concentration cell discussed in Section 16.5.1. Ion selective electrodes are a very useful devices for measuring the concentration of a "selected" ion in solution. The heart of the device is a membrane which is selectively permeable to the analyte ion. On one side of the membrane is sealed a solution of the analyte ion having known concentration; since the solution is sealed in, this concentration is fixed. The solution of fixed concentration is in contact with a redox electrode, and this half cell therefore has a fixed potential. The other side of the membrane is exposed to the solution of the analyte whose concentration is to be measured (Figure 4). A potential develops at the membrane whenever the solutions on either side differ in concentration.

The electrode in contact with the solution of fixed concentration, together with the membrane, make up one half cell. To complete the cell, a reference electrode such as a Saturated Calomel Electrode must be provided. The equipment for measuring the concentration of an analyte to which the membrane of the selective ion electrode is permeable consists of two electrodes: the ion-selective electrode and the reference electrode. The whole apparatus develops a potential which is proportional to the logarithm of the concentration of the analyte, as expected from the Nernst equation.

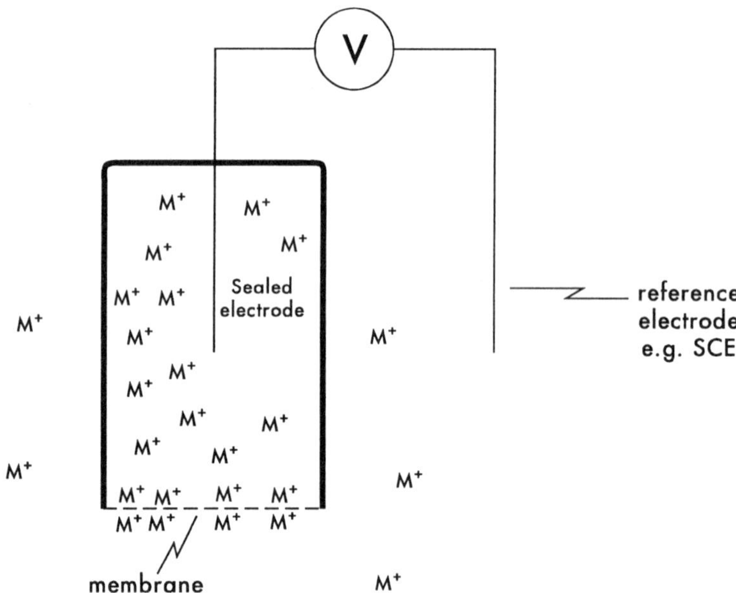

Figure 16.4: Schematic of an electrochemical cell utilizing an ion selective electrode

The following argument shows how the potential of the apparatus depends upon the concentration of the analyte. The reference electrode has a fixed potential. The potential of the ion selective electrode is composed of two parts: the potential of the sealed-in electrode (fixed) and the potential at the membrane, $\mathscr{E}_{memb}$. The latter depends upon the concentrations of the analyte (A) on the two sides of the membrane, since this is effectively a concentration cell.

$$\mathscr{E}_{memb} = (0.0257/n)\ln\{[A]_{unknown}/[A]_{fixed}\}$$
$$\text{or} \quad \mathscr{E}_{memb} = (0.0591/n)\log_{10}\{[A]_{unknown}/[A]_{fixed}\}$$

Since $[A]_{unknown}$ is the only variable quantity, the overall cell potential is given by the equation below.

$$\mathscr{E}_{cell} = \mathscr{E}(\text{reference electrode}) - \mathscr{E}(\text{sealed in electrode}) - \mathscr{E}_{memb}$$
$$\mathscr{E}_{cell} = \text{Constant} + (0.0591/n)\log_{10}[A]_{unkown}$$

pH meter

The pH meter employs an ion selective electrode whose membrane is a very thin piece of a special glass which is unusually permeable to H^+. Inside the membrane is a solution of HCl(aq), which is the solution of fixed H^+ concentration. A reference electrode completes the circuit. The pH meter changes its overall cell potential according to the concentration of H^+ in the analyte solution. Since $n = 1$, we can write:

$$\mathscr{E}_{cell} = \text{Constant} + 0.0591\log_{10}[H^+]$$

Since $pH = -\log_{10}[H^+]$, this corresponds to:

$$\mathscr{E}_{cell} = \text{Constant} - 0.0591\text{pH}$$

The pH meter changes its potential by 59 mV for each change of 1 pH unit. The read-out of the meter is set to read directly in pH units, and the meter itself is calibrated in buffer solutions of known pH. Note the following point about the pH meter, and about ion selective electrodes in general. The meter responds linearly to log(concentration), not to concentration itself. This feature means that the pH meter is useful for measuring $[H^+]$ over an exceptionally wide range of values — say, 1 mol L^{-1} to 10^{-14} mol L^{-1} (pH 0 - 14).

Very compact pH meters can be made, and they can also be made rugged by protecting the fragile glass membrane with a plastic shield. The associated electronic voltmeter can be made to operate using small batteries, and as a result, pH meters are well suited to work "in the field". For example, the pH of rainwater, and of remote streams and lakes can easily be monitored, and in addition, no special skill is required to operate the equipment.

Other ion selective electrodes have been developed to monitor the concentrations of a wide variety of analytes. They vary only in the nature of the membrane, and of course in the identity of the solution on the fixed concentration side of the membrane. Like the pH meter, ion selective electrodes are suited to making rapid concentration measurements over a wide range of [Analyte].

Ion selective electrodes find application in environmental work for monitoring industrial waste streams for toxic elements (examples: Cd^{2+}, Hg^{2+}, S^{2-}) to ensure that discharges are within regulatory standards. They have also been miniaturized, and it is now possible to measure the concentrations of ions such as H^+, Na^+, and Ca^{2+} directly inside living tissues. Modified electrodes also permit the analysis of gases such as CO_2, NH_3, HCN, and H_2S using hand-held equipment. Such sensors are extremely valuable in monitoring work-places for unacceptable concentrations of toxic gases. Since the output of a selected ion monitor is an electrical signal, these devices can also provide on-line monitoring of the composition of an industrial product or waste stream, can be connected to a computer for continuous monitoring, and can even be set to sound an alarm or shut off a process if the concentration of the analyte moves outside a specified range.

16.6 Batteries

16.6.1 Flashlight battery (dry cell)

About 10^9 of these disposable batteries are used each year in the U.S. alone. They are the familiar batteries used to power flashlights and transistor radios (and even portable pH meters). Each has $\mathscr{E} = 1.5$ V when new; the voltage gradually falls during use as the reactants are consumed. The anode is a cylindrical "cup" of zinc which forms the bottom and sides of the battery; the bottom of the battery is the negative (anode) connection to an electrical circuit. The central cathode (positive connection) is a graphite rod[1]; this is actually inert — the cathode reaction involves the reduction of solid MnO_2. The container is filled with a paste containing water, MnO_2(s), and NH_4Cl, besides inert materials. Although the paste is not completely dry, the dry cell got its name because earlier batteries[2] all used aqueous solutions and could not be tipped up without spilling

---

[1] The following definitions are helpful.
Anode: the negative connection of a conventional battery (galvanic cell); the electrode to which anions migrate.
Cathode: the positive connection of a conventional battery (galvanic cell); the electrode to which cations migrate.

[2] The dry cell was invented in 1866 by Leclanché, a French engineer. It is also known as the Leclanché cell.

their contents.

Anode reaction: $Zn(s) + 4\ NH_3 \rightarrow Zn(NH_3)_4^{2+} + 2\ e^-$

The role of the NH_3 is to convert Zn^{2+} into $Zn(NH_3)_4^{2+}$, which keeps the concentration of Zn^{2+} low.

Cathode reaction: $MnO_2(s) + H_2O + e^- \rightarrow MnO_2H + OH^-$

The hydroxide ion neutralizes NH_4^+ in the electrolyte paste; this simultaneously keeps the concentration of OH^- low, and generates the NH_3 needed in the anode reaction.

16.6.2 Alkaline batteries

"Alkaline" batteries are simply a variant of the conventional dry cell. They use NaOH in place of NH_4Cl — hence the term "alkaline battery". The anode is usually a zinc rod (the cylindrical shape of the battery is achieved by use of a steel casing). The cell diagram is:

$$Zn(s) \mid Zn(OH)_2(s), NaOH, MnO_2(s) \mid C(s)$$

Anode reaction: $Zn(s) + 2\ OH^-(aq) \rightarrow Zn(OH)_2\ (s) + 2\ e^-$

Cathode reaction: $MnO_2\ (s) + H_2O + e^- \rightarrow MnO_2H + OH^-$

The hydroxide ions produced at the cathode are consumed at the anode, thus maintaining $c(OH^-)$ constant. Because hydroxide ions travel much faster in solution than NH_4^+, the alkaline battery has a steadier voltage output than the dry cell, which is easily "polarized" — that is, its voltage drops because of a build up of products in the vicinity of the electrodes.

16.6.3 Mercury batteries

Mercury batteries are used widely in everyday life, in applications such as cameras, hearing aids, and heart pacemakers. The advantages of the mercury battery are its small size and the constancy of the voltage of the mercury battery, almost to the point of complete discharge. Constant voltage is vital in the applications mentioned, and is understandable from thermodynamic considerations.

The mercury cell is represented as follows.

Steel, Zn/Hg(s) | ZnO(s), 40% KOH, HgO(s) | Hg(ℓ), C(s)

The anode is made of zinc amalgam (an *amalgam* is a mixture of mercury and another metal); the cathode is a paste of HgO, Hg, and graphite.

Anode reaction: Zn/Hg(s) + 2 OH^-(aq) → ZnO(s) + $H_2O(\ell)$ + 2 e^-

Cathode reaction: HgO(s) + H_2O + 2 e^- → Hg(ℓ) + 2 OH^-

The cell reaction is:

Zn(s) + HgO(s) → ZnO(s) + Hg(ℓ)

Notice that all these substances are solids or liquids, and hence their concentrations/activities are all unity. From the Nernst equation:

$$[3] \qquad \mathscr{E} = \mathscr{E}^\circ - \left\{\frac{RT}{n\mathscr{F}}\right\} \ln Q$$

$$Q = \frac{\{a(Hg, \ell)a(ZnO, s)}{a(Zn, s)a(HgO, s)} = 1$$

Hence $\mathscr{E} = \mathscr{E}^\circ$ = 1.35V throughout the life of the battery.

About 30% of U.S. production of mercury is used in the manufacture of mercury batteries. The disposal of spent mercury batteries in garbage has become a source of mercury pollution, especially for communities which incinerate garbage — the volatile element mercury passes out with the flue gases.

16.6.4 Nickel-cadmium batteries

Unlike the batteries discussed so far, the nickel-cadmium battery is rechargeable. In this battery, the anode is composed of cadmium metal, and the cathode is made of NiO(OH) supported on a nickel grid. The electrolyte is KOH(aq).

Cd(s) | $Cd(OH)_2$ (s), OH^-, NiO(OH) | $Ni(OH)_2$ (s), Ni

Anode reaction: Cd(s) + 2 OH^-(aq) → $Cd(OH)_2$ (s) + 2 e^-

Cathode reaction: NiO(OH)(s) + H_2O + e^- → $Ni(OH)_2$(s) + OH^-(aq)

When the nickel-cadmium battery is recharged, these two reactions are reversed. However, note that during recharge, the identities of the anode and the cathode are transposed, because now reduction occurs at the cadmium electrode, and oxidation occurs at the nickel oxide electrode.

Cathode reaction: $Cd(OH)_2\ (s) + 2\ e^- \rightarrow Cd(s) + 2\ OH^-\ (aq)$

Anode reaction: $Ni(OH)_2\ (s) + OH^-\ (aq) \rightarrow NiO(OH)(s) + H_2O + e^-$

The nickel-cadmium battery thus functions as a galvanic cell when it powers, for example, an electric razor, but is an electrolytic cell when it is being recharged.

16.6.5 The lead-acid battery

The lead-acid battery (also called a storage battery) is familiar as the battery in your car. Its great strength is that it can be discharged and recharged a large number of times unlike, say, a dry cell, which is thrown away when it is spent. Like the nickel-cadmium battery, the lead acid battery operates galvanically when you use it to start your car, but electrolytically when it is being recharged.

Both the electrodes of the lead-acid battery are made of lead, but the electrochemically active substance on the cathode is a layer of lead dioxide. The following electrochemical reactions occur when the lead-acid battery functions as a galvanic cell.

$Pb(s)\ |\ PbSO_4\ (s), Pb^{2+}\ (aq), H_2SO_4(aq), PbO_2\ (s)\ |\ Pb(s)$

Anode reaction: $Pb(s) + SO_4^{2-}\ (aq) \rightarrow PbSO_4\ (s) + 2\ e^-$

The concentration of Pb^{2+} (aq) is kept low because the insoluble $PbSO_4$ precipitates out on the anode.

Cathode reaction:

$$PbO_2(s) + 4\ H^+\ (aq) + SO_4^{2-}\ (aq) + 2\ e^- \rightarrow PbSO_4(s) + 2\ H_2O(\ell)$$

Overall chemistry:

$$Pb(s) + PbO_2\ (s) + 2\ H_2SO_4\ (aq) \rightarrow 2\ PbSO_4\ (s) + 2\ H_2O(\ell)$$

Charging reverses this reaction (and, of course, the identities of the anode and the cathode). A convenient method of checking the charge of a lead-acid battery is to measure the density of the electrolyte, since sulfuric acid (charged) is denser than water (discharged).

Used car batteries distribute a lot of lead into the environment; despite recycling, they are the major source of lead in municipal waste. Left lying about, for example on the farm, they are the cause of numerous poisonings, especially of inquisitive cattle. Recycled, they result in local lead pollution when the old electrodes are redistilled (lead has m.p. 327 °C and b.p. 1740 °C) to recover the metal. Lead recycling plants in Canada fall under the definition of secondary lead smelters, emission standards for which limit the amount of lead emitted to 46 mg m^{-3}.

16.6.6 *Batteries and transportation*

Much research has been carried out into the development of electric vehicles, since they do not emit air pollutants as do gasoline and diesel-powered vehicles (Chapter 9). Electric trains, streetcars and trolleys are limited to tracks or overhead wires to supply their electrical power. Battery powered cars and vans have been available for many years, with the lead-acid battery being used as the power source. These vehicles must be "plugged in" at night for recharging.

Unfortunately, the high density of lead (density 11.3 g cm^{-3}) severely limits the use of the lead-acid battery for transportation. The batteries add greatly to the mass of the vehicle, because the lead-acid cell has a low energy density, meaning that the ratio of the extractable energy to the mass of the battery is low. This limits the range of the vehicle or, equivalently, the time between recharging. In order to extract 4 mol of electrons from a lead-acid battery, 2 mol of $PbSO_4$ (0.6 kg) must undergo chemical change.

Current research involves finding alternatives which would employ lighter metals such as lithium or aluminum. The successful development of the all-electric car seems to depend upon producing a lightweight power source which can either be recharged simply, or for which a spent electrode can be easily and cheaply replaced.

Lithium is the lightest metal. Interest in using it for transportation batteries is high because the same 4 mol electrons which require 0.6 kg of $PbSO_4$ to react could be produced from a mere 28 g of Li metal. So far, no completely satisfactory lithium battery has been devised. The lithium-chlorine battery is capable of delivering high power, but operates at very high temperature (> 600 °C); although it can be recharged successfully, it suffers from rapid corrosion.

$$Li(\ell) \mid LiCl(\ell) \mid Cl_2(g) \mid C(s)$$

Li metal is oxidized to Li^+ at the anode, and Cl_2 (g) is reduced to Cl^- at the carbon cathode.

A similar concept led to the development of the sodium-sulfur battery.

$$Na(\ell) \mid Al_2O_3\,(s) \mid \{Na_2S + S\}(\ell) \mid C(s)$$

Na metal is oxidized to Na^+ at the anode, and elemental sulfur is reduced to S^{2-} at the carbon cathode. The potential of this cell ranges from 2.07 V when the electrolyte composition is Na_2S_5 (Na_2S + 4 S) to 1.74 V at $Na_2S_{2.7}$. The sodium-sulfur cell is rechargeable, and operates at 350 °C.

16.6.7 Fuel cells

Conventional batteries convert chemical energy directly into electricity. The electrical power which can be drawn from them is limited by the amounts of chemicals available to drive the electrode reactions.

A fuel cell is like a battery in that it converts chemical energy directly into electricity. It differs in that the chemical reactants are continuously fed to the electrodes. This opens up the very attractive possibility of using a fuel (such as a hydrocarbon) to produce electricity directly. Suppose the following reaction could be run in a fuel cell:

$$\text{gasoline } (C_{10}H_{20}) + 20\ O_2(g) \rightarrow 10\ CO_2 + 10\ H_2O$$

Gasoline and oxygen (air) could be continuously fed in, just as in a conventional engine, and CO_2 and water exhausted. Because a fuel cell releases much of its energy in the form of electricity rather than heat, it is not subject to the thermodynamic restrictions on heat engines (Section 14.3.3)[3]. Electric power stations and vehicles could be run much more efficiently, given an appropriate fuel cell.

Unfortunately, no such fuel cell is yet available. The two chief problems to be surmounted are (i) finding a suitable chemical reaction and (ii) obtaining the reactants in a sufficiently high state of purity that they do not "poison" (inactivate) the electrodes in long term use.

To date, the only practical fuel cell is the H_2/O_2 fuel cell. Hydrogen is oxidized to water at the anode, and oxygen is reduced to OH^- at the cathode. The electrodes are composed of inert metals, or of graphite.

$$C(s), H_2(g) \mid \text{concentrated KOH(aq)} \mid O_2(g), C(s)$$

Anode reaction: $H_2(g) + 2\ OH^-(aq) \rightarrow 2\ H_2O(\ell) + 2\ e^-$

Cathode reaction: $O_2(g) + 2\ H_2O(\ell) + 4\ e^- \rightarrow 4\ OH^-(aq)$

This fuel cell is too expensive for common use, but is very reliable. It is used in the U.S. space program.

Scientists at Alcan have experimented with a cell that is something of a hybrid between a conventional battery and a fuel cell. This cell uses a

---

[3] Not all the free energy of the fuel cell can be converted to electricity however; the ($T\Delta S$) contribution to ΔG will ultimately be converted to the heat of the surroundings: see Section 14.3.3)

replaceable aluminum anode (oxidized to Al^{3+}). When the electrode is used up, it can be replaced with a new one, and the electrolyte reprocessed to metallic aluminum. The concept is that such a battery could be used for transportation, with quick replacement at a service station. Unfortunately, the prototypes require an extremely pure grade of aluminum for long term use. Commercialization will only be possible if the cheaper refinery grade of aluminum can be used.

16.7 Electrolytic reactions

16.7.1 Faraday's laws of electrolysis

Electrolysis involves using electricity to drive an otherwise non-spontaneous reaction. Michael Faraday deduced the quantitative "laws of electrolysis", which relate the extent of chemical reaction to the amount of electricity which has passed through the solution. These are summarized in modern form below.

Law of electrolysis: *The number of moles of product formed in an electrolytic reaction is equivalent to the number of moles of electrons that have passed through the solution*

We have to say "equivalent to" rather than "equal to" in the definition above, because not all chemical changes involve 1 mol electrons per mole of product. Compare the two cathodic reactions below.

(i) $Ag^+(aq) + e^- \rightarrow Ag(s)$
(ii) $Cu^{2+}(aq) + 2\,e^- \rightarrow Cu(s)$

Reaction (i) produces 1 mol Ag per mole of electrons, but reaction (ii) produces only 0.5 mol Cu per mole of electrons.

Rule: *The **Faraday constant** ($\mathscr{F}$) relates the amount of charge which passes through a solution to the number of moles of electrons that have been involved in the redox reaction.*

As noted in Section 16.4, $\mathscr{F}$ has the value 96,487 C mol^{-1}. The charge in coulombs (C) is the product of the current (amperes, A, units coulombs per second) and time.

Q (charge) = I x t

16.7.2 Analytical applications of Faraday's laws

Let's first look at a very simple process, **electrodeposition**. Suppose we had a solution containing an unknown quantity of Cu^{2+}. We could set up an electrochemical cell in which passing an electric current would deposit copper on a cathode.

e.g. $\| Cu^{2+}$ (aq), [c] = ? | Cu(s)

Current could be passed until all the copper was removed from solution as Cu metal. Then the analysis could be performed in one of two ways: (i) weigh the dry cathode before and after reaction and determine the mass increase; (ii) measure the amount of charge passed (C = current x time). In the first case, one could determine the number of moles of Cu(s) formed. In the second case, one would determine the number of moles of electrons needed to deposit all the copper. Method (ii) is much more precise, and is most commonly done today. The technique is called **coulometry**, because the number of coulombs of charge is measured experimentally.

Example: *Calculate the increase in mass of a copper cathode when 154.3 mA is passed through a solution of $CuSO_4$ in contact with the cathode for a period of 240.38 s.*

Answer:

Charge	$= \text{Current x time} = (154.3 \times 10^{-3}\ A)(240.38\ s) = 37.09\ C$
$\mathcal{F}$	$= 96{,}487\ C\ mol^{-1}$
$n(e^-)$	$= 37.09\ C/96{,}487\ C\ mol^{-1} = 3.844 \times 10^{-4}\ mol$
$n(Cu)$	$= n(e^-) \times (1\ mol\ Cu/2\ mol\ e^-) = 1.922 \times 10^{-4}\ mol$
mass (Cu)	$= (1.922 \times 10^{-4}\ mol) \times \mathbf{M}(Cu)$
	$= (1.922 \times 10^{-4}\ mol) \times 63.55\ g\ mol^{-1}$
	$= 1.221 \times 10^{-2}\ g$

A useful branch of coulometry is **coulometric titration**. In this method the solution to be analysed is subjected to an electrolytic reaction. The amount of charge passed to reach the equivalence point is measured. An indicator is often needed to assist in locating the endpoint of the titration. Coulometry has been likened to titrating the solution with electrons, instead of with a chemical reagent from a burette. For example, coulometric titration could be used to estimate Fe^{2+}, by arranging to oxidize it quantitatively (at the anode of an electrochemical cell) to Fe^{3+}.

16.7.3 Electroplating

Electroplating involves covering the surface of one metal with a layer of a different metal, for the sake of improved appearance and/or improved corrosion resistance. Some examples follow.

- silver plating and gold plating of ornaments and cutlery
- chromium plating of automobile bumpers and accessories
- cadmium plating of steel screws

In each of these reactions, the object to be electroplated is made the cathode of an electrochemical cell, and current is passed through the cell until a layer of the desired thickness is obtained. The surface of the object being plated must be carefully prepared, and the rate of deposition carefully controlled so that the deposited layer will adhere properly to the surface.

In most cases, the anode is made of the metal to be deposited; the electrolyte is always a salt of the same metal. The anode gradually dissolves and must be periodically replaced. Many different formulations exist for the preparation of satisfactory plating baths; those described here do not include the various special additives that are used.

Acidic plating baths: These are used for the plating of "bright" copper (as opposed to dull ("matte") copper, and for plating nickel. The relevant metal is used as the anode. For nickel plating, the bath contains $NiCl_2$ /$NiSO_4$ /H_2SO_4; in the case of copper plating, $CuSO_4$ /H_2SO_4 is used.

Cyanide plating baths: These are used for the deposition of Cd, matte Cu, Ag, and Zn. In all these cases, deposition of the metal from acidic solution is impractical, because it gives a finish which is easily abraded. The cyanide serves the specific purpose of keeping the concentration of M^{n+} low by means of the following equilibria, whose equilibrium constants are large in the forward direction.

$$Ag^+(aq) + 2\ CN^- \rightleftharpoons Ag(CN)_2^-(aq)$$

$$M^{2+}(aq) + 4\ CN^- \rightleftharpoons M(CN)_4^{2-}(aq) \qquad M = Cd, Cu, Zn$$

The cathode reaction can be viewed as dissociation of the cyanide complex followed by deposition of the free metal, or can be written as:

$$(e.g)\ \ Cd(CN)_4^{2-}(aq) + 2\ e^- \rightleftharpoons Cd(s) + 4\ CN^-(aq)$$

Care is needed to protect workers from the high concentrations (up to 2 mol L^{-1}) of the highly toxic cyanide solutions.

Chromium plating baths: Chromium is plated from Cr(VI) in the form of chromic acid H_2CrO_4. Exceptionally, chromium is not used as the anode; instead, the chromic acid used up is replaced by adding more of it to the plating bath. The anode is made of lead, which rapidly develops a protective layer of PbO_2. Water is therefore oxidized to elemental oxygen, which is evolved as bubbles around the anode.

Cathode reaction: $H_2CrO_4\,(aq) + 6\,e^- + 6\,H^+(aq) \rightleftharpoons Cr(s) + 4\,H_2O(\ell)$

Anode reaction: $2\,H_2O(\ell) \rightleftharpoons O_2(g) + 4\,H^+(aq) + 4\,e^-$

A major concern in chromium plating is exposure of the workers to Cr(VI) salts, which are carcinogenic.

Gold plating baths: The electroactive substance is $Au(CN)_2^-$, analogous to silver. However, the anodes are made of either platinum or stainless steel rather than gold. As with chromium, the electrolyte concentration is maintained by chemical replacement of $KAu(CN)_2$, and the anode reaction is the oxidation of water to O_2. Since this reaction is carried out in alkaline solution, the anode reaction is different from the case of chromium plating.

Anode reaction: $4\,OH^-(aq) \rightarrow O_2(g) + 2\,H_2O(\ell) + 4\,e^-$

Environmental problems in the electroplating industry

These arise from two sources: spent plating baths, and "anode sludges". The spent plating solutions contain metals and often cyanide. Metals such as Cd, Cu, Ag, Ni, and Cr are toxic; however, they may be precipitated as hydroxides at high pH, leaving a sludge for further treatment or disposal. Cyanide is destroyed by treatment of the solution with chlorine (OCl^- in the alkaline solution: see Chapter 10). This oxidizes cyanide to cyanate (OCN^-), and then to innocuous products.

$$CN^-(aq) + OCl^-(aq) \rightleftharpoons Cl^-(aq) + OCN^-(aq)$$

$$2\,OCN^-(aq) + 3\,ClO^- + 2\,OH^-(aq) \rightleftharpoons 2\,CO_3^{2-}(aq) + N_2(g) + 3\,Cl^-(aq) + H_2O(\ell)$$

Anode sludges are formed as the anode dissolves. Typically, they consist of impurities in the anode, including metals less easily oxidized than the metal being plated. Since many poorly oxidizable metals are toxic, the sludges present a disposal problem. In the past, these toxic sludges have been carelessly disposed of in municipal landfill sites, with the potential for contamination of ground water.

16.7.4 Electrolysis of water: hydrogen storage

Electrolysis breaks down water into elemental hydrogen and oxygen. The reaction is usually carried out in acidic solution.

cell reaction: $2\ H_2O(\ell) \xrightarrow{\text{electrolysis}} 2\ H_2(g) + O_2(g)$

anode reaction: $2\ H_2O(\ell) \rightarrow O_2(g) + 4\ H^+(aq) + 4\ e$

cathode reaction: $2\ \{2\ H^+(aq) + 2\ e \rightarrow H_2(g)\}$

Electrolysis of water has been considered by power utilities, particularly those employing hydraulic or nuclear power, as a way of using surplus off-peak electricity production. The concept is to store hydrogen as an energy "source", and then to regenerate electricity during periods of peak power demand, perhaps using fuel cell technology. In Canada, Hydro Quebec has invested considerable research effort into water electrolysis.

In acidic solution, the electrolysis of water requires input of $\mathscr{E}° = -1.23$ V. In principle, it should be possible to convert water to hydrogen and oxygen with an electrical input of 1.23 V, and to recapture 1.23 V from a fuel cell. Problems arises in practice. First, $\mathscr{E}°$ values apply to cells operating with all reactants and products at unit activity and under conditions of thermodynamic reversibility: almost at equilibrium. High power electrolysis moves the reaction far from equilibrium, and some of the free energy is dissipated as heat instead of electricity. Second, gas evolution at an electrode is generally accompanied by an **overvoltage**, which is a voltage in addition to the calculated value. It may be likened to an activation energy to make the electrode reaction "go". In practice, 1.5-1.6 V is usually needed for electrolysis, rather than the theoretical 1.23 V.

16.7.5 The chlor-alkali process

The chlor-alkali process is the electrolysis of brine (NaCl, aq) to produce chlorine and sodium hydroxide (Chapter 4).

$$NaCl(aq) + 2\ H_2O(\ell) \xrightarrow{\text{electrolysis}} 2\ NaOH(aq) + H_2(g) + Cl_2(g)$$

In the past, the chlor-alkali industry has been the source of serious environmental pollution. This problem was recognized about 1970, and new technology has allowed successful elimination of the problem. Let us trace the development of the story.

In the traditional flowing mercury cell, the anodes are made of graphite or of platinum-coated titanium, and a layer of liquid mercury acts as the cathode. The mercury cathode flows slowly through the cell (Figure 5).

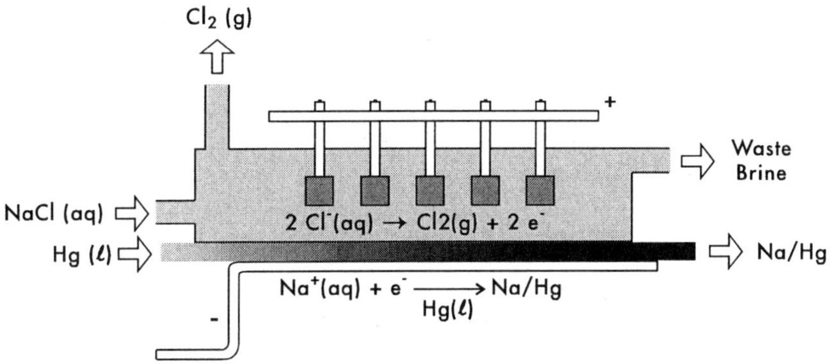

Figure 16.5: Flowing cell for electrolyzing brine

Chlorine is formed at the anode:

$$Cl^-(aq) \rightarrow \tfrac{1}{2}\, Cl_2(g) + e^-$$

Na/Hg (sodium amalgam) is formed at the cathode.

$$Na^+(aq) + e^- \rightarrow Na/Hg$$

This reaction is interesting, because $\mathscr{E}°$ for the conversion of Na^+(aq) to Na(s) is -2.71 V, and one would normally expect the conversion of H^+(aq) to H_2(g) instead[4]. This does not happen because of: (i) the exceptionally large overvoltage for reduction of H^+ at a mercury cathode (1.2 - 1.3 V); (ii) the lowering of the free energy for sodium formation through generation of the amalgam rather than the free metal, which removes the sodium from the aqueous solution with which it would react. The dilute sodium amalgam is led into a separate chamber and reacted with water.

$$2\, Na/Hg + 2\, H_2O(\ell) \rightarrow 2\, NaOH(aq) + H_2(g) + 2\, Hg(\ell)$$

The sodium hydroxide is formed as a concentrated solution and the mercury is

---

[4] The reaction $2\,H^+(aq) + 2\,e \rightarrow H_2(g)$ has $\mathscr{E}° = 0.000$ V, but this value applies when $a(H^+) = 1$. This corresponds to pH = 0. At pH near 7, $\mathscr{E} = -0.414$ V (for explanation, review Section 16.5.2). When the overvoltage for evolution of hydrogen on mercury is considered as well, we see that evolution of H_2(g) will not commence until $\mathscr{E}$ reaches about - 1.6 V at near neutral pH.

recycled. Further evaporation gives pure solid NaOH. An advantage of the mercury cell is that hydrogen and chlorine are produced in separate chambers, thus minimizing the risk of explosion from reaction between gaseous hydrogen and chlorine.

In the late 1960's it was discovered that chlor-alkali plants were leaking large amounts of elemental mercury into the environment. The need for abundant cooling water necessitates siting these plants on rivers; poor maintenance and poor inventory control led to losses of mercury, causing the river sediments downstream to become contaminated with liquid mercury. As much as 200 g of elemental mercury was being lost per tonne of chlorine produced. After pollution by the chloralkali industry was recognized, better housekeeping and government regulation greatly reduced losses of mercury to the environment, and by 1980 mercury discharge had been cut to *ca.* 0.15 g mercury per tonne of chlorine.

Besides these improvements in the operation of mercury cells, new mercury-free technology has been developed. The new cells use a special membrane to separate the anode and the cathode compartments of the cell. The special property of the membrane is that it is an ion-exchanger (compare Chapter 10), which allows cations to pass through, but excludes anions and neutral molecules (Figure 6).

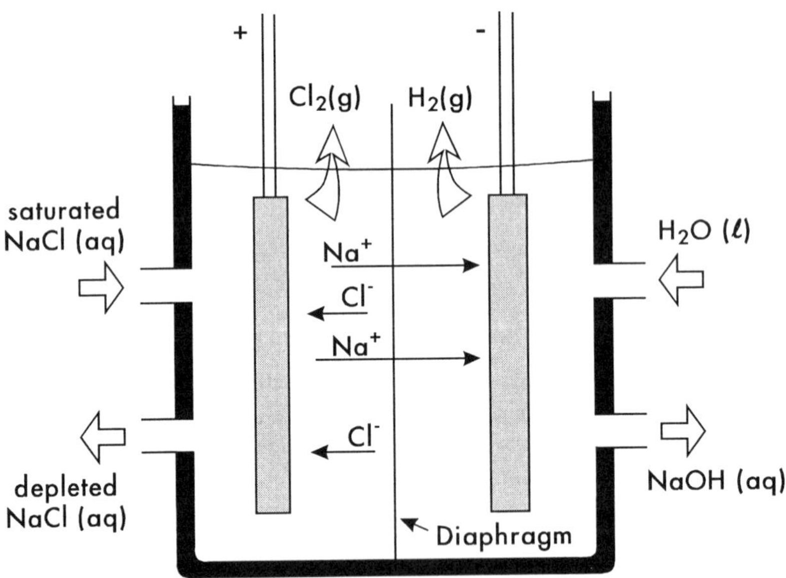

Figure 16.6: Membrane cell for chlor-alkali production

The electrochemistry of the membrane cell is different from that of the mercury cathode cell in that H^+ rather than Na^+ is reduced at the cathode. When H^+

from water is discharged at the cathode, $H_2(g)$ is formed, leaving behind OH^- in the aqueous solution. The conversion of $H^+(aq)$ to $H_2(g)$ leaves an excess of OH^- in the solution, and Na^+ migrates through the membrane to restore the charge balance. A pure grade of NaOH is obtained from the membrane cell because chloride anions cannot pass from the anodic to the cathodic compartment, hence the NaOH is uncontaminated by NaCl.

Anode reaction: $Cl^-(aq) \rightarrow \frac{1}{2} Cl_2(g) + e^-$

Cathode reaction: $H_2O(\ell) + e^- \rightarrow \frac{1}{2} H_2(g) + OH^-(aq)$

Recognizing that removal of H^+ from water leaves OH^- behind, the overall electrochemical reaction is:

$$H_2O(\ell) + Cl^-(aq) \rightarrow \tfrac{1}{2} H_2(g) + \tfrac{1}{2} Cl_2(g) + OH^-(aq)$$

In terms of actual reactants:

$$H_2O(\ell) + NaCl(aq) \rightarrow \tfrac{1}{2} H_2(g) + \tfrac{1}{2} Cl_2(g) + NaOH(aq)$$

All chlor-alkali plants commissioned in North America since the early 1970's have employed this newer technology; there are still mercury cathode cells in use, but they are less polluting than formerly, and the last of them will probably be taken out of service shortly after the year 2000. Elsewhere, as for example in Europe, mercury cells are still in widespread use.

A major outlet for both Cl_2 and NaOH is the bleaching of wood pulp to make paper. Many pulp mills in more remote locations (*e.g.*, Northern Canada) have their own chlor-alkali plants. Concerns arise from the use of chlorine as a bleach since chlorine is a chlorinating agent, and hence pulp mills discharge large amounts of chlorinated, and potentially toxic, byproducts into rivers. There is great pressure on the industry to replace chlorine by other bleaches such as ClO_2, peroxides, or ozone, which do not produce chlorinated byproducts. Reduction in the demand for Cl_2 without a compensating reduced demand for NaOH upsets the stoichiometry of the chlor-alkali process, as discussed in Chapter 4.

Problems

Section 16.1 - 16.2

1. Complete the following sentences.
   a) The electrode at which oxidation occurs is called the ...
   b) The electrode at which reduction occurs is called the ...
   c) A(n) .... cell is one that requires the input of electrical energy to drive the chemical reaction.
   d) A(n) .... cell is one in which the energy of a chemical reaction is released in the form of electrical energy.
   e) Reduction is the process in which electrons are (added to/removed from) a chemical substance.
   f) Oxidation is the process in which electrons are (added to/removed from) a chemical substance.

2. State whether oxidation or reduction occurred in each of the (unbalanced) chemical changes below.
   a) $Cu^{2+} \rightarrow Cu$
   b) $CH_4 \rightarrow CO_2$
   c) $O_2 \rightarrow H_2O$
   d) $MnO_4^- \rightarrow Mn^{2+}$
   e) $BrCl_3 \rightarrow Br^- + 3Cl^-$
   f) $CH_2O \rightarrow CH_3OH$
   g) $2Hg^{2+} \rightarrow Hg_2^{2+}$
   h) $ClO_2 \rightarrow ClO_2^-$

3. Identify the anode, the cathode, and the chemical reaction occurring at each electrode in the cells below.
   a) $Cu(s) \mid Cu^{2+}, aq \parallel Hg^{2+}, aq \mid Hg(\ell)$
   b) $Pt(s) \mid Fe^{2+} (aq); Fe^{3+} (aq) \parallel Zn^{2+} (aq) \mid Zn(s)$
   c) $Pt(s), H_2(g) \mid H^+ (aq) \parallel Pb^{2+} (aq) \mid Pb(s)$
   d) $Pt(s) \mid CH_2O (aq); HCHO_2 (aq) \parallel Ni^{2+} (aq) \mid Ni(s)$
   e) $C(s), Cl_2 (g) \mid Cl^- (aq) \parallel Zn^{2+} (aq) \mid Zn(s)$

4. Write cell diagrams for the following cells
   a) Cu^{2+} is reduced to Cu(s) at the cathode; Fe(s) is oxidized to Fe^{2+} at the anode.
   b) Zinc is oxidized to Zn^{2+} at the anode; hydrogen gas is reduced to H^+, aq at an inert platinum electrode.
   c) Formic acid is oxidized to CO(g) at an inert carbon anode; Cl_2 (aq) is reduced to Cl^- (aq) at an inert carbon cathode.
   d) Copper metal is oxidized to Cu^{2+} at the anode; Cu^{2+} is reduced to Cu(s) at the cathode.

Sections 16.3 - 16.4

1. True or False?
   a) A reduction reaction for which $\mathscr{E}°$ is positive is a spontaneous process.
   b) A reduction reaction for which $\mathscr{E}$ is positive is a spontaneous process.
   c) For a half reaction having a given value of $\mathscr{E}°$, $\mathscr{E}°$ is equal to $\mathscr{E}$(cell) when the other electrode of the cell is a standard hydrogen electrode.

2. Calculate the standard cell potential for each of the following cells.
   a) Cd(s) | Cd^{2+} (aq) ‖ Cu^{2+} (aq) | Cu(s)
   b) Zn(s) | Zn^{2+} (aq) ‖ Fe^{2+} (aq); Fe^{3+} (aq) | Pt(s)
   c) Hg(ℓ) | Hg^{2+} (aq) ‖ Ag^+ (aq) | Ag(s)

3. For the cells in question 2, what were the likely values of the concentrations of the aqueous solution if $\mathscr{E} = \mathscr{E}°$?

4. Calculate the standard free energy change (kJ mol^{-1}) for each of the cells in question 2. Assume 298 K.

5. Calculate the potentials for the following half cells.
   a) ‖ Cu^{2+}, 0.0055 mol L^{-1} | Cu(s)
   b) Cu(s) | Cu^{2+}, 0.012 mol L^{-1} ‖
   c) C(s), Cl_2 (g, p = 1.00 atm) | Cl^-, 0.15 mol L^{-1} ‖
   d) ‖ Fe^{3+}, 0.115 mol L^{-1}; Fe^{2+}, 0.012 mol L^{-1} | Pt(s)
   e) ‖ Cd^{2+}, 0.072 mol L^{-1} | Cd(s)

6. Calculate the potentials of the cells below.
   a) Zn(s) | Zn^{2+}, 0.10 mol L^{-1} ‖ Cu^{2+}, 0.050 mol L^{-1} | Cu(s)
   b) Zn(s) | Zn^{2+}, 0.10 mol L^{-1} ‖ Cu^{2+}, 1.0 mol L^{-1} | Cu(s)
   c) Zn(s) | Zn^{2+}, 0.80 mol L^{-1} ‖ Cu^{2+}, 1.4 x 10^{-5} mol L^{-1} | Cu(s)
   d) Pt(s), I_2 (s) | I^-, 0.070 mol L^{-1} ‖ Fe^{3+}, 0.112 mol L^{-1}; Fe^{2+}, 0.46 mol L^{-1} | Pt(s)

7. Identify the half reactions in each of the reactions below. (Remember to neglect spectator ions.)
   a) Zn(s) + 2HCl (aq) → $ZnCl_2$ (aq) + H_2(g)
   b) $CuSO_4$ (aq) + Ni(s) → Cu(s) + $NiSO_4$ (aq)
   c) $Br_2(\ell)$ + $2FeBr_2$ (aq) → $2FeBr_3$ (aq)
   d) 3Cu(s) + $8HNO_3$ (aq) → $3Cu(NO_3)_2$ (aq) + 2NO(g) + $4H_2O(\ell)$

8. Calculate equilibrium constants for each of the reactions in question 7, at 25°C in each case.

9. Calculate standard reduction potentials for the equilibria below at 298K.
   a) AgCl(s) ⇌ Ag^+ (aq) + Cl^- (aq)
   b) HAc (aq) ⇌ H^+ (aq) + Ac^- (aq)

10. Use the results of question 9 and the rule of multiple equilibria to calculate $\mathscr{E}°$ for each of the following half cells.
    a) ‖ Cl^- (a =1) | AgCl(s); Ag(s)
    b) ‖ Ac^- (a = 1), HAc (a = 1) | H_2 (g, 1 atm); Pt(s)

Sections 16.5 - 16.6

1. For the cell:
   Cu(s) | Cu^{2+} (a=1) ‖ Cu^{2+} (x mol L^{-1}) | Cu(s)
   calculate the concentration of Cu^{2+} in the cathode compartment if $\mathscr{E}$(cell) = 0.081 V.

2. Calculate the change in potential when the solution in which a glass electrode is immersed is changed from pH 8.10 to pH 10.27.

3. Calculate the change in potential when a calcium ion selective electrode is taken from a solution having [Ca^{2+}] = 1.5 x 10^{-2} mol L^{-1} and placed in a solution having [Ca^{2+}] = 4.2 x 10^{-4} mol L^{-1}.

4. Calculate the change in potential of a fluoride ion selective electrode (initially placed in 0.100 L of a KF solution of concentration 0.025 mol L^{-1}), when 1.0×10^{-3} mol of $BaCl_2$ is added to the solution. Assume that BaF_2 is precipitated quantitatively.

5. Calculate $\mathscr{E}$ for a hydrogen electrode having $p(H_2,g) = 1$ atm, and pH = 3.24. Assume that the electrode acts as the anode of a cell.

6. a) Calculate $\mathscr{E}°$ for the lead-acid battery when it is being used as a galvanic cell.

   b) Calculate the concentration of H_2SO_4 in the lead-acid battery when it is fully charged ($\mathscr{E}$= 2.0 V) and when it is almost discharged ($\mathscr{E}$= 1.75 V).

Section 16.7

1. The following experiment was carried out to determine precisely the value of the Faraday constant. A current of 0.2036390 amps was passed through a cell containing an extremely pure silver anode. Current was passed for 18,000.075s, and the loss of mass of the anode was 4.097900g. Calculate the value of the Faraday constant, taking M(Ag) = 107.868 g mol^{-1}.

2. A current of 0.02196 A was passed through the cathode below for 4.65h.

   $\| H^+, aq \mid H_2$ (g), Pt(s)

   What volume of H_2 (g), collected at 22.5°C and 100.8 kPa, was produced?

3. A solution of H_2S (aq) was analysed using the reaction below

   $H_2S + I_2 \rightarrow S(s) + 2H^+ + 2I^-$

   To 50.00 mL of the H_2S solution was added excess KI, and oxidation of I^- to I_2 occurred at the anode of an electrochemical cell. Electrolysis at a constant current of 47.6 mA was carried out for 776 s, at which time excess I_2 was just beginning to be present in solution. Calculate the concentration of H_2S (aq).

4. A solution of $FeCl_2$ (50.00 mL) was oxidized electrochemically for 685 s at a constant current of 51.7 mA. This completed the oxidation of Fe^{2+} to Fe^{3+}. What was the concentration of $FeCl_2$?

5. A solution of $Na_2Cr_2O_7$ (25.00 mL) was reduced electrochemically to Cr^{3+}

   $$Cr_2O_7^{2-} + 14H^+ + 6e^- \rightarrow 2Cr^{3+} + 7H_2O$$

   A current of 64.2 mA was passed for 1543 s in order to reduce the $Cr_2O_7^{2-}$ quantitatively. Calculate the concentration of $Cr_2O_7^{2-}$.

6. A pilot scale reactor for electrorefining copper is operated at 2.00 A. Calculate the mass of copper that can be purified per hour.

7. The mass of Cl_2 produced in Canada by electrolyzing brine was 1.4×10^6 tonnes. Calculate the cost of the electricity used to make this amount of chlorine, if the price of electricity is 8.4¢ per kWh.

8. Assume that in the 1960's 200 g of mercury was lost to the environment per tonne of Cl_2 produced. What mass of mercury was lost at that time, assuming 1.5×10^6 t annual production of Cl_2?

17 METALS AND MINING

17.1 Occurrence and use of metals

The history of civilization parallels the exploitation of the Earth's metallic resources. Early civilizations were familiar with the precious metals gold and silver, as well as copper, tin and lead. Alloys of these metals such as bronze (copper and tin, or copper and lead) were also used. The "Bronze Age" was followed by the "Iron Age", and indeed iron was the chief metal used for construction up to the middle nineteenth century, when technology was developed for the production of steel (the term steel refers both to iron with lower impurity content than was available previously, and also to alloys of iron with small amounts of other metals). Light non-ferrous alloys did not become commercially available until the twentieth century, when the technology for their electrochemical production had been developed.

The development in the use of metals has nothing to do with their abundance in the Earth's crust. Gold, silver, and copper are rather rare elements, while aluminum, iron, and magnesium are very common: Table 17.1.

Table 17.1: Occurrence of metals and other elements in the Earth's crust

Metals		Non-metals	
Element	**c, ppm**[a]	**Element**	**c, ppm**
Al	83,000	O	460,000
Fe	56,000	Si	280,000
Ca	42,000	H	1,400
Na	24,000	P	1,000
Mg	23,000	F	630
K	21,000	S	260
Ti	5,700	C	200
Mn	950	Cl	130

Minor elements

Cr, 100; Ni, 75; Zn, 70; Cu, 55; Pb, 12; Sn, 2; Cd, 0.2; Hg, 0.08; Ag, 0.07; Au, 0.004

a average concentration in the Earth's crust

From Table 17.1, we see that the familiarity of the elements has nothing to do with their relative abundance. Localization of metals in deposits of high concentration, and ease of isolation of the metal from the ore are more important than high average abundance. Thus it was the ease — or difficulty — of obtaining a particular metal that determined whether it came into use early or late in history.

Gold was known to the ancient world, because it is found in nature as the free metal. Apart from its use in decoration, it was a rather useless metal to the ancients, because it is very soft and hence has little strength. Compounds of copper, silver, and mercury are easily reduced to the metal, and so these metals were available by the use of very simple technology, such as heating appropriate compounds in a charcoal fire.

Metals such as iron, lead, tin, and zinc can also be produced using charcoal, with the additional innovation of mixing the charcoal with the compound that is to be reduced. This technology takes advantage of the thermodynamically spontaneous oxidation of carbon (charcoal) to CO or CO_2, coupling this reaction with the non-spontaneous reduction to the metal. For example, in the case of iron:

$$2\ C(s) + O_2(g) \rightarrow 2\ CO(g)$$
$$3\ CO(g) + Fe_2O_3(s) \rightarrow Fe(\ell) + 3\ CO_2(g)$$

Reduction of a metal oxide with carbon or CO is non-spontaneous for the most reactive metals, and these are produced electrochemically. This technology was not possible until after 1800, following the invention of galvanic cells to supply electricity. Consequently, metallic potassium, sodium, aluminum etc were unknown before that time. These metals are produced by the electrolysis of molten salts rather than aqueous solutions, because the metals are so reactive with water that they cannot be formed in aqueous solution.

17.2 Electrochemical series

The relative ease of isolating a metal from its ores is illustrated by a table of reduction potentials — often called the "electrochemical series". An electrochemical series is shown in Table 17.2, which gives the potentials for the reduction of the aqueous metal cation to the element. This table is not meant to imply that the technology for the recovery of these metals involves reduction in aqueous solution; it is given to allow comparison between the different metals. We shall discuss in Section 17.3 the actual methods used for the production of some of these metals.

Table 17.2: Standard reduction potentials for the conversion of M^{n+} (aq) to the metal

Ion	$\mathscr{E}^{o}$	Ion	$\mathscr{E}^{o}$
K^{+}	-2.93	Ca^{2+}	-2.87
Na^{+}	-2.71	Mg^{2+}	-2.37
Al^{3+}	-1.67	Ti^{2+}	-1.63
Mn^{2+}	-1.18	Zn^{2+}	-0.76
Cr^{3+}	-0.74	Ni^{2+}	-0.26
Pb^{2+}	-0.13	Fe^{3+}	-0.04
	H^{+}	0.00	
Cu^{2+}	+0.34	Ag^{+}	+0.80
Hg^{2+}	+0.85	Au^{3+}	+1.50

Of the metals listed, potassium is seen to be the most electropositive (most easily oxidized to K^{+}; or conversely, K^{+} is least easily reduced to the metal), and gold is the least electropositive. Remember that when $\mathscr{E}^{o}$ is positive, the reduction of the cation to the metal is spontaneous ($K > 1$) when the aqueous ion is present at unit activity. For the electropositive metals, which have negative standard reduction potentials, the spontaneous reaction ($K > 1$) is oxidation of the metal to the aqueous cation when the cation is present at unit activity.

The electrochemical series is useful for explaining the ability of certain metals to displace other metals from their salts. For example, zinc reacts with aqueous $CuSO_4$ to produce $ZnSO_4$ and metallic copper.

$$Zn(s) + CuSO_4(aq) \rightarrow Cu(s) + ZnSO_4(aq)$$

This reaction was met in Section 16.3 as the basis for an electrochemical cell. In that case the anode and cathode compartments were separated and the energy of the reaction was captured as an electric potential. If the components are simply mixed, the free energy of the reaction is dissipated as heat. The reaction above illustrates a general theme: that a metal cation can be displaced from one of its salts by reaction with a more electropositive metal. In this case, the less electropositive copper is forced out of solution by the more electropositive zinc. Replacement of one metal by another is readily explained in terms of reduction potentials.

Cu^{2+} (aq) + 2 e^- → Cu(s) $\mathscr{E}° = + 0.34$ V
Zn^{2+}(aq) + 2 e^- → Zn(s) $\mathscr{E}° = - 0.76$ V

The replacement reaction is therefore spontaneous to the extent of $\mathscr{E}° = + 1.10$ V (+0.34V - (-0.76V)).

Hydrogen occupies an important place in the electrochemical series. The reduction H^+ (aq) → H_2(g) has $\mathscr{E}° = 0.00$ V. Consequently, any metal for which the corresponding cation has a negative reduction potential can displace hydrogen gas from an acidic solution. For example:

Fe(s) + 2 HCl(aq) → $FeCl_2$(aq) + H_2(g)

or: Fe(s) + 2 H^+ (aq) → Fe^{2+}(aq) + H_2(g)

For this reaction $\mathscr{E}° = 0.00$ V - (-0.45 V) = + 0.45 V. The oxidation of Fe metal to Fe^{2+} (aq) in acidic solution is accompanied by the reduction of H^+ (aq) to gaseous H_2. This reaction is spontaneous when the reactants and products are present under standard conditions.

Metals for which the reduction potential M(s) → M^{n+} (aq) + n e^- is positive cannot displace H_2(g) from an acidic solution (remember that the *oxidation* potential for these metals M → M^{n+} is negative: non-spontaneous). These metals, such as Cu, Ag, Hg, and Au, dissolve in acid only under oxidizing conditions. For example:

Cu(s) + HCl(aq) → no reaction

but 3 Cu(s) + 8 HNO_3(aq) → 3 $Cu(NO_3)_2$(aq) + 4 $H_2O(\ell)$ + 2 NO(g)

In order to oxidize Cu metal to Cu^{2+} (aq), nitric acid had to be reduced to NO(g), a half reaction for which $\mathscr{E}° = + 0.96$ V. Nitric acid can therefore oxidize any metal whose oxidation potential (reverse of reduction potential) is < + 0.96 V. Among the metals in Table 17.2, this includes all but gold.

Example: *Although copper is insoluble in dilute hydrochloric acid, it does dissolve in moderately concentrated sulfuric acid.*

Cu(s) + 2 H_2SO_4 (aq) → $CuSO_4$ (aq) + SO_2 (aq) + 2 $H_2O(\ell)$

Calculate $\mathscr{E}°$ for this reaction (it is negative). Why therefore does copper dissolve in sulfuric acid, and evolve SO_2(g) from the solution?

Answer: *Break the reaction into two half reactions (omit one SO_4^{2-} from reactants and products as it is a spectator ion).*

Oxidation: *Cu(s) → Cu^{2+} (aq) + 2 e^-*

Reduction: *SO_4^{2-} (aq) + 4 H^+ (aq) → SO_2(aq) + 2 $H_2O(\ell)$*

Look up standard reduction potentials in Appendix 5.

Oxidation reaction (reverse of reduction) $\mathscr{E}°$ = -(+ 0.337 V)

Reduction reaction: $\mathscr{E}°$ = + 0.172 V

Combine these two half reactions: $\mathscr{E}°$ = - 0.337 + 0.172 = - 0.165 V

The standard reduction potential is -0.165 V (apparently non-spontaneous). The reaction succeeds in practice because SO_2(g) is evolved from the solution, therefore displacing the equilibrium to the right, because p(SO_2,g) << 1 atm.

17.3 Isolation of metals from their ores and uses of these metals

We shall discuss a few representative metals only. They will be discussed in terms of increasing difficulty of isolating the metal.

17.3.1 Gold

As already noted, gold is found in the elemental state. It is very widely distributed, but its concentration is almost always very low. Free gold is separated from other rocks by virtue of its great density (18.9 g cm $^{-3}$). "Panning" by hand is effective for alluvial (river) deposits, where the action of water has already separated the gold physically from (some) of the rock. Commercial operations employ a comparable mechanized process, after crushing the ore finely to obtain the maximum degree of physical separation of the gold particles from the associated rock. About 15 ppm of gold is needed in order for ore to be worked economically by this method. Lower grade ores can be worked by extracting the gold using aqueous cyanide solutions, in which gold dissolves chemically in the presence of atmospheric oxygen.

$$2\ Au(s) + 4\ CN^-(aq) + \tfrac{1}{2}\ O_2(g) + H_2O(\ell) \rightarrow 2\ Au(CN)_2^-(aq) + 2\ OH^-(aq)$$

Gold extraction can be a serious source of water pollution by cyanide, especially where "heap leaching" is practised. Heap leaching involves percolating a dilute solution of NaCN through an above-ground mound of gold ore; the extracted gold is recovered from dilute aqueous solution.

Besides its use in jewellery and ornamentation, a major use of gold is in the microelectronics industry, due to its exceptional electrical conductivity and chemical inertness (*i.e.*, resistance to corrosion). Used computer circuitry is reprocessed in order to recover its gold content.

Gold is an example of a "noble metal", a term inherited from alchemy to indicate metals that are not attacked by the common acids. Gold is insoluble even in concentrated nitric acid (the "aqua fortis", or strong acid, of alchemy); it may be dissolved in a mixture of nitric and hydrochloric acids ("aqua regia", or royal acid), which is even more highly oxidizing than nitric acid alone.

17.3.2 Copper

Although a little copper is found in the native (uncombined) state, the common ores of copper are the carbonate and various sulfides (notably Cu_2S and

$CuFeS_2$). The sulfides may be converted to the metal by roasting in air, a technology known as "pyrometallurgy": obtaining the metal through fire (pyros).

$$Cu_2S + O_2 \rightarrow 2\ Cu + SO_2$$
$$CuFeS_2 + 2\ O_2 \rightarrow Cu + FeO + 2\ SO_2$$

These reactions can be the source of considerable acidic emissions, unless pollution control is practised (see Chapter 13).

Low grade copper ores can be worked by an aqueous process ("hydrometallurgy"), in which the ore is extracted with dilute sulfuric acid to give Cu^{2+}(aq).

e.g. $CuCO_3(s) + 2\ H^+(aq) \rightarrow Cu^{2+}(aq) + H_2CO_3(aq)$

The extract can be reduced with any convenient reducing agent; scrap iron is often used.

$$Fe(s) + Cu^{2+}(aq) \rightarrow Cu(s) + Fe^{2+}(aq)$$

This last reaction is an example of the displacement of the less electropositive copper from solution by the more electropositive iron.

$$Cu^{2+}(aq) + 2\ e^- \rightarrow Cu(s) \qquad \mathscr{E}^\circ = +\ 0.34\ V$$
$$Fe^{2+}(aq) + 2\ e^- \rightarrow Fe(s) \qquad \mathscr{E}^\circ = -\ 0.45\ V$$

The replacement reaction is therefore spontaneous to the extent of $\mathscr{E}^\circ = +\ 0.79$ V. In the presence of air, at pH > 3.5, Fe(II) spontaneously and rapidly oxidizes to Fe(III), which precipitates as hydrated Fe_2O_3.

The major use of copper is in electrical applications, on account of its excellent electrical conductivity. These applications require copper of very high purity (> 99.99%), and so the copper produced as already described must be further purified. This is accomplished by **electrorefining**, a process identical in concept to electroplating (Section 16.8). The impure copper is made the anode of an electrochemical cell, with a small piece of pure copper serving as the cathode. Passage of an electric current causes the anode to dissolve, while pure copper is deposited on the cathode. The anode sludges contain small amounts of precious metals such as silver and gold, in addition to worthless non-metallic impurities, and it is economic to recover the precious metals as byproducts of electrorefining.

17.3.3 Mercury

Mercury was known in the ancient world, and through history it has been ascribed somewhat magical properties, on account of it being a liquid metal: latin name hydrargyrum, liquid silver; old English name quicksilver, "living silver." Mercury is a relatively rare element; it occurs mainly as HgS, cinnabar[1], from which the metal is recovered by roasting the sulfide in air.

$$HgS(s) + O_2(g) \rightarrow Hg(g) + SO_2(g)$$

Purification differs from that of copper; the low boiling point of mercury (357 °C) allows it to be purified by distillation rather than by electrorefining.

Mercury has many uses. We have met the mercury battery and the flowing mercury cell for the chlor-alkali reaction (Chapter 16). Organomercurials (compounds containing Hg-C covalent bonds) are used in agriculture as fungicides, *e.g.*, for seed dressings. The electrical uses of mercury include its application as a seal to exclude air when tungsten light bulb filaments are manufactured; in fluorescent light tubes and mercury arc lamps, which are used for street lighting and as germicidal lamps. The property of mercury being a liquid metal is exploited in certain electrical switching gear; the mercury is sealed into a glass container with electrical contacts at one end. The assembly is balanced so that under conditions of load the mercury completes the circuit; if the load is removed, the mercury runs away to the other end of the container and the electrical circuit is broken.

Although the vapour pressure of mercury is very low (see Table 17.3), its high toxicity and its nature as a cumulative poison lead to serious risks of exceeding the regulatory standard for mercury (0.05 mg m^{-3}) in the workplace.

The use of mercury diffusion pumps and similar equipment in scientific laboratories presents the same hazard in case of spills, and it is not unknown for puddles of mercury to be found beneath the floors when old laboratories are renovated! Other occupations having the potential for exposure include thermometer manufacture (the end is closed by hand after the mercury is introduced), the mining and refining of mercury, and dental surgery (the use of Ag-Hg amalgams for tooth fillings). More information on mercury toxicity is presented in Section 17.5.

---

[1] Mercury sulfide comes in two forms: cinnabar, which is black, and **vermillion**, which has for centuries been used as a red pigment for oil based paint. Mercury poisoning among artists has occurred as a result of licking the paint brush to get a fine point.

Table 17.3: Vapour pressure of elemental mercury

t, °C	p(Hg), atm	t, °C	p(Hg), atm
0	2.7×10^{-7}	10	7.1×10^{-7}
20	1.7×10^{-6}	30	4.0×10^{-6}
40	8.6×10^{-6}	50	1.8×10^{-5}
60	3.5×10^{-5}	70	6.7×10^{-5}
80	1.2×10^{-4}	90	2.2×10^{-4}
100	3.8×10^{-4}		

17.3.4 Nickel, lead, and zinc

These elements all occur principally as sulfides, and their ores frequently occur in combination with each other, as well as with sulfides of iron and copper, and small amounts of silver. Their metallurgy involves reduction of the metal oxide with coke (a porous form of carbon which is produced from coal by driving off all the volatile constituents as coal tar). The process may be divided into four steps.

1. Ore purification: The ore is finely crushed and subjected to oil flotation (Section 13.6) to separate the metal sulfides from inert rock (gangue) and, as far as possible, from each other, on the basis of the different densities of the various fractions. The rejected material is disposed of as **tailings**.

2. Roasting: The partly purified ore is heated in air. Details of this reaction and the associated problems of acidic emissions were described in Chapter 13 for the case of nickel.

e.g. $2\ ZnS + 3\ O_2 \rightarrow 2\ ZnO + 2\ SO_2$

3. Smelting: The oxide is heated with coke in the presence of air. Reduction involves partial oxidation of carbon to CO, which is the actual reducing agent for the metal oxide (ZnO shown as an example).

$$2\ C(s) + O_2(g) \rightarrow 2\ CO(g)$$
$$CO(g) + ZnO(s) \rightarrow Zn(\ell) + CO_2(g)$$

These reactions are exothermic, and the high temperatures required commercially are therefore self-sustaining.

4. **Purification**: All these metals may be purified by electrorefining, as in the case of copper, although zinc is most commonly purified by distillation, b.p. 907 °C. The anode sludges from electrorefining afford commercially useful quantities of silver and indeed, most of the commercial production of silver is as a byproduct of the smelting of lead and zinc.

In the case of nickel, two different purification methods are used: electrorefining, and the Mond process in which nickel is separated from non-volatile impurities by the reversible formation of $Ni(CO)_4(g)$. This reaction was described in detail in Chapter 8.

Uses of lead, nickel and zinc

Lead was used from ancient times because of its resistance to corrosion (this is due to the formation of a chemically resistant oxide coating on the surface). The Romans used lead in plumbing (Section 10.8), and through the Middle Ages, lead was used as a roofing material on important buildings such as cathedrals. Lead oxide is used as a component of crystal glass, to give the high refractive index and hence its ability to disperse light spectrally. Concern has been raised about high levels of lead in wines stored for long periods in crystal decanters.

Major modern uses of metallic lead are in automotive storage batteries (Section 16.7) and as an absorber of high energy radiation (X-rays and γ-rays) in medical facilities and the nuclear industry. Lead is used in the manufacture of tetraethyllead ($Pb(C_2H_5)_4$) as an octane enhancer for gasoline (Section 9.7). This use, which at one time accounted for 70% of lead use in Canada, is now discontinued in Canada with the move to unleaded gasoline. Leaded gasoline is still available in many other countries, for example in Europe and in Mexico. Large amounts of lead were formerly used as the white pigment in paints ("white lead", $2PbCO_3.Pb(OH)_2$); the toxicity of these paints has led to white lead being replaced by ZnO and TiO_2 as pigmentation agents. Chrome yellow (lead chromate) is the familiar yellow pigment used to paint North American school buses, and "red lead" (Pb_3O_4) is used as the base of rustproofing paints for iron and steel.

Nickel belongs to the same chemical family as iron, but is much less susceptible to corrosion. It is used to coat iron and steel with a covering that is corrosion-resistant, hard, and capable of taking high polish. Nickel is mostly used in the production of alloys which are stain and corrosion resistant (*e.g.*, stainless steel). Some nickel steels are very hard and are used as armour plate. Copper-nickel alloys are used in coinage (cupronickel) and in equipment that is resistant to corrosion by sea water. An aluminum-nickel-cobalt alloy (Alnico) is used to make magnets.

A major use of zinc is in the protection of steel. The zinc may be applied either as a coating of the liquid metal (galvanized steel) or by electrodeposition, as in the production of rustproof electrical conduits such as those used in commercial buildings. Zinc is also used to make dye castings for the automotive industry. Alloys of zinc include brass (with copper) and solder (with lead). Zinc oxide is widely used as a white pigment.

17.3.5 *Iron and steel*

Iron is one of the most abundant elements. It occurs as oxides (Fe_2O_3: hematite, and Fe_3O_4: magnetite, or magnetic iron oxide), and as various sulfides, of which pyrite (FeS_2: fool's gold) is the commonest. Iron oxide is most convenient for producing the metal, because no roasting step is needed (see also Section 13.6).

Iron is still made by a variant of the method used in the ancient world: reduction with carbon. Today, the reaction is carried out in a large kiln called a blast furnace. Coke is prepared in coke ovens for use as the reducing agent[2]. Coke, limestone and impure iron oxide are loaded into the top of the blast furnace, and preheated air or pure oxygen is forced into the bottom of the furnace, where it reacts with coke to form CO(g), which is the reducing agent. Reduction occurs at about 1000-1050 °C, which is below the melting point of pure iron (1535 °C), but above the melting point of an alloy of approximate composition Fe, 96%; C, 4% which forms in the blast furnace.

$$2\,C(s) + O_2(g) \rightarrow 2\,CO(g)$$
$$3\,CO(g) + Fe_2O_3(s) \rightarrow Fe(\ell) + 3\,CO_2(g)$$

The limestone ($CaCO_3$) decomposes into CaO, which has the function of removing any residual silicate materials present in the ore in the form of "slag" ($CaSiO_3$).

$$CaCO_3(s) \rightarrow CaO(s) + CO_2(g)$$
$$CaO(s) + SiO_2(s) \rightarrow CaSiO_3(\ell)$$

Liquid iron (density 7.9 g cm $^{-3}$) collects at the bottom of the furnace, where it is covered by a layer of molten slag. These two liquids are drawn off at the bottom of the furnace, while new reactants are added at the top, permitting continuous operation.

The newly formed iron (pig iron), which contains about 4% carbon and

---

[2] "Coking" of coal produces significant quantities of polycyclic aromatic hydrocarbons, which are carcinogenic, and represent a potential health hazard to coke oven workers.

smaller amounts of other elements, is very hard and brittle. Its carbon content must be lowered to make steel, which is much more resistant to mechanical shock. Conversion to steel is accomplished by blowing pure oxygen through the molten pig iron: this oxidizes most of the impurities away (carbon as CO_2, silicon as SiO_2, and phosphorus as P_2O_5). The product is low-carbon ("mild") steel which is used in most non-specialty applications, such as steel cans and automotive bodies. Alloys of iron with metals such as chromium, manganese, or nickel give extra-hard steels for use as armour plate and railway lines. Stainless steel contains iron, chromium, and nickel.

17.3.6 Tin

Tin occurs mostly as the oxide SnO_2, and is reduced to the metal with coal at high temperatures. Major deposits of tin occur in Asia and Africa.

A familiar use of tin is in the production of "tin cans", which are actually mild steel coated with a thin layer of tin (see Section 17.4). Tin is used in this application because of the low mammalian toxicity of tin salts. Other uses are in the production of window glass and plate glass, where the glass is allowed to float on a layer of $Sn(\ell)$, melting point 232 °C, giving a flatness to the surface that was previously unattainable. Electrically conducting glass is made by spraying tin salts on to glass, to make frost-free windshields. Certain organotin compounds are used as biocides (Section 17.5), and as stabilizers to protect plastics such as polyvinyl chloride (PVC) from breakdown.

Tin forms two allotropes. "White tin" is the metallic form, which is stable above 13 °C, and has the familiar highly polished surface. Below this temperature, tin slowly changes into "grey tin", which is non-metallic, dull, and has low structural strength. The formation of grey tin in unheated buildings in winter was formerly known as "tin pest". Tin organ pipes in cold churches were occasionally known to disintegrate in winter because of tin pest.

17.3.7 Aluminum, magnesium and the alkali metals

These metals are all produced electrochemically. Only in the late nineteenth century and twentieth century have the metals used in the manufacture of light alloys become available, in particular aluminum, magnesium, and titanium. The story is told that at the Court of the French Emperor Napoleon III in the 1850s, the most important guests ate with utensils made of the newly discovered metal aluminum, while the lesser nobles had to make do with mere gold!

Aluminum is the most abundant metal in the lithosphere. Most of it occurs in clays, from which isolation is uneconomic. The only commercial ore is bauxite (hydrated aluminum oxide), which is purified by dissolution in sodium hydroxide solution followed by reprecipitation with $CO_2(g)$. This procedure takes advantage of the amphoteric behaviour of aluminum oxide, allowing it to be separated from iron oxide, the chief impurity, since iron oxide is insoluble in aqueous base.

$$Al_2O_3(s) + 2\ OH^-(aq) + 3\ H_2O(\ell) \rightarrow 2\ Al(OH)_4^-(aq)$$
$$Al(OH)_4^-(aq) + CO_2(g) \rightarrow Al(OH)_3(s) + HCO_3^-(aq)$$

The precipitated aluminum hydroxide is heated to give pure Al_2O_3, which is electrolyzed in a solution of molten fluoride salts at about 1000 °C, using carbon anodes; the steel tank in which the reaction is carried out serves as the cathode.

anode reaction: $2\ O^{2-} + C(s) \rightarrow CO_2(g) + 4\ e^-$ (x 3)

cathode reaction: $Al^{3+} + 3\ e^- \rightarrow Al(\ell)$ (x 4)

overall reaction: $2\ Al_2O_3 + 3\ C \rightarrow 2\ Al + 3\ CO_2$

There are several interesting features of this process. The carbon anodes are consumed in the reaction. Aluminum production is very energy intensive, and aluminum smelters are normally sited close to abundant hydroelectricity. For example, Alcan operates a large aluminum smelter at Kitimat on the west coast of British Columbia, where rainfall averages nearly 500 cm per year. A "human interest" point is that the reaction was discovered in 1886 by Charles Hall, while he was an undergraduate student at Oberlin College, Ohio.

Aluminum has low density (2.7 g cm^{-3}) and good electrical conductivity. Hydroelectric transmission lines are made of aluminum, strengthened with steel. The strength of aluminum is greatly improved by alloying with small amounts of other metals, such as copper and magnesium; the combination of high strength and low density make aluminum alloys the ideal materials for aircraft construction.

The position of aluminum in the electrochemical series would lead one to expect that it would corrode rapidly. It does not do so because the metal surface is protected by a tightly-held layer of Al_2O_3, preventing access of air and water to the metal below. Corrosion occurs rapidly if the oxide layer is destroyed, as for example in strong acid, or even dilute base (never use washing soda ($Na_2CO_3.10H_2O$) on aluminum pots and pans!). Additional protection can be obtained by "anodizing" the surface: that is, making the aluminum object the anode of an electrochemical cell, in order to increase the thickness of the oxide layer. Coloured anodized aluminum is produced by the addition of dyes to the electrolyte solution during anodization.

Magnesium is produced electrochemically by the electrolysis of molten $MgCl_2$, which can be obtained from sea water. This process takes advantage of the low solubility of $Mg(OH)_2$ compared with the hydroxides of the other common cations in sea water (Chapter 10).

$$Mg^{2+}(aq) + Ca(OH)_2(aq) \rightarrow Ca^{2+}(aq) + Mg(OH)_2(s)$$
$$Mg(OH)_2 + 2\ HCl \rightarrow MgCl_2 + 2\ H_2O$$

In the electrolysis, Mg^{2+} is reduced to the metal at the cathode, and Cl^- is oxidized to elemental chlorine at the anode.

With a density 1.7 g cm^{-3}, magnesium is even lighter than aluminum; magnesium alloys are used extensively in the aerospace industry, for example as aircraft wheels. A longtime problem has been that the highly reactive magnesium could not be machined, because friction causes the machinings to take fire in air. Until recently, this limited magnesium alloys to parts that could be stamped or pressed; however, modern technology allows magnesium parts to be machined under a blanket of inert gas such as nitrogen or argon.

Magnesium is used in old-fashioned flash bulbs, and in fireworks and other incendiary devices. The chemical reaction is the formation of magnesium oxide; the reaction is so highly exothermic that the MgO becomes white hot.

Among the **alkali metals** sodium, like magnesium, is produced by electrolysis of molten NaCl; potassium is produced by electrolysis of molten KOH. Sodium is used in the production of tetraethyllead (Section 9.7); a sodium-potassium alloy is used as a liquid heat-exchange medium in some types of nuclear power station.

17.3.8 Chromium and titanium

Both these metals are produced by chemical reactions different from those encountered so far: by displacement from a salt using a more reactive metal.

Chromium is produced by the reaction of Cr_2O_3 with powdered aluminum; the reaction is so exothermic that the chromium is formed in the liquid state.

$$Cr_2O_3(s) + Al(s) \rightarrow Al_2O_3(s) + Cr(\ell)$$

Uses of chromium are in chrome plating (Section 16.8) and in the production of stainless steel.

Titanium is produced from TiO_2, the main ore, by conversion to $TiCl_4(\ell)$, which is reduced with metallic magnesium. Titanium (density 4.5 g cm^{-3}) is used in the aerospace industry in applications requiring greater strength than aluminum alloys can provide. Purified TiO_2 is important as a white pigment for paints.

17.4 Corrosion

17.4.1 Mechanism of corrosion

Corrosion is the process whereby a metal reacts during use in the environment. Very inert metals such as gold do not corrode away; others such as lead and aluminum do not corrode due to the formation of a protective coating on their surface. Copper gradually forms a green "patina" which is desired architecturally on roofs and domes of public buildings.

Corrosion is a particular problem in the case of iron and steel. Attack by atmospheric oxygen and moisture yields rust — hydrated Fe_2O_3. Unlike Al_2O_3 on aluminum, rust does not provide protection to the metal underneath. The rust layer does not adhere tightly to the metal below, and eventually the metal rusts away completely.

Corrosion of iron and steel is an electrochemical process, in which impurity sites (carbon or less electropositive metals) act as cathodes, and iron acts as the anode.

The reduction of O_2 at a cathode has $\mathscr{E}° = + 1.23$ V, and can therefore make dissolution of iron spontaneous.

$$O_2(g) + 4\,H^+(aq) + 4\,e^- \rightarrow 2\,H_2O(\ell)$$

$$Fe(s) \rightarrow Fe^{2+}(aq) + 2\,e^-$$

Once in solution, Fe^{2+} is oxidized to Fe^{3+} which precipitates onto the metal surface as hydrated Fe_2O_3. This mechanism explains why the rust layer adheres poorly to the metal, and offers no protection to further attack. The greater tendency of iron to rust in sea water, or in areas that are exposed to road salt, is explained by the higher electrolyte concentration of these media. Rusting is almost non-existent in desert climates, where water is not available to complete the electrochemical cell.

17.4.2 Cathodic protection

The cost of corrosion in North America has been estimated at tens of billions of dollars annually, in damage to bridges, ships, oil rigs, etc. The only way to inhibit corrosion is to protect the metal surface — by painting or by coating with another metal. The metal used most often for this purpose is zinc. Zinc may be applied by electrodeposition, as is done to make rust-resistant electrical conduits in public buildings, or by dipping the object to be protected into liquid zinc (hot galvanization). "Galvanized" steel is used for roofing sheets, rust-resistant nails and fencing wire. It offers protection in two ways: first, by physically limiting access of air to the iron underneath; second, because zinc is more electropositive than iron. In the event that an electrochemical reaction should begin, zinc will form the anode (and thus dissolve), while iron will act as the cathode. This is sometimes called "cathodic protection"; alternatively, the zinc may be described as a "sacrificial anode", meaning that the zinc dissolves away in order to protect the steel (cathode) underneath. In desert climates, galvanized steel is predicted to last for hundreds of years; however, in moist climates protection may last as little as five years, especially if corrosion is accelerated by air pollution by acidic gases.

anode reaction: $Zn(s) \rightarrow Zn^{2+}(aq) + 2\,e^-$

cathode reaction: $O_2(g) + 4\,H^+(aq) + 4\,e^- \rightarrow 2\,H_2O(\ell)$

Cathodic protection is also used on ocean-going ships. A slab of zinc or magnesium may be bolted to the hull of the ship so that this slab of metal may act as a sacrificial anode, and thus protect the steel hull of the ship. The anode may be replaced when it has dissolved. Cathodic protection is especially important in sea water, because the high concentrations of electrolytes greatly favour electrochemical corrosion.

17.4.3 Tin plating

Modern tin cans are made from mild steel coated with a thin layer of tin. A disadvantage of tin plating is that tin is less electropositive than iron. This situation is opposite to that encountered with zinc plating. Therefore, if the tin layer becomes damaged, the iron below rusts rapidly, with the iron acting as anode and oxygen goes into solution at the tin cathode.

anode reaction: $Fe(s) \rightarrow Fe^{2+}(aq) + 2\,e^-$

This reaction is then followed by oxidation of Fe(II) and precipitation of hydrated $Fe_2O_3(s)$ as described earlier.

cathode reaction: $O_2(g) + 4\,H^+(aq) + 4\,e^- \rightarrow 2\,H_2O(\ell)$

17.5 Toxicity of certain metals

Toxicity of metals presents a different problem from the toxicity of, for example, pesticides. Pesticides are organic compounds, whose toxicity results from the particular arrangement of carbon, hydrogen, nitrogen, oxygen, and other non-metallic elements in the molecular structure. These non-metallic elements are not intrinsically toxic. Consequently, conversion to non-toxic material can be achieved by chemical transformation, for example by incineration or by biological degradation. This is not possible for toxic metals, where the element itself is intrinsically toxic. Chemical structure may modify toxicity, in the sense that one compound of a given metal may be more or less toxic than another. For example, organo-mercury compounds (those containing Hg-C covalent bonds) are more toxic to mammals than simple Hg(II) salts, because organomercurials are non-polar and can be bioconcentrated from water into living tissue.

Many metals of commercial importance are toxic. These include Hg, Pb, Cu, Ag, Cd, and Ni. Iron is toxic, even though it is a necessary trace nutrient, at elevated concentrations; however, toxicity caused by environmental iron is rarely a problem because it precipitates from solution as Fe_2O_3 (see Section 17.6). Toxic metals in drinking water were discussed in Section 10.8.

17.5.1 Toxicity of mercury

The toxicity of mercury has long been known. It is a neurotoxin, which causes symptoms such as quarrelsome behaviour, headache and depression, and muscle tremors. The expression "mad as a hatter" derives from the use of mercury(II) nitrate in making felt for hats, a use which continued until about 1940. Workers in the industry suffered long term exposure to mercury salts, hence their symptoms.

Other occupations which formerly led to mercury exposure were gilding and mirror making. Gilding was the method by which objects were plated with gold and silver, before electroplating was developed in the mid-nineteenth century. An amalgam (alloy) of 10 parts of mercury to 1 part of gold was painted on to the object to be gilded, and then the mercury was evaporated away - without proper ventilation. Until the end of the last century, mirrors were made by applying a mercury/tin amalgam to glass and then evaporating the mercury[3]. Mercury was also an important ingredient in alchemy; it has been suggested that the peculiar behaviour of Isaac Newton around 1690 was occasioned by his interest in alchemy at the time.

In some places mercury occurs as the free element: in the mercury mines in Sicily, where the mercury occurs in shales, the miners are exposed to mercury vapour, levels of which in the air may reach ≈ 5 mg m^{-3}. Figure 1 is an example of the handwriting of a Sicilian miner suffering from the tremors of mercury poisoning. Even where free mercury is absent, care has to be taken to minimize the workers' exposure to dust.

---

[3] The modern silver mirror is made by reducing silver nitrate with a weak reducing agent such as glucose; if the surface of the glass is clean, a thin layer of silver adheres to the glass, and a protective finish is applied on top of the silver layer.

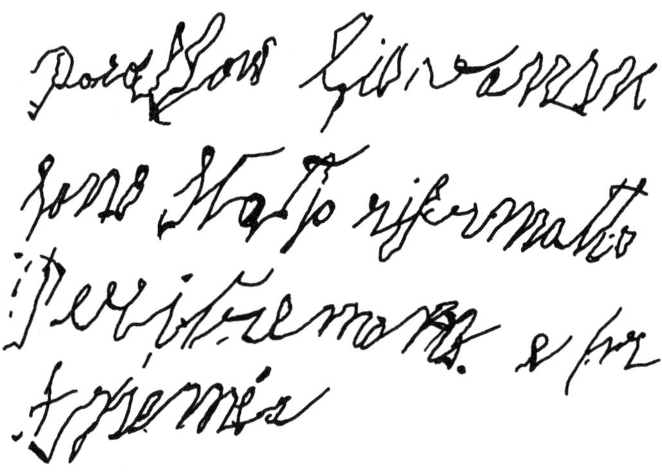

Figure 17.1: Handwriting of a Sicilian miner suffering from mercury poisoning

Mercury poisoning associated with environmental contamination

1. <u>The Reed Paper controversy at Dryden, Ontario</u>

The Reed paper company operated a pulp mill at Dryden, Ontario, together with a chlor-alkali plant to supply the chemicals needed to bleach the pulp. Between 1962 and 1970 it is estimated that about 10 tonnes of mercury was lost into the Wabigoon-English River systems. Use of mercury was limited after 1970 and discontinued in 1975, but minor losses from the plant continue even today, because of the large amount of mercury still dispersed on the site.

A serious situation developed because two aboriginal bands used the waters of the Wabigoon-English River systems for fishing. The fish caught there were said to comprise a major part of the diet of the bands as well as their principal source of livelihood. Tests taken on the fish showed them to be contaminated by mercury far in excess of the 0.5 ppm which had been set as the standard for human consumption. Some members of the bands were found to have tissue levels of up to 600 ppb of mercury, which is at the lower end of the range of clinical mercury poisoning. After years of legal wrangling, the bands were eventually compensated to the extent of $8 million in 1985.

What should be done about rehabilitating the river in a case like this? In this as in other similar cases, the losses of mercury were so great that pools of liquid mercury may be found in the sediments. Dredging has been rejected as an option, since it would be prohibitively expensive and would also disturb the river sediments. Stirring up the mud would pose a short term threat to aquatic life and would resuspend the mercury in the biotic zone. Doing nothing allows the mercury to become ever more deeply buried in the sediment. A point often forgotten in discussions about environmental contamination by metals is that these elements occur naturally in the environment; it is from the natural environment that they were extracted in the first place.

2. Minamata disease

The largest episode of mass poisoning due to mercury occurred in the 1950's in the Japanese fishing village of Minamata. About 1300 people were afflicted with physical signs of mercury poisoning over the period 1953-1960, and about 200 died. The patients showed signs of anorexia, irritability, and other psychiatric symptoms, but considerable research was needed to make the link to mercury.

The affected residents all ate large amounts of fish and shellfish. Cats fed on fish scraps showed similar symptoms. At first, it was believed that food poisoning was responsible, but the patients showed no fever or gastrointestinal disturbance, and all bacteriology was negative. Chemical poisoning was then considered, and a chemical plant manufacturing acetaldehyde came under suspicion because both the number of victims and the severity of their symptoms increased with proximity to the plant. The actual toxic agent was not traced for some time, but was eventually identified as mercury, used as a catalyst in the conversion of acetylene (C_2H_2) into acetaldehyde ($CH_3CH{=}O$). Mercury was detected in the fish and shellfish, the amount decreasing with distance from the acetaldehyde plant, and the methylmercury cation CH_3Hg^+ was detected in the waters of Minamata Bay. Minamata disease was thus caused by the loss of mercury residues from the acetaldehyde plant into Minamata Bay, where they were taken up by the fish and shellfish, and bioconcentrated in the form of lipophilic methylmercury derivatives (see below for the methylation of mercury).

The health of the Minamata residents was monitored by analysis of their hair for CH_3HgCl, since the concentration of CH_3HgCl in hair is proportional to the concentration of mercury in the patients' blood, and hence to the average amount of mercury in the diet. Up to 500 ppm of CH_3HgCl was detected in the hair of clinically affected patients, compared with 5 ppm among Japanese in general.

Regular consumption of mercury in the diet leads to the accumulation of mercury in the body, because the rate of clearance of mercury from the body is slow. For this reason, mercury is described as a cumulative poison. A single meal contaminated by mercury at a specified level may cause no ill effects, but the same concentration in a steady diet can lead to sickness or death. This situation stands in contrast to a readily excreted or metabolized poison such as cyanide, which does not accumulate.

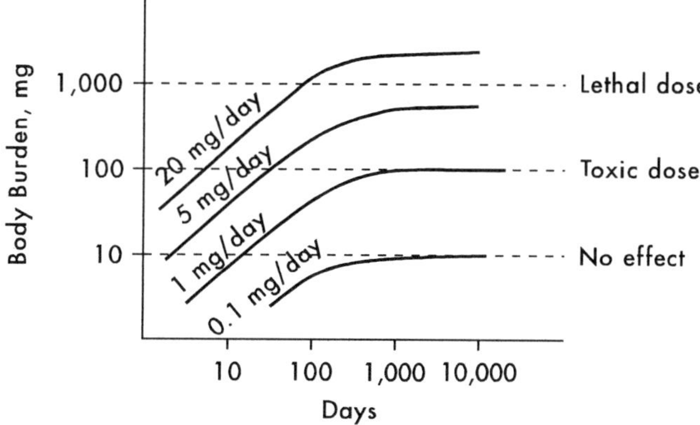

Figure 17.2: Accumulation curves for different levels of mercury in the diet.

Speciation of aqueous mercury

The usual form of mercury in aqueous solution is the Hg^{2+} ion. Mercury has two oxidation states, Hg(I) and Hg(II), although mercury(I) — which exists as the unusual ion $^{+}Hg\text{-}Hg^{+}$ — is stable only as insoluble salts such as Hg_2Cl_2. It disproportionates in solution.

$$Hg_2^{2+}\,(aq) \rightarrow Hg^{2+}\,(aq) + Hg(\ell)$$

The major complicating factor in the environmental chemistry of mercury is its biological methylation to CH_3Hg^+ and $(CH_3)_2Hg$, which converts inorganic mercury to forms which are both more toxic and more lipophilic. Mercury is responsible for neurological conditions because methylmercury compounds are lipid soluble (and hence bioconcentrate) and are able to cross the blood-brain barrier.

The carbon-mercury bond is intrinsically weak (about 200 kJ mol^{-1}), but is almost completely non-polar. As a result, organomercurials tend to be kinetically unreactive, even though not especially stable thermodynamically.

Environmental regulation of mercury

As noted in Chapter 10, standards for mercury in drinking water are typically 1.0 $\mu g\ L^{-1}$ (1 ppb). The Great Lakes joint water quality agreement between Canada and the United States sets a target of 0.2 μg of mercury per liter of unfiltered lake water. This agreement implies that states and provinces bordering the Great Lakes will control their discharges; Ontario's standards are 1.0 ppb of mercury in waste water and 0.1 ppb for water discharged into sanitary sewers. In Canada, fish should not be eaten if their mercury content exceeds 0.5 ppm.

Typical workplace air quality standards are 0.05 mg m^{-3} for inorganic mercury and 0.01 mg m^{-3} for organic mercury. The lower standard for organic mercury is in consideration of the greater toxic potential of the lipophilic and bioconcentratable organic forms of mercury. Mercury salts cause kidney damage, while elemental mercury — like the alkylmercurials — affects the central nervous system, giving rise to symptoms such as tremors, irritability, and sleeplessness. Elemental mercury is mainly hazardous as the vapour; there is less danger of absorbing the metal from the digestive tract.

17.5.2 Toxicity of lead

The toxicity of lead paint (paint containing white lead, $PbCO_3.Pb(OH)_2$) towards children is well recognized. Children chewing on lead-painted toys or household surfaces may ingest significant quantities of the metal, which is recognized as a cumulative poison, responsible for mental retardation in children. Children are more susceptible than adults, because they absorb a higher proportion of environmental lead from the digestive tract. Lead competes with calcium for target biomolecules, since Pb^{2+} and Ca^{2+} are very similar in size. More affluent children seem to be less at risk from lead poisoning, perhaps because their diet is better, and thus their stores of Ca^{2+} are higher.

The half-life of lead in humans is estimated to be about 6 yr (whole body) and about 15-20 years (skeletal). Movement out of the skeleton is thus very slow and lead, like mercury, is a cumulative poison. However, it is not true to say, as appears sometimes in the news media, that lead is accumulated through the lifetime and never eliminated. What has been shown in several studies is that skeletal burdens of lead increase almost linearly with age; this suggests that the steady state with respect to lead is not normally reached. For patients clinically affected, reaction of Pb^{2+} with ethylenediaminetetraacetic acid (EDTA) has been found beneficial in reducing body burdens of lead, through excretion of the water-soluble complex $Pb(EDTA)^{2-}$.

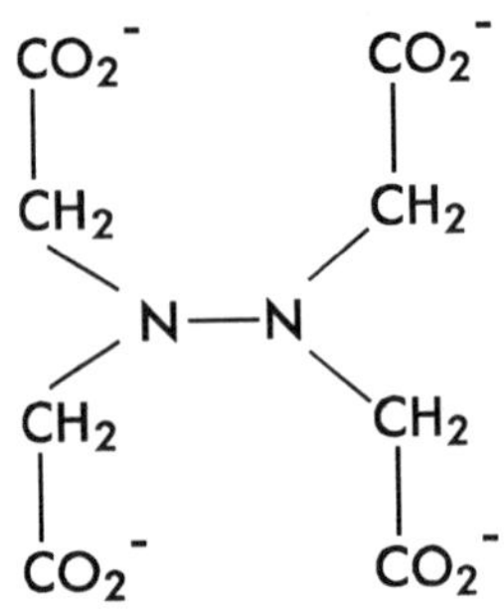

EDTA^{4-}

Lead, like mercury, causes neurological problems. Children can suffer mental retardation, lower performance on IQ tests, and hyperactivity. Severe exposure in adults causes irritability, sleeplessness, and irrational behaviour. The appetite is depressed, which can lead to emaciation, and death can ensue due to starvation. Again like mercury, organolead compounds are more toxic than simple lead salts because they are non-polar, lipid-soluble, and can more readily cross the blood-brain barrier.

Lead poisoning is believed to have been the fate of the Franklin expedition, which left England in May 1845 in a bid to discover the elusive Northwest Passage through the Canadian Arctic. It was never heard from again. Ultimately, the graves of several of the crew members were located, and in 1984 and 1986, the bodies of three crew members were subjected to autopsy, having lain perfectly preserved for 130 years in the permafrost. The emaciation of the bodies, plus high levels of lead in bone and hair, pointed to lead poisoning. The source of the lead? Lead solder, which was used to seal the tin cans which contained the crew's provisions.

The limit for lead in drinking water is 50 ppb, currently under downward revision in many jurisdictions to 5 ppb. Problems associated with the use of lead plumbing were discussed in Chapter 10.

The common compounds of lead are Pb(II) derivatives. As a member of the periodic group IVA, lead also forms tetravalent compounds, which are covalent. Of these, the most important commercially are the tetraalkylleads, which are, or have been, used as gasoline additives (Section 9.7). At the temperature of the engine, the weak Pb-C bonds cleave, allowing reactive C_2H_5 radicals to initiate smooth combustion in the engine. In other respects, the non-polar Pb-C bond is unreactive, and organolead compounds tend to be kinetically inert, like organomercurials.

$$Pb(C_2H_5)_4(g) \rightarrow Pb(g) + 4\ C_2H_5(g)$$

17.5.3 Marine antifoulants

An important economic consideration in the shipping industry is the growth of barnacles on ships' hulls. Both the extra mass of a thick layer of barnacles and the extra friction between the hull and the water reduce speed and increase fuel consumption. Removal of barnacles is very expensive, since the vessel must be taken into dry dock, and the barnacles scraped off physically. Antifoulant paints have therefore been developed in order to inhibit the growth of barnacles.

Three considerations determine the suitability of a material as the active biocide in the paint: toxicity towards the target organism (barnacles), lack of toxicity to non-target organisms, and long effective lifetime between paintings.

Incorporation of metal salts into paint has been used for ship protection for about 100 years. Copper salts are still used to some extent, while the use of compounds of mercury and lead has been discontinued because of their toxicity. Organotins of the general formula R_3SnX came into use in the 1960s. In the formula R_3SnX, R represents an organic group such as butyl (C_4H_9) and X is an anion. These tributyltin compounds are known as TBT compounds for short. They contain three Sn-C bonds.

It is thought that the biocide works by releasing tiny concentrations of the organotin into the boundary layer of water adjacent to the hull. The solubility properties of the organotin are chosen so as to maintain effectiveness at the smallest possible release rate, so that the paint will remain effective as long as possible. Typical release rates are ≈ 1 μg cm^{-2} day^{-1}. TBT compounds show moderate bioconcentration factors of 10^2-10^3. In heavy shipping areas such as harbours, TBT compounds can pose a threat to inshore fisheries such as oysters, shrimp, and crab at concentrations 0.1-0.01 ppb, and as a result, many countries now restrict the use of TBT compounds to ocean going vessels only, and prohibit its use on most recreational boats. Targets for TBT in coastal waters have been set in many jurisdictions at 2-20 ppt (2-20 ng L^{-1}).

17.6 Acid mine drainage

The mining industry is the source of very visible environmental problems. First, the scope of its operations is enormous; a commercially viable mine will exploit an ore-body covering an area of many square kilometers. Strip mining is the most economical means of mining near-surface minerals. In this method, the soil and rock overlying the ore-body (overburden) is removed, and the ore is removed from a huge pit. In the past, little attention has been paid to reclaiming the land after the ore-body has been worked out, and unsightly scars result on the landscape where frequently nothing will grow for many years.

Acid mine drainage involves the acidification of streams in the vicinity downstream of a mine, and the deposition of iron, which forms an unsightly slimy orange precipitate of iron(III) hydroxide, on the rocks of the stream bed. It is an environmental problem wherever sulfide-containing minerals are mined.

Acid mine drainage is actually a biological problem. As noted earlier in this chapter, many important metals are mined as sulfides, while coal also contains metal sulfides such as FeS_2. Some of these materials will be present in the waste material ("spoil") that is discarded at the mine site. Bacteria of the *Thiobacillus* family are able to use these sulfur compounds for the reduction of molecular oxygen, and some of these bacteria oxidize the sulfur to sulfuric acid.

$$H_2S(aq) + 2\ O_2(g) \rightarrow SO_4^{2-}(aq) + 2\ H^+(aq)$$

or $$S^{2-}(aq) + 2\ O_2(g) \rightarrow SO_4^{2-}(aq)$$

In this reaction, the strongly basic anion S^{2-} is exchanged for the neutral anion SO_4^{2-}; or alternatively, the weak acid H_2S is exchanged for the strongly acidic H_2SO_4. Either description is compatible with a large increase in the acidity of the solution.

In the immediate vicinity of the mine pH values as low as 1-2 may be reached, and under these conditions, ferric oxide (in the mine spoil or in the abandoned mine itself) may also dissolve.

$$Fe_2O_3(s) + 6\,H^+(aq) \rightarrow 2\,Fe^{3+}(aq) + 3\,H_2O(\ell)$$

Oxidation of the Fe(II) in FeS_2 to Fe(III) occurs simultaneously with oxidation of sulfur. This reaction occurs both chemically and microbially. The chemical reaction follows the rate law shown below.

$$-d[Fe(II)]/dt = k[Fe(II)][OH^-]^2 p(O_2)$$

This rate equation suggests that the rate increases as $[OH^-]^2$, or equivalently that the $\log_{10}$(rate) increases proportionately with (2 x pH). The oxidation of Fe(II) is a chemical process above about pH 5, but mostly biological below about pH 4: see Figure 3. The straight line portion of the curve, and its dotted extrapolation, correspond to the kinetics shown above; at low pH, the rate of oxidation is higher than predicted from the chemical rate law, and represents the incursion of biological oxidation.

As the stream travels away from the mine site, it becomes diluted with uncontaminated water. The pH rises, and the dissolved iron precipitates when the pH exceeds about 3.5, because $Fe(OH)_3$ is extremely insoluble ($K_{sp} \approx 10^{-38}$ (mol $L^{-1})^4$). Consequently, Fe(III) salts are only stable in acidic solutions.

$$Fe^{3+}(aq) + 3\,H_2O(\ell) \rightarrow Fe(OH)_3(s)\ [\textit{or}\ \tfrac{1}{2}\,Fe_2O_3.nH_2O] + 3\,H^+(aq)$$

Thus the manifestations of acid mine drainage are (i) the loss of aquatic life close to the mine site due to excessive acidity, (ii) the deposition of hydrated Fe_2O_3 some distance downstream.

In the case of ores containing metals other than iron, these other metals (*e.g.*, Cu, Pb, Zn) become leached in soluble form in the vicinity of the mine. Recall from Section 12.3 that metal salts tend to be more soluble at low pH. Since their concentrations are often > 100 ppm, it may be economically feasible to recover them by hydrometallurgy. At some large mines, the quantities of metals in solution may amount to millions of tonnes, at concentrations a few ppm, but spread over thousands of hectares. A major thrust in environmental clean-up is site remediation of present and former mine sites. Contamination at such sites by toxic metal ions is predicted to continue for hundreds of years in the absence of remediation technology.

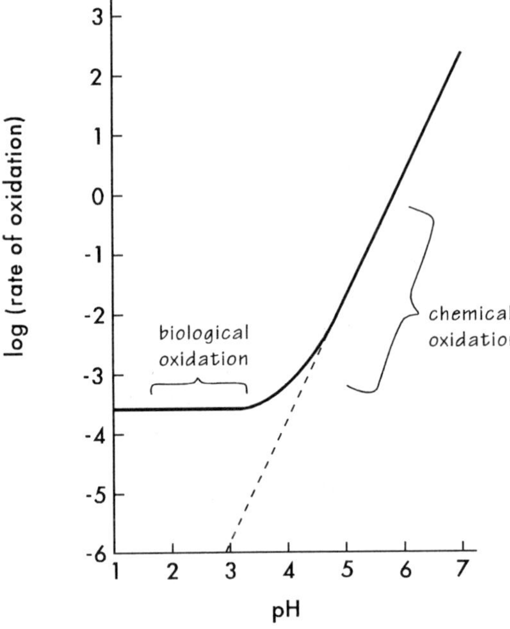

Figure 17.3: Oxidation of Fe(II) as a function of pH

It is reasonable to ask why acidification should be a problem at a mine site, but not when the same sulfide minerals lie undisturbed in the ground. The source of the problem lies in the way the ores are treated. In order to concentrate the valuable minerals from the gangue, the ore is crushed very thoroughly to a consistency of fine sand before being subjected to oil flotation. The rejected material, known as "tailings", is present as an aqueous slurry, and is dumped in "tailings ponds", where there is free access to atmospheric oxygen. Because the material is finely crushed, the surface to volume ratio is greatly increased compared with the original rock. Bacterial oxidation can therefore proceed rapidly, producing a large amount of H^+ per unit time. In the same way, tunnelling in deep pit mines perforates the ore body, again allowing access to oxygen and microorganisms.

At abandoned mine sites, the mine pit may become filled with water through seepage from the water table. This may produce a large lake having pH 3 or so. It is not practical to neutralize the acid, for example with lime, because further biological oxidation of exposed sulfide surfaces would soon re-acidify the lake. The management and restoration of former mine sites promises to be one of the greatest environmental challenges for future decades, since no practical solution has yet been found. The day is rapidly coming when it will not be possible for mining companies to fulfil completely their responsibilities for a site even after site reclamation; a residual responsibility for the site will likely remain in perpetuity.

Problems

Sections 17.1 - 17.2

1. The mass of the lithosphere is estimated to be 2.4×10^{19} t. Calculate the total masses of gold, silver, copper, and iron in the lithosphere, using the data of Table 17.1.

2. Tell what happens (in words) when the following reactants are mixed. If a chemical reaction takes place, write a net ionic equation.
   a) Zinc metal is placed in a copper(II) sulfate solution.
   b) Copper metal is placed in zinc sulfate solution.
   c) Powdered nickel is placed in mercury(II) chloride solution.
   d) Powdered silver is placed in calcium chloride solution.
   e) Copper metal is placed in hydrochloric acid.
   f) Manganese metal is placed in sulfuric acid.

3. Write a plausible equation for the reaction which occurs when silver dissolves in moderately concentrated nitric acid.

4. Calculate $\mathscr{E}$ for the following cells:
   a) $Fe(s) \mid H^+ (a = 1), Fe^{2+}, (a = 1) \mid H_2$ (g, 1 atm), M(s)
   b) $Fe(s) \mid H^+$, pH 6.2, Fe^{2+} (0.58 mol L^{-1}) $\mid H_2$(g, 1 atm)), M(s)
   M(s) is an impurity site on the surface of the iron.

5. Repeat Question 4 where the cathode reaction is the reduction of O_2 to H_2O rather than the release of H_2(g). Assume $p(O_2\ (g))$ = 1.00 atm (part a) and 0.04 atm (part b).

Section 17.3

1. In a heap leaching process for gold, assumed to take place at 298 K, a solution of NaCN (1000 ppm) is percolated through a heap of gold ore.
   a) Calculate the pH of the solution
   b) Calculate K for the reaction below at 298 K
   $$2Au(s) + 4CN^-(aq) + \tfrac{1}{2}O_2(g) + H_2O(\ell) \rightarrow 2\,Au(CN)_2^-(aq) + 2OH^-(aq)$$
   c) Calculate the percentage of gold which would be leached at equilibrium using 1000 ppm of NaCN, and assuming that $p(O_2\ (g))$ = 0.0090 atm inside the ore body.

2. Calculate the cost of electricity per tonne of copper refined if electricity costs 9.6 c/kWh.

3. Calculate the maximum theoretical removal of copper from a leachate containing 145 ppm of Cu^{2+} by the use of scrap iron
   a) if Cu^{2+} is the only cation present
   b) if the solution also contains 350 ppm of Fe^{2+}

4. A student breaks a thermometer and 0.5 g of mercury is left uncollected on the floor of a laboratory 3 m x 4 m x 3 m high. At equilibrium at 20°C, is the regulatory standard for mercury (0.05 mg m^{-3}) exceeded?

5. The maximum allowable concentration of mercury in drinking water is 1 ppb. Is this guideline exceeded if a drinking water source is at equilibrium with a rock containing HgS(s)?

6. A model for mercury cycling in the atmosphere used the following data. Steady state of Hg(g) in the atmosphere 1.2×10^9 g; rates of input in g yr^{-1}: volcanoes, 2×10^7; biomass decomposition, 4×10^7; volatilization from oceans, 4×10^9; volatilization from the land 1.8×10^{10}; anthropogenic, 1.0×10^{10}. Calculate the residence time of mercury in the atmosphere.

7. For each tonne of aluminum produced by electrolysis calculate the mass of carbon anodes oxidized.

Sections 17.4 - 17.5

1. a) Calculate $\mathscr{E}°$ for the reaction of cathodic protection.

   $$Zn(s) + \tfrac{1}{2}O_2(g) + 2H^+(aq) \rightarrow Zn^{2+}(aq) + H_2O(\ell)$$

   b) Calculate $\mathscr{E}$, and hence ΔG for the above reaction at 298 K if $[Zn^{2+}] = 0.001$ mol L^{-1} $p(O_2(g)) = 0.21$ atm at pH 5.1 and at pH 8.1.

2. At an intake of 1 mg Hg per day in the diet, humans experience a steady state burden of 100 mg Hg in the body. Calculate the residence time of mercury in the human body, and the half life for excretion of mercury.

3. a) The World Health Organization sets 0.2 mg per week as an acceptable intake of mercury for a 60 kg person. In Canada, Great Lakes fish are considered edible if they contain no more than 0.5 ppm of mercury. What is the maximum mass of Great Lakes fish containing 0.5 ppm of mercury that a Canadian should eat weekly?

   b) Calculate the mass of mercury in a 1.5 kg lake trout contaminated by 0.45 ppm of Hg^{2+}.

4. A "low lead" paint contains 0.5% lead by weight and loses 60% of its weight upon drying. An 11 kg child chews on an object painted with this paint. What mass of dried paint needs to be ingested for the child to take up the World Health Organization's recommended daily lead intake of no more than 6 μg kg^{-1}?

5. The movement of lead in the blood of adult males may be summarized:

   Blood lead = 140 μg L^{-1} Net transfer to bone = 7.5 μg day^{-1}

   Blood volume = 4.8 L Net excretion rate = 24 μg day^{-1}

   Calculate the residence time of lead in the blood.

6. A lead recycling plant begins operation on the shores of a hitherto clean lake of capacity 3.0×10^6 m^3. It discharges into the lake 12 m^3 per hour of waste containing 15 ppm of Pb^{2+}. The other inflow and outflow of the lake is a river with a flow rate 8400 m^3 h^{-1}.

   a) Calculate the steady state concentration of Pb^{2+} in the lake, which is well mixed, and has no other source or sink for Pb^{2+}.

   b) Calculate the residence time of Pb^{2+} in the lake at the steady state.

   c) How long does it take for the Pb^{2+} level to reach 50%?, 90%?, of its steady state value?

7. At 20°C the oxidation of Fe^{2+} to Fe^{3+} follows the following rate law

$$\frac{-d[Fe^{2+}]}{dt} = 8 \times 10^{13}\,[Fe^{2+}]\,[OH^-]^2 p(O_2) \text{ mol L}^{-1}\text{ min}^{-1}$$

   a) What are the units of the rate constant?

   b) What is the order of the reaction at constant pH, and constant $p(O_2)$?

   c) Calculate the rate of reaction of Fe^{2+} in mol L^{-1} min^{-1} and the half-life of the reaction under the following conditions

      i) pH 4.30 $p(O_2)$ = 0.21 atm $[Fe^{2+}]_o$ = 75 ppm

      ii) pH 7.45 $p(O_2)$ = 0.21 atm $[Fe^{2+}]_o$ = 75 ppm

   d) In each of the cases (c) calculate how long the reaction will proceed before $Fe(OH)_3$ begins to precipitate.

Appendix 1

Standard Molar Enthalpies of Formation (298 K), Standard Molar Free Energies of Formation (298 K), and Absolute Entropies

Formula	ΔH°_f kJ mol^{-1}	ΔG°_f kJ mol^{-1}	S° J K^{-1} mol^{-1}
Aluminum			
Al (s)	0	0	28.33
Al (g)	326.4	285.7	164.43
Al^{3+} (aq)	-531	-485	-
Al_2O_3 (s)	-1675.7	-1582.3	50.92
$Al_2O_3.3H_2O$ (s)	-2586.67	-2310.41	136.90
$Al_2(SO_4)_3$ (s)	-3440.84	-3100.14	239.3
$AlPO_4$ (s)	-1733.8	-1618.0	90.79
Argon			
Ar (g)	0	0	154.734
Barium			
Ba (s)	0	0	62.8
Ba^{2+} (aq)	-537.64	-560.77	-
BaO (s)	-553.5	-525.1	70.42
$Ba(OH)_2$ (s)	-944.47	-	-
BaF_2 (s)	-1207.1	-1156.9	96.36
$BaCl_2$ (s)	-858.6	-810.4	123.68
$BaSO_4$ (s)	-1473.2	-1362.3	132.2
$BaCO_3$ (s)	-1216.3	-1137.6	112.1
Boron			
B (s)	0	0	5.86
B (g)	562.7	518.8	153.34
B_2O_3 (s)	-1272.77	-1193.70	53.97
B_2H_6 (g)	35.6	86.6	232.00
H_3BO_3 (s)	-1094.33	-969.02	88.83

Formula	ΔH°_f kJ mol^{-1}	ΔG°_f kJ mol^{-1}	S° J K^{-1} mol^{-1}
Boron (continued)			
BF_3 (g)	-1137.00	-1120.38	254.01
BCl_3 (ℓ)	-427.2	-387.4	206.3
BCl_3 (g)	-403.76	-388.77	289.99
Bromine			
Br_2 (ℓ)	0	0	152.231
Br_2 (g)	30.907	3.110	245.354
Br (g)	111.884	82.412	174.913
Br^- (aq)	-121.55	-103.96	-
HBr (g)	-36.40	-53.45	198.586
BrF (g)	-93.85	-109.18	228.97
BrF_3 (ℓ)	-300.8	-240.5	178.2
BrF_3 (g)	-255.60	-299.45	292.42
BrF_5 (ℓ)	-458.6	-351.9	225.1
BrF_5 (g)	-428.9	-350.6	320.08
BrCl (g)	14.64	-0.98	239.99
Cadmium			
Cd (s)	0	0	51.76
Cd^{2+} (aq)	-75.90	-77.612	-
CdO (s)	-258.2	-228.4	54.8
$Cd(OH)_2$ (s)	-560.7	-473.7	96
CdS (s)	-161.9	-156.5	64.9
Calcium			
Ca (s)	0	0	41.42
Ca^{2+} (aq)	-542.83	-553.58	-
CaO (s)	-635.09	-604.05	39.75
$Ca(OH)_2$ (s)	-986.09	-898.56	83.39
CaF_2 (s)	-1219.6	-1167.3	68.87
$CaCl_2$ (s)	-795.8	-748.1	104.6
$CaSO_4$ (s)	-1425.24	-1313.49	108.4
$CaSO_4.2H_2O$ (s)	-2022.63	-1797.44	194.1
$CaCO_3$ (s)	-1206.92	-1128.84	92.6

Formula	ΔH°_f kJ mol^{-1}	ΔG°_f kJ mol^{-1}	S° J K^{-1} mol^{-1}
<u>Carbon</u>			
C (s, graphite)	0	0	5.740
C (s, diamond)	1.895	2.900	2.377
C (g)	716.682	671.290	157.987
CO (g)	-110.525	-137.152	197.565
CO_2 (g)	-393.509	-394.359	213.63
CO_3^{2-} (aq)	-677.14	-527.86	-
CH_4 (g)	-74.81	-50.75	186.155
HCO_2^- (aq), formate	-425.55	-351.0	-
HCO_3^- (aq), bicarbonate	-691.99	-586.84	-
HCHO (g)	-108.57	-102.55	218.66
HCOOH (ℓ)	-424.72	-361.42	128.95
HCOOH (aq, undissociated)	-425.43	-372.4	-
CH_3OH (ℓ)	-238.66	-166.35	126.8
CH_3OH (g)	-200.66	-162.01	239.70
CF_4 (g)	-925	-879	261.50
CF_3Br (g)	-642.7	-616.3	297.6
CCl_4 (ℓ)	-135.44	-65.28	216.40
CCl_4 (g)	-102.9	-60.62	309.74
$COCl_2$ (g)	-218.8	-204.6	283.42
CH_3Cl (g)	-80.83	-57.40	234.47
CH_2Cl_2 (ℓ)	-121.46	-67.29	177.8
CH_2Cl_2 (g)	-92.47	-65.9	270.12
$CHCl_3$ (ℓ)	-134.47	-73.69	201.7
$CHCl_3$ (g)	-103.14	-70.37	295.60
CF_2Cl_2 (g)	-477	-439	300.7
$CFCl_3$ (g)	-301.33	-236.86	225.35
CS_2 (ℓ)	89.70	65.27	151.34
CS_2 (g)	117.36	67.15	237.73
CN^- (aq)	150.6	172.4	-
HCN (ℓ)	108.87	124.94	112.84
HCN (aq, undissociated)	107.1	119.7	-
$CO(NH_2)_2$ (s)	-333.51	-197.44	104.60

Formula	ΔH°_f kJ mol^{-1}	ΔG°_f kJ mol^{-1}	S° J K^{-1} mol^{-1}
Carbon (continued)			
$C_2O_4^{2-}$ (aq), oxalate	-825.1	-674.0	-
C_2H_2 (g)	226.73	209.17	200.83
C_2H_4 (g)	52.26	68.08	219.45
C_2H_6 (g)	-84.68	-32.92	229.49
CH_3COO^- (aq), acetate	-486.01	-369.39	-
CH_3CHO (ℓ)	-192.30	-128.20	160.2
CH_3CHO (g)	-166.19	-128.91	250.2
CH_3COOH (ℓ)	-484.5	-390.0	159.8
CH_3COOH (aq, undissociated)	-485.76	-396.56	-
C_2H_5OH (ℓ)	-277.69	-174.89	160.7
C_3H_8 (g)	-104.5	-23.4	296.9
n-C_8H_{18} (ℓ), octane	-250.0	6.36	361.2
C_8H_{18} (ℓ), isooctane	-225.0	12.8	423.0
Chlorine			
Cl_2 (g)	0	0	222.957
Cl (g)	121.679	105.696	165.089
Cl^- (aq)	-167.159	-131.244	-
ClO (g)	101.84	98.11	226.52
ClO^- (aq)	-107.1	-36.8	-
ClO_2 (g)	102.5	120.5	256.73
ClO_3^- (aq)	-103.97	-7.98	-
ClO_4^- (aq)	-129.33	-8.60	-
Cl_2O (g)	80.3	97.9	266.10
HCl (g)	-92.307	-95.299	186.799
HClO (g)	-78.7	-66.1	236.56
HClO (aq, undissociated)	-120.9	-79.9	-
ClF (g)	-54.48	-55.94	217.78
ClF_3 (g)	-163.2	-123.0	281.50
$COCl_2$ (g)	-219	-205	283.4

Formula	ΔH°_f kJ mol^{-1}	ΔG°_f kJ mol^{-1}	S° J K^{-1} mol^{-1}
Chromium			
Cr (s)	0	0	23.77
Cr (g)	396.6	351.8	174.39
CrO_4^{2-} (aq)	-881.15	-727.82	-
$Cr_2O_7^{2-}$ (aq)	-1490.3	-1301.2	-
Copper			
Cu (s)	0	0	33.150
Cu (g)	338.32	298.61	166.27
Cu^+ (aq)	71.67	49.98	-
Cu^{2+} (aq)	64.77	65.49	-
CuO (s)	-157.3	-129.7	42.63
Cu_2O (s)	-168.6	-146.0	93.14
$Cu(OH)_2$ (s)	-449.8	-	-
$CuCl_2$ (s)	-220.1	-175.7	108.07
Cu_2S (s)	-79.5	-86.2	120.9
$CuSO_4$ (s)	-771.36	-661.9	109
$CuSO_4.5H_2O$ (s)	-2279.65	-1880.055	300.4
Flourine			
F_2 (g)	0	0	202.67
F (g)	78.99	61.93	158.645
F^- (aq)	-332.63	-278.81	-
F^- (g)	-255.39	-	-
HF (g)	-271.1	-273.2	173.670
HF (aq, undissociated)	-320.08	-296.85	-
Gold			
Au (s)	0	0	47.40
$Au(CN)_2^-$ (aq)	242.3	285.8	-
Helium			
He (g)	0	0	126.041

Formula	ΔH°_f kJ mol^{-1}	ΔG°_f kJ mol^{-1}	S° J K^{-1} mol^{-1}
Hydrogen			
H_2 (g)	0	0	130.575
H (g)	217.965	203.263	114.604
H^+ (aq)	0	0	-
H^+ (g)	1536.202	-	-
OH^- (aq)	-229.994	-157.277	-
H_2O (ℓ)	-285.830	-237.178	69.91
H_2O (g)	-241.818	-228.588	188.716
OH (g)	38.95	34.23	183.6
H_2O_2 (ℓ)	-187.78	-120.42	109.6
H_2O_2 (g)	-136.31	-105.60	232.6
HO_2 (g)	21	-	-
Iodine			
I_2 (s)	0	0	116.135
I_2 (g)	62.438	19.360	260.58
I (g)	106.838	70.283	180.682
I^- (aq)	-55.19	-51.57	-
IO_3^- (aq)	-221.3	-128.0	-
HI (g)	26.48	1.72	206.485
HIO_3 (aq, undissociated)	-211.3	-132.7	-
Iron			
Fe (s)	0	0	27.28
Fe (g)	416.3	370.7	180.381
Fe^{2+} (aq)	-89.1	-78.90	-
Fe^{3+} (aq)	-48.5	-4.7	-
Fe_2O_3 (s)	-824.2	-742.2	87.40
Fe_3O_4 (s)	-1118.4	-1015.4	146.4
$FeCl_2$ (s)	-341.79	-302.33	117.95
$FeCl_3$ (s)	-399.49	-334.05	142.3
FeS (s)	-100.0	-100.4	60.29
FeS_2 (s)	-178.2	-166.9	52.93
$FeSO_4$ (s)	-928.4	-820.9	107.5

Formula	ΔH°_f kJ mol^{-1}	ΔG°_f kJ mol^{-1}	S° J K^{-1} mol^{-1}
Iron (continued)			
$FeSO_4.7H_2O$ (s)	-3014.57	-2510.28	409.2
$Fe(CO)_5$ (g)	-733.9	-697.26	445.3
Lead			
Pb (s)	0	0	64.81
Pb (g)	195.0	161.9	175.264
Pb^{2+} (aq)	-1.7	-24.43	-
PbO (s, yellow)	-217.32	-187.91	68.70
PbO (s, red)	-218.99	-188.95	66.5
PbO_2 (s)	-277.4	-217.36	68.6
Pb_3O_4 (s)	-718.4	-601.3	211.3
PbF_2 (s)	-664.0	-617.1	110.5
$PbCl_2$ (s)	-359.41	-314.13	136.0
$PbBr_2$ (s)	-278.7	-261.92	161.5
PbI_2 (s)	-175.48	-173.64	174.85
PbS (s)	-100.4	-98.7	91.2
$PbSO_4$ (s)	-919.94	-813.21	148.57
$Pb(C_2H_5)_4$ (ℓ)	52.7	-	-
Lithium			
Li (s)	0	0	29.12
Li (g)	159.37	126.69	138.66
Li^+ (g)	685.783	-	-
Li^+ (aq)	-278.49	-293.31	13.4
Li_2O (s)	-597.94	-561.20	37.57
LiOH (s)	-484.93	-438.98	42.80
Magnesium			
Mg (s)	0	0	32.68
Mg (g)	147.70	113.13	148.650
Mg^{2+} (aq)	-466.85	-454.8	-
MgO (s)	-601.70	-569.45	26.94

Formula	ΔH°_f kJ mol^{-1}	ΔG°_f kJ mol^{-1}	S° J K^{-1} mol^{-1}
Magnesium (continued)			
$Mg(OH)_2$ (s)	-924.54	-833.58	63.18
MgF_2 (s)	-1123.4	-1070.2	57.24
$MgCl_2$ (s)	-641.32	-591.82	89.62
$MgSO_4$ (s)	-1284.9	-1170.7	91.6
$MgSO_4.7H_2O$ (s)	-3388.71	-2871.9	372.
$3MgO.2SiO_2.2H_2O$ (s, chrysotile)	-4365.6	-4038.0	221.3
$MgCO_3.CaCO_3$ (dolomite)	-2326	-2164	155.2
Manganese			
Mn (s)	0	0	32.01
Mn (g)	280.7	238.5	173.59
Mn^{2+} (aq)	-220.75	-228.1	-
MnO (s)	-385.22	-362.92	59.71
MnO_2 (s)	-520.03	-465.17	53.05
MnO_4^- (aq)	-541.4	-447.3	-
$MnCO_3$ (s)	-481.29	-440.53	118.24
Mercury			
Hg (ℓ)	0	0	76.02
Hg (g)	61.317	31.853	174.85
Hg^{2+} (aq)	171.1	164.40	-
Hg_2^{2+} (aq)	172.4	153.52	-
HgO (s, red)	-90.83	-58.555	70.29
HgO (s, yellow)	-90.46	-58.425	71.1
$HgCl_2$ (s)	-224.3	-178.6	146.0
Hg_2Cl_2 (s)	-265.22	-210.778	192.5
HgS (s, red)	-58.2	-50.6	82.4
HgS (s, black)	-53.6	-47.7	88.3
Hg_2SO_4 (s)	-743.12	-625.880	200.66
$Hg(CH_3)_2$ (ℓ)	59.8	140.3	209
$Hg(CH_3)_2$ (g)	94.39	146.0	306

Formula	$\Delta H°_f$ kJ mol^{-1}	$\Delta G°_f$ kJ mol^{-1}	S° J K^{-1} mol^{-1}
Neon			
Ne (g)	0	0	146.219
Nickel			
Ni (s)	0	0	29.87
Ni (g)	429.7	384.5	182.084
Ni^{2+} (aq)	-54.0	-45.6	-
NiO (s)	-239.7	-211.7	37.99
NiS (s)	-82.0	-79.5	52.97
$Ni(CO)_4$ (ℓ)	-633.0	-588.3	313.4
$Ni(CO)_4$ (g)	-602.91	-587.26	410.5
Nitrogen			
N_2 (g)	0	0	191.50
N (g)	472.704	455.579	153.189
N_3^- (aq)	275.14	348.2	-
NO (g)	90.25	86.55	210.652
NO_2 (g)	33.18	51.29	239.95
NO_2^- (aq)	-104.6	-32.2	123.0
NO_3^- (aq)	-205.0	-108.81	-
N_2O (g)	82.05	104.20	219.74
N_2O_4 (ℓ)	-19.50	97.44	209.2
N_2O_4 (g)	9.16	97.89	304.29
N_2O_5 (s)	-43.1	113.8	178.2
N_2O_5 (g)	11.3	115.0	355.7
NH_3 (g)	-46.11	-16.42	192.34
NH_3 (aq)	-80.29	-26.57	-
N_2H_4 (ℓ)	50.63	149.24	121.21
N_2H_4 (g)	95.40	159.28	238.36
HNO_3 (ℓ)	-174.10	-80.79	155.60
HNO_3 (g)	-135.1	-74.77	226.3
NH_4NO_3 (s)	-365.56	-184.02	151.08
NH_4Cl (s)	-314.43	-202.97	94.6
$(NH_4)_2SO_4$ (s)	-1180.85	-901.90	220.1

Formula	ΔH°_f kJ mol^{-1}	ΔG°_f kJ mol^{-1}	S° J K^{-1} mol^{-1}
Oxygen			
O_2 (g)	0	0	205.029
O (g)	249.170	231.747	160.946
O_3 (g)	149.7	163.2	238.82
OH (g)	38.95	34.23	183.6
OH^- (aq)	-229.994	-157.277	-
Phosphorus			
P (s, white)	0	0	41.09
P (s, red)	-17.6	-12.1	22.80
P (g)	314.64	278.28	168.084
P_4 (g)	58.91	24.47	279.87
PO_4^{3-} (aq)	-1277.4	-1018.8	-
P_4O_{10} (s)	-2984.0	-2697.9	228.86
HPO_4^{2-} (aq)	-1292.14	-1089.23	-
$H_2PO_4^-$ (aq)	-1296.29	-1130.38	-
H_3PO_4 (s)	-1279.0	-1119.2	110.50
H_3PO_4 (aq, undissociated)	-1288.34	-1142.65	-
PF_3 (g)	-918.8	-897.5	273.13
PCl_3 (ℓ)	-319.7	-272.3	217.1
PCl_3 (g)	-287.0	-267.8	311.67
PCl_5 (g)	-374.9	-305.1	364.47
Potassium			
K (s)	0	0	64.18
K (g)	89.24	60.62	160.336
K^+ (g)	514.26	-	-
K^+ (aq)	-252.38	-283.27	-
KOH (s)	-424.764	-379.11	78.9
KF (s)	-567.27	-537.77	66.57
KCl (s)	-436.747	-409.16	82.59
$KClO_3$ (s)	-397.73	-296.32	143.1
$KClO_4$ (s)	-432.75	-303.17	151.0
KBr (s)	-393.798	-380.68	95.90

Formula	ΔH°_f kJ mol^{-1}	ΔG°_f kJ mol^{-1}	S° J K^{-1} mol^{-1}
Potassium (continued)			
KI (s)	-327.900	-324.892	106.32
KIO_3 (s)	-501.37	-418.40	151.46
KNO_3 (s)	-494.63	-394.93	133.05
K_2CO_3 (s)	-1151.02	-1063.5	155.52
$KMnO_4$ (s)	-837.2	-737.7	171.71
K_2CrO_4 (s)	-1403.7	-1295.8	200.12
$K_2Cr_2O_7$ (s)	-2061.5	-1881.9	291.2
Silicon			
Si (s)	0	0	18.83
Si (g)	455.6	411.3	167.86
SiO_2 (s, quartz)	-910.94	-856.67	41.84
SiH_4 (g)	34.3	56.9	204.51
Silver			
Ag (s)	0	0	42.55
Ag (g)	284.55	245.68	172.888
Ag^+ (aq)	105.579	77.107	-
AgCl (s)	-127.068	-109.805	96.2
$AgCl_2^-$ (aq)	-245.2	-215.4	-
AgBr (s)	-100.37	-96.90	107.1
Ag_2S (s)	-32.59	-40.67	144.01
$AgNO_3$ (s)	-124.39	-33.48	140.92
$Ag(NH_3)_2^+$ (aq)	-111.29	-17.25	-
AgCN (s)	146.0	156.9	107.19
$Ag(CN)_2^-$ (aq)	270.3	305.5	-
Sodium			
Na (s)	0	0	51.21
Na (g)	107.32	76.794	153.603
Na^+ (g)	609.358	-	-
Na^+ (aq)	-240.12	-261.905	-

Formula	ΔH°_f kJ mol^{-1}	ΔG°_f kJ mol^{-1}	S° J K^{-1} mol^{-1}
Sodium (continued)			
$NaOH$ (s)	-425.609	-379.527	62.455
NaF (s)	-573.647	-543.510	51.46
$NaCl$ (s)	-411.153	-384.154	72.13
$NaClO_4$ (s)	-383.30	-254.93	142.3
$NaBr$ (s)	-361.062	-348.983	86.82
NaI (s)	-287.78	-286.06	98.53
Na_2SO_4 (s)	-1387.08	-1270.23	149.58
$Na_2SO_4.10H_2O$ (s)	-4327.26	-3627.40	592.0
$NaHSO_4$ (s)	-1125.5	-992.9	113.0
$NaSO_4$ (s)	-358.65	-284.60	103.8
$NaNO_3$ (s)	-467.85	-367.07	116.52
Na_2CO_3 (s)	-1130.68	-1044.49	134.98
$NaHCO_3$ (s)	-950.81	-851.0	101.7
$Na_2B_4O_7.10H_2O$ (s)	-6288.6	-5516.6	586
Na_2CrO_4 (s)	-1342.2	-1235.00	176.61
Sulfur			
S (s, rhombic)	0	0	31.80
S (s, monoclinic)	0.33	-	-
S (g)	278.805	238.283	167.712
S^{2-} (aq)	33.1	85.8	-
S_2 (g)	128.37	79.33	228.07
S_8 (g)	102.30	49.66	430.87
SO_2 (g)	-296.830	-300.194	248.11
SO_3 (ℓ)	-441.04	-373.80	113.8
SO_3 (g)	-395.72	-371.08	256.65
SO_3^{2-} (aq)	-635.5	-486.5	-
SO_4^{2-} (aq)	-909.27	-744.60	-
HS^- (aq)	-17.6	12.08	-
H_2S (g)	-20.63	-33.59	205.68
H_2S (aq, undissociated)	-39.7	-27.86	-
HSO_3^- (aq)	-626.22	-527.80	-

Formula	ΔH°_f kJ mol^{-1}	ΔG°_f kJ mol^{-1}	S° J K^{-1} mol^{-1}
Sulfur (continued)			
HSO_4^- (aq)	-887.34	-755.99	-
H_2SO_4 (ℓ)	-813.989	-690.101	156.795
SF_4 (g)	-774.9	-731.4	291.92
SF_6 (g)	-1209	-1105.4	291.71
SO_2Cl_2 (g)	-212.5	-320.0	311.83
Titanium			
Ti (s)	0	0	30.63
Ti (g)	469.9	425.1	180.189
$TiCl_3$ (s)	-720.9	-653.5	139.7
$TiCl_4$ (ℓ)	-804.2	-737.2	252.34
$TiCl_4$ (g)	-763.2	-726.7	354.8
Tungsten			
W (s)	0	0	32.64
W (g)	849.4	807.1	173.841
WO_2 (s)	-589.69	-533.92	50.54
WO_3 (s)	-842.87	-764.08	75.90
Uranium			
U (s)	0	0	50.21
U (g)	535.6	491.2	199.66
UO_2 (s)	-1084.9	-1031.7	77.03
UO_3 (s)	-1223.8	-1145.9	96.11
U_3O_8 (s)	-3574.8	-3369.8	282.59
UF_6 (s)	-2197.0	-2068.6	227.6
UF_6 (g)	-2147.4	-2063.8	377.8
Vanadium			
V (s)	0	0	28.91
V (g)	514.21	754.46	182.189
VO^{2+} (aq)	-486.6	-446.4	-

Formula	ΔH°_f kJ mol^{-1}	ΔG°_f kJ mol^{-1}	S° J K^{-1} mol^{-1}
Vanadium (continued)			
VO_2^+ (aq)	-649.8	-587.0	-
VO_3^- (aq)	-888.3	-783.6	-
V_2O_3 (s)	-1218.8	-1139.3	98.3
V_2O_4 (s)	-1427.2	-1318.4	102.5
V_2O_5 (s)	-1550.6	-1419.6	131.0
Xenon			
Xe (g)	0	0	169.574
Zinc			
Zn (s)	0	0	41.63
Zn (g)	130.729	95.178	160.875
Zn^{2+} (aq)	-153.89	-147.06	-
ZnO (s)	-348.28	-318.32	43.64
$Zn(OH)_2$ (s)	-643.25	-555.14	81.6
ZnF_2 (s)	-764.4	-713.3	73.68
$ZnCl_2$ (s)	-415.05	-369.431	111.46
$ZnBr_2$ (s)	-328.65	-312.13	138.5
ZnS (s)	-205.98	-201.29	57.7

Appendix 2

Single Bond Energies (kJ/mol) at 25°C

	H	C	N	O	S	F	Cl	Br	I
H	436	414	389	464	339	565	431	368	297
C		347	293	351	259	485	331	276	238
N			159	222	-	272	201	243	-
O				138	-	184	205	201	201
S					226	285	255	213	-
F						153	255	255	-
Cl							243	218	209
Br								193	180
I									151

Double Bond Energies (kJ/mol) at 25°C

	C	N	O	S
C	620	615	730	575
N		420	590	-
O			495	-
S				-

Triple Bond Energies (kJ/mol) at 25°C

	C	N	O
C	860	890	1076
N		946	-

Appendix 3

Acid dissociation constants at 25°C in order of decreasing acid strength.

Acid	Formula	K_a(mol L^{-1})
Benzenesulfonic	$C_6H_5SO_3H$	0.2
Iodic	HIO_3	0.17
Sulfurous	H_2SO_3	1.7×10^{-2}
Hydrogen sulfate	HSO_4^-	1.2×10^{-2}
Chlorous	$HClO_2$	1.0×10^{-2}
Phosphoric	H_3PO_4	7.5×10^{-3}
Lactic	$HC_3H_5O_3$	8.4×10^{-4}
Nitrous	HNO_2	4.6×10^{-4}
Hydrofluoric	HF	3.5×10^{-4}
Formic	$HCHO_2$	1.8×10^{-4}
Benzoic	$HC_7H_5O_2$	6.5×10^{-5}
Anilinium	$HC_6H_7N^+$	2.3×10^{-5}
Hydrazoic	HN_3	1.9×10^{-5}
Acetic	$HC_2H_3O_2$	1.8×10^{-5}
Pyridinium	$HC_5H_5N^+$	5.6×10^{-6}
Carbonic	H_2CO_3	4.2×10^{-7}
Hydrogen sulfite	HSO_3^-	1.0×10^{-7}
Hydrogen sulfide	H_2S	9.1×10^{-8}
Dihydrogen phosphate	$H_2PO_4^-$	6.3×10^{-8}
Hypochlorous	HClO	3.0×10^{-8}
Hydrazinium	$HN_2H_4^+$	5.9×10^{-9}
Boric	H_3BO_3	7.2×10^{-10}
Ammonium	NH_4^+	5.6×10^{-10}
Hydrocyanic	HCN	4.9×10^{-10}
Phenol	HC_6H_5O	1.3×10^{-10}
Hydrogen carbonate	HCO_3^-	4.8×10^{-11}
Methylammonium	HCH_5N^+	2.8×10^{-11}
Ethylammonium	$HC_2H_7N^+$	1.5×10^{-11}
Hydrogen peroxide	H_2O_2	2.4×10^{-12}
Hydrogen sulfide anion	HS^-	1.1×10^{-12}
Hydrogen phosphate	HPO_4^{2-}	2.2×10^{-13}

Appendix 4

Solubility product constants at 25°C by cation, in alphabetical order.

Substance	Formula	K_{sp}
Aluminium hydroxide	$Al(OH)_3$	1.7×10^{-33} $(mol\ L^{-1})^4$
Aluminum phosphate	$AlPO_4$	1.0×10^{-21} $(mol\ L^{-1})^2$
Barium carbonate	$BaCO_3$	2.6×10^{-9} $(mol\ L^{-1})^2$
Barium flouride	BaF_2	1.0×10^{-6} $(mol\ L^{-1})^3$
Barium sulfate	$BaSO_4$	1.1×10^{-10} $(mol\ L^{-1})^2$
Cadmium hydroxide	$Cd(OH)_2$	5.3×10^{-15} $(mol\ L^{-1})^3$
Cadmium sulfide	CdS	1.4×10^{-29} $(mol\ L^{-1})^2$
Calcium carbonate	$CaCO_3$	6.0×10^{-9} $(mol\ L^{-1})^2$
Calcium fluoride	CaF_2	1.6×10^{-10} $(mol\ L^{-1})^3$
Calcium hydroxide	$Ca(OH)_2$	7.9×10^{-6} $(mol\ L^{-1})^3$
Calcium phosphate	$Ca_3(PO_4)_2$	2.0×10^{-33} $(mol\ L^{-1})^5$
Calcium phosphate (hydroxylapatite)	$Ca_5(PO_4)_3OH$	1.0×10^{-56} $(mol\ L^{-1})^9$
Calcium sulfate	$CaSO_4$	3.7×10^{-5} $(mol\ L^{-1})^2$
Calcium magnesium carbonate	$CaCO_3.MgCO_3$	2.6×10^{-13} $(mol\ L^{-1})^4$
Copper (II) hydroxide	$Cu(OH)_2$	2.2×10^{-20} $(mol\ L^{-1})^3$
Copper (II) iodate	$Cu(IO_3)_2$	1.4×10^{-7} $(mol\ L^{-1})^3$
Copper (II) sulfide	CuS	1.3×10^{-36} $(mol\ L^{-1})^2$
Iron (II) carbonate	$FeCO_3$	3.1×10^{-11} $(mol\ L^{-1})^2$
Iron (II) hydroxide	$Fe(OH)_2$	4.8×10^{-17} $(mol\ L^{-1})^3$
Iron (III) hydroxide	$Fe(OH)_3$	1.0×10^{-38} $(mol\ L^{-1})^4$
Lead chloride	$PbCl_2$	1.6×10^{-5} $(mol\ L^{-1})^3$
Lead chromate	$PbCrO_4$	2.8×10^{-13} $(mol\ L^{-1})^2$
Lead fluoride	PbF_2	1.1×10^{-7} $(mol\ L^{-1})^3$
Lead hydroxide	$Pb(OH)_2$	1.4×10^{-20} $(mol\ L^{-1})^3$
Lead sulfate	$PbSO_4$	1.8×10^{-8} $(mol\ L^{-1})^2$
Lead sulfide	PbS	8.8×10^{-29} $(mol\ L^{-1})^2$
Lithium carbonate	Li_2CO_3	1.1×10^{-3} $(mol\ L^{-1})^3$
Magnesium carbonate	$MgCO_3$	3.5×10^{-8} $(mol\ L^{-1})^2$

Substance	Formula	K_{sp}
Magnesium hydroxide	$Mg(OH)_2$	5.6×10^{-12} (mol $L^{-1})^3$
Mercury (II) sulfide	HgS	6.4×10^{-53} (mol $L^{-1})^2$
Mercury (I) chloride	Hg_2Cl_2	1.5×10^{-18} (mol $L^{-1})^3$
Mercury (I) sulfate	Hg_2SO_4	8.0×10^{-7} (mol $L^{-1})^2$
Silver bromide	AgBr	5.3×10^{-13} (mol $L^{-1})^2$
Silver chloride	AgCl	1.8×10^{-10} (mol $L^{-1})^2$
Silver cyanide	AgCN	2.2×10^{-16} (mol $L^{-1})^2$
Silver sulfate	Ag_2SO_4	1.2×10^{-5} (mol $L^{-1})^3$
Silver sulfide	Ag_2S	6.6×10^{-50} (mol $L^{-1})^3$
Zinc sulfide	ZnS	2.9×10^{-25} (mol $L^{-1})^2$

Appendix 5

Standard reduction potentials in aqueous solution at 25 °C, in decreasing order

Half-reaction	E°,V
$F_2\ (g) + 2e \rightleftharpoons 2F^-$	2.866
$O_3\ (g) + 2H^+ + 2e \rightleftharpoons O_2\ (g) + H_2O$	2.076
$OH\ (aq) + e \rightleftharpoons OH^-$	2.02
$N_2O\ (g) + 2H^+ + 2e \rightleftharpoons N_2 + H_2O$	1.766
$Au^+ + e \rightleftharpoons Au$	1.692
$PbO_2\ (s) + SO_4^{2-} + 4H^+ + 2e \rightleftharpoons PbSO_4\ (s) + 2H_2O$	1.691
$MnO_4^- + 4H^+ + 3e \rightleftharpoons MnO_2 + 2H_2O$	1.679
$HClO + H^+ + e \rightleftharpoons \frac{1}{2}Cl_2\ (g) + H_2O$	1.611
$MnO_4^- + 8H^+ + 5e \rightleftharpoons Mn^{2+} + 4H_2O$	1.507
$HO_2\ (aq) + H^+ + e \rightleftharpoons H_2O_2$	1.495
$HClO + H^+ + 2e \rightleftharpoons H_2O + Cl^-$	1.482
$Cl_2\ (g) + 2e \rightleftharpoons 2Cl^-$	1.358
$ClO_2\ (aq) + H^+ + e \rightleftharpoons HClO_2$	1.277
$O_3\ (g) + H_2O + 2e \rightleftharpoons O_2 + 2OH^-$	1.24
$Cr_2O_7^{2-} + 14H^+ + 6e \rightleftharpoons 2Cr^{3+} + 7H_2O$	1.232
$O_2\ (g) + 4H^+ + 4e \rightleftharpoons 2H_2O$	1.229
$MnO_2\ (s) + 4H^+ + 2e \rightleftharpoons Mn^{2+} + 2H_2O$	1.224
$Br_2\ (\ell) + 2e \rightleftharpoons 2Br^-$	1.066
$NO_3^- + 4H^+ + 3e \rightleftharpoons NO\ (g) + 2H_2O$	0.957
$ClO_2\ (aq) + e \rightleftharpoons ClO_2^-$	0.954
$2Hg^{2+} + 2e \rightleftharpoons Hg_2^{2+}$	0.920
$Hg^{2+} + 2e \rightleftharpoons Hg\ (\ell)$	0.851
$Ag^+ + e \rightleftharpoons Ag\ (s)$	0.799
$Fe^{3+} + e \rightleftharpoons Fe^{2+}$	0.771
$O_2\ (g) + 2H^+ + 2e \rightleftharpoons H_2O_2$	0.695
$I_2 + 2e \rightleftharpoons 2I^-$	0.536
$Cu^+ + e \rightleftharpoons Cu\ (s)$	0.521
$O_2\ (g) + 2H_2O + 4e \rightleftharpoons 4OH^-$	0.401
$Cu^{2+} + 2e \rightleftharpoons Cu\ (s)$	0.337

Half-reaction	E°,V
$Hg_2Cl_2 + 2e \rightleftharpoons 2Hg\ (\ell) + 2Cl^-$	0.268
Saturated calomel electrode	0.241
$AgCl + e \rightleftharpoons Ag\ (s) + Cl^-$	0.222
$SO_4^{2-} + 4H^+ + 2e \rightleftharpoons H_2SO_3 + H_2O$	0.172
$Cu^{2+} + e \rightleftharpoons Cu^+$	0.153
$SO_4^{2-} + 8H^+ + 8e \rightleftharpoons S^{2-} + 4H_2O$	0.145
$S_4O_6^{2-} + 2e \rightleftharpoons 2S_2O_3^{2-}$	0.08
$2H^+ + 2e \rightleftharpoons H_2\ (g)$	0.000
$Fe^{3+} + 3e \rightleftharpoons Fe\ (s)$	-0.037
$Pb^{2+} + 2e \rightleftharpoons Pb\ (s)$	-0.126
$O_2 + H_2O + 2e \rightleftharpoons H_2O_2 + 2OH^-$	-0.146
$Ni^{2+} + 2e \rightleftharpoons Ni\ (s)$	-0.257
$PbSO_4\ (s) + 2e \rightleftharpoons Pb\ (s) + SO_4^{2-}$	-0.359
$Cd^{2+} + 2e \rightleftharpoons Cd\ (s)$	-0.403
$Fe^{2+} + 2e \rightleftharpoons Fe\ (s)$	-0.443
$Cr^{3+} + 3e \rightleftharpoons Cr\ (s)$	-0.744
$Zn^{2+} + 2e \rightleftharpoons Zn\ (s)$	-0.762
$2H_2O + 2e \rightleftharpoons H_2\ (g) + 2OH^-$	-0.828
$Mn^{2+} + 2e \rightleftharpoons Mn\ (s)$	-1.185
$Al^{3+} + 3e \rightleftharpoons Al\ (s)$	-1.662
$Mg^{2+} + 2e \rightleftharpoons Mg\ (s)$	-2.372
$Na^+ + e \rightleftharpoons Na\ (s)$	-2.71

INDEX

D

Q, R

T

U, V, W, X, Y, Z

About the Author

Nigel J. Bunce was born in Sutton Coldfield, England. He earned a B.A. (1964) and D. Phil. (1967) degrees from Oxford University. He moved to Canada in 1967, and held a Killam Memorial fellowship at the University of Alberta for two years. Since 1969 he has been with the Department of Chemistry and Biochemistry of the University of Guelph. Dr. Bunce's researches are in environmental chemistry generally, and his specific areas of interest at present are the toxicology of dioxins, and organic photochemistry. Dr. Bunce is also the author of *Environmental Chemistry*.

BACK COVER

On a clear day in the Himalayas, one can hardly see the mountains. The smoggy view is caused by air-borne particles (originating in agricultural areas in India), loose soil and dust (from deforestation in Nepal), and smoke from cooking fires throughout the region.